T0350461

Graduate Texts in Mathematics 119

Editorial Board
S. Axler F.W. Gehring K.A. Ribet

Springer
New York
Berlin
Heidelberg
Barcelona
Budapest
Hong Kong
London
Milan
Paris
Singapore
Tokyo

Graduate Texts in Mathematics

continued after index

Joseph J. Rotman

An Introduction to Algebraic Topology

With 92 Illustrations

Springer

Joseph J. Rotman
Department of Mathematics
University of Illinois
Urbana, IL 61801
USA

Editorial Board

S. Axler
Mathematics Department
San Francisco State University
San Francisco, CA 94132
USA

F.W. Gehring
Mathematics Department
East Hall
University of Michigan
Ann Arbor, MI 48109
USA

K.A. Ribet
Department of Mathematics
University of California at Berkeley
Berkeley, CA 94720-3840
USA

Mathematics Subject Classification (1991): 55-01

Library of Congress Cataloging-in-Publication Data
Rotman, Joseph J.,
An introduction to algebraic topology.
(Graduate texts in mathematics; 119)
Bibliography: p.
Includes index.
1. Algebraic topology. I. Title. II. Series.
QA612.R69 1988 514'.2 87-37646

© 1988 by Springer-Verlag New York Inc.
All rights reserved. This work may not be translated or copied in whole or in part without the written permission of the publisher (Springer-Verlag, 175 Fifth Avenue, New York, NY 10010, USA), except for brief excerpts in connection with reviews or scholarly analysis. Use in connection with any form of information storage and retrieval, electronic adaptation, computer software, or by similar or dissimilar methodology now known or hereafter developed is forbidden.
The use of general descriptive names, trade names, trademarks, etc. in this publication, even if the former are not especially identified, is not to be taken as a sign that such names, as understood by the Trade Marks and Merchandise Marks Act, may accordingly be used freely by anyone.

Typeset by Asco Trade Typesetting Ltd., Hong Kong.
Printed and bound by R. R. Donnelley & Sons, Harrisonburg, Virginia.

9 8 7 6 5 4 (Fourth corrected printing, 1998)

ISBN 978-0-387-96678-6 ISBN 978-1-4612-4576-6 (eBook)
DOI 10.1007/978-1-4612-4576-6

To my wife Marganit
and my children Ella Rose and Daniel Adam
without whom this book would have
been completed two years earlier

Preface

There is a canard that every textbook of algebraic topology either ends with the definition of the Klein bottle or is a personal communication to J. H. C. Whitehead. Of course, this is false, as a glance at the books of Hilton and Wylie, Maunder, Munkres, and Schubert reveals. Still, the canard does reflect some truth. Too often one finds too much generality and too little attention to details.

There are two types of obstacle for the student learning algebraic topology. The first is the formidable array of new techniques (e.g., most students know very little homological algebra); the second obstacle is that the basic definitions have been so abstracted that their geometric or analytic origins have been obscured. I have tried to overcome these barriers. In the first instance, new definitions are introduced only when needed (e.g., homology with coefficients and cohomology are deferred until after the Eilenberg–Steenrod axioms have been verified for the three homology theories we treat—singular, simplicial, and cellular). Moreover, many exercises are given to help the reader assimilate material. In the second instance, important definitions are often accompanied by an informal discussion describing their origins (e.g., winding numbers are discussed before computing $\pi_1(S^1)$, Green's theorem occurs before defining homology, and differential forms appear before introducing cohomology).

We assume that the reader has had a first course in point-set topology, but we do discuss quotient spaces, path connectedness, and function spaces. We assume that the reader is familiar with groups and rings, but we do discuss free abelian groups, free groups, exact sequences, tensor products (always over **Z**), categories, and functors.

I am an algebraist with an interest in topology. The basic outline of this book corresponds to the syllabus of a first-year's course in algebraic topology designed by geometers and topologists at the University of Illinois, Urbana; other expert advice came (indirectly) from my teachers, E. H. Spanier and S. Mac Lane, and from J. F. Adams's *Algebraic Topology: A Student's Guide*. This latter book is strongly recommended to the reader who, having finished this book, wants direction for further study.

I am indebted to the many authors of books on algebraic topology, with a special bow to Spanier's now classic text. My colleagues in Urbana, especially Ph. Tondeur, H. Osborn, and R. L. Bishop, listened and explained. M.-E. Hamstrom took a particular interest in this book; she read almost the entire manuscript and made many wise comments and suggestions that have improved the text; my warmest thanks to her. Finally, I thank Mrs. Dee Wrather for a superb job of typing and Springer-Verlag for its patience.

Joseph J. Rotman

Addendum to Second Corrected Printing

Though I did read the original galleys carefully, there were many errors that eluded me. I thank all who apprised me of mistakes in the first printing, especially David Carlton, Monica Nicolau, Howard Osborn, Rick Rarick, and Lewis Stiller.

November 1992 Joseph J. Rotman

Addendum to Fourth Corrected Printing

Even though many errors in the first printing were corrected in the second printing, some were unnoticed by me. I thank Bernhard J. Elsner and Martin Meier for apprising me of errors that persisted into the the second and third printings. I have corrected these errors, and the book is surely more readable because of their kind efforts.

April, 1998 Joseph Rotman

To the Reader

Doing exercises is an essential part of learning mathematics, and the serious reader of this book should attempt to solve all the exercises as they arise. An asterisk indicates only that an exercise is cited elsewhere in the text, sometimes in a proof (those exercises used in proofs, however, are always routine).

I have never found references of the form 1.2.1.1 convenient (after all, one decimal point suffices for the usual description of real numbers). Thus, Theorem 7.28 here means the 28th theorem in Chapter 7.

Contents

CHAPTER 0

Introduction

One expects algebraic topology to be a mixture of algebra and topology, and that is exactly what it is. The fundamental idea is to convert problems about topological spaces and continuous functions into problems about algebraic objects (e.g., groups, rings, vector spaces) and their homomorphisms; the method may succeed when the algebraic problem is easier than the original one. Before giving the appropriate setting, we illustrate how the method works.

Notation

Let us first introduce notation for some standard spaces that is used throughout the book.

$\mathbf{Z}$ = integers (positive, negative, and zero).
$\mathbf{Q}$ = rational numbers.
$\mathbf{C}$ = complex numbers.
$\mathbf{I} = [0, 1]$, the (closed) unit interval.
$\mathbf{R}$ = real numbers.
$\mathbf{R}^n = \{(x_1, x_2, \ldots, x_n) | x_i \in \mathbf{R} \text{ for all } i\}$.

$\mathbf{R}^n$ is called **real n-space** or **euclidean space** (of course, $\mathbf{R}^n$ is the cartesian product of n copies of $\mathbf{R}$). Also, $\mathbf{R}^2$ is homeomorphic to $\mathbf{C}$; in symbols, $\mathbf{R}^2 \approx \mathbf{C}$. If $x = (x_1, \ldots, x_n) \in \mathbf{R}^n$, then its **norm** is defined by $\|x\| = \sqrt{\sum_{i=1}^n x_i^2}$ (when $n = 1$, then $\|x\| = |x|$, the absolute value of x). We regard $\mathbf{R}^n$ as the subspace of $\mathbf{R}^{n+1}$ consisting of all $(n + 1)$-tuples having last coordinate zero.

$$S^n = \{x \in \mathbf{R}^{n+1} : \|x\| = 1\}.$$

S^n is called the **n-sphere** (of radius 1 and center the origin). Observe that $S^n \subset \mathbf{R}^{n+1}$(as the circle $S^1 \subset \mathbf{R}^2$); note also that the 0-sphere S^0 consists of the two points $\{1, -1\}$ and hence is a discrete two-point space. We may regard S^n as the **equator** of S^{n+1}:

$$S^n = \mathbf{R}^{n+1} \cap S^{n+1} = \{(x_1, \ldots, x_{n+2}) \in S^{n+1}: x_{n+2} = 0\}.$$

The **north pole** is $(0, 0, \ldots, 0, 1) \in S^n$; the **south pole** is $(0, 0, \ldots, 0, -1)$. The **antipode** of $x = (x_1, \ldots, x_{n+1}) \in S^n$ is the other endpoint of the diameter having one endpoint x; thus the antipode of x is $-x = (-x_1, \ldots, -x_{n+1})$, for the distance from $-x$ to x is 2.

$$D^n = \{x \in \mathbf{R}^n: \|x\| \leq 1\}.$$

D^n is called the **n-disk** (or **n-ball**). Observe that $S^{n-1} \subset D^n \subset \mathbf{R}^n$; indeed S^{n-1} is the boundary of D^n in $\mathbf{R}^n$.

$$\Delta^n = \{(x_1, x_2, \ldots, x_{n+1}) \in \mathbf{R}^{n+1}: \text{each } x_i \geq 0 \text{ and } \textstyle\sum x_i = 1\}.$$

Δ^n is called the **standard n-simplex**. Observe that Δ^0 is a point, Δ^1 is a closed interval, Δ^2 is a triangle (with interior), Δ^3 is a (solid) tetrahedron, and so on. It is obvious that $\Delta^n \approx D^n$, although the reader may not want to construct[1] a homeomorphism until Exercise 2.11.

There is a standard homeomorphism from $S^n - \{\text{north pole}\}$ to $\mathbf{R}^n$, called **stereographic projection**. Denote the north pole by N, and define $\sigma: S^n - \{N\} \to \mathbf{R}^n$ to be the intersection of $\mathbf{R}^n$ and the line joining x and N. Points on the latter line have the form $tx + (1 - t)N$; hence they have coordinates $(tx_1, \ldots, tx_n, tx_{n+1} + (1 - t))$. The last coordinate is zero for $t = (1 - x_{n+1})^{-1}$; hence

$$\sigma(x) = (tx_1, \ldots, tx_n),$$

where $t = (1 - x_{n+1})^{-1}$. It is now routine to check that σ is indeed a homeomorphism. Note that $\sigma(x) = x$ if and only if x lies on the equator S^{n-1}.

Brouwer Fixed Point Theorem

Having established notation, we now sketch a proof of the **Brouwer fixed point theorem**: if $f: D^n \to D^n$ is continuous, then there exists $x \in D^n$ with $f(x) = x$. When $n = 1$, this theorem has a simple proof. The disk D^1 is the closed interval $[-1, 1]$; let us look at the graph of f inside the square $D^1 \times D^1$.

---

[1] It is an exercise that a compact convex subset of $\mathbf{R}^n$ containing an interior point is homeomorphic to D^n (convexity is defined in Chapter 1); it follows that Δ^n, D^n, and $\mathbf{I}^n$ are homeomorphic.

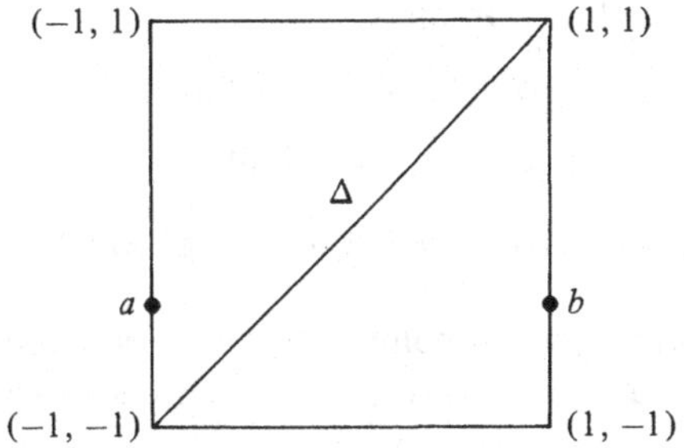

Theorem 0.1. *Every continuous $f: D^1 \to D^1$ has a fixed point.*

PROOF. Let $f(-1) = a$ and $f(1) = b$. If either $f(-1) = -1$ or $f(1) = 1$, we are done. Therefore, we may assume that $f(-1) = a > -1$ and that $f(1) = b < 1$, as drawn. If G is the graph of f and Δ is the graph of the identity function (of course, Δ is the diagonal), then we must prove that $G \cap \Delta \neq \varnothing$. The idea is to use a connectedness argument to show that every path in $D^1 \times D^1$ from a to b must cross Δ. Since f is continuous, $G = \{(x, f(x)): x \in D^1\}$ is connected [G is the image of the continuous map $D^1 \to D^1 \times D^1$ given by $x \mapsto (x, f(x))$]. Define $A = \{(x, f(x)): f(x) > x\}$ and $B = \{(x, f(x)): f(x) < x\}$. Note that $a \in A$ and $b \in B$, so that $A \neq \varnothing$ and $B \neq \varnothing$. If $G \cap \Delta = \varnothing$, then G is the disjoint union

$$G = A \cup B.$$

Finally, it is easy to see that both A and B are open in G, and this contradicts the connectedness of G. □

Unfortunately, no one knows how to adapt this elementary topological argument when $n > 1$; some new idea must be introduced. There is a proof using the *simplicial approximation theorem* (see [Hirsch]). There are proofs by analysis (see [Dunford and Schwartz, pp. 467–470] or [Milnor (1978)]); the basic idea is to approximate a continuous function $f: D^n \to D^n$ by smooth functions $g: D^n \to D^n$ in such a way that f has a fixed point if all the g do; one can then apply analytic techniques to smooth functions.

Here is a proof of the Brouwer fixed point theorem by algebraic topology. We shall eventually prove that, for each $n \geq 0$, there is a *homology functor* H_n with the following properties: for each topological space X there is an abelian group $H_n(X)$, and for each continuous function $f: X \to Y$ there is a homomorphism $H_n(f): H_n(X) \to H_n(Y)$, such that:

$$H_n(g \circ f) = H_n(g) \circ H_n(f) \tag{1}$$

whenever the composite $g \circ f$ is defined;

$$H_n(1_X) \text{ is the identity function on } H_n(X), \tag{2}$$

where 1_X is the identity function on X;

$$H_n(D^{n+1}) = 0 \quad \text{for all } n \geq 1; \tag{3}$$

$$H_n(S^n) \neq 0 \quad \text{for all } n \geq 1. \tag{4}$$

Using these H_n's, we now prove the Brouwer theorem.

Definition. A subspace X of a topological space Y is a **retract** of Y if there is a continuous map[2] $r: Y \to X$ with $r(x) = x$ for all $x \in X$; such a map r is called a **retraction**.

Remarks. (1) Recall that a topological space X contained in a topological space Y is a **subspace** of Y if a subset V of X is open in X if and only if $V = X \cap U$ for some open subset U of Y. Observe that this guarantees that the inclusion $i: X \hookrightarrow Y$ is continuous, because $i^{-1}(U) = X \cap U$ is open in X whenever U is open in Y. This parallels group theory: a group H contained in a group G is a **subgroup** of G if and only if the inclusion $i: H \hookrightarrow G$ is a homomorphism (this says that the group operations in H and in G coincide).

(2) One may rephrase the definition of retract in terms of functions. If $i: X \hookrightarrow Y$ is the inclusion, then a continuous map $r: Y \to X$ is a retraction if and only if

$$r \circ i = 1_X.$$

(3) For abelian groups, one can prove that a subgroup H of G is a retract of G if and only if H is a **direct summand** of G; that is, there is a subgroup K of G with $K \cap H = 0$ and $K + H = G$ (see Exercise 0.1).

Lemma 0.2. *If $n \geq 0$, then S^n is not a retract of D^{n+1}.*

PROOF. Suppose there were a retraction $r: D^{n+1} \to S^n$; then there would be a "commutative diagram" of topological spaces and continuous maps

$$\begin{array}{ccc}
 & D^{n+1} & \\
 {\scriptstyle i}\nearrow & & \searrow{\scriptstyle r} \\
 S^n & \xrightarrow{\ \ 1\ \ } & S^n
\end{array}$$

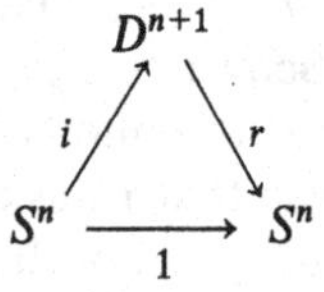

(here commutative means that $r \circ i = 1$, the identity function on S^n). Applying H_n gives a diagram of abelian groups and homomorphisms:

$$\begin{array}{ccc}
 & H_n(D^{n+1}) & \\
 {\scriptstyle H_n(i)}\nearrow & & \searrow{\scriptstyle H_n(r)} \\
 H_n(S^n) & \xrightarrow[{H_n(1)}]{} & H_n(S^n).
\end{array}$$

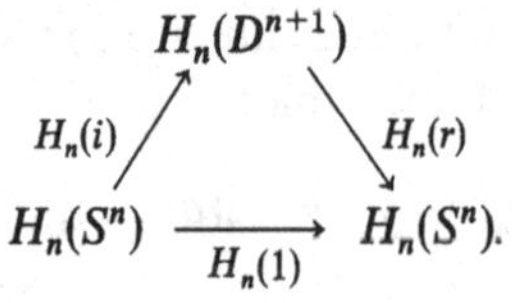

[2] We use the words *map* and *function* interchangeably.

By property (1) of the homology functor H_n, the new diagram commutes: $H_n(r) \circ H_n(i) = H_n(1)$. Since $H_n(D^{n+1}) = 0$, by (3), it follows that $H_n(1) = 0$. But $H_n(1)$ is the identity on $H_n(S^n)$, by (2). This contradicts (4) because $H_n(S^n) \neq 0$. □

Note how homology functors H_n have converted a topological problem into an algebraic one.

We mention that Lemma 0.2 has an elementary proof when $n = 0$. It is plain that a retraction $r: Y \to X$ is surjective. In particular, a retraction $r: D^1 \to S^0$ would be a continuous map from $[-1, 1]$ onto the two-point set $\{\pm 1\}$, and this contradicts the fact that a continuous image of a connected set is connected.

Theorem 0.3 (Brouwer). *If $f: D^n \to D^n$ is continuous, then f has a fixed point.*

PROOF. Suppose that $f(x) \neq x$ for all $x \in D^n$; the distinct points x and $f(x)$ thus determine a line. Define $g: D^n \to S^{n-1}$ (the boundary of D^n) as the function

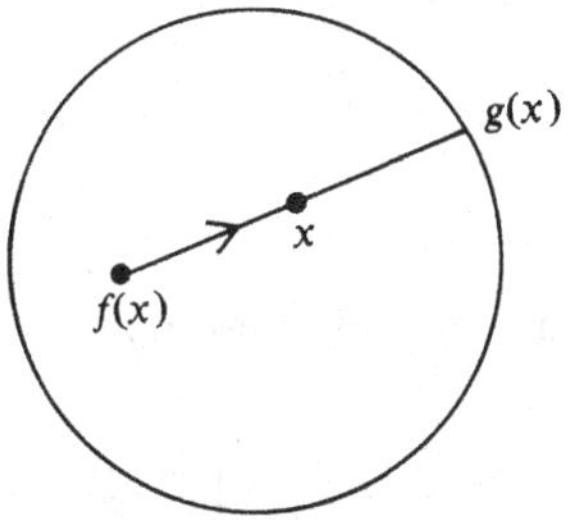

assigning to x that point where the ray from $f(x)$ to x intersects S^{n-1}. Obviously, $x \in S^{n-1}$ implies $g(x) = x$. The proof that g is continuous is left as an exercise in analytic geometry. We have contradicted the lemma. □

There is an extension of this theorem to infinite-dimensional spaces due to Schauder (which explains why there is a proof of the Brouwer fixed point theorem in [Dunford and Schwartz]): if D is a compact convex subset of a Banach space, then every continuous $f: D \to D$ has a fixed point. The proof involves approximating $f - 1_D$ by a sequence of continuous functions each of which is defined on a finite-dimensional subspace of D where Brouwer's theorem applies.

EXERCISES

*0.1. Let H be a subgroup of an abelian group G. If there is a homomorphism $r: G \to H$ with $r(x) = x$ for all $x \in H$, then $G = H \oplus \ker r$. (*Hint*: If $y \in G$, then $y = r(y) + (y - r(y))$.)

0.2. Give a proof of Brouwer's fixed point theorem for $n = 1$ using the proof of Theorem 0.3 and the remark preceding it.

0.3. Assume, for $n \geq 1$, that $H_i(S^n) = \mathbf{Z}$ if $i = 0$, n, and that $H_i(S^n) = 0$ otherwise. Using the technique of the proof of Lemma 0.2, prove that the equator of the n-sphere is not a retract.

0.4. If X is a topological space homeomorphic to D^n, then every continuous $f: X \to X$ has a fixed point.

0.5. Let $f, g: \mathbf{I} \to \mathbf{I} \times \mathbf{I}$ be continuous; let $f(0) = (a, 0)$ and $f(1) = (b, 1)$, and let $g(0) = (0, c)$ and $g(1) = (1, d)$ for some $a, b, c, d \in \mathbf{I}$. Show that $f(s) = g(t)$ for some s, $t \in \mathbf{I}$; that is, the paths intersect. (*Hint*: Use Theorem 0.3 for a suitable map $\mathbf{I} \times \mathbf{I} \to \mathbf{I} \times \mathbf{I}$.) (There is a proof in [Maehara]; this paper also shows how to derive the Jordan curve theorem from the Brouwer theorem.)

0.6. (Perron). Let $A = [a_{ij}]$ be a real $n \times n$ matrix with $a_{ij} > 0$ for every i, j. Prove that A has a positive eigenvalue λ; moreover, there is a corresponding eigenvector $x = (x_1, x_2, \ldots, x_n)$ (i.e., $Ax = \lambda x$) with each coordinate $x_i > 0$. (*Hint*: First define $\sigma: \mathbf{R}^n \to \mathbf{R}$ by $\sigma(x_1, x_2, \ldots, x_n) = \sum_{i=1}^{n} x_i$, and then define $g: \Delta^{n-1} \to \Delta^{n-1}$ by $g(x) = Ax/\sigma(Ax)$, where $x \in \Delta^{n-1} \subset \mathbf{R}^n$ is regarded as a column vector. Apply the Brouwer fixed point theorem after showing that g is a well defined continuous function.)

Categories and Functors

Having illustrated the technique, let us now give the appropriate setting for algebraic topology.

Definition. A **category** $\mathscr{C}$ consists of three ingredients: a class of **objects**, obj $\mathscr{C}$; sets of **morphisms** $\mathrm{Hom}(A, B)$, one for every ordered pair $A, B \in \text{obj } \mathscr{C}$; **composition** $\mathrm{Hom}(A, B) \times \mathrm{Hom}(B, C) \to \mathrm{Hom}(A, C)$, denoted by $(f, g) \mapsto g \circ f$, for every $A, B, C \in \text{obj } \mathscr{C}$, satisfying the following axioms:

(i) the family of $\mathrm{Hom}(A, B)$'s is pairwise disjoint;
(ii) composition is associative when defined;
(iii) for each $A \in \text{obj } \mathscr{C}$, there exists an **identity** $1_A \in \mathrm{Hom}(A, A)$ satisfying $1_A \circ f = f$ for every $f \in \mathrm{Hom}(B, A)$, all $B \in \text{obj } \mathscr{C}$, and $g \circ 1_A = g$ for every $g \in \mathrm{Hom}(A, C)$, all $C \in \text{obj } \mathscr{C}$.

Remarks. (1) The associativity axiom stated more precisely is: if f, g, h are morphisms with either $h \circ (g \circ f)$ or $(h \circ g) \circ f$ defined, then the other is also defined and both composites are equal.

(2) We distinguish class from set: a **set** is a class that is small enough to have a cardinal number. Thus, we may speak of the *class* of all topological spaces, but we cannot say the *set* of all topological spaces. (The set theory we accept has primitive undefined terms: class, element, and the membership relation $\in$. All the usual constructs (e.g., functions, subclasses, Boolean opera-

tions, relations) are permissible except that the statement $x \in A$ is always false whenever x is a class that is not a set.)

(3) The only restriction on $\operatorname{Hom}(A, B)$ is that it be a set. In particular, $\operatorname{Hom}(A, B) = \varnothing$ is allowed, although axiom (iii) shows that $\operatorname{Hom}(A, A) \neq \varnothing$ because it contains 1_A.

(4) Instead of writing $f \in \operatorname{Hom}(A, B)$, we usually write $f: A \to B$.

EXAMPLE 0.1. $\mathscr{C}$ = **Sets**. Here obj $\mathscr{C}$ = all sets, $\operatorname{Hom}(A, B) = \{$all functions $A \to B\}$, and composition is the usual composition of functions.

This example needs some discussion. Our requirement, in the definition of category, that Hom sets are pairwise disjoint is a reflection of our insistence that a function $f: A \to B$ is given by its **domain** A, its **target** B, and its **graph**: $\{$all $(a, f(a))$: $a \in A\} \subset A \times B$. In particular, if A is a proper subset of B, we distinguish the inclusion $i: A \hookrightarrow B$ from the identity 1_A even though both functions have the same domain and the same graph; $i \in \operatorname{Hom}(A, B)$ and $1_A \in \operatorname{Hom}(A, A)$, and so $i \neq 1_A$. This distinction is essential. For example, in the proof of Lemma 0.2, $H_n(i) = 0$ and $H_n(1_A) \neq 0$ when $A = S^n$ and $B = D^{n+1}$. Here are two obvious consequences of this distinction: (1) If $B \subset B'$ and $f: A \to B$ and $g: A \to B'$ are functions with the same graph (and visibly the same domain), then $g = i \circ f$, where $i: B \hookrightarrow B'$ is the inclusion. (2) One may form the composite $h \circ g$ only when target g = domain h. Others may allow one to compose $g: A \to B$ with $h: C \to D$ when $B \subset C$; we insist that the only composite defined here is $h \circ i \circ g$, where $i: B \hookrightarrow C$ is the given inclusion.

Now that we have explained the fine points of the definition, we continue our list of examples of categories.

EXAMPLE 0.2. $\mathscr{C}$ = **Top**. Here obj $\mathscr{C}$ = all topological spaces, $\operatorname{Hom}(A, B) = \{$all continuous functions $A \to B\}$, and composition is usual composition.

Definition. Let $\mathscr{C}$ and $\mathscr{A}$ be categories with obj $\mathscr{C} \subset$ obj $\mathscr{A}$. If $A, B \in$ obj $\mathscr{C}$, let us denote the two possible Hom sets by $\operatorname{Hom}_{\mathscr{C}}(A, B)$ and $\operatorname{Hom}_{\mathscr{A}}(A, B)$. Then $\mathscr{C}$ is a **subcategory** of $\mathscr{A}$ if $\operatorname{Hom}_{\mathscr{C}}(A, B) \subset \operatorname{Hom}_{\mathscr{A}}(A, B)$ for all $A, B \in$ obj $\mathscr{C}$ and if composition in $\mathscr{C}$ is the same as composition in $\mathscr{A}$; that is, the function $\operatorname{Hom}_{\mathscr{C}}(A, B) \times \operatorname{Hom}_{\mathscr{C}}(B, C) \to \operatorname{Hom}_{\mathscr{C}}(A, C)$ is the restriction of the corresponding composition with subscripts $\mathscr{A}$.

EXAMPLE 0.2′. The category **Top** has many interesting subcategories. First, we may restrict objects to be subspaces of euclidean spaces, or Hausdorff spaces, or compact spaces, and so on. Second, we may restrict the maps to be differentiable or analytic (assuming that these make sense for the objects being considered).

EXAMPLE 0.3. $\mathscr{C}$ = **Groups**. Here obj $\mathscr{C}$ = all groups, $\operatorname{Hom}(A, B) = \{$all homomorphisms $A \to B\}$, and composition is usual composition (Hom sets are so called because of this example).

EXAMPLE 0.4. $\mathscr{C} = \mathbf{Ab}$. Here obj $\mathscr{C}$ = all abelian groups, and $\operatorname{Hom}(A, B) = \{$all homomorphisms $A \to B\}$; **Ab** is a subcategory of **Groups**.

EXAMPLE 0.5. $\mathscr{C} = \mathbf{Rings}$. Here obj $\mathscr{C}$ = all rings (always with a two-sided identity element), $\operatorname{Hom}(A, B) = \{$all ring homomorphisms $A \to B$ that preserve identity elements$\}$, and usual composition.

EXAMPLE 0.6. $\mathscr{C} = \mathbf{Top}^2$. Here obj $\mathscr{C}$ consists of all ordered pairs (X, A), where X is a topological space and A is a subspace of X. A morphism $f: (X, A) \to (Y, B)$ is an ordered pair (f, f'), where $f: X \to Y$ is continuous and $fi = jf'$ (where i and j are inclusions),

$$\begin{array}{ccc} A & \xrightarrow{i} & X \\ {\scriptstyle f'}\downarrow & & \downarrow{\scriptstyle f} \\ B & \xrightarrow[j]{} & Y; \end{array}$$

and composition is coordinatewise (usually one is less pedantic, and one says that a morphism is a continuous map $f: X \to Y$ with $f(A) \subset B$). $\mathbf{Top}^2$ is called the category of **pairs** (of topological spaces).

EXAMPLE 0.7. $\mathscr{C} = \mathbf{Top}_*$. Here obj $\mathscr{C}$ consists of all ordered pairs (X, x_0), where X is a topological space and x_0 is a point of X. $\mathbf{Top}_*$ is a subcategory of $\mathbf{Top}^2$ (subspaces here are always one-point subspaces), and it is called the category of **pointed spaces**; x_0 is called the **basepoint** of (X, x_0), and morphisms are called **pointed maps** (or **basepoint preserving maps**). The category $\mathbf{Sets}_*$ of pointed sets is defined similarly.

Of course, there are many other examples of categories, and others arise as we proceed.

EXERCISES

0.7. Let $f \in \operatorname{Hom}(A, B)$ be a morphism in a category $\mathscr{C}$. If f has a left inverse g ($g \in \operatorname{Hom}(B, A)$ and $g \circ f = 1_A$) and a right inverse h ($h \in \operatorname{Hom}(B, A)$ and $f \circ h = 1_B$), then $g = h$.

0.8. (i) Let $\mathscr{C}$ be a category and let $A \in \operatorname{obj} \mathscr{C}$. Prove that $\operatorname{Hom}(A, A)$ has a unique identity 1_A.

(ii) If $\mathscr{C}'$ is a subcategory of $\mathscr{C}$, and if $A \in \operatorname{obj} \mathscr{C}'$, then the identity of A in $\operatorname{Hom}_{\mathscr{C}'}(A, A)$ is the identity 1_A in $\operatorname{Hom}_{\mathscr{C}}(A, A)$.

*0.9. A set X is called **quasi-ordered** (or **pre-ordered**) if X has a transitive and reflexive relation $\leq$. (Of course, such a set is partially ordered if, in addition, $\leq$ is antisymmetric.) Prove that the following construction gives a category $\mathscr{C}$. Define obj $\mathscr{C} = X$; if $x, y \in X$ and $x \not\leq y$, define $\operatorname{Hom}(x, y) = \varnothing$; if $x \leq y$, define $\operatorname{Hom}(x, y)$ to be a set with exactly one element, denoted by i_y^x; if $x \leq y \leq z$, define composition by $i_z^y \circ i_y^x = i_z^x$.

*0.10. Let G be a **monoid**, that is, a semigroup with 1. Show that the following construction gives a category $\mathscr{C}$. Let obj $\mathscr{C}$ have exactly one element, denoted by $*$; define $\text{Hom}(*, *) = G$, and define composition $G \times G \to G$ as the given multiplication in G. (This example shows that morphisms may not be functions.)

0.11. Show that one may regard **Top** as a subcategory of $\mathbf{Top}^2$ if one identifies a space X with the pair $(X, \varnothing)$.

Definition. A **diagram** in a category $\mathscr{C}$ is a directed graph whose vertices are labeled by objects of $\mathscr{C}$ and whose directed edges are labeled by morphisms in $\mathscr{C}$. A **commutative diagram** in $\mathscr{C}$ is a diagram in which, for each pair of vertices, every two paths (composites) between them are equal as morphisms.

This terminology comes from the particular diagram

$$\begin{array}{ccc} A & \xrightarrow{g'} & A' \\ {\scriptstyle f}\downarrow & & \downarrow{\scriptstyle f'} \\ B & \xrightarrow[g]{} & B' \end{array},$$

which commutes if $g \circ f = f' \circ g'$. Of course, we have already encountered commutative diagrams in the proof of Lemma 0.2.

EXERCISES

*0.12. Given a category $\mathscr{C}$, show that the following construction gives a category $\mathscr{M}$. First, an object of $\mathscr{M}$ is a morphism of $\mathscr{C}$. Next, if $f, g \in \text{obj } \mathscr{M}$, say, $f: A \to B$ and $g: C \to D$, then a morphism in $\mathscr{M}$ is an ordered pair (h, k) of morphisms in $\mathscr{C}$ such that the diagram

$$\begin{array}{ccc} A & \xrightarrow{f} & B \\ {\scriptstyle h}\downarrow & & \downarrow{\scriptstyle k} \\ C & \xrightarrow[g]{} & D \end{array}$$

commutes. Define composition coordinatewise:

$$(h', k') \circ (h, k) = (h' \circ h, k' \circ k).$$

0.13. Show that $\mathbf{Top}^2$ is a subcategory of a suitable morphism category (as constructed in Exercise 0.12). (*Hint*: Take $\mathscr{C} = \mathbf{Top}$, and let $\mathscr{M}$ be the corresponding morphism category; regard a pair (X, A) as an inclusion $i: A \to X$.)

The next simple construction is useful.

Definition. A **congruence** on a category $\mathscr{C}$ is an equivalence relation $\sim$ on the class $\bigcup_{(A,B)} \text{Hom}(A, B)$ of all morphisms in $\mathscr{C}$ such that:

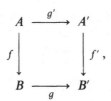

(i) $f \in \mathrm{Hom}(A, B)$ and $f \sim f'$ implies $f' \in \mathrm{Hom}(A, B)$;
(ii) $f \sim f'$, $g \sim g'$, and the composite $g \circ f$ exists imply that

$$g \circ f \sim g' \circ f'.$$

Theorem 0.4. *Let $\mathscr{C}$ be a category with congruence $\sim$, and let $[f]$ denote the equivalence class of a morphism f. Define $\mathscr{C}'$ as follows:*

$$\mathrm{obj}\ \mathscr{C}' = \mathrm{obj}\ \mathscr{C};$$

$$\mathrm{Hom}_{\mathscr{C}'}(A, B) = \{[f]: f \in \mathrm{Hom}_{\mathscr{C}}(A, B)\};$$

$$[g] \circ [f] = [g \circ f].$$

Then $\mathscr{C}'$ is a category.

Proof. Property (i) in the definition of congruence shows that $\sim$ partitions each set $\mathrm{Hom}_{\mathscr{C}}(A, B)$, and this implies that $\mathrm{Hom}_{\mathscr{C}'}(A, B)$ is a set; moreover, the family of these sets is pairwise disjoint. Property (ii) in the definition of congruence shows that composition in $\mathscr{C}'$ is well defined, and it is routine to see that composition in $\mathscr{C}'$ is associative and that $[1_A]$ is the identity morphism on A. □

The category $\mathscr{C}'$ just constructed is called a **quotient category** of $\mathscr{C}$; one usually denotes $\mathrm{Hom}_{\mathscr{C}'}(A, B)$ by $[A, B]$.

The most important quotient category for us is the *homotopy category* described in Chapter 1. Here is a lesser example. Let $\mathscr{C}$ be the category of groups and let $f, f' \in \mathrm{Hom}(G, H)$. Define $f \sim f'$ if there exists $a \in H$ with $f(x) = af'(x)a^{-1}$ for all $x \in G$ (one may say that f and f' are conjugate). It is routine to check that $\sim$ is an equivalence relation on each $\mathrm{Hom}(G, H)$. To see that $\sim$ is a congruence, assume that $f \sim f'$, that $g \sim g'$, and that $g \circ f$ exists. Thus f and $f' \in \mathrm{Hom}(G, H)$, g and $g' \in \mathrm{Hom}(H, K)$, there is $a \in H$ with $f(x) = af'(x)a^{-1}$ for all $x \in G$, and there is $b \in K$ with $g(y) = bg'(y)b^{-1}$ for all $y \in H$. It is easy to see that $g(f(x)) = [g(a)b]g'(f'(x))[g(a)b]^{-1}$ for all $x \in G$, that is, $g \circ f \sim g' \circ f'$. Thus the quotient category is defined. If G and H are groups, then $[G, H]$ is the set of all "conjugacy classes" $[f]$, where $f: G \to H$ is a homomorphism.

Exercise

0.14. Let G be a group and let $\mathscr{C}$ be the one-object category it defines (Exercise 0.10 applies because every group is a monoid): $\mathrm{obj}\ \mathscr{C} = \{*\}$, $\mathrm{Hom}(*, *) = G$, and composition is the group operation. If H is a normal subgroup of G, define $x \sim y$ to mean $xy^{-1} \in H$. Show that $\sim$ is a congruence on $\mathscr{C}$ and that $[*, *] = G/H$ in the corresponding quotient category.

Just as topological spaces are important because they carry continuous functions, so categories are important because they carry functors.

Definition. If $\mathscr{A}$ and $\mathscr{C}$ are categories, a **functor** $T: \mathscr{A} \to \mathscr{C}$ is a function, that is,

(i) $A \in \text{obj } \mathscr{A}$ implies $TA \in \text{obj } \mathscr{C}$,
and
(ii) if $f: A \to A'$ is a morphism in $\mathscr{A}$, then $Tf: TA \to TA'$ is a morphism in $\mathscr{C}$,
such that
(iii) if f, g are morphisms in $\mathscr{A}$ for which $g \circ f$ is defined, then

$$T(g \circ f) = (Tg) \circ (Tf);$$

(iv) $T(1_A) = 1_{TA}$ for every $A \in \text{obj } \mathscr{A}$.

Our earlier discussion of homology functors H_n can now be rephrased: for each $n \geq 0$, we shall construct a functor H_n: **Top** $\to$ **Ab** with $H_n(D^{n+1}) = 0$ and $H_n(S^n) \neq 0$.

EXAMPLE 0.8. The **forgetful functor** F: **Top** $\to$ **Sets** assigns to each topological space its underlying set and to each continuous function itself ("forgetting" its continuity). Similarly, there are forgetful functors **Groups** $\to$ **Sets**, **Ab** $\to$ **Groups**, **Ab** $\to$ **Sets**, and so on.

EXAMPLE 0.9. If $\mathscr{C}$ is a category, the **identity functor** $J: \mathscr{C} \to \mathscr{C}$ is defined by $JA = A$ for every object A and $Jf = f$ for every morphism f.

EXAMPLE 0.10. If M is a fixed topological space, then T_M: **Top** $\to$ **Top** is a functor, where $T_M(X) = X \times M$ and, if $f: X \to Y$ is continuous, then $T_M(f): X \times M \to Y \times M$ is defined by $(x, m) \mapsto (f(x), m)$.

EXAMPLE 0.11. Fix an object A in a category $\mathscr{C}$. Then $\text{Hom}(A, \quad): \mathscr{C} \to$ **Sets** is a functor assigning to each object B the set $\text{Hom}(A, B)$ and to each morphism $f: B \to B'$ the **induced map** $\text{Hom}(A, f): \text{Hom}(A, B) \to \text{Hom}(A, B')$ defined by $g \mapsto f \circ g$. One usually denotes the induced map $\text{Hom}(A, f)$ by f_*.

Functors as just defined are also called *covariant functors* to distinguish them from *contravariant functors* that reverse the direction of arrows. Thus the functor of Example 0.11 is sometimes called a **covariant Hom functor**.

Definition. If $\mathscr{A}$ and $\mathscr{C}$ are categories, a **contravariant functor** $S: \mathscr{A} \to \mathscr{C}$ is a function, that is,

(i) $A \in \text{obj } \mathscr{A}$ implies $SA \in \text{obj } \mathscr{C}$,
and
(ii) if $f: A \to A'$ is a morphism in $\mathscr{A}$, then $Sf: SA' \to SA$ is a morphism in $\mathscr{C}$,
such that:
(iii) if f, g are morphisms in $\mathscr{A}$ for which $g \circ f$ is defined, then

$$S(g \circ f) = S(f) \circ S(g);$$

(iv) $S(1_A) = 1_{SA}$ for every $A \in \text{obj } \mathscr{A}$.

EXAMPLE 0.12. Fix an object B in a category $\mathscr{C}$. Then $\mathrm{Hom}(\quad, B)$: $\mathscr{C} \to$ **Sets** is a contravariant functor assigning to each object A the set $\mathrm{Hom}(A, B)$ and to each morphism $g: A \to A'$ the **induced map** $\mathrm{Hom}(g, B)$: $\mathrm{Hom}(A', B) \to \mathrm{Hom}(A, B)$ defined by $h \mapsto h \circ g$. One usually denotes the induced map $\mathrm{Hom}(g, B)$ by g^*; $\mathrm{Hom}(\quad, B)$ is called a **contravariant Hom functor**.

EXAMPLE 0.13. Let F be a field and let $\mathscr{C}$ be the category of all finite-dimensional vector spaces over F. Define S: $\mathscr{C} \to \mathscr{C}$ by $S(V) = V^* = \mathrm{Hom}(V, F)$ and $Sf = f^*$. Thus S is the **dual space functor** that assigns to each vector space V its dual space V^* consisting of all linear functionals on V and to each linear transformation f its transpose f^*. Note that this example is essentially a special case of the preceding one, since F is a vector space over itself.

For quite a while, we shall deal exclusively with covariant functors, but contravariant functors are important and will eventually arise.

When working with functors, one is forced to state problems in a form recognizable by them. Thus, in our proof of the Brouwer fixed point theorem, we had to rephrase the definition of retraction from the version using elements, "$r(x) = x$ for all $x \in X$", to an equivalent version using functions: "$r \circ i = 1_X$". Similarly, one must rephrase the definition of bijection.

Definition. An **equivalence** in a category $\mathscr{C}$ is a morphism f: $A \to B$ for which there exists a morphism g: $B \to A$ with $f \circ g = 1_B$ and $g \circ f = 1_A$.

Theorem 0.5. *If $\mathscr{A}$ and $\mathscr{C}$ are categories and T: $\mathscr{A} \to \mathscr{C}$ is a functor of either variance, then f an equivalence in $\mathscr{A}$ implies that Tf is an equivalence in $\mathscr{C}$.*

PROOF. Apply T to the equations $f \circ g = 1$ and $g \circ f = 1$. □

EXERCISES

0.15. Let $\mathscr{A}$ and $\mathscr{C}$ be categories, and let T: $\mathscr{A} \to \mathscr{C}$ be a functor of either variance. If D is a commutative diagram in $\mathscr{A}$, then $T(D)$ (i.e., relabel all vertices and (possibly reversed) arrows) is a commutative diagram in $\mathscr{C}$.

0.16. Check that the following are the equivalences in the specified category: (i) **Sets**: bijections; (ii) **Top**: homeomorphisms; (iii) **Groups**: isomorphisms; (iv) **Rings**: isomorphisms; (v) quasi-ordered set: all i_y^x, where $x \le y$ and $y \le x$; (vi) $\mathbf{Top}^2$: all f: $(X, A) \to (X', A')$, where f: $X \to X'$ is a homeomorphism for which $f(A) = A'$; (vii) monoid G: all elements having a two-sided inverse.

*0.17. Let $\mathscr{C}$ and $\mathscr{A}$ be categories, and let $\sim$ be a congruence on $\mathscr{C}$. If T: $\mathscr{C} \to \mathscr{A}$ is a functor with $T(f) = T(g)$ whenever $f \sim g$, then T defines a functor T': $\mathscr{C}' \to \mathscr{A}$ (where $\mathscr{C}'$ is the quotient category) by $T'(X) = T(X)$ for every object X and $T'([f]) = T(f)$ for every morphism f.

0.18. For an abelian group G, let

$$tG = \{x \in G\colon x \text{ has finite order}\}$$

denote its **torsion subgroup**.

(i) Show that t defines a functor **Ab** $\to$ **Ab** if one defines $t(f) = f|tG$ for every homomorphism f.
(ii) If f is injective, then $t(f)$ is injective.
(iii) Give an example of a surjective homomorphism f for which $t(f)$ is not surjective.

0.19. Let p be a fixed prime in **Z**. Define a functor F: **Ab** $\to$ **Ab** by $F(G) = G/pG$ and $F(f)\colon x + pG \mapsto f(x) + pH$ (where $f\colon G \to H$ is a homomorphism).
(i) Show that if f is a surjection, then $F(f)$ is a surjection.
(ii) Give an example of an injective homomorphism f for which $F(f)$ is not injective.

*0.20. (i) If X is a topological space, show that $C(X)$, the set of all continuous real-valued functions on X, is a commutative ring with 1 under pointwise operations:

$$f + g\colon x \mapsto f(x) + g(x) \quad \text{and} \quad f \cdot g\colon x \mapsto f(x)g(x)$$

for all $x \in X$.
(ii) Show that $X \mapsto C(X)$ gives a (contravariant) functor **Top** $\to$ **Rings**.

One might expect that the functor C: **Top** $\to$ **Rings** of Exercise 0.20 is as valuable as the homology functors. Indeed, a theorem of Gelfand and Kolmogoroff (see [Dugundji, p. 289]) states that for X and Y compact Hausdorff, $C(X)$ and $C(Y)$ isomorphic as rings implies that X and Y are homeomorphic. Paradoxically, a less accurate translation of a problem from topology to algebra is usually more interesting than a very accurate one. The functor C is not as useful as other functors precisely because of the theorem of Gelfand and Kolmogoroff: the translated problem is exactly as complicated as the original one and hence cannot be any easier to solve (one can hope only that the change in viewpoint is helpful). Aside from homology, other functors to be introduced are cohomology groups, cohomology rings, and homotopy groups, one of which is the fundamental group.

CHAPTER 1

Some Basic Topological Notions

Homotopy

One often replaces a complicated function by another, simpler function that somehow approximates it and shares an important property of the original function. An allied idea is the notion of "deforming" one function into another: "perturbing" a function a bit may yield a new simpler function similar to the old one.

Definition. If X and Y are spaces and if f_0, f_1 are continuous maps from X to Y, then f_0 is **homotopic** to f_1, denoted by $f_0 \simeq f_1$, if there is a continuous map $F: X \times \mathbf{I} \to Y$ with

$$F(x, 0) = f_0(x) \quad \text{and} \quad F(x, 1) = f_1(x) \quad \text{for all } x \in X.$$

Such a map F is called a **homotopy**. One often writes $F: f_0 \simeq f_1$ if one wishes to display a homotopy.

If $f_t: X \to Y$ is defined by $f_t(x) = F(x, t)$, then a homotopy F gives a one-parameter family of continuous maps deforming f_0 into f_1. One thinks of f_t as describing the deformation at time t.

We now present some basic properties of homotopy, and we prepare the way with an elementary lemma of point-set topology.

Lemma 1.1 (Gluing lemma). *Assume that a space X is a finite union of closed subsets: $X = \bigcup_{i=1}^{n} X_i$. If, for some space Y, there are continuous maps $f_i: X_i \to Y$ that agree on overlaps ($f_i|X_i \cap X_j = f_j|X_i \cap X_j$ for all i, j), then there exists a unique continuous $f: X \to Y$ with $f|X_i = f_i$ for all i.*

PROOF. It is obvious that f defined by $f(x) = f_i(x)$ if $x \in X_i$ is the unique well defined function $X \to Y$ with restrictions $f|X_i = f_i$ for all i; only the continuity of f need be established. If C is a closed set in Y, then

$$\begin{aligned} f^{-1}(C) &= X \cap f^{-1}(C) = \left(\bigcup X_i\right) \cap f^{-1}(C) \\ &= \bigcup (X_i \cap f^{-1}(C)) \\ &= \bigcup (X_i \cap f_i^{-1}(C)) = \bigcup f_i^{-1}(C). \end{aligned}$$

Since each f_i is continuous, $f_i^{-1}(C)$ is closed in X_i; since X_i is closed in X, $f_i^{-1}(C)$ is closed in X. Therefore $f^{-1}(C)$ is closed in X, being a finite union of closed sets, and so f is continuous. □

There is another version of the gluing lemma, using open sets, whose proof is that of Lemma 1.1, *mutatis mutandis.*

Lemma 1.1′ (Gluing lemma). *Assume that a space X has a (possibly infinite) open cover: $X = \bigcup X_i$. If, for some space Y, there are continuous maps $f_i: X_i \to Y$ that agree on overlaps, then there exists a unique continuous $f: X \to Y$ with $f|X_i = f_i$ for all i.*

Theorem 1.2. *Homotopy is an equivalence relation on the set of all continuous maps $X \to Y$.*

PROOF. *Reflexivity.* If $f: X \to Y$, define $F: X \times \mathbf{I} \to Y$ by $F(x, t) = f(x)$ for all $x \in X$ and all $t \in \mathbf{I}$; clearly $F: f \simeq f$.

Symmetry: Assume that $f \simeq g$, so there is a continuous $F: X \times \mathbf{I} \to Y$ with $F(x, 0) = f(x)$ and $F(x, 1) = g(x)$ for all $x \in X$. Define $G: X \times \mathbf{I} \to Y$ by $G(x, t) = F(x, 1 - t)$, and note that $G: g \simeq f$.

Transitivity: Assume that $F: f \simeq g$ and $G: g \simeq h$. Define $H: X \times \mathbf{I} \to Y$ by

$$H(x, t) = \begin{cases} F(x, 2t) & \text{if } 0 \le t \le \frac{1}{2} \\ G(x, 2t - 1) & \text{if } \frac{1}{2} \le t \le 1. \end{cases}$$

Because these functions agree on the overlap $\{(x, \frac{1}{2}): x \in X\}$, the gluing lemma applies to show that H is continuous. Therefore $H: f \simeq h$. □

Definition. If $f: X \to Y$ is continuous, its **homotopy class** is the equivalence class

$$[f] = \{\text{continuous } g: X \to Y: g \simeq f\}.$$

The family of all such homotopy classes is denoted by $[X, Y]$.

Theorem 1.3. *Let $f_i: X \to Y$ and $g_i: Y \to Z$, for $i = 0, 1$, be continuous. If $f_0 \simeq f_1$ and $g_0 \simeq g_1$, then $g_0 \circ f_0 \simeq g_1 \circ f_1$; that is, $[g_0 \circ f_0] = [g_1 \circ f_1]$.*

PROOF. Let $F: f_0 \simeq f_1$ *and* $G: g_0 \simeq g_1$ *be homotopies. First, we show that*

$$g_0 \circ f_0 \simeq g_1 \circ f_0. \tag{$*$}$$

Define $H: X \times \mathbf{I} \to Z$ by $H(x, t) = G(f_0(x), t)$. Clearly, H is continuous; moreover, $H(x, 0) = G(f_0(x), 0) = g_0(f_0(x))$ and $H(x, 1) = G(f_0(x), 1) = g_1(f_0(x))$. Next, observe that

$$K: g_1 \circ f_0 \simeq g_1 \circ f_1, \tag{**}$$

where $K: X \times \mathbf{I} \to Z$ is the composite $g_1 \circ F$. Finally, use (*) and (**) together with the transitivity of the homotopy relation. □

Corollary 1.4. *Homotopy is a congruence on the category* **Top**.

PROOF. Immediate from Theorems 1.2 and 1.3. □

It follows at once from Theorem 0.4 that there is a quotient category whose objects are topological spaces X, whose Hom sets are $\text{Hom}(X, Y) = [X, Y]$, and whose composition is $[g] \circ [f] = [g \circ f]$.

Definition. The quotient category just described is called the **homotopy category**, and it is denoted by **hTop**.

All the functors T: **Top** $\to \mathscr{A}$ that we shall construct, where $\mathscr{A}$ is some "algebraic" category (e.g., **Ab**, **Groups**, **Rings**), will have the property that $f \simeq g$ implies $T(f) = T(g)$. This fact, aside from a natural wish to identify homotopic maps, makes homotopy valuable, because it guarantees that the algebraic problem in $\mathscr{A}$ arising from a topological problem via T is simpler than the original problem. Furthermore, Exercise 0.17 shows that every such functor gives a functor **hTop** $\to \mathscr{A}$, and so the homotopy category is actually quite fundamental.

What are the equivalences in **hTop**?

Definition. A continuous map $f: X \to Y$ is a **homotopy equivalence** if there is a continuous map $g: Y \to X$ with $g \circ f \simeq 1_X$ and $f \circ g \simeq 1_Y$. Two spaces X and Y have the **same homotopy type** if there is a homotopy equivalence $f: X \to Y$.

If one rewrites this definition, one sees that f is a homotopy equivalence if and only if $[f] \in [X, Y]$ is an equivalence in **hTop**. Thus the passage from **hTop** to the more familiar **Top** is accomplished by removing brackets and by replacing $=$ by $\simeq$.

Clearly, homeomorphic spaces have the same homotopy type, but the converse is false, as we shall see (Theorem 1.12).

The next two results show that homotopy is related to interesting questions.

Definition. Let X and Y be spaces, and let $y_0 \in Y$. The **constant map** at y_0 is the function $c: X \to Y$ with $c(x) = y_0$ for all $x \in X$. A continuous map $f: X \to Y$ is **nullhomotopic** if there is a constant map $c: X \to Y$ with $f \simeq c$.

Theorem 1.5. *Let* $\mathbf{C}$ *denote the complex numbers, let* $\Sigma_\rho \subset \mathbf{C} \approx \mathbf{R}^2$ *denote the circle with center at the origin* 0 *and radius* ρ, *and let* $f_\rho^n\colon \Sigma_\rho \to \mathbf{C} - \{0\}$ *denote the restriction to* Σ_ρ *of* $z \mapsto z^n$. *If none of the maps* f_ρ^n *is nullhomotopic* ($n \geq 1$ *and* $\rho > 0$), *then the fundamental theorem of algebra is true (i.e., every nonconstant complex polynomial has a complex root).*

PROOF. Consider the polynomial with complex coefficients:

$$g(z) = z^n + a_{n-1}z^{n-1} + \cdots + a_1 z + a_0.$$

Choose $\rho > \max\{1, \sum_{i=0}^{n-1} |a_i|\}$, and define $F\colon \Sigma_\rho \times \mathbf{I} \to \mathbf{C}$ by

$$F(z, t) = z^n + \sum_{i=0}^{n-1} (1 - t)a_i z^i.$$

It is obvious that $F\colon g|\Sigma_\rho \simeq f_\rho^n$ if we can show that the image of F is contained in $\mathbf{C} - \{0\}$; that is, $F(z, t) \neq 0$ (this restriction is crucial because, as we shall see in Theorem 1.13, every continuous function having values in a "contractible" space, e.g., in $\mathbf{C}$, is nullhomotopic). If, on the contrary, $F(z, t) = 0$ for some $t \in \mathbf{I}$ and some z with $|z| = \rho$, then $z^n = -\sum_{i=0}^{n-1} (1 - t)a_i z^i$. The triangle inequality gives

$$\rho^n \leq \sum_{i=0}^{n-1} (1 - t)|a_i|\rho^i \leq \sum_{i=0}^{n-1} |a_i|\rho^i \leq \left(\sum_{i=0}^{n-1} |a_i|\right)\rho^{n-1},$$

for $\rho > 1$ implies that $\rho^i \leq \rho^{n-1}$. Canceling ρ^{n-1} gives $\rho \leq \sum_{i=0}^{n-1} |a_i|$, a contradiction.

Assume now that g has no complex roots. Define $G\colon \Sigma_\rho \times \mathbf{I} \to \mathbf{C} - \{0\}$ by $G(z, t) = g((1 - t)z)$. (Since g has no roots, the values of G do lie in $\mathbf{C} - \{0\}$.) Visibly, $G\colon g|\Sigma_\rho \simeq k$, where k is the constant function at a_0. Therefore $g|\Sigma_\rho$ is nullhomotopic and, by transitivity, f_ρ^n is nullhomotopic, contradicting the hypothesis. □

Remark. We shall see later (Corollary 1.23) that $\mathbf{C} - \{0\}$ is essentially the circle $S^1 = \Sigma_1$; more precisely, $\mathbf{C} - \{0\}$ and S^1 have the same homotopy type.

A common problem involves extending a map $f\colon X \to Z$ to a larger space Y; the picture is

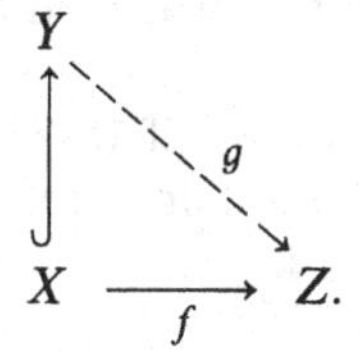

Homotopy itself raises such a problem: if $f_0, f_1\colon X \to Z$, then $f_0 \simeq f_1$ if we can extend $f_0 \cup f_1\colon X \times \{0\} \cup X \times \{1\} \to Z$ to all of $X \times \mathbf{I}$.

Theorem 1.6. *Let* $f\colon S^n \to Y$ *be a continuous map into some space* Y. *The following conditions are equivalent*:

(i) *f is nullhomotopic;*
(ii) *f can be extended to a continuous map $D^{n+1} \to Y$;*
(iii) *if $x_0 \in S^n$ and $k: S^n \to Y$ is the constant map at $f(x_0)$, then there is a homotopy $F: f \simeq k$ with $F(x_0, t) = f(x_0)$ for all $t \in \mathbf{I}$.*

Remark. Condition (iii) is a technical improvement on (i) that will be needed later; using terminology not yet introduced, it says that "F is a homotopy rel$\{x_0\}$".

PROOF. (i) $\Rightarrow$ (ii). Assume that $F: f \simeq c$, where $c(x) = y_0$ for all $x \in S^n$. Define $g: D^{n+1} \to Y$ by

$$g(x) = \begin{cases} y_0 & \text{if } 0 \le \|x\| \le \frac{1}{2} \\ F(x/\|x\|, 2 - 2\|x\|) & \text{if } \frac{1}{2} \le \|x\| \le 1. \end{cases}$$

Note that all makes sense: if $x \neq 0$, then $x/\|x\| \in S^n$; if $\frac{1}{2} \le \|x\| \le 1$, then $2 - 2\|x\| \in \mathbf{I}$; if $\|x\| = \frac{1}{2}$, then $2 - 2\|x\| = 1$ and $F(x/\|x\|, 1) = c(x/\|x\|) = y_0$. The gluing lemma shows that g is continuous. Finally, g does extend f: if $x \in S^n$, then $\|x\| = 1$ and $g(x) = F(x, 0) = f(x)$.

(ii) $\Rightarrow$ (iii). Assume that $g: D^{n+1} \to Y$ extends f. Define $F: S^n \times \mathbf{I} \to Y$ by $F(x, t) = g((1 - t)x + tx_0)$; note that $(1 - t)x + tx_0 \in D^{n+1}$, since this is just a point on the line segment joining x and x_0. Visibly, F is continuous. Now $F(x, 0) = g(x) = f(x)$ (since g extends f), while $F(x, 1) = g(x_0) = f(x_0)$ for all $x \in S^n$; hence $F: f \simeq k$, where $k: S^n \to Y$ is the constant map at $f(x_0)$. Finally, $F(x_0, t) = g(x_0) = f(x_0)$ for all $t \in \mathbf{I}$.

(iii) $\Rightarrow$ (i). Obvious. □

Compare this theorem with Lemma 0.2. If $Y = S^n$ and f is the identity, then Lemma 0.2 (not yet officially known!) implies that f is not nullhomotopic (otherwise S^n would be a retract of D^{n+1}).

Convexity, Contractibility, and Cones

Let us name a property of D^{n+1} that was used in the last proof.

Definition. A subset X of $\mathbf{R}^m$ is **convex** if, for each pair of points $x, y \in X$, the line segment joining x and y is contained in X. In other words, if $x, y \in X$, then $tx + (1 - t)y \in X$ for all $t \in \mathbf{I}$.

It is easy to give examples of convex sets; in particular, $\mathbf{I}^n$, $\mathbf{R}^n$, D^n, and Δ^n are convex. The sphere S^n considered as a subset of $\mathbf{R}^{n+1}$ is not convex.

Definition. A space X is **contractible** if 1_X is nullhomotopic.

Theorem 1.7. *Every convex set X is contractible.*

PROOF. Choose $x_0 \in X$, and define $c: X \to X$ by $c(x) = x_0$ for all $x \in X$. Define $F: X \times \mathbf{I} \to X$ by $F(x, t) = tx_0 + (1 - t)x$. It is easy to see that $F: 1_X \simeq c$. □

A hemisphere is contractible but not convex, so that the converse of Theorem 1.7 is not true. After proving Theorem 1.6, we observed that Lemma 0.2 implies that S^n is not contractible.

EXERCISES

1.1. Let $x_0, x_1 \in X$ and let $f_i: X \to X$ for $i = 0, 1$ denote the constant map at x_i. Prove that $f_0 \simeq f_1$ if and only if there is a continuous $F: \mathbf{I} \to X$ with $F(0) = x_0$ and $F(1) = x_1$.

1.2. (i) If $X \approx Y$ and X is contractible, then Y is contractible.
(ii) If X and Y are subspaces of euclidean space, $X \approx Y$, and X is convex, show that Y may not be convex.

*1.3. Let $R: S^1 \to S^1$ be rotation by α radians. Prove that $R \simeq 1_S$, where 1_S is the identity map of S^1. Conclude that every continuous map $f: S^1 \to S^1$ is homotopic to a continuous map $g: S^1 \to S^1$ with $g(1) = 1$ (where $1 = e^{2\pi i 0} \in S^1$).

1.4. (i) If X is a convex subset of $\mathbf{R}^n$ and Y is a convex subset of $\mathbf{R}^m$, then $X \times Y$ is a convex subset of $\mathbf{R}^{n+m}$.
(ii) If X and Y are contractible, then $X \times Y$ is contractible.

*1.5. Let $X = \{0\} \cup \{1, \frac{1}{2}, \frac{1}{3}, \ldots, 1/n, \ldots\}$ and let Y be a countable discrete space. Show that X and Y do not have the same homotopy type. (*Hint*: Use the compactness of X to show that every map $X \to Y$ takes all but finitely many points of X to a common point of Y.)

1.6. Contractible sets and hence convex sets are connected.

1.7. Let X be **Sierpinski space**: $X = \{x, y\}$ with topology $\{X, \varnothing, \{x\}\}$. Prove that X is contractible.

1.8. (i) Give an example of a continuous image of a contractible space that is not contractible.
(ii) Show that a retract of a contractible space is contractible.

1.9. If $f: X \to Y$ is nullhomotopic and if $g: Y \to Z$ is continuous, then $g \circ f$ is nullhomotopic.

The coming construction of a "cone" will show that every space can be imbedded in a contractible space. Before giving the definition, let us recall the construction of a quotient space.

Definition. Let X be a topological space and let $X' = \{X_j: j \in J\}$ be a partition of X (each X_j is nonempty, $X = \bigcup X_j$, and the X_j are pairwise disjoint). The **natural map** $\nu: X \to X'$ is defined by $\nu(x) = X_j$, where X_j is the (unique) subset in the partition containing x. The **quotient topology** on X' is the family of all subsets U' of X' for which $\nu^{-1}(U')$ is open in X.

It is easy to see that $\nu: X \to X'$ is a continuous map when X' has the quotient topology. There are two special cases that we wish to mention. If A is a subset of X, then we write X/A for X', where the partition of X consists of A together with all the one-point subsets of $X - A$ (this construction collapses A to a point but does not identify any other points of X; therefore, this construction differs from the quotient group construction for X a group and A a normal subgroup). The second special case arises from an equivalence relation $\sim$ on X; in this case, the partition consists of the equivalence classes, the natural map is given by $\nu: x \mapsto [x]$ (where $[x]$ denotes the equivalence class containing x), and the quotient space is denoted by $X/\sim$. The natural map is always a continuous surjection, but it may not be an open map [see Exercise 1.23(iii)].

Example 1.1. Consider the space $\mathbf{I} = [0, 1]$ and let A be the two-point subset $A = \{0, 1\}$. Intuitively, the quotient space $\mathbf{I}/A$ identifies 0 and 1 and ought to be the circle S^1; we let the reader supply the details that it is.

Example 1.2. As an example of the quotient topology using an equivalence relation, let $X = \mathbf{I} \times \mathbf{I}$

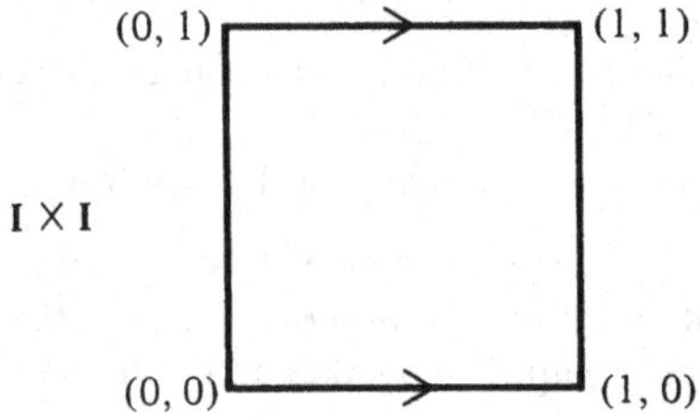

and define $(x, 0) \sim (x, 1)$ for every $x \in \mathbf{I}$. We let the reader show that $X/\sim$ is homeomorphic to the cylinder $S^1 \times \mathbf{I}$. As a further example, suppose we define a second equivalence relation on $\mathbf{I} \times \mathbf{I}$ by $(x, 0) \sim (x, 1)$ for all $x \in \mathbf{I}$ and $(0, y) \sim (1, y)$ for all $y \in \mathbf{I}$. Now $\mathbf{I} \times \mathbf{I}/\sim$ is the **torus** $S^1 \times S^1$ (first one has a cylinder and then one glues the circular ends together).

Example 1.3. If $h: X \to Y$ is a function, then **ker** $\boldsymbol{h}$ is the equivalence relation on X defined by $x \sim x'$ if $h(x) = h(x')$. The corresponding quotient space is denoted by $X/\ker h$. Note that, given $h: X \to Y$, there always exists an injection $\varphi: X/\ker h \to Y$ making the following diagram commute:

$$\begin{array}{ccc} X & \xrightarrow{h} & Y \\ & {\scriptstyle \nu} \searrow \quad \nearrow {\scriptstyle \varphi} & \\ & X/\ker h, & \end{array}$$

namely, $\varphi([x]) = h(x)$.

If $h: X \to Y$ is continuous, it is a natural question whether the map $\varphi: X/\ker h \to Y$ of Example 1.3 is continuous.

Definition. A continuous surjection $f: X \to Y$ is an **identification** if a subset U of Y is open if and only if $f^{-1}(U)$ is open in X.

EXAMPLE 1.4. If $\sim$ is an equivalence relation on X and $X/\sim$ is given the quotient topology, then the natural map $v: X \to X/\sim$ is an identification.

EXAMPLE 1.5. If $f: X \to Y$ is a continuous surjection that is either open or closed, then f is an identification.

EXAMPLE 1.6. If $f: X \to Y$ is a continuous map having a **section** (i.e., there is a continuous $s: Y \to X$ with $fs = 1_Y$), then f is an identification (note that f must be a surjection).

Theorem 1.8. *Let $f: X \to Y$ be a continuous surjection. Then f is an identification if and only if, for all spaces Z and all functions $g: Y \to Z$, one has g continuous if and only if gf is continuous.*

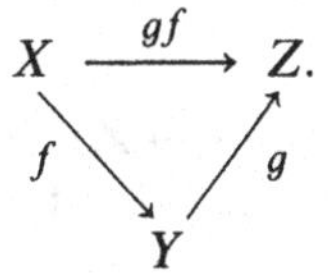

PROOF. Assume f is an identification. If g is continuous, then gf is continuous. Conversely, let gf be continuous and let V be an open set in Z. Then $(gf)^{-1}(V) = f^{-1}(g^{-1}(V))$ is open in X; since f is an identification, $g^{-1}(V)$ is open in Y, hence g is continuous.

Assume the condition. Let $Z = X/\ker f$, let $v: X \to X/\ker f$ be the natural map, and let $\varphi: X/\ker f \to Y$ be the injection of Example 1.3. Note that φ is surjective because f is. Consider the commutative diagram

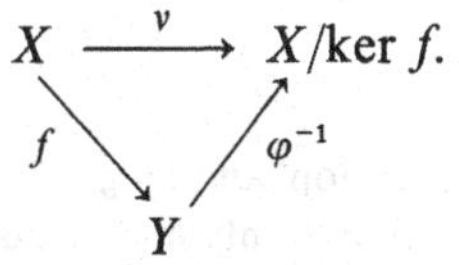

That $\varphi^{-1}f = v$ is continuous implies that φ^{-1} is continuous, by hypothesis. Also, φ is continuous because v is an identification. We conclude that φ is a homeomorphism, and the result follows at once. □

Definition. Let $f: X \to Y$ be a function and let $y \in Y$. Then $f^{-1}(y)$ is called the **fiber** over y.

If $f: X \to Y$ is a homomorphism between groups, then the fiber over 1 is the (group-theoretic) kernel of f, while the fiber over an arbitrary point y is a coset of the subgroup $\ker f$. More generally, fibers are the equivalence classes of the equivalence relation $\ker f$ on X.

Corollary 1.9. *Let $f: X \to Y$ be an identification and, for some space Z, let $h: X \to Z$ be a continuous function that is constant on each fiber of f. Then $hf^{-1}: Y \to Z$ is continuous.*

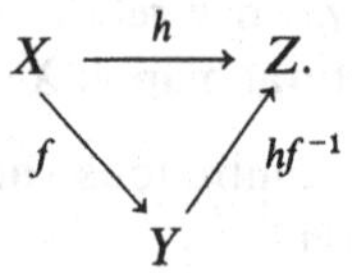

Moreover, hf^{-1} is an open map (or a closed map) if and only if $h(U)$ is open (or closed) in Z whenever U is an open (or closed) set in X of the form $U = f^{-1}f(U)$.[1]

PROOF. That h is constant on each fiber of f implies that $hf^{-1}: Y \to Z$ is a well defined function; hf^{-1} is continuous because $(hf^{-1})f = h$ is continuous, and Theorem 1.8 applies. Finally, if V is an open set in Y, then $f^{-1}(V)$ is an open set of the stated form: $f^{-1}(V) = f^{-1}f(f^{-1}(V))$; the result now follows easily. □

Remark. If A is a subset of X and $h: X \to Z$ is constant on A, then h is constant on the fibers of the natural map $\nu: X \to X/A$.

Corollary 1.10. *Let X and Z be spaces, and let $h: X \to Z$ be an identification. Then the map $\varphi: X/\ker h \to Z$, defined by $[x] \mapsto h(x)$, is a homeomorphism.*

PROOF. It is plain that the function $\varphi: X/\ker h \to Z$ is a bijection; φ is continuous, by Corollary 1.9. Let $\nu: X \to X/\ker h$ be the natural map. To see that φ is an open map, let U be an open set in $X/\ker h$. Then $h^{-1}\varphi(U) = \nu^{-1}(U)$ is an open set in X, because ν is continuous, and hence $\varphi(U)$ is open, because h is an identification. □

EXERCISES

*1.10. Let $f: X \to Y$ be an identification, and let $g: Y \to Z$ be a continuous surjection. Then g is an identification if and only if gf is an identification.

*1.11. Let X and Y be spaces with equivalence relations $\sim$ and $\square$, respectively, and let $f: X \to Y$ be a continuous map preserving the relations (if $x \sim x'$, then $f(x) \,\square\, f(x')$). Prove that the induced map $\bar{f}: X/\!\sim \;\to Y/\square$ is continuous; moreover, if f is an identification, then so is $\bar{f}$.

1.12. Let X and Z be compact Hausdorff spaces, and let $h: X \to Z$ be a continuous surjection. Prove that $\varphi: X/\ker h \to Z$, defined by $[x] \mapsto h(x)$, is a homeomorphism.

---

[1] Recall elementary set theory: if $f: X \to Y$ is a function and $U \subset \operatorname{im} f$, then $ff^{-1}(U) = U$ and $U \subset f^{-1}f(U)$; in general, there is no equality $U = f^{-1}f(U)$.

Definition. If X is a space, define an equivalence relation on $X \times \mathbf{I}$ by $(x, t) \sim (x', t')$ if $t = t' = 1$. Denote the equivalence class of (x, t) by $[x, t]$. The **cone** over X, denoted by CX, is the quotient space $X \times \mathbf{I}/\sim$.

One may also regard CX as the quotient space $X \times \mathbf{I}/X \times \{1\}$. The identified point $[x, 1]$ is called the **vertex**; we have essentially introduced a new point v not in X (the vertex) and joined each point in X to v by a line segment.

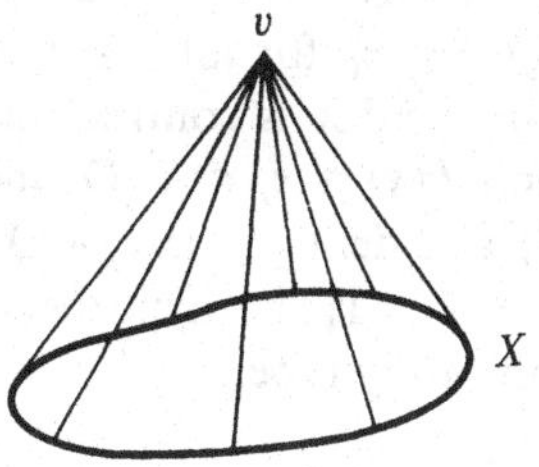

This picture is fine when X is compact Hausdorff, but it may be misleading otherwise: the quotient topology may have more open sets than expected.[2]

Example 1.7. For spaces X and Y, every continuous map $f: X \times \mathbf{I} \to Y$ with $f(x, 1) = y_0$, say, for all $x \in X$, induces a continuous map $\bar{f}: CX \to Y$, namely, $\bar{f}: [x, t] \mapsto f(x, t)$. In particular, let $f: S^n \times \mathbf{I} \to D^{n+1}$ be the map $(u, t) \mapsto (1 - t)u$; since $f(u, 1) = 0$ for all $u \in S^n$, there is a continuous map $\bar{f}: CS^n \to D^{n+1}$ with $[u, t] \mapsto (1 - t)u$. The reader may check that $\bar{f}$ is a homeomorphism (thus D^{n+1} is the cone over S^n with vertex 0).

Exercises

*1.13. For fixed t with $0 \le t < 1$, prove that $x \mapsto [x, t]$ defines a homeomorphism from a space X to a subspace of CX.

1.14. Prove that $X \mapsto CX$ defines a functor **Top** $\to$ **Top** (the reader must define the behavior on morphisms). (*Hint*: Use Exercise 1.11.)

Theorem 1.11. *For every space X, the cone CX is contractible.*

Proof. Define $F: CX \times \mathbf{I} \to CX$ by $F([x, t], s) = [x, (1 - s)t + s]$. □

Combining Theorem 1.11 with Exercise 1.13 shows that every space can be embedded in a contractible space.

---

[2] Let X be the set of positive integers regarded as points on the x-axis in $\mathbf{R}^2$; let $C'X$ denote the subspace of $\mathbf{R}^2$ obtained by joining each $(n, 0) \in X$ to $v = (0, 1)$ with a line segment. There is a continuous bijection $CX \to C'X$, but CX is not homeomorphic to $C'X$ (see [Dugundji, p. 127]).

The next result shows that contractible spaces are the simplest objects in **hTop**.

Theorem 1.12. *A space X has the same homotopy type as a point if and only if X is contractible.*

PROOF. Let $\{a\}$ be a one-point space, and assume that X and $\{a\}$ have the same homotopy type. There are thus maps $f: X \to \{a\}$ (visibly constant) and $g: \{a\} \to X$ (with $g(a) = x_0 \in X$, say) with $g \circ f \simeq 1_X$ and $f \circ g \simeq 1_{\{a\}}$ (actually, $f \circ g = 1_{\{a\}}$). But $gf(x) = g(a) = x_0$ for all $x \in X$, so that $g \circ f$ is constant. Therefore 1_X is nullhomotopic and X is contractible.

Assume that $1_X \simeq k$, where $k(x) \equiv x_0 \in X$. Define $f: X \to \{x_0\}$ as the constant map at x_0 (no choice!), and define $g: \{x_0\} \to X$ by $g(x_0) = x_0$. Note that $f \circ g = 1_{\{x_0\}}$ and that $g \circ f = k \simeq 1_X$, by hypothesis. We have shown that X and $\{x_0\}$ have the same homotopy type. □

This theorem suggests that contractible spaces may behave as singletons, especially when homotopy is in sight.

Theorem 1.13. *If Y is contractible, then any two maps $X \to Y$ are homotopic (indeed they are nullhomotopic).*

PROOF. Assume that $1_Y \simeq k$, where there is $y_0 \in y$ with $k(y) = y_0$ for all $y \in Y$. Define $g: X \to Y$ as the constant map $g(x) = y_0$ for all $x \in X$. If $f: X \to Y$ is any continuous map, we claim that $f \simeq g$. Consider the diagram

$$X \longrightarrow Y \overset{k}{\underset{1_Y}{\rightrightarrows}} Y.$$

Since $1_Y \simeq k$, Theorem 1.3 gives $f = 1_Y \circ f \simeq k \circ f = g$. □

If X is contractible (instead of Y), this result is false (indeed this result is false for X a singleton). However, the result is true when combined with a connectivity hypothesis (Exercise 1.19). This hypothesis also answers the question whether two nullhomotopic maps $X \to Y$ are necessarily homotopic (as they are in Theorem 1.13).

Paths and Path Connectedness

Definition. A **path** in X is a continuous map $f: \mathbf{I} \to X$. If $f(0) = a$ and $f(1) = b$, one says that f is a path **from** a **to** b.

Do not confuse a path f with its image $f(\mathbf{I})$, but do regard a path as a parametrized curve in X. Note that if f is a path in X from a to b, then $g(t) = f(1 - t)$ defines a path in X from b to a (of course, $g(\mathbf{I}) = f(\mathbf{I})$).

Definition. A space X is **path connected** if, for every $a, b \in X$, there exists a path in X from a to b.

Theorem 1.14. *If X is path connected, then X is connected.*

PROOF. If X is disconnected, then X is the disjoint union $X = A \cup B$, where A and B are nonempty open subsets of X. Choose $a \in A$ and $b \in B$, and let $f: \mathbf{I} \to X$ be a path from a to b. Now $f(\mathbf{I})$ is connected, yet

$$f(\mathbf{I}) = (A \cap f(\mathbf{I})) \cup (B \cap f(\mathbf{I}))$$

displays $f(\mathbf{I})$ as disconnected, a contradiction. □

The converse of Theorem 1.14 is false.

EXAMPLE 1.8. The **sin(1/x) space** X is the subspace $X = A \cup G$ of $\mathbf{R}^2$, where $A = \{(0, y): -1 \le y \le 1\}$ and $G = \{(x, \sin(1/x)): 0 < x \le 1/2\pi\}$.

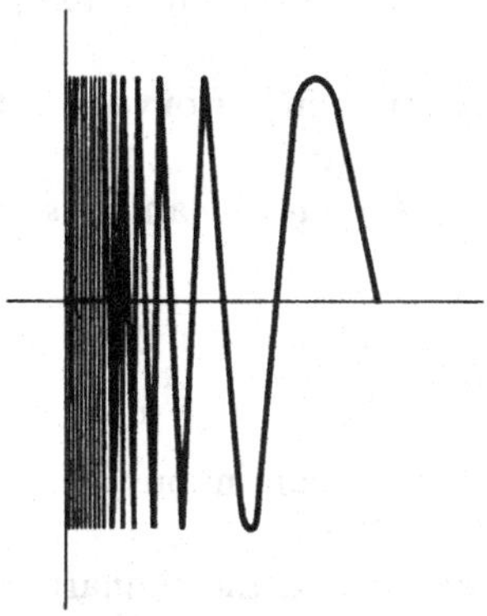

It is easy to see that X is connected, because the component of X that contains G is closed (components are always closed) and A is contained in the closure of G. Exercise 1.15 contains a hint toward proving that X is not path connected.

EXERCISES

*1.15. Show that the sin(1/x) space X is not path connected. (*Hint*: Assume that $f: \mathbf{I} \to X$ is a path from (0, 0) to $(1/2\pi, 0)$. If $t_0 = \sup\{t \in \mathbf{I}: f(t) \in A\}$, then $a = f(t_0) \in A$ and $f(s) \notin A$ for all $s > t_0$. One may thus assume that there is a path $g: \mathbf{I} \to X$ with $g(0) \in A$ and with $g(t) \in G$ for all $t > 0$.)

1.16. Show that S^n is path connected for all $n \ge 1$.

1.17. If $U \subset \mathbf{R}^n$ is open, then U is connected if and only if U is path connected. (This is false if "open" is replaced by "closed": the sin(1/x) space is a (compact) subset of $\mathbf{R}^2$.)

1.18. Every contractible space is path connected.

*1.19. (i) A space X is path connected if and only if every two constant maps $X \to X$ are homotopic.
(ii) If X is contractible and Y is path connected, then any two continuous maps $X \to Y$ are homotopic (and each is nullhomotopic).

1.20. Let A and B be path connected subspaces of a space X. If $A \cap B \neq \emptyset$ is path connected, then $A \cup B$ is path connected.

*1.21. If X and Y are path connected, then $X \times Y$ is path connected.

*1.22. If $f: X \to Y$ is continuous and X is path connected, then $f(X)$ is path connected.

Let us now analyze path connectedness as one analyzes connectedness.

Theorem 1.15. *If X is a space, then the binary relation $\sim$ on X defined by "$a \sim b$ if there is a path in X from a to b" is an equivalence relation.*

PROOF. *Reflexivity*: If $a \in X$, the constant function $f: \mathbf{I} \to X$ with $f(t) = a$ for all $t \in \mathbf{I}$ is a path from a to a.

Symmetry: If $f: \mathbf{I} \to X$ is a path in X from a to b, then $g: \mathbf{I} \to X$ defined by $g(t) = f(1 - t)$ is a path from b to a.

Transitivity: If f is a path from a to b and g is a path from b to c, define $h: \mathbf{I} \to X$ by

$$h(t) = \begin{cases} f(2t) & \text{if } 0 \le t \le \frac{1}{2} \\ g(2t - 1) & \text{if } \frac{1}{2} \le t \le 1. \end{cases}$$

The gluing lemma shows that h is continuous. □

The reader has probably noticed the similarity of this proof to that of Theorem 1.2: homotopy is an equivalence relation on the set of all continuous maps $X \to Y$. This will be explained in Chapter 12 when we discuss function spaces.

Definition. The equivalence classes of X under the relation $\sim$ in Theorem 1.15 are called the **path components** of X.

We now can see that every space is the disjoint union of path connected subspaces, namely, its path components.

EXERCISES

*1.23. (i) The $\sin(1/x)$ space X has exactly two path components: the vertical line A and the graph G.
(ii) Show that the graph G is not closed. Conclude that, in contrast to components (which are always closed), path components may not be closed.
(iii) Show that the natural map $\nu: X \to X/A$ is not an open map. (*Hint*: Let U

be the open disk with center $(0, \frac{1}{2})$ and radius $\frac{1}{4}$; show that $\nu(X \cap U)$ is not open in X/A ($\approx [0, \frac{1}{2\pi}]$).)

*1.24. The path components of a space X are maximal path connected subspaces; moreover, every path connected subset of X is contained in a unique path component of X.

1.25. Prove that the $\sin(1/x)$ space is not homeomorphic to **I**.

Let us use this notion to construct a (simple-minded) functor.

Definition. Define $\pi_0(X)$ to be the set of path components of X. If $f: X \to Y$, define $\pi_0(f): \pi_0(X) \to \pi_0(Y)$ to be the function taking a path component C of X to the (unique) path component of Y containing $f(C)$ (Exercises 1.24 and 1.22).

Theorem 1.16. π_0: **Top** $\to$ **Sets** *is a functor. Moreover, if* $f \simeq g$, *then* $\pi_0(f) = \pi_0(g)$.

PROOF. It is an easy exercise to check that π_0 preserves identities and composition; that is, π_0 is a functor.

Assume that $F: f \simeq g$, where $f, g: X \to Y$. If C is a path component of X, then $C \times \mathbf{I}$ is path connected (Exercise 1.21), hence $F(C \times \mathbf{I})$ is path connected (Exercise 1.22). Now

$$f(C) = F(C \times \{0\}) \subset F(C \times \mathbf{I})$$

and

$$g(C) = F(C \times \{1\}) \subset F(C \times \mathbf{I});$$

the unique path component of Y containing $F(C \times \mathbf{I})$ thus contains both $f(C)$ and $g(C)$. This says that $\pi_0(f) = \pi_0(g)$. □

Corollary 1.17. *If X and Y have the same homotopy type, then they have the same number of path components.*

PROOF. Assume that $f: X \to Y$ and $g: Y \to X$ are continuous with $g \circ f \simeq 1_X$ and $f \circ g \simeq 1_Y$. Then $\pi_0(g \circ f) = \pi_0(1_X)$ and $\pi_0(f \circ g) = \pi_0(1_Y)$, by Theorem 1.16. Since π_0 is a functor, it follows that $\pi_0(f)$ is a bijection. □

Here is a more conceptual proof. One may regard π_0 as a functor **hTop** $\to$ **Sets**, by Exercise 0.17. If $f: X \to Y$ is a homotopy equivalence, then $[f]$ is an equivalence in **hTop**, and so $\pi_0([f])$ (which is $\pi_0(f)$, by definition) is an equivalence in **Sets**, by Theorem 0.5.

π_0 is not a very thrilling functor since its values lie in **Sets**, and the only thing one can do with a set is count it. Still, it is as useful as counting ordinary components (which is how one proves that S^1 and **I** are not homeomorphic (after deleting a point)). π_0 is the first (zeroth?) of a sequence of functors. The next is π_1, the fundamental group, which takes values in **Groups**; the others, $\pi_2, \pi_3, \ldots$, are called (higher) homotopy groups and take values in **Ab** (we shall study these functors in Chapter 11).

Definition. A space X is **locally path connected** if, for each $x \in X$ and every open neighborhood U of x, there is an open V with $x \in V \subset U$ such that any two points in V can be joined by a path in U.

Corollary 1.19 will show that one can choose V so that every two points in V can be joined by a path in V; that is, V is path connected.

EXAMPLE 1.9. Let X be the subspace of $\mathbf{R}^2$ obtained from the $\sin(1/x)$ space by adjoining a curve from $(0, 1)$ to $(\frac{1}{2\pi}, 0)$. It is easy to see that X is path connected but not locally path connected.

Theorem 1.18. *A space X is locally path connected if and only if path components of open subsets are open. In particular, if X is locally path connected, then its path components are open.*

PROOF. Assume that X is locally path connected and that U is an open subset of X. Let C be a path component of U, and let $x \in C$. There is an open V with $x \in V \subset U$ such that every point of V can be joined to x by a path in U. Hence each point of V lies in the same path component as x, and so $V \subset C$. Therefore C is open.

Conversely, let U be an open set in X, let $x \in U$, and let V be the path component of x in U. By hypothesis, V is open. Therefore X is locally path connected. □

Corollary 1.19. *X is locally path connected if and only if, for each $x \in X$ and each open neighborhood U of x, there is an open path connected V with $x \in V \subset U$.*

PROOF. If X is locally path connected, then choose V to be the path component of U containing x. The converse is obvious. □

Corollary 1.20. *If X is locally path connected, then the components of every open set coincide with its path components. In particular, the components of X coincide with the path components of X.*

PROOF. Let C be a component of an open set U in X, and let $\{A_j: j \in J\}$ be the path components of C; then C is the disjoint union of the A_j: by Theorem 1.18, each A_j is open in C, hence each A_j is closed in C (its complement being the open set, which is the union of the other A's). Were there more than one A_j, then C would be disconnected. □

Corollary 1.21. *If X is connected and locally path connected, then X is path connected.*

PROOF. Since X is connected, X has only one component; since X is locally path connected, this component is a path component. □

EXERCISES

*1.26. A locally path connected space is locally connected. (Recall that a space is **locally connected** if every point has a connected open neighborhood.) (*Hint*: A space is locally connected if and only if components of open sets are open.)

1.27. If X and Y are locally path connected, then so is $X \times Y$.

*1.28. Every open subset of a locally path connected space is itself locally path connected.

Definition. Let A be a subspace of X and let $i: A \hookrightarrow X$ be the inclusion. Then A is a **deformation retract** of X if there is a continuous $r: X \to A$ such that $r \circ i = 1_A$ and $i \circ r \simeq 1_X$.

Of course, every deformation retract is a retract. One can rephrase the definition as follows: there is a continuous $F: X \times \mathbf{I} \to X$ such that $F(x, 0) = x$ for all $x \in X$, $F(x, 1) \in A$ for all $x \in X$, and $F(a, 1) = a$ for all $a \in A$ (in this formulation, we have $r(x) = F(x, 1)$). The next result is immediate.

Theorem 1.22. *If A is a deformation retract of X, then A and X have the same homotopy type.*

Corollary 1.23. *S^1 is a deformation retract of $\mathbf{C} - \{0\}$, and so these spaces have the same homotopy type.*

PROOF. Write each nonzero complex number z in polar coordinates:

$$z = \rho e^{i\theta}, \qquad \rho > 0, \quad 0 \leq \theta < 2\pi.$$

Define $F: (\mathbf{C} - \{0\}) \times \mathbf{I} \to \mathbf{C} - \{0\}$ by

$$F(\rho e^{i\theta}, t) = [(1 - t)\rho + t]e^{i\theta}.$$

It is clear that F is never 0 and that F satisfies the requirements making $S^1 = \{e^{i\theta}: 0 \leq \theta < 2\pi\}$ a deformation retract of $\mathbf{C} - \{0\}$. □

EXERCISES

*1.29. For $n \geq 1$, show that S^n is a deformation retract of $\mathbf{R}^{n+1} - \{0\}$.

1.30. For $n \geq 1$, show that S^n is a deformation retract of the "punctured disk" $D^{n+1} - \{0\}$.

*1.31. Let $a = (0, \ldots, 0, 1)$ and $b = (0, \ldots, 0, -1)$ be the north and south poles, respectively, of S^n. Show that the equator S^{n-1} is a deformation retract of $S^n - \{a, b\}$, hence S^{n-1} and $S^n - \{a, b\}$ have the same homotopy type.

1.32. Assume that X, Y, and Z are spaces with $X \subset Y$. If X is a retract, then every continuous map $f: X \to Z$ can be extended to a continuous map $\tilde{f}: Y \to Z$, namely, $\tilde{f} = fr$, where $r: Y \to X$ is a retraction. Prove that if X is a retract of Y and if f_0 and f_1 are homotopic continuous maps $X \to Z$, then $\tilde{f}_0 \simeq \tilde{f}_1$.

Definition. Let $f: X \to Y$ be continuous and define[3]

$$M_f = ((X \times \mathbf{I}) \coprod Y)/\sim,$$

where $(x, t) \sim y$ if $y = f(x)$ and $t = 1$. Denote the class of (x, t) in M_f by $[x, t]$ and the class of y in M_f by $[y]$ (so that $[x, 1] = [f(x)]$). The space M_f is called the **mapping cylinder** of f.

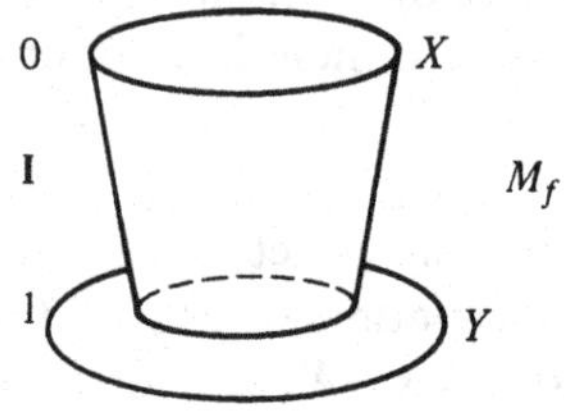

EXERCISES

1.33. If Y is a one-point space, then $f: X \to Y$ must be constant. Prove that the mapping cylinder in this case is CX, the cone on X.

1.34. (i) Define $i: X \to M_f$ by $i(x) = [x, 0]$ and $j: Y \to M_f$ by $j(y) = [y]$. Show that i and j are homeomorphisms to subspaces of M_f.

(ii) Define $r: M_f \to Y$ by $r[x, t] = f(x)$ for all $(x, t) \in X \times \mathbf{I}$ and $r[y] = y$. Prove that r is a retraction: $rj = 1_Y$.

(iii) Prove that Y is a deformation retract of M_f. (*Hint*: Define $F: M_f \times \mathbf{I} \to M_f$ by

$$F([x, t], s) = [x, (1 - s)t + s] \quad \text{if } x \in X, t, s \in \mathbf{I};$$

$$F([y], s) = [y] \quad \text{if } y \in Y, s \in \mathbf{I}.)$$

(iv) Show that every continuous map $f: X \to Y$ is homotopic to $r \circ i$, where i is an injection and r is a homotopy equivalence.

[3] If A and B are topological spaces, then $A \coprod B$ denotes their disjoint union topologized so that both A and B are open sets.

CHAPTER 2
Simplexes

Affine Spaces

Many interesting spaces are constructed from certain familiar subsets of euclidean space, called simplexes. This brief chapter is devoted to describing these sets and maps between them.

Definition. A subset A of euclidean space is called **affine** if, for every pair of distinct points $x, x' \in A$, the line determined by x, x' is contained in A.

Observe that affine subsets are convex (convexity requires only that the line *segment* between x and x' lies in the set). Note also that, by default, $\varnothing$ and one-point subsets are affine.

Theorem 2.1. *If $\{X_j: j \in J\}$ is a family of convex (or affine) subsets of $\mathbf{R}^n$, then $\bigcap X_j$ is also convex (or affine).*

PROOF. Immediate from the definitions. □

It thus makes sense to speak of the **convex** (or **affine**) **set** in $\mathbf{R}^n$ **spanned** by a subset X of $\mathbf{R}^n$ (also called the **convex hull** of X), namely, the intersection of all convex (or affine) subsets of $\mathbf{R}^n$ containing X. We denote the convex set spanned by X by $[X]$ (note that $[X]$ does exist, for $\mathbf{R}^n$ itself is affine, hence convex). It is hopeless to try to describe arbitrary convex subsets of $\mathbf{R}^n$: for example, for every subset K of S^1, the set $D^2 - K$ is convex. Even closed convex sets exist in abundance. However, we can describe $[X]$ for finite X.

Definition. An **affine combination** of points $p_0, p_1, \ldots, p_m$ in $\mathbf{R}^n$ is a point x with

$$x = t_0 p_0 + t_1 p_1 + \cdots + t_m p_m,$$

where $\sum_{i=0}^{m} t_i = 1$. A **convex combination** is an affine combination for which $t_i \geq 0$ for all i.

For example, a convex combination of x, x' has the form $tx + (1 - t)x'$ for $t \in \mathbf{I}$.

Theorem 2.2. *If $p_0, p_1, \ldots, p_m \in \mathbf{R}^n$, then $[p_0, p_1, \ldots, p_m]$, the convex set spanned by these points, is the set of all convex combinations of $p_0, p_1, \ldots, p_m$.*

PROOF. Let S denote the set of all convex combinations.

$[p_0, \ldots, p_m] \subset S$: It suffices to show that S is a convex set containing $\{p_0, \ldots, p_m\}$. First, if we set $t_j = 1$ and the other $t_i = 0$, then we see that $p_j \in S$ for every j. Second, let $\alpha = \sum a_i p_i$ and $\beta = \sum b_i p_i \in S$, where a_i, $b_i \geq 0$ and $\sum a_i = 1 = \sum b_i$. We claim that $t\alpha + (1 - t)\beta \in S$ for $t \in \mathbf{I}$. Now

$$t\alpha + (1 - t)\beta = \sum_{i=0}^{m} [ta_i + (1 - t)b_i]p_i.$$

This is a convex combination of $p_0, \ldots, p_m$, hence lies in S:

(i) $\sum [ta_i + (1 - t)b_i] = t\sum a_i + (1 - t)\sum b_i = t + (1 - t) = 1$;
(ii) $ta_i + (1 - t)b_i \geq 0$ because each term is nonnegative.

$S \subset [p_0, \ldots, p_m]$: If X is any convex set containing $\{p_0, \ldots, p_m\}$, we show that $S \subset X$ by induction on $m \geq 0$. If $m = 0$, then $S = \{p_0\}$ and we are done. Let $m > 0$. If $t_i \geq 0$ and $\sum t_i = 1$, is $p = \sum t_i p_i$ in X? We may assume that $t_0 \neq 1$ (otherwise $p = p_0 \in X$); by induction,

$$q = \left(\frac{t_1}{1 - t_0}\right)p_1 + \cdots + \left(\frac{t_m}{1 - t_0}\right)p_m \in X$$

(for this is a convex combination), and so

$$p = t_0 p_0 + (1 - t_0)q \in X,$$

because X is convex. □

Corollary 2.3. *The affine set spanned by $\{p_0, p_1, \ldots, p_m\} \subset \mathbf{R}^n$ consists of all affine combinations of these points.*

PROOF. A minor variation of the proof just given. □

Definition. An ordered set of points $\{p_0, p_1, \ldots, p_m\} \subset \mathbf{R}^n$ is **affine independent** if $\{p_1 - p_0, p_2 - p_0, \ldots, p_m - p_0\}$ is a linearly independent subset of the real vector space $\mathbf{R}^n$.

Any linearly independent subset of $\mathbf{R}^n$ is an affine independent set; the converse is not true, because any linearly independent set together with the origin is affine independent. Any one point set $\{p_0\}$ is affine independent (there are no points of the form $p_i - p_0$ with $i \neq 0$, and $\varnothing$ is linearly independent); a set $\{p_0, p_1\}$ is affine independent if $p_1 - p_0 \neq 0$, that is, if $p_1 \neq p_0$; a set $\{p_0, p_1, p_2\}$ is affine independent if it is not collinear; a set $\{p_0, p_1, p_2, p_3\}$ is affine independent if it is not coplanar.

Theorem 2.4. *The following conditions on an ordered set of points* $\{p_0, p_1, \ldots, p_m\}$ *in* $\mathbf{R}^n$ *are equivalent.*

(i) $\{p_0, p_1, \ldots, p_m\}$ *is affine independent;*
(ii) *if* $\{s_0, s_1, \ldots, s_m\} \subset \mathbf{R}$ *satisfies* $\sum_{i=0}^m s_i p_i = 0$ *and* $\sum_{i=0}^m s_i = 0$, *then* $s_0 = s_1 = \cdots = s_m = 0$;
(iii) *each* $x \in A$, *the affine set spanned by* $\{p_0, p_1, \ldots, p_m\}$, *has a unique expression as an affine combination:*

$$x = \sum_{i=0}^{m} t_i p_i \quad \textit{and} \quad \sum_{i=0}^{m} t_i = 1.$$

Proof. (i) $\Rightarrow$ (ii). Assume that $\sum s_i = 0$ and that $\sum s_i p_i = 0$. Then

$$\sum_{i=0}^{m} s_i p_i = \sum_{i=0}^{m} s_i p_i - \left(\sum_{i=0}^{m} s_i\right) p_0 = \sum_{i=0}^{m} s_i(p_i - p_0) = \sum_{i=1}^{m} s_i(p_i - p_0)$$

(because $p_i - p_0 = 0$ when $i = 0$). Affine independence of $\{p_0, \ldots, p_m\}$ gives linear independence of $\{p_1 - p_0, \ldots, p_m - p_0\}$, hence $s_i = 0$ for $i = 1, 2, \ldots, m$. Finally, $\sum s_i = 0$ implies that $s_0 = 0$ as well.

(ii) $\Rightarrow$ (iii). Assume that $x \in A$. By Corollary 2.3,

$$x = \sum_{i=0}^{m} t_i p_i,$$

where $\sum_{i=0}^m t_i = 1$. If, also,

$$x = \sum_{i=0}^{m} t_i' p_i,$$

where $\sum_{i=0}^m t_i' = 1$, then

$$0 = \sum_{i=0}^{m} (t_i - t_i')p_i.$$

Since $\sum (t_i - t_i') = \sum t_i - \sum t_i' = 1 - 1 = 0$, it follows that $t_i - t_i' = 0$ for all i, and $t_i = t_i'$ for all i, as desired.

(iii) $\Rightarrow$ (i). We may assume that $m \neq 0$. Assume that each $x \in A$ has a unique expression as an affine combination of $p_0, \ldots, p_m$. We shall reach a contradiction by assuming that $\{p_1 - p_0, \ldots, p_m - p_0\}$ is linearly dependent. If so, there would be real numbers r_i, not all zero, with

$$0 = \sum_{i=1}^{m} r_i(p_i - p_0).$$

Let $r_j \neq 0$; indeed, multiplying the equation by r_j^{-1} if necessary, we may suppose that $r_j = 1$. Now $p_j \in A$ has two expressions as an affine combination of $p_0, \ldots, p_m$:

$$p_j = 1p_j;$$

$$p_j = -\sum_{i \neq j} r_i p_i + \left(1 + \sum_{i \neq j} r_i\right) p_0,$$

where $1 \leq i \leq m$ in the summations (recall that $r_j = 1$). □

Corollary 2.5. *Affine independence is a property of the set $\{p_0, \ldots, p_m\}$ that is independent of the given ordering.*

PROOF. The characterizations of affine independence in the theorem do not depend on the given ordering. □

Corollary 2.6. *If A is the affine set in $\mathbf{R}^n$ spanned by an affine independent set $\{p_0, \ldots, p_m\}$, then A is a translate of an m-dimensional sub-vector-space V of $\mathbf{R}^n$, namely,*

$$A = V + x_0$$

for some $x_0 \in \mathbf{R}^n$.

PROOF. Let V be the sub-vector-space with basis $\{p_1 - p_0, \ldots, p_m - p_0\}$, and set $x_0 = p_0$. □

Definition. A set of points $\{a_1, a_2, \ldots, a_k\}$ in $\mathbf{R}^n$ is in **general position** if every $n + 1$ of its points forms an affine independent set.

Observe that the property of being in general position depends on n. Thus, assume that $\{a_1, a_2, \ldots, a_k\} \subset \mathbf{R}^n$ is in general position. If $n = 1$, we are saying that every pair $\{a_i, a_j\}$ is affine independent; that is, all the points are distinct. If $n = 2$, we are saying that no three points are collinear, and if $n = 3$, that no four points are coplanar.

Let $r_0, r_1, \ldots, r_m$ be real numbers. Recall that the $(m + 1) \times (m + 1)$ *Vandermonde matrix* V has as its ith column $[1, r_i, r_i^2, \ldots, r_i^m]$; moreover, $\det V = \prod_{j<i}(r_i - r_j)$, hence V is nonsingular if all the r_i are distinct. If one subtracts column 0 from each of the other columns of V, then the ith column (for $i > 0$) of the new matrix is

$$[0, r_i - r_0, r_i^2 - r_0^2, \ldots, r_i^m - r_0^m].$$

If V^* is the southeast $m \times m$ block of this new matrix, then $\det V^* = \det V$ (consider Laplace expansion across the first row).

Theorem 2.7. *For every $k \geq 0$, euclidean space $\mathbf{R}^n$ contains k points in general position.*

Proof. We may assume that $k > n + 1$ (otherwise, choose the origin together with $k - 1$ elements of a basis). Select k distinct reals $r_1, r_2, \ldots, r_k$, and for each $i = 1, 2, \ldots, k$, define

$$a_i = (r_i, r_i^2, \ldots, r_i^n) \in \mathbf{R}^n.$$

We claim that $\{a_1, a_2, \ldots, a_k\}$ is in general position. If not, there are $n + 1$ points $\{a_{i_0}, a_{i_1}, \ldots, a_{i_n}\}$ not affine independent, hence $\{a_{i_1} - a_{i_0}, a_{i_2} - a_{i_0}, \ldots, a_{i_n} - a_{i_0}\}$ is linearly dependent. There are thus real numbers $s_1, s_2, \ldots, s_n$, not all zero, with

$$0 = \sum s_j(a_{i_j} - a_{i_0}) = (\sum s_j(r_{i_j} - r_{i_0}), \sum s_j(r_{i_j}^2 - r_{i_0}^2), \ldots, \sum s_j(r_{i_j}^n - r_{i_0}^n)).$$

If V^* is the $n \times n$ southeast block of the $(n + 1) \times (n + 1)$ Vandermonde matrix obtained from $r_{i_0}, r_{i_1}, \ldots, r_{i_n}$, and if σ is the column vector $\sigma = (s_1, s_2, \ldots, s_n)$, then the vector equation above is $V^*\sigma = 0$. But since all the r_i are distinct, V^* is nonsingular and $\sigma = 0$, contradicting our hypothesis that not all the s_i are zero. □

There are other proofs of this theorem using induction on k. The key geometric observation needed is that $\mathbf{R}^n$ is not the union of only finitely many (proper) affine subsets (the reader may take this observation as an exercise).

Exercises

2.1. Every affine subset A of $\mathbf{R}^n$ is spanned by a finite subset. (*Hint*: Choose a maximal affine independent subset of A.) Conclude that every nonempty affine subset of $\mathbf{R}^n$ is as described in Corollary 2.6.

*2.2. Assume that $n < k$ and that the vector space $\mathbf{R}^n$ is isomorphic to a subspace of $\mathbf{R}^k$ (not necessarily the subspace of all those vectors whose last $k - n$ coordinates are 0). If X is a subset of $\mathbf{R}^n$, then the affine set spanned by X in $\mathbf{R}^n$ is the same as the affine set spanned by X in $\mathbf{R}^k$.

2.3. Show that S^n contains an affine independent set with $n + 2$ points. (*Hint*: Theorem 2.7.)

Definition. Let $\{p_0, p_1, \ldots, p_m\}$ be an affine independent subset of $\mathbf{R}^n$, and let A be the affine set spanned by this subset. If $x \in A$, then Theorem 2.4 gives a unique $(m + 1)$-tuple $(t_0, t_1, \ldots, t_m)$ with $\sum t_i = 1$ and $x = \sum_{i=0}^m t_i p_i$. The entries of this $(m + 1)$-tuple are called the **barycentric coordinates** of x (relative to the ordered set $\{p_0, p_1, \ldots, p_m\}$).

In light of Exercise 2.2, the barycentric coordinates of a point relative to $\{p_0, p_1, \ldots, p_m\} \subset \mathbf{R}^n$ do not depend on the ambient space $\mathbf{R}^n$.

Definition. Let $\{p_0, p_1, \ldots, p_m\}$ be an affine independent subset of $\mathbf{R}^n$. The convex set spanned by this set, denoted by $[p_0, p_1, \ldots, p_m]$, is called the (affine) ***m*-simplex** with **vertices** $p_0, p_1, \ldots, p_m$.

Theorem 2.8. *If $\{p_0, p_1, \dots, p_m\}$ is affine independent, then each x in the m-simplex $[p_0, p_1, \dots, p_m]$ has a unique expression of the form*

$$x = \sum t_i p_i, \quad \text{where } \sum t_i = 1 \text{ and each } t_i \geq 0.$$

PROOF. Theorem 2.2 shows that every $x \in [p_0, \dots, p_m]$ is such a convex combination. Were this expression not unique, the barycentric coordinates of x would not be unique. □

Definition. If $\{p_0, \dots, p_m\}$ is affine independent, the **barycenter** of $[p_0, \dots, p_m]$ is $(1/m + 1)(p_0 + p_1 + \cdots + p_m)$.

Barycenter comes from the Greek *barys* meaning heavy; thus, barycenter is just "center of gravity". Let us consider some low-dimensional examples; we assume that $\{p_0, \dots, p_m\}$ is affine independent.

EXAMPLE 2.1. $[p_0]$ is a 0-simplex and consists of one point, which is its own barycenter.

EXAMPLE 2.2. The 1-simplex $[p_0, p_1] = \{tp_0 + (1 - t)p_1 : t \in \mathbf{I}\}$ is the closed line segment with endpoints p_0, p_1. The barycenter $\frac{1}{2}(p_0 + p_1)$ is the midpoint of the line segment.

EXAMPLE 2.3. The 2-simplex $[p_0, p_1, p_2]$ is a triangle (with interior) with vertices p_0, p_1, p_2; the barycenter $\frac{1}{3}(p_0 + p_1 + p_2)$ is the center of gravity (this is easy to see in the special case of an equilateral triangle). Note that the three edges are $[p_0, p_1]$, $[p_1, p_2]$, and $[p_0, p_2]$. Now $[p_0, p_1]$ is the edge opposite

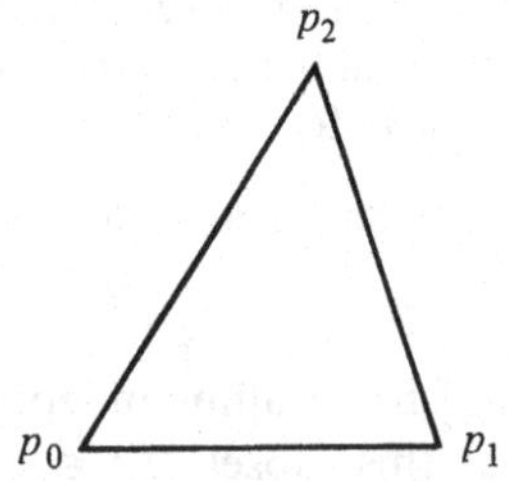

p_2 and is the 1-simplex obtained by deleting p_2. Thus, a point on this edge has barycentric coordinates $(t, 1 - t, 0)$; that is, the coordinate t_2 is 0. More generally, (t_0, t_1, t_2) lies on an edge if and only if one of its coordinates is zero (after all, such points are convex combinations of the endpoints of their respective edges).

EXAMPLE 2.4. The 3-simplex $[p_0, p_1, p_2, p_3]$ is the (solid) tetrahedron with vertices p_0, p_1, p_2, p_3. The triangular face opposite p_i consists of all those points whose ith barycentric coordinate is zero.

EXAMPLE 2.5. For $i = 0, 1, \ldots, n$, let e_i denote the point in $\mathbf{R}^{n+1}$ having (cartesian) coordinates all zeros except for 1 in the $(i + 1)$st position. Clearly, $\{e_0, e_1, \ldots, e_n\}$ is affine independent (it is even linearly independent). Now $[e_0, e_1, \ldots, e_n]$ consists of all convex combinations $x = \sum t_i e_i$. In this case, barycentric and cartesian coordinates $(t_0, t_1, \ldots, t_n)$ coincide, and $[e_0, e_1, \ldots, e_n] = \Delta^n$, the standard n-simplex.

The next definition gives names to what was seen in the examples.

Definition. Let $[p_0, p_1, \ldots, p_m]$ be an m-simplex. The **face opposite** p_i is

$$[p_0, \ldots, \hat{p}_i, \ldots, p_m] = \{\textstyle\sum t_j p_j \colon t_j \geq 0, \sum t_j = 1, \text{ and } t_i = 0\}$$

(circumflex ^ means "delete"). The **boundary** of $[p_0, p_1, \ldots, p_m]$ is the union of its faces.

Clearly, an m-simplex has $m + 1$ faces. For an integer k with $0 \leq k \leq m - 1$, one sometimes speaks of a ***k*-face** of $[p_0, p_1, \ldots, p_m]$, namely, a k-simplex spanned by $k + 1$ of the vertices $\{p_0, p_1, \ldots, p_m\}$. In this terminology, the faces defined above are $(m - 1)$-faces.

The following theorem will be needed when we discuss barycentric subdivision.

Theorem 2.9. *Let S denote the n-simplex $[p_0, \ldots, p_n]$.*

(i) *If $u, v \in S$, then* $\|u - v\| \leq \sup_i \|u - p_i\|$.
(ii) diam $S = \sup_{i,j} \|p_i - p_j\|$.
(iii) *If b is the barycenter of S, then* $\|b - p_i\| \leq (n/n + 1)$ diam S.

PROOF. (i) $v = \sum t_i p_i$, where $t_i \geq 0$ and $\sum t_i = 1$. Therefore

$$\begin{aligned}\|u - v\| &= \|u - \textstyle\sum t_i p_i\| = \|(\sum t_i)u - \sum t_i p_i\| \\ &\leq \textstyle\sum t_i \|u - p_i\| \leq \sum t_i \sup_i \|u - p_i\| = \sup_i \|u - p_i\|.\end{aligned}$$

(ii) By (i), $\|u - p_i\| \leq \sup_j \|p_j - p_i\|$.

(iii) Since $b = (1/n + 1)\sum p_i$, we have

$$\begin{aligned}\|b - p_i\| &= \left\| \sum_{j=0}^{n} (1/n + 1)p_j - p_i \right\| = \left\| \sum_{j=0}^{n} (1/n + 1)p_j - \left(\sum_{j=0}^{n} (1/n + 1)\right) p_i \right\| \\ &= \left\| \sum_{j=0}^{n} (1/n + 1)(p_j - p_i) \right\| \\ &\leq (1/n + 1) \sum_{j=0}^{n} \|p_j - p_i\| \\ &\leq (n/n + 1) \sup_{i,j} \|p_j - p_i\| \quad (\text{for } \|p_j - p_i\| = 0 \text{ when } j = i) \\ &= (n/n + 1) \operatorname{diam} S.\end{aligned}$$

□

Affine Maps

Definition. Let $\{p_0, p_1, \ldots, p_m\} \subset \mathbf{R}^n$ be affine independent and let A denote the affine set it spans. An **affine map** $T: A \to \mathbf{R}^k$ (for some $k \geq 1$) is a function satisfying

$$T(\sum t_j p_j) = \sum t_j T(p_j)$$

whenever $\sum t_j = 1$. The restriction of T to $[p_0, p_1, \ldots, p_m]$ is also called an **affine map**.

Thus affine maps preserve affine combinations, hence convex combinations. It is clear that an affine map is determined by its values on an affine independent subset; its restriction to a simplex is thus determined by its values on the vertices. Moreover, uniqueness of barycentric coordinates relative to $\{p_0, \ldots, p_m\}$ shows that such an affine T exists, since the formula in the definition is well defined.

Theorem 2.10. *If $[p_0, \ldots, p_m]$ is an m-simplex, $[q_0, \ldots, q_n]$ an n-simplex, and $f: \{p_0, \ldots, p_m\} \to [q_0, \ldots, q_n]$ any function, then there exists a unique affine map $T: [p_0, \ldots, p_m] \to [q_0, \ldots, q_n]$ with $T(p_i) = f(p_i)$ for $i = 0, 1, \ldots, m$.*

PROOF. Define $T(\sum t_i p_i) = \sum t_i f(p_i)$, where $\sum t_i p_i$ is a convex combination. Uniqueness is obvious. □

EXERCISES

*2.4. If $T: \mathbf{R}^n \to \mathbf{R}^k$ is affine, then $T(x) = \lambda(x) + y_0$, where $\lambda: \mathbf{R}^n \to \mathbf{R}^k$ is a linear transformation and $y_0 \in \mathbf{R}^k$ is fixed. (*Hint*: Define $y_0 = T(0)$.)

2.5. Every affine map is continuous.

*2.6. Prove that any two m-simplexes are homeomorphic via an affine map.

*2.7. Give an explicit formula for the affine map $\theta: \mathbf{R} \to \mathbf{R}$ carrying $[s_1, s_2] \to [t_1, t_2]$ with $\theta(s_i) = t_i$, $i = 1, 2$. In particular, give a formula for the affine map taking [32, 212] onto [0, 100]. (*Hint*: $\theta(x) = \lambda x + x_0$, by Exercise 2.4.)

*2.8. Let $A \subset \mathbf{R}^n$ be an affine set and let $T: A \to \mathbf{R}^k$ be an affine map. If $X \subset A$ is affine (or convex), then $T(X) \subset \mathbf{R}^k$ is affine (or convex). In particular, if a, b are distinct points in A and if ℓ is the line segment with endpoints a, b, then $T(\ell)$ is the line segment with endpoints $T(a)$, $T(b)$ if $T(a) \neq T(b)$, and $T(\ell)$ collapses to the point $T(a)$ if $T(a) = T(b)$.

2.9. If $\{p_0, p_1, \ldots, p_m\}$ is affine independent with barycenter b, then $\{b, p_0, \ldots, \hat{p}_i, \ldots, p_m\}$ (i.e., delete p_i) is affine independent for each i.

*2.10. Show that, for $0 \leq i \leq m$, $[p_0, \ldots, p_m]$ is homeomorphic to the cone $C[p_0, \ldots, \hat{p}_i, \ldots, p_m]$ with vertex p_i.

*2.11. Give an explicit homeomorphism from an n-simplex $[p_0, \ldots, p_n]$ to D^n. (*Hint*: Any n-simplex is homeomorphic to Δ^n, by Exercise 2.6, and $\Delta^n \approx D^n$ by radial stretching.)

CHAPTER 3

The Fundamental Group

The first functor we have constructed on **Top** (actually, on **hTop**), namely, π_0, takes values in **Sets**; it is of limited use because it merely counts the number of path components. The functor to be constructed in this chapter takes values in **Groups**, the category of (not necessarily abelian) groups. The basic idea is that one can "multiply" two paths f and g if f ends where g begins.

The Fundamental Groupoid

Definition. Let $f, g\colon \mathbf{I} \to X$ be paths with $f(1) = g(0)$. Define a path $f * g\colon \mathbf{I} \to X$ by

$$(f * g)(t) = \begin{cases} f(2t) & \text{if } 0 \le t \le \frac{1}{2} \\ g(2t - 1) & \text{if } \frac{1}{2} \le t \le 1. \end{cases}$$

The gluing lemma shows that $f * g$ is continuous (for $f(1) = g(0)$), and so $f * g$ is a path in X. Our aim is to construct a group whose elements are certain homotopy classes of paths in X with binary operation $[f][g] = [f * g]$. Now if we impose the rather mild condition that X be path connected, then contractibility of $\mathbf{I}$ implies that all maps $\mathbf{I} \to X$ are homotopic (Exercise 1.19(ii)); thus, there is only one homotopy class of maps. Since groups of order 1 carry little information, we modify our earlier definition of homotopy.

Definition. Let $A \subset X$ and let $f_0, f_1\colon X \to Y$ be continuous maps with $f_0|A = f_1|A$. We write

$$f_0 \simeq f_1 \text{ rel } A$$

if there is a continuous map $F: X \times \mathbf{I} \to Y$ with $F: f_0 \simeq f_1$ and

$$F(a, t) = f_0(a) = f_1(a) \quad \text{for all } a \in A \text{ and all } t \in \mathbf{I}.$$

The homotopy F above is called a **relative homotopy** (more precisely, a homotopy rel A); in contrast, the original definition (which may be viewed as a homotopy rel $A = \varnothing$) is called a **free homotopy**. We leave to the reader the routine exercise that, for fixed $A \subset X$, homotopy rel A is an equivalence relation on the set of continuous maps $X \to Y$.

Definition. Let $\dot{\mathbf{I}} = \{0, 1\}$ be the boundary of $\mathbf{I}$ in $\mathbf{R}$. The equivalence class of a path $f: \mathbf{I} \to X$ rel $\dot{\mathbf{I}}$ is called the **path class** of f and is denoted by $[f]$.

No confusion should arise from using the same notation for the homotopy class of a path as for its path class, because we have remarked that the (free) homotopy class is always trivial.

Theorem 3.1. *Assume that f_0, f_1, g_0, g_1 are paths in X with*

$$f_0 \simeq f_1 \text{ rel } \dot{\mathbf{I}} \quad \textit{and} \quad g_0 \simeq g_1 \text{ rel } \dot{\mathbf{I}}.$$

If $f_0(1) = f_1(1) = g_0(0) = g_1(0)$, then $f_0 * g_0 \simeq f_1 * g_1$ rel $\dot{\mathbf{I}}$.

Remark. In path class notation, if $[f_0] = [f_1]$ and $[g_0] = [g_1]$, then $[f_0 * g_0] = [f_1 * g_1]$ (assuming that the stars are defined).

PROOF. If $F: f_0 \simeq f_1$ rel $\dot{\mathbf{I}}$ and $G: g_0 \simeq g_1$ rel $\dot{\mathbf{I}}$, then one checks easily that $H: \mathbf{I} \times \mathbf{I} \to X$ defined by

$$H(t, s) = \begin{cases} F(2t, s) & \text{if } 0 \le t \le \frac{1}{2} \\ G(2t - 1, s) & \text{if } \frac{1}{2} \le t \le 1 \end{cases}$$

is a continuous map (the gluing lemma applies because both functions agree on $\{\frac{1}{2}\} \times \mathbf{I}$) that is a relative homotopy $f_0 * g_0 \simeq f_1 * g_1$ rel $\dot{\mathbf{I}}$. □

EXERCISES

*3.1. Generalize Theorem 1.3 as follows. Let $A \subset X$ and $B \subset Y$ be given. Assume that $f_0, f_1: X \to Y$ with $f_0|A = f_1|A$ and $f_i(A) \subset B$ for $i = 0.1$; assume $g_0, g_1: Y \to Z$ with $g_0|B = g_1|B$. If $f_0 \simeq f_1$ rel A and $g_0 \simeq g_1$ rel B, then $g_0 \circ f_0 \simeq g_1 \circ f_1$ rel A.

*3.2. (i) If $f: \mathbf{I} \to X$ is a path with $f(0) = f(1) = x_0 \in X$, then there is a continuous $f': S^1 \to X$ given by $f'(e^{2\pi i t}) = f(t)$. If $f, g: \mathbf{I} \to X$ are paths with $f(0) = f(1) = x_0 = g(0) = g(1)$ and if $f \simeq g$ rel $\dot{\mathbf{I}}$, then $f' \simeq g'$ rel$\{1\}$ (of course, $1 = e^0 \in S^1$).

(ii) If f and g are as above, then $f \simeq f_1$ rel $\dot{\mathbf{I}}$ and $g \simeq g_1$ rel $\dot{\mathbf{I}}$ implies that $f' * g' \simeq f_1' * g_1'$ rel$\{1\}$.

3.3. Using Theorem 1.6, show (with the notation of Exercise 3.2) that if f and g are paths with g constant, then $f' \simeq g'$ rel$\{1\}$ if and only if there is a free homotopy $f' \simeq g'$.

Definition. If $f: \mathbf{I} \to X$ is a path from x_0 to x_1, call x_0 the **origin** of f and write $x_0 = \alpha(f)$; call x_1 the **end** of f and write $x_1 = \omega(f)$. A path f in X is **closed** at x_0 if $\alpha(f) = x_0 = \omega(f)$.

Observe that if f and g are paths with $f \simeq g$ rel $\dot{\mathbf{I}}$, then $\alpha(f) = \alpha(g)$ and $\omega(f) = \omega(g)$; therefore we may speak of the **origin** and **end** of a path class and write $\alpha[f]$ and $\omega[f]$.

Definition. If $p \in X$, then the constant function $i_p: \mathbf{I} \to X$ with $i_p(t) = p$ for all $t \in \mathbf{I}$ is called the **constant path** at p. If $f: \mathbf{I} \to X$ is a path, its **inverse** path $f^{-1}: \mathbf{I} \to X$ is defined by $t \mapsto f(1-t)$.

Exercises

*3.4. Let $\sigma: \Delta^2 \to X$ be continuous, where $\Delta^2 = [e_0, e_1, e_2]$.

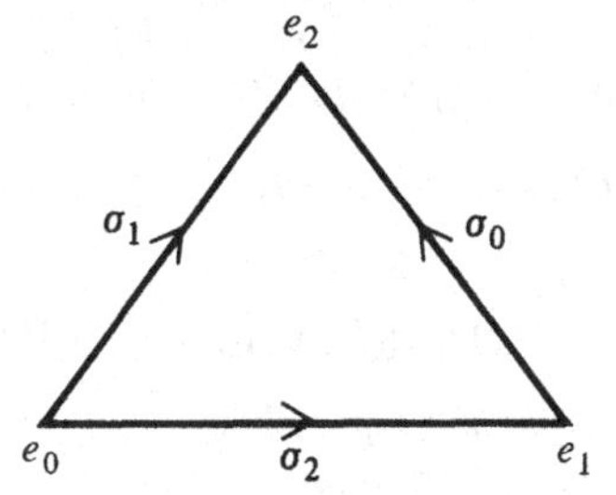

Define $\varepsilon_0: \mathbf{I} \to \Delta^2$ as the affine map with $\varepsilon_0(0) = e_1$ and $\varepsilon_0(1) = e_2$; similarly, define ε_1 by $\varepsilon_1(0) = e_0$ and $\varepsilon_1(1) = e_2$, and define ε_2 by $\varepsilon_2(0) = e_0$ and $\varepsilon_2(1) = e_1$. Finally, define $\sigma_i = \sigma \circ \varepsilon_i$ for $i = 0, 1, 2$.

(i) Prove that $(\sigma_0 * \sigma_1^{-1}) * \sigma_2$ is nullhomotopic rel $\dot{\mathbf{I}}$. (*Hint*: Theorem 1.6.)

(ii) Prove that $(\sigma_1 * \sigma_0^{-1}) * \sigma_2^{-1}$ is nullhomotopic rel $\dot{\mathbf{I}}$.

(iii) Let $F: \mathbf{I} \times \mathbf{I} \to X$ be continuous, and define paths α, β, γ, δ in X as indicated in the figure.

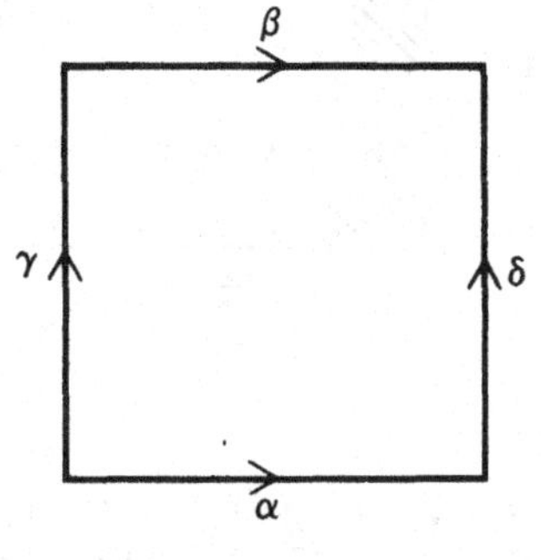

Thus, $\alpha(t) = F(t, 0)$, $\beta(t) = F(t, 1)$, $\gamma(t) = F(0, t)$, and $\delta(t) = F(1, t)$. Prove that $\alpha \simeq \gamma * \beta * \delta^{-1}$ rel $\dot{\mathbf{I}}$.

*3.5. Let $f_0 \simeq f_1$ rel $\dot{\mathbf{I}}$ and $g_0 \simeq g_1$ rel $\dot{\mathbf{I}}$ be paths in X and Y, respectively. If, for $i = 0$, 1, (f_i, g_i) is the path in $X \times Y$ defined by $t \mapsto (f_i(t), g_i(t))$, prove that $(f_0, g_0) \simeq (f_1, g_1)$ rel $\dot{\mathbf{I}}$.

*3.6. (i) If $f \simeq g$ rel $\dot{\mathbf{I}}$, then $f^{-1} \simeq g^{-1}$ rel $\dot{\mathbf{I}}$, where f, g are paths in X.
(ii) If f and g are paths in X with $\omega(f) = \alpha(g)$, then

$$(f * g)^{-1} = g^{-1} * f^{-1}.$$

(iii) Give an example of a closed path f with $f * f^{-1} \neq f^{-1} * f$.
(iv) Show that if $\alpha(f) = p$ and f is not constant, then $i_p * f \neq f$.

Exercise 3.6 shows that it is hopeless to force paths to form a group under $*$ unless we can somehow identify, for example, $f * f^{-1}$ with $f^{-1} * f$ (of course, there are other obstacles as well). The next theorem shows that replacing paths by path classes resolves most problems.

Theorem 3.2. *If X is a space, then the set of all path classes in X under the (not always defined) binary operation $[f][g] = [f * g]$ forms an algebraic system (called a* **groupoid***) satisfying the following properties:*

(i) *each path class $[f]$ has an origin $\alpha[f] = p \in X$ and an end $\omega[f] = q \in X$, and*

$$[i_p][f] = [f] = [f][i_q];$$

(ii) *associativity holds whenever possible;*
(iii) *if $p = \alpha[f]$ and $q = \omega[f]$, then*

$$[f][f^{-1}] = [i_p] \quad \text{and} \quad [f^{-1}][f] = [i_q].$$

PROOF. (i) We show only that $i_p * f \simeq f$ rel $\dot{\mathbf{I}}$; the other half is similar.

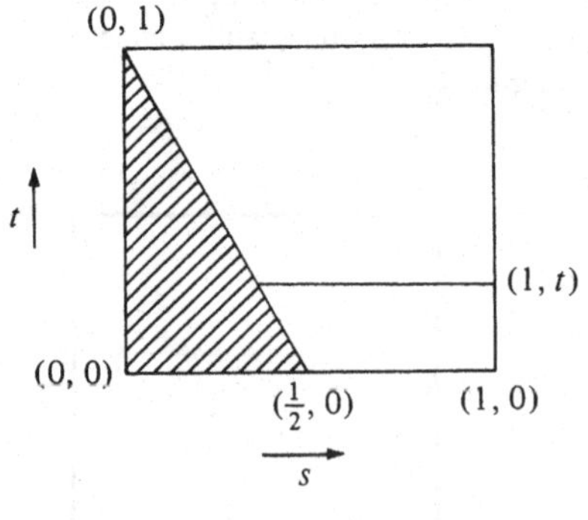

First, draw the line in $\mathbf{I} \times \mathbf{I}$ joining (0, 1) to $(\frac{1}{2}, 0)$; its equation is $2s = 1 - t$. For fixed t, define θ_t: $[(1 - t)/2, 1] \to [0, 1]$ as the affine map matching the endpoints of these intervals. By Exercise 2.7,

$$\theta_t(s) = \frac{s - (1-t)/2}{1 - (1-t)/2}.$$

Define $H\colon \mathbf{I} \times \mathbf{I} \to X$ by

$$H(s, t) = \begin{cases} p & \text{if } 2s \le 1 - t \quad ((s, t) \in \text{shaded triangle}) \\ f(\theta_t(s)) = f((2s - 1 + t)/(1 + t)) & \text{if } 2s \ge 1 - t. \end{cases}$$

One sees easily that H is continuous (using the gluing lemma),[1] that $H\colon i_p * f \simeq f$, and that $\dot{\mathbf{I}}$ remains fixed during the homotopy.

(ii) To prove associativity, use the picture below.

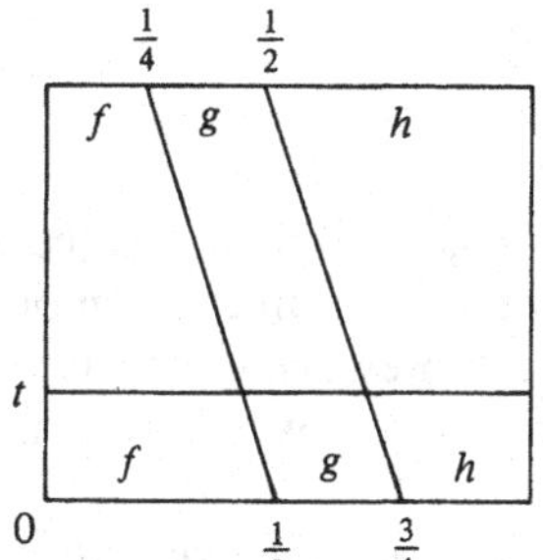

First, draw the slanted lines in $\mathbf{I} \times \mathbf{I}$ and write their equations. On each of the three pieces, construct a continuous function whose formula is, for each fixed t, the affine map from the bottom 0th interval (e.g., from $[0, \frac{1}{2}]$) to the upper tth interval (e.g., to $[0, (2 - t)/4]$). It suffices to show that the continuous map obtained by gluing maps together, as in part (i), is a homotopy $f * (g * h) \simeq (f * g) * h$ rel $\dot{\mathbf{I}}$, and this is routine.

(iii) We show only that $f * f^{-1} \simeq i_p$ rel $\dot{\mathbf{I}}$; the other half is similar. One proceeds as in the first two cases, subdividing $\mathbf{I} \times \mathbf{I}$; here are the formulas. Define $H\colon \mathbf{I} \times \mathbf{I} \to X$ by

$$H(s, t) = \begin{cases} f(2s(1 - t)) & \text{if } 0 \le s \le \frac{1}{2} \\ f(2(1 - s)(1 - t)) & \text{if } \frac{1}{2} \le s \le 1. \end{cases}$$

That H is the desired relative homotopy is left to the reader. □

The groupoid in Theorem 3.2 is not a group because multiplication is not always defined; we remedy this defect in the most naive possible way, namely, by restricting our attention to closed paths. See [Brown] for uses of groupoids in topology.

[1] $\mathbf{I} \times \mathbf{I}$ is divided into two pieces: a triangle and a quadrilateral. The affine maps on each tth interval give the formula for a function of two variables defined on the quadrilateral; this formula is used to show that this function is continuous.

Definition. Fix a point $x_0 \in X$ and call it the **basepoint**. The **fundamental group** of X with basepoint x_0 is

$$\pi_1(X, x_0) = \{[f]: [f] \text{ is a path class in } X \text{ with } \alpha[f] = x_0 = \omega[f]\}$$

with binary operation

$$[f][g] = [f * g].$$

Theorem 3.3. *$\pi_1(X, x_0)$ is a group for each $x_0 \in X$.*

PROOF. This follows at once from Theorem 3.2. □

The Functor π_1

We have been led to the category **Top**$_*$ of pointed spaces and pointed maps that we introduced in Chapter 0. Recall that a morphism $f: (X, x_0) \to (Y, y_0)$ is a continuous map $f: X \to Y$ preserving the basepoint: $f(x_0) = y_0$. In **Top**$_*$, one usually chooses 0 as the basepoint of **I** and 1 as the basepoint of S^1.

Theorem 3.4. *π_1: **Top**$_* \to$ **Groups** is a (covariant) functor. Moreover, if h, $k: (X, x_0) \to (Y, y_0)$ and $h \simeq k \operatorname{rel}\{x_0\}$, then $\pi_1(h) = \pi_1(k)$.*

PROOF. If $[f] \in \pi_1(X, x_0)$, define $\pi_1(h)$ by $[f] \mapsto [h \circ f]$. Note that the composite $h \circ f: \mathbf{I} \to Y$ is defined, is continuous, and is a closed path in Y at y_0; thus $[h \circ f] \in \pi_1(Y, y_0)$. Also, $\pi_1(h)$ is well defined: if $f \simeq f' \operatorname{rel} \dot{\mathbf{I}}$, then $h \circ f \simeq h \circ f' \operatorname{rel} \dot{\mathbf{I}}$ (Exercise 3.1). If f and g are closed paths in X at x_0, then evaluation of both sides shows that there is equality (not merely homotopy)

$$h \circ (f * g) = (h \circ f) * (h \circ g);$$

it follows that $\pi_1(h)$ is a homomorphism.

It is routine to check that π_1 preserves composition and identities in **Top**$_*$, so that π_1 is indeed a functor.

Finally, Exercise 3.1 shows that $h \simeq k \operatorname{rel}\{x_0\}$ implies that $h \circ f \simeq k \circ f \operatorname{rel} \dot{\mathbf{I}}$ whenever f is a closed path in X at x_0. Thus $[h \circ f] = [k \circ f]$ for all such f; that is, $\pi_1(h) = \pi_1(k)$. □

Remarks. (1) One usually writes h_* instead of $\pi_1(h)$ and calls h_* the map **induced** by h.

(2) We have shown that $h_* = k_*$ if there is a relative homotopy $h \simeq k \operatorname{rel}\{x_0\}$. We have not shown that $h_* = k_*$ if there is a free homotopy $h \simeq k$ (between pointed maps h and k), and this may not be true (we shall return to this point in Lemma 3.8).

(3) There is a category appropriate to the fundamental group functor π_1. Define the **pointed homotopy category, hTop**$_*$, as the quotient category arising

from the congruence of relative homotopy: if $f_0, f_1: (X, x_0) \to (Y, y_0)$, then $f_0 \simeq f_1 \text{ rel}\{x_0\}$. The objects of **hTop**$_*$ are pointed spaces (X, x_0), morphisms $(X, x_0) \to (Y, y_0)$ are relative homotopy classes $[f]$, where $f: (X, x_0) \to (Y, y_0)$ is a pointed map, and composition is given by $[h][f] = [h \circ f]$ (when h, f can be composed in **Top**$_*$). By Exercise 3.2, each closed path $f: (\mathbf{I}, \dot{\mathbf{I}}) \to (Y, y_0)$ may be viewed as a pointed map $f': (S^1, 1) \to (Y, y_0)$. If Hom sets in **hTop**$_*$ are denoted by $[(X, x_0), (Y, y_0)]$, then $[f] \mapsto [f']$ is a bijection

$$\pi_1(Y, y_0) \rightarrow [(S^1, 1), (Y, y_0)].$$

Using Exercise 3.2(ii), one may introduce a multiplication in the Hom set, namely, $[f'][g'] = [(f * g)']$, and the bijection is now an isomorphism. Therefore π_1 is an instance of a covariant Hom functor (Example 0.11). Roughly speaking, the fundamental group of a space Y is just the set of morphisms $S^1 \to Y$. We shall elaborate on this theme when we introduce the higher homotopy group functors π_n (which, roughly speaking, are the morphisms of S^n into a space). These remarks are designed to place π_1 in its proper context, to whet the reader's appetite for the π_n's, and to indicate that paying attention to categories is worthwhile. On the other hand, we must say that the fundamental group was invented and used (by Poincaré) 50 years before anyone dreamt of categories!

Let us return to properties of fundamental groups. The next result shows that one may as well assume that spaces are path connected.

Let x_0 be a basepoint of a space X, and let A be a subspace of X containing x_0; the inclusion $j: (A, x_0) \hookrightarrow (X, x_0)$ is a pointed map, and hence it induces a homomorphism $j_*: \pi_1(A, x_0) \to \pi_1(X, x_0)$, namely, $[f] \mapsto [jf]$ (where f is a closed path in A at x_0). The path jf is the path f now regarded as a path in X. It is possible that f is not nullhomotopic in A, yet f (really, jf) is nullhomotopic in X (e.g., take X to be a contractible space containing A—the cone CA will do for X); the extra room in X may allow f to be contracted to a point in X even though this is impossible in A. The homomorphism j_* may thus have a kernel.

Theorem 3.5. *Let $x_0 \in X$, and let X_0 be the path component of X containing x_0. Then*

$$\pi_1(X_0, x_0) \cong \pi_1(X, x_0).$$

PROOF. Let $j: (X_0, x_0) \hookrightarrow (X, x_0)$ be the inclusion. If $[f] \in \ker j_*$, then $jf \simeq c \text{ rel } \dot{\mathbf{I}}$, where $c: \mathbf{I} \to X$ is the constant path at x_0. If $F: \mathbf{I} \times \mathbf{I} \to X$ is a homotopy, then $F(0, 0) = x_0$; as $F(\mathbf{I} \times \mathbf{I})$ is path connected, it follows that $F(\mathbf{I} \times \mathbf{I}) \subset X_0$. It is now a simple matter to see that f is nullhomotopic in X_0. Hence j_* is injective. To see that j_* is surjective, observe that if $f: \mathbf{I} \to X$ is a closed path at x_0, then $f(\mathbf{I}) \subset X_0$. Be fussy and define $f': \mathbf{I} \to X_0$ by $f'(t) = f(t)$ for all $t \in \mathbf{I}$; note that $jf' = f$. □

What happens when the basepoint is changed?

Theorem 3.6. *If X is path connected and $x_0, x_1 \in X$, then*

$$\pi_1(X, x_0) \cong \pi_1(X, x_1).$$

PROOF. Let γ be a path in X from x_0 to x_1. Define $\varphi: \pi_1(X, x_0) \to \pi_1(X, x_1)$ by $[f] \mapsto [\gamma^{-1}][f][\gamma]$ (note that the multiplication occurs in the groupoid of X). Using Theorem 3.2, one sees easily that φ is an isomorphism (with inverse $[g] \mapsto [\gamma][g][\gamma^{-1}]$). □

It follows that the fundamental group of a space X is independent of the choice of basepoint when X is path connected.

Let us establish notation. In a cartesian product $H \times K$, there are two **projections**: $p: H \times K \to H$ and $q: H \times K \to K$ defined by $p(h, k) = h$ and $q(h, k) = k$. Also, if $\alpha: L \to H$ and $\beta: L \to K$ are functions from some set L, then there is a function $(\alpha, \beta): L \to H \times K$ defined by $(\alpha, \beta)(x) = (\alpha(x), \beta(x))$. Of course, $p \circ (\alpha, \beta) = \alpha$ and $q \circ (\alpha, \beta) = \beta$.

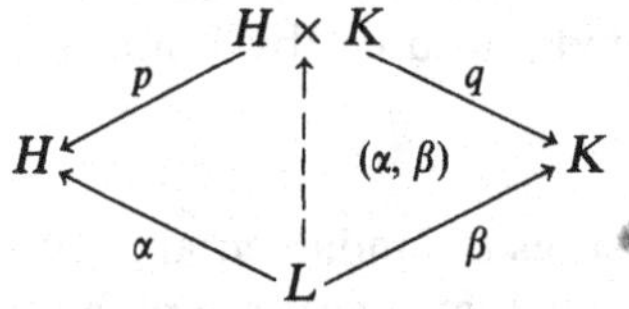

If the sets are groups and the functions are homomorphisms, then (α, β) is a homomorphism; if the sets are topological spaces and the functions are continuous, then (α, β) is continuous.

Theorem 3.7. *If (X, x_0) and (Y, y_0) are pointed spaces, then*

$$\pi_1(X \times Y, (x_0, y_0)) \cong \pi_1(X, x_0) \times \pi_1(Y, y_0).$$

PROOF. Let $p: (X \times Y, (x_0, y_0)) \to (X, x_0)$ and $q: (X \times Y, (x_0, y_0)) \to (Y, y_0)$ be the projections. Then $(p_*, q_*): \pi_1(X \times Y, (x_0, y_0)) \to \pi_1(X, x_0) \times \pi_1(Y, y_0)$ is a homomorphism. In more detail, if $f: \mathbf{I} \to X \times Y$ is a closed path at (x_0, y_0), then $(p_*, q_*): [f] \mapsto (p_*[f], q_*[f]) = ([pf], [qf])$. We show that (p_*, q_*) is an isomorphism by displaying its inverse. Let g be a closed path in X at x_0, and let h be a closed path in Y at y_0; define $\theta: \pi_1(X, x_0) \times \pi_1(Y, y_0) \to \pi_1(X \times Y, (x_0, y_0))$ by

$$\theta: ([g], [h]) \mapsto [(g, h)],$$

where $(g, h): \mathbf{I} \to X \times Y$ is defined by $t \mapsto (g(t), h(t))$; Exercise 3.5 shows that θ is well defined. It is routine to check that (p_*, q_*) and θ are inverse. □

Remark. Often it is not enough to know that two groups are isomorphic; one needs to know an explicit isomorphism. For example, we shall use the isomorphisms (p_*, q_*) and θ in the proof of Theorem 3.20.

EXERCISES

3.7. If X is the $\sin(1/x)$ space, prove that $\pi_1(X, x_0) = \{1\}$ for every $x_0 \in X$.

*3.8. Give an example of a contractible space that is not locally path connected. (*Hint*: Take the cone on a suitable space.)

*3.9. Let X be a space. Show that there is a category $\mathscr{C}$ with obj $\mathscr{C} = X$, with $\text{Hom}(p, q) = \{$all path classes $[f]$ with $\alpha[f] = p$ and $\omega[f] = q\}$, and with composition $\text{Hom}(p, q) \times \text{Hom}(q, r) \to \text{Hom}(p, r)$ defined by $([f], [g]) \mapsto [f * g]$. Show that every morphism in $\mathscr{C}$ is an equivalence.

3.10. If (X, x_0) is a pointed space, let the path component of X containing x_0 be the basepoint of $\pi_0(X)$; show that π_0 defines a functor $\mathbf{Top}_* \to \mathbf{Sets}_*$ (pointed sets).

*3.11. If $X = \{x_0\}$ is a one-point space, then $\pi_1(X, x_0) = \{1\}$.

Choosing a basepoint in X is only an artifice to extract a group from a groupoid. On this minor point, we have constructed new categories $\mathbf{Top}_*$ and $\mathbf{hTop}_*$; eventually, we shall see that we have not overreacted. Nevertheless, these constructions raise an honest question: Do spaces having the same homotopy type have isomorphic fundamental groups?

Lemma 3.8. *Assume that $F\colon \varphi_0 \simeq \varphi_1$ is a (free) homotopy, where $\varphi_i\colon X \to Y$ is continuous for $i = 0, 1$. Choose $x_0 \in X$ and let λ denote the path $F(x_0, \ \ )$ in Y from $\varphi_0(x_0)$ to $\varphi_1(x_0)$. Then there is a commutative diagram*

$$\begin{array}{ccc} \pi_1(X, x_0) & \xrightarrow{\varphi_{1*}} & \pi_1(Y, \varphi_1(x_0)) \\ & \searrow^{\varphi_{0*}} & \downarrow \psi \\ & & \pi_1(Y, \varphi_0(x_0)), \end{array}$$

where ψ is the isomorphism $[g] \mapsto [\lambda * g * \lambda^{-1}]$.

PROOF. Let $f\colon \mathbf{I} \to X$ be a closed path at x_0, and define $G\colon \mathbf{I} \times \mathbf{I} \to Y$ by

$$G(t, s) = F(f(t), s).$$

Note that $G\colon \varphi_0 \circ f \simeq \varphi_1 \circ f$ (of course, $\varphi_0 \circ f$ and $\varphi_1 \circ f$ are closed paths in Y at $\varphi_0(x_0)$ and $\varphi_1(x_0)$, respectively). Consider the two triangulations of the square $\mathbf{I} \times \mathbf{I}$ pictured below.

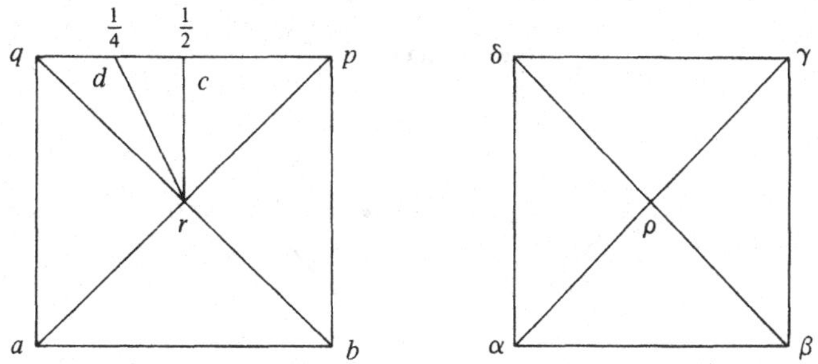

Define a continuous map $H: \mathbf{I} \times \mathbf{I} \to \mathbf{I} \times \mathbf{I}$ by first defining it on each triangle and then invoking the gluing lemma. On each triangle (2-simplex), H shall be an affine map; it thus suffices to evaluate H on each vertex (observe that agreement on overlaps is automatic here). Define $H(a) = H(q) = \alpha$, $H(b) = H(p) = \beta$; $H(c) = \gamma$; $H(d) = \delta$; $H(r) = \rho$. By Exercise 2.8, the vertical edge $[a, q]$ collapses to α, and the vertical edge $[b, p]$ collapses to β. Also, $[q, d]$ goes to $[\alpha, \delta]$, $[d, c]$ goes to $[\delta, \gamma]$, and $[c, p]$ goes to $[\gamma, \beta]$. The map $J = G \circ H: \mathbf{I} \times \mathbf{I} \to Y$ is easily seen to be a relative homotopy;

$$J: \varphi_0 \circ f \simeq (\lambda * (\varphi_1 \circ f)) * \lambda^{-1} \text{ rel } \dot{\mathbf{I}}.$$

Therefore $\varphi_{0*}[f] = [\varphi_0 \circ f] = [\lambda * \varphi_1 \circ f * \lambda^{-1}]$ (using homotopy associativity). On the other hand, $\psi\varphi_{1*}[f] = \psi[\varphi_1 \circ f] = [\lambda * \varphi_1 \circ f * \lambda^{-1}]$, as desired. □

This lemma shows that freely homotopic maps φ_0 and φ_1 may not induce the same homomorphism between fundamental groups, because they differ by the isomorphism ψ.

Corollary 3.9. *Assume that $\varphi_i: (X, x_0) \to (Y, y_0)$, for $i = 0, 1$, are freely homotopic.*

(i) *φ_{0*} and φ_{1*} are conjugate; that is, there is $[\lambda] \in \pi_1(Y, y_0)$ with $\varphi_{0*}[f] = [\lambda]\varphi_{1*}([f])[\lambda]^{-1}$ for every $[f] \in \pi_1(X, x_0)$.*
(ii) *If $\pi_1(Y, y_0)$ is abelian, then $\varphi_{0*} = \varphi_{1*}$.*

Proof. In the notation of the lemma, we have $\varphi_0(x_0) = y_0 = \varphi_1(x_0)$, and the path λ in Y is now a closed path at y_0; therefore $[\lambda]$ lies in $\pi_1(Y, y_0)$. The path class $[\lambda * \varphi_1 * f * \lambda^{-1}]$, which can always be factored in the groupoid of Y, now factors in the group $\pi_1(Y, y_0)$:

$$\begin{aligned}[\lambda * \varphi_1 \circ f * \lambda^{-1}] &= [\lambda][\varphi_1 \circ f][\lambda^{-1}] \\ &= [\lambda]\varphi_{1*}([f])[\lambda]^{-1}.\end{aligned}$$

This proves (i), and the second statement is immediate from this. □

Theorem 3.10. *If $\beta: X \to Y$ is a homotopy equivalence, then the induced homomorphism $\beta_*: \pi_1(X, x_0) \to \pi_1(Y, \beta(x_0))$ is an isomorphism for every $x_0 \in X$.*

Proof. Choose a continuous map $\alpha: Y \to X$ with $\alpha \circ \beta \simeq 1_X$ and $\beta \circ \alpha \simeq 1_Y$. By the lemma, the lower triangle of the diagram below commutes.

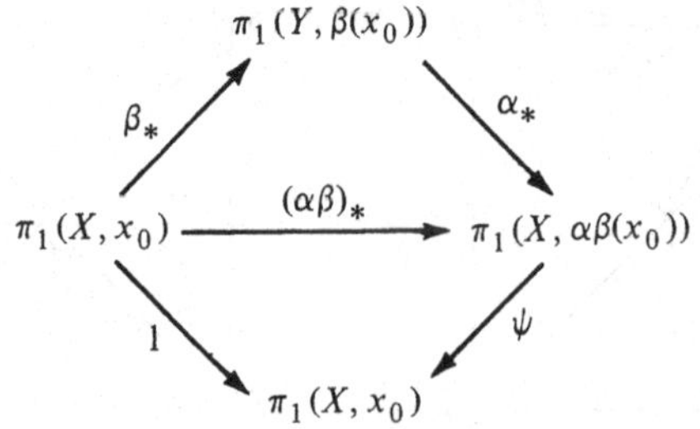

Since ψ is an isomorphism, it follows that $(\alpha\beta)_*$ is an isomorphism. Now the top triangle commutes because π_1 is a functor: $(\alpha\beta)_* = \alpha_*\beta_*$. It follows that β_* is injective and α_* is surjective. A similar diagram arising from $\beta\alpha \simeq 1_Y$ shows that β_* is surjective; that is, β_* is an isomorphism. □

Corollary 3.11. *Let X and Y be path connected spaces having the same homotopy type. Then, for every $x_0 \in X$ and $y_0 \in Y$, we have*

$$\pi_1(X, x_0) \cong \pi_1(Y, y_0).$$

Proof. The theorem gives $\pi_1(X, x_0) \cong \pi_1(Y, \beta(x_0))$ if $\beta: X \to Y$ is a homotopy equivalence, and Theorem 3.6 shows that the isomorphism class of either side is independent of the choice of basepoint. □

Corollary 3.12. *If X is a contractible space and $x_0 \in X$, then*

$$\pi_1(X, x_0) = \{1\}.$$

Proof. Corollary 3.11 and Exercise 3.11. (This result also follows from Theorem 1.13.) □

Definition. A space X is called **simply connected** if it is path connected and $\pi_1(X, x_0) = \{1\}$ for every $x_0 \in X$.

According to this definition, all simply connected spaces are path connected; that is, both π_1 and π_0 are trivial. The reader should be aware that some authors allow simply connected spaces that are not path connected; this means that every path component is simply connected in our sense.

Remark. In complex variables, one calls an open connected subset U of $\hat{\mathbf{C}}$ ($= \mathbf{C} \cup \{\infty\}$, homeomorphic to S^2) simply connected if its complement is connected. This agrees with our definition, but it requires some work to prove it: $\pi_1(U, u_0) = \{1\}$ if and only if $S^2 - U$ is connected.

We have just shown, in Corollary 3.12, that contractible spaces are simply connected. The converse is false; for example, we will see eventually that S^n is simply connected whenever $n \geq 2$, yet these spheres are not contractible.

Here is another consequence of Theorem 3.10.

Corollary 3.13. *If $\beta: (X, x_0) \to (Y, y_0)$ is (freely) nullhomotopic, then the induced homomorphism $\beta_*: \pi_1(X, x_0) \to \pi_1(Y, y_0)$ is trivial.*[2]

Proof. If $k: X \to Y$ is a constant map at y_1, say, then it is easy to see that $k_*: \pi_1(X, x_0) \to \pi_1(Y, y_1)$ is trivial ($k_*[f] = [k \circ f]$, and $k \circ f$ is a constant

[2] If G and H are groups, a homomorphism $\varphi: G \to H$ is called **trivial** if $\varphi(x) = 1$ for all $x \in G$, where 1 is the identity element of H.

path). Suppose that $\beta \simeq k$, as in the hypothesis. By Lemma 3.8, there is an isomorphism ψ with $\psi\beta_* = k_*$; it follows that $\beta_* = \psi^{-1}k_*$ is trivial. □

$\pi_1(S^1)$

We have yet to exhibit a space that is not simply connected, that is, a space with a nontrivial fundamental group. Since $\pi_1(X, x_0)$ consists of relative homotopy classes of maps $S^1 \to X$, the space $X = S^1$ suggests itself for consideration.

EXERCISE

3.12. If $\pi_1(Y, y_0) \neq \{1\}$ for some pointed space (Y, y_0), then $\pi_1(S^1, 1) \neq \{1\}$. (*Hint*: Otherwise 1_S is nullhomotopic, where 1_S is the identity map on S^1, and this implies that $f = f \circ 1_S$ is nullhomotopic for every closed path f in Y at y_0.)

To compute $\pi_1(S^1, 1)$, let us view S^1 as the set of all complex numbers z with $\|z\| = 1$. One feels that $z \mapsto z^2$, which wraps **I** around S^1 twice, ought not to be homotopic to the constant map $z \mapsto z^0 = 1$, and so we seek a way to distinguish these two functions (of course, we must even distinguish their homotopy classes). Recall from complex variables that these functions can be distinguished by a certain line integral called the **winding number**:

$$W(f) = \frac{1}{2\pi i}\oint_f \frac{dz}{z}$$

(here $f: (\mathbf{I}, \dot{\mathbf{I}}) \to (S^1, 1)$ is a parametrization of the circle by some "nice", e.g., differentiable, function f). Evaluate $W(f)$ by rewriting $f(t) = \exp \tilde{f}(t)$ for some real-valued function $\tilde{f}$ [$\exp s$ denotes $e^{2\pi i s}$]. With this rewriting, one can convert the line integral into an ordinary integral via the substitution $z = f(t) = \exp \tilde{f}(t)$. Thus $dz = z2\pi i\tilde{f}'(t)\,dt$ and

$$W(f) = \frac{1}{2\pi i}\oint_f \frac{dz}{z} = \int_0^1 \tilde{f}'(t)\,dt = \tilde{f}(1) - \tilde{f}(0).$$

For example, let $f(t) = e^{2\pi i m t}$ be the function wrapping **I** around S^1 $|m|$ times (counterclockwise if $m \geq 0$ and clockwise if $m < 0$). Here we may let $\tilde{f}(t) = mt$, and so

$$W(f) = \tilde{f}(1) - \tilde{f}(0) = m.$$

(Note that there are other possible choices for $\tilde{f}$, namely, $\tilde{f}(t) = mt + k$ for any fixed integer k. This multitude of choices is easily explained: $\tilde{f}$ is essentially $\log f$, and the complex logarithm is not single-valued.) Here is the point of these remarks. Investigation of $\pi_1(S^1)$ in the spirit of the winding number suggests constructing maps $\tilde{f}: \mathbf{I} \to \mathbf{R}$ with $f(t) = e^{2\pi i \tilde{f}(t)}$ (for every closed path f in S^1); moreover, attention should be paid to $\tilde{f}(1)$ and $\tilde{f}(0)$.

Lemma 3.14. *Let X be a compact convex subset of some $\mathbf{R}^k$, let $f: (X, x_0) \to (S^1, 1)$ be continuous, let $t_0 \in \mathbf{Z}$, and let $\exp t$ denote $e^{2\pi i t}$. Then there exists a unique continuous $\tilde{f}: (X, x_0) \to (\mathbf{R}, t_0)$ with $\exp \tilde{f} = f$.*

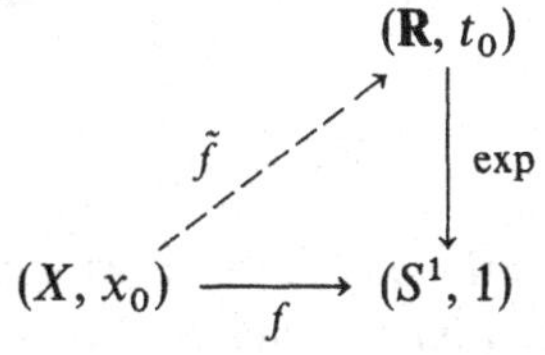

Remarks. (1) $\tilde{f}$ is called a **lifting** of f.

(2) In order that $\exp \tilde{f}(x_0) = f(x_0) = 1$, t_0 must be an integer.

Proof. Since X is compact metric, f must be uniformly continuous. There is thus $\varepsilon > 0$ such that whenever $\|x - x'\| < \varepsilon$, then $\|f(x) - f(x')\| < 2$ (we choose $2 = \operatorname{diam} S^1$ to guarantee that $f(x)$ and $f(x')$ are not antipodal, i.e., $f(x)f(x')^{-1} \neq -1$). Now X bounded implies the existence of a positive integer n with $\|x - x_0\|/n < \varepsilon$ for all $x \in X$.

For each $x \in X$, subdivide the line segment having endpoints x_0 and x (which is contained in X by convexity) into n intervals of equal length using (uniquely determined) points $x_0, x_1, \ldots, x_n = x$. Thus $\|x_j - x_{j+1}\| = \|x - x_0\|/n < \varepsilon$, hence $f(x_j)^{-1}f(x_{j+1}) \neq -1$. For each j with $0 \le j \le n-1$, the function $g_j: X \to S^1 - \{-1\}$ defined by

$$g_j(x) = f(x_j)^{-1}f(x_{j+1})$$

is easily seen to be continuous (for multiplication $S^1 \times S^1 \to S^1$ and inversion $S^1 \to S^1$ are continuous); note that $g_j(x_0) = 1$ for all j. Since S^1 is a multiplicative group, there is a "telescoping product" in S^1:

$$\begin{aligned} f(x) &= f(x_0)[f(x_0)^{-1}f(x_1)][f(x_1)^{-1}f(x_2)]\cdots[f(x_{n-1})^{-1}f(x_n)] \\ &= f(x_0)g_0(x)g_1(x)\cdots g_{n-1}(x). \end{aligned}$$

Now the restriction of exp to $(-\frac{1}{2}, \frac{1}{2})$ is a homeomorphism from $(-\frac{1}{2}, \frac{1}{2})$ to $S^1 - \{-1\}$; let us call its inverse λ (actually, $\lambda = (1/2\pi i)\log$); note that $\lambda(1) = 0$. Since $\operatorname{im} g_j \subset S^1 - \{-1\}$ for all j, each $\lambda \circ g_j$ is defined and continuous. Define $\tilde{f}: X \to \mathbf{R}$ by

$$\tilde{f}(x) = t_0 + \lambda(g_0(x)) + \lambda(g_1(x)) + \cdots + \lambda(g_{n-1}(x)).$$

Now $\tilde{f}$ is continuous (it is a sum of continuous functions), $\tilde{f}(x_0) = t_0$ (because $g_j(x_0) = 1$ for all j and $\lambda(1) = 0$), and $\exp \tilde{f} = f$ (because exp is a homomorphism).

To prove uniqueness of $\tilde{f}$, assume that $\tilde{g}: X \to \mathbf{R}$ is a continuous function with $\exp \tilde{g} = f$ and $\tilde{g}(x_0) = t_0$. Define $h: X \to \mathbf{R}$ by $h(x) = \tilde{f}(x) - \tilde{g}(x)$; it is clear that h is continuous. Now

$$\exp h(x) = \exp(\tilde{f}(x) - \tilde{g}(x)) = \exp \tilde{f}(x)/\exp \tilde{g}(x) = 1,$$

because $\exp \tilde{f} = f = \exp \tilde{g}$. But $\exp: \mathbf{R} \to S^1$ is a homomorphism with kernel $\mathbf{Z}$. Therefore $h: X \to \mathbf{R}$ is integer-valued. Since X is connected (it is convex), it follows from the discreteness of $\mathbf{Z}$ that h is constant. Finally, $h(x_0) = \tilde{f}(x_0) - \tilde{g}(x_0) = t_0 - t_0 = 0$ shows that the constant is zero; that is, $\tilde{f} = \tilde{g}$. □

Corollary 3.15. *Let $f: (\mathbf{I}, \dot{\mathbf{I}}) \to (S^1, 1)$ be continuous.*

(i) *There exists a unique continuous $\tilde{f}: \mathbf{I} \to \mathbf{R}$ with $\exp \tilde{f} = f$ and $\tilde{f}(0) = 0$.*
(ii) *If $g: (\mathbf{I}, \dot{\mathbf{I}}) \to (S^1, 1)$ is continuous and $f \simeq g$ rel $\dot{\mathbf{I}}$, then $\tilde{f} \simeq \tilde{g}$ rel $\dot{\mathbf{I}}$ (where $\exp \tilde{g} = g$ and $\tilde{g}(0) = 0$); moreover, $\tilde{f}(1) = \tilde{g}(1)$.*

Proof. (i) This follows from the lemma because $\mathbf{I}$ is compact convex.

(ii) Note that $\mathbf{I} \times \mathbf{I}$ is compact convex; choose $(0, 0)$ as a basepoint. If $F: \mathbf{I} \times \mathbf{I} \to S^1$ is a relative homotopy, $F: f \simeq g$ rel $\dot{\mathbf{I}}$, then the lemma provides a continuous map $\tilde{F}: \mathbf{I} \times \mathbf{I} \to \mathbf{R}$ with $\exp \tilde{F} = F$ and with $\tilde{F}(0, 0) = 0$. We claim that $\tilde{F}: \tilde{f} \simeq \tilde{g}$ rel $\dot{\mathbf{I}}$; that is, the homotopy F can be lifted. If $\varphi_0: \mathbf{I} \to \mathbf{R}$ is defined by $\varphi_0(t) = \tilde{F}(t, 0)$, then $\exp \varphi_0(t) = \exp \tilde{F}(t, 0) = F(t, 0) = f(t)$; since $\varphi_0(0) = \tilde{F}(0, 0) = 0$, uniqueness of lifting gives $\varphi_0 = \tilde{f}$. Define $\theta_0: \mathbf{I} \to \mathbf{R}$ by $\theta_0(t) = \tilde{F}(0, t)$; a similar argument shows that θ_0 is the constant function $\theta_0(t) \equiv 0$; it follows that $\tilde{F}(0, 1) = 0$. Define $\varphi_1: \mathbf{I} \to \mathbf{R}$ by $\varphi_1(t) = \tilde{F}(t, 1)$; as above, $\exp \varphi_1(t) = F(t, 1) = g(t)$ and $\varphi_1(0) = \tilde{F}(0, 1) = 0$, hence $\varphi_1(t) = \tilde{g}$. Finally, define $\theta_1: \mathbf{I} \to \mathbf{R}$ by $\theta_1(t) = \tilde{F}(1, t)$. Now $\exp \theta_1$ is the constant function c with value $f(1)$, and $\theta_1(0) = \tilde{f}(1)$. Therefore the constant function at $\tilde{f}(1)$ is a lifting of c, and uniqueness gives $\theta_1(t) \equiv \tilde{f}(1)$ for all $t \in \mathbf{I}$. Hence $\tilde{g}(1) = \tilde{f}(1)$ and $\tilde{F}$ is a relative homotopy $\tilde{F}: \tilde{f} \simeq \tilde{g}$ rel $\dot{\mathbf{I}}$. □

Part (ii) of this corollary shows that differentiable functions $f, g: (\mathbf{I}, \dot{\mathbf{I}}) \to (S^1, 1)$ which are homotopic rel $\dot{\mathbf{I}}$ have the same winding number: $W(f) = W(g)$ because $\tilde{f}(1) - \tilde{f}(0) = \tilde{f}(1) = \tilde{g}(1) = \tilde{g}(1) - \tilde{g}(0)$.

Definition. If $f: (\mathbf{I}, \dot{\mathbf{I}}) \to (S^1, 1)$ is continuous, define the **degree** of f by

$$\deg f = \tilde{f}(1),$$

where $\tilde{f}$ is the unique lifting of f with $\tilde{f}(0) = 0$.

Observe that $\exp \tilde{f}(1) = f(1) = 1$ hence $\tilde{f}(1)$ lies in the kernel of the homomorphism exp, namely, $\mathbf{Z}$. Thus, $\deg f \in \mathbf{Z}$ for every $f: (\mathbf{I}, \dot{\mathbf{I}}) \to (S^1, 1)$. Also, if $f(z) = z^m$ (more precisely, if $f(t) = \exp(mt)$), we saw above that $\tilde{f}(1) = m$; this explains the term degree.

Theorem 3.16. *The function $d: \pi_1(S^1, 1) \to \mathbf{Z}$ given by $[f] \mapsto \deg f$ is an isomorphism. In particular, $\deg(f * g) = \deg f + \deg g$.*

Proof. First, Corollary 3.15(ii) shows that d is a well defined function. Second, d is a surjection because, for each $m \in \mathbf{Z}$, the function $f(z) = z^m$ has degree m

(as we have just observed above). Assume that $\deg f = 0$, where f is a closed path in S^1 at 1. Thus $\tilde{f}(1) = 0$, which says that $\tilde{f}$ is a closed path in $\mathbf{R}$ at 0. Now $\exp: (\mathbf{R}, 0) \to (S^1, 1)$ induces a homomorphism $\pi_1(\mathbf{R}, 0) \to \pi_1(S^1, 1)$ with $[\tilde{f}] \mapsto [\exp \tilde{f}] = [f]$. But $\mathbf{R}$ contractible implies that $\pi_1(\mathbf{R}, 0) = \{1\}$, so that $[\tilde{f}] = 1$ and $[f] = 1$ (the identity element of $\pi_1(R^1, 1)$). It remains to show that d is a homomorphism, for then we can conclude that $\ker d$ is trivial and d is injective.

Assume that f and g are closed paths in S^1 at 1 of degrees m and n, respectively. To compute $\deg(f * g)$, we must find a path $\tilde{h}: \mathbf{I} \to \mathbf{R}$ with $\exp \tilde{h} = f * g$ and with $\tilde{h}(0) = 0$; then $\deg(f * g) = \tilde{h}(1)$. Let $\tilde{g}$ be the lifting of g with $\tilde{g}(0) = 0$. Define $\tilde{\gamma}: \mathbf{I} \to \mathbf{R}$ by $\tilde{\gamma}(t) = m + \tilde{g}(t)$, so that $\tilde{\gamma}$ is a path in $\mathbf{R}$ from m to $m + n$. Now let $\tilde{f}$ be the lifting of f with $\tilde{f}(0) = 0$ (and $\tilde{f}(1) = m$). Then $\tilde{f} * \tilde{\gamma}$ is a path in $\mathbf{R}$ with $(\tilde{f} * \tilde{\gamma})(0) = 0$ and $(\tilde{f} * \tilde{\gamma})(1) = m + n$. We claim that $\tilde{f} * \tilde{\gamma}$ is a lifting of $f * g$:

$$\exp(\tilde{f} * \tilde{\gamma})(t) = \begin{cases} \exp \tilde{f}(2t) & \text{if } 0 \leq t \leq \frac{1}{2} \\ \exp \tilde{\gamma}(2t - 1) & \text{if } \frac{1}{2} \leq t \leq 1. \end{cases}$$

Now $\exp \tilde{f}(s) = f(s)$ for $s \in \mathbf{I}$, because $\tilde{f}$ is a lifting of f; also, $\exp \tilde{\gamma}(s) = \exp(m + \tilde{g}(s)) = e^{2\pi i m} \exp \tilde{g}(s) = g(s)$, because $m \in \mathbf{Z}$ and $\tilde{g}$ is a lifting of g (incidentally, this shows that $\tilde{\gamma}$ is the lifting of g with $\tilde{\gamma}(0) = m$). Hence $\exp(\tilde{f} * \tilde{\gamma}) = f * g$. Therefore

$$\deg(f * g) = (\tilde{f} * \tilde{\gamma})(1) = m + n = \deg f + \deg g.$$

It follows that $d: \pi_1(S^1, 1) \to \mathbf{Z}$ is a homomorphism and hence is an isomorphism. □

Corollary 3.17. *S^1 is not simply connected.*

Corollary 3.18. *Two closed paths in S^1 at 1 are homotopic* rel $\dot{\mathbf{I}}$ *if and only if they have the same degree.*

Proof. If $f \simeq g$ rel $\dot{\mathbf{I}}$, then $\deg f = \deg g$, for we have already shown that $d: \pi_1(S^1, 1) \to \mathbf{Z}$ is well defined. Conversely, $\deg f = \deg g$ implies that $[f] = [g]$ because d is injective. □

Theorem 3.19 (Fundamental Theorem of Algebra). *Every nonconstant polynomial with complex coefficients has a complex root.*

Proof. Let Σ_ρ denote the circle in $\mathbf{C}$ of radius ρ and center at the origin and, for $n \geq 1$, let $f_\rho^n: \Sigma_\rho \to \mathbf{C} - \{0\}$ be the restriction to Σ_ρ of $z \mapsto z^n$. By Theorem 1.5, it suffices to prove that f_ρ^n is not (freely) nullhomotopic. Consider the composite $h: S^1 \to \Sigma_\rho \to \mathbf{C} - \{0\} \to S^1$, where the maps are $z \mapsto \rho z$, $z \mapsto z^n$, and $z \mapsto z/\|z\|$; one checks that $h(z) = z^n$. Were f_ρ^n nullhomotopic, then it would follow that h is nullhomotopic. Corollary 3.13 now says that $h_*: \pi_1(S^1, 1) \to \pi_1(S^1, 1)$ is trivial. In particular, $h_*[\exp] = [h \exp] = [\exp^n]$ is trivial; that is,

$\exp^n$ is nullhomotopic rel $\dot{\mathbf{I}}$, and so $\exp^n$ has degree 0. But we know that $\exp^n$ has degree $n \geq 1$, and this is a contradiction. □

There are other proofs of the fundamental theorem of algebra (one of the simplest is E. Artin's variation of a proof by Gauss, which requires only two facts, both following from the intermediate value theorem: every positive real number has a positive square root; every real polynomial of odd degree has a real root (see [Jacobson, p. 293]). The proof of Theorem 3.19, however, still illustrates that the ideas we are developing are powerful. Later, we shall investigate methods of computing fundamental groups, one of which (covering spaces) generalizes the computation of $\pi_1(S^1, 1)$ just given. We shall also see that $\pi_1(X, x_0)$ may not be abelian; indeed, given any group G, there exists a space X with $\pi_1(X, x_0) \cong G$.

EXERCISES

3.13. Let $u: (\mathbf{I}, \dot{\mathbf{I}}) \to (S^1, 1)$ be the closed path $t \mapsto \exp(t)$. Show that $[u]$ is a generator of $\pi_1(S^1, 1)$.

*3.14. If f is a closed path in S^1 at 1 and if $m \in \mathbf{Z}$, then $t \mapsto f(t)^m$ is a closed path in S^1 at 1 and

$$\deg(f^m) = m \deg f.$$

3.15. Let $f: (\mathbf{I}, \dot{\mathbf{I}}) \to (S^1, a)$ be a closed path in S^1 at $a = \exp(\alpha)$. Define **degree** $f =$ degree $R \circ f$, where $R: S^1 \to S^1$ is rotation by $-2\pi\alpha$ radians. Prove that two closed paths f and g in S^1 (with $f(0) = a$ and $g(0) = b$) are homotopic (with closed paths at every time t of the homotopy) if and only if they have the same degree. (*Hint*: Corollary 3.18, Exercise 1.3, and Theorem 1.6.)

3.16. Compute $\pi_1(T, t_0)$, where T is the torus $S^1 \times S^1$.

3.17. Prove that S^1 is not a retract of D^2.

3.18. Prove the Brouwer fixed point theorem for continuous maps $D^2 \to D^2$.

3.19. Let f be a closed path in S^1 at 1.
(i) If f is not surjective, then $\deg f = 0$.
(ii) Give an example of a surjective f with $\deg f = 0$.

*3.20. Let X be a space with basepoint x_0, and let $\{U_j: j \in J\}$ be an open cover of X by path connected subspaces such that:
(i) $x_0 \in U_j$ for all j;
(ii) $U_j \cap U_k$ is path connected for all j, k.
(It follows that X is path connected.) Prove that $\pi_1(X, x_0)$ is generated by the subgroups im i_{j*}, where $i_j: (U_j, x_0) \hookrightarrow (X, x_0)$ is the inclusion. (*Hint*: If $f: \mathbf{I} \to X$ is a closed path in X at x_0, use a Lebesgue number of the open cover $\{f^{-1}(U_j): j \in J\}$ of $\mathbf{I}$.)

*3.21. If $n \geq 2$, prove that S^n is simply connected. (*Hint*: Use Exercise 3.20 with the open cover $\{U_1, U_2\}$ of S^n, where U_1 is the complement of the north pole and U_2 is the complement of the south pole.)

3.22. If $n \geq 2$, then S^n and S^1 do not have the same homotopy type.

Definition. A topological group is a group G whose underlying set is equipped with a topology[3] such that:

(i) the multiplication map $\mu: G \times G \to G$, given by $(x, y) \mapsto xy$, is continuous if $G \times G$ has the product topology;
(ii) the inversion map $i: G \to G$, given by $x \mapsto x^{-1}$, is continuous.

Both $\mathbf{R}^n$ (under addition) and S^1 (under multiplication) are topological groups.

EXERCISES

*3.23. Let G be a topological group and let H be a normal subgroup. Prove that G/H is a topological group, where G/H is regarded as the quotient space of G by the kernel of the natural map.

*3.24. Let G be a simply connected topological group and let H be a discrete closed normal subgroup. Prove that $\pi_1(G/H, 1) \cong H$. (*Hint*: Adapt the proof of Theorem 3.16 with exp: $\mathbf{R} \to S^1$ replaced by the natural map $v: G \to G/H$, and with the open neighborhood $(-\frac{1}{2}, \frac{1}{2})$ of 0 in $\mathbf{R}$ replaced by a suitable open neighborhood of the identity element 1 in G.) (*Remark*: If G is T_0, then every discrete subgroup of G is necessarily closed.)

3.25. Let $GL(n, \mathbf{R})$ denote the multiplicative group of all $n \times n$ nonsingular real matrices. Regard $GL(n, \mathbf{R})$ as a subspace of $\mathbf{R}^{n^2}$, and show that it and its subgroups are topological groups.

3.26. A discrete normal subgroup H of a connected topological group G is contained in the center of G (i.e., each $h \in H$ commutes with every $x \in G$), hence is abelian. (*Hint*: Fix $h \in H$ and show that $\varphi: G \to H$ defined by $\varphi(x) = xhx^{-1}h^{-1}$ is constant.) Conclude that $\pi_1(G/H, 1)$ is abelian when G is simply connected and H is a discrete closed normal subgroup.

The next result is a vast generalization of the conclusion of the last exercise.

Definition. A pointed space (X, x_0) is called an ***H*-space** (after H. Hopf) if there is a pointed map $m: (X \times X, (x_0, x_0)) \to (X, x_0)$ such that each of the (necessarily pointed) maps $m(x_0,\ )$ and $m(\ , x_0)$ on (X, x_0) is homotopic to $1_X \operatorname{rel}\{x_0\}$. One calls x_0 a **homotopy identity**.

Clearly, every topological group X with identity x_0 and multiplication m is an H-space (one even has equality instead of relative homotopy).

To help us evaluate the induced map $m(x_0,\ )_*$, let us restate the definition of H-space so that it is phrased completely in terms of maps. If $k: X \to X$ is the constant map at x_0 and $(k, 1_X): X \to X \times X$ is the map $x \mapsto (x_0, x)$, then $m(x_0,\ )$ is the composite $m \circ (k, 1_X)$. Similarly, $m(\ , x_0)$ is the composite

[3] One often assumes as part of the definition that G has some separation property. It is known (see [Hewitt and Ross, p. 70]) that if G is T_0, then it is completely regular.

$m \circ (1_X, k)$. In an H-space, therefore, each of these composites is homotopic to 1_X rel$\{x_0\}$.

Recall an elementary property of direct products of groups: if $x \in G$ and $y \in H$, then in $G \times H$,

$$(x, 1)(1', y) = (x, y) = (1', y)(x, 1),$$

where 1 denotes the identity element in H and $1'$ denotes the identity element of G.

Theorem 3.20. *If (X, x_0) is an H-space, then $\pi_1(X, x_0)$ is abelian.*

PROOF. In Theorem 3.7, we have proved that $\theta: \pi_1(X, x_0) \times \pi_1(X, x_0) \to \pi_1(X \times X, (x_0, x_0))$, defined by $([f], [g]) \mapsto [(f, g)]$, is an isomorphism, where (f, g) is the path in $X \times X$ given by $t \mapsto (f(t), g(t))$. Choose $[f], [g] \in \pi_1(X, x_0)$. Now

$$\begin{aligned}
[g] &= (m \circ (k, 1_X))_*[g] && \text{(definition of } H\text{-space)} \\
&= m_*(k, 1_X)_*[g] && (\pi_1 \text{ is a functor}) \\
&= m_*[(k, 1_X) \circ g] && \text{(definition of induced map)} \\
&= m_*[(kg, g)] \\
&= m_*\theta([kg], [g]) && \text{(definition of } \theta) \\
&= m_*\theta(e, [g]),
\end{aligned}$$

where $e = [k]$ is the identity element of $\pi_1(X, x_0)$. Similarly,

$$[f] = m_*\theta([f], e),$$

because $m \circ (1_X, k) \simeq 1_X$ rel$\{x_0\}$. Since $m_*\theta: \pi_1(X, x_0) \times \pi_1(X, x_0) \to \pi_1(X, x_0)$ is a homomorphism, we have

$$\begin{aligned}
m_*\theta([f], [g]) &= m_*\theta((e, [g])([f], e)) \\
&= m_*\theta((e, [g]))m_*\theta(([f], e)) = [g][f].
\end{aligned}$$

If instead one factors $([f], [g]) = ([f], e)(e, [g])$, one obtains $m_*\theta([f], [g]) = [f][g]$. We conclude that $[g][f] = [f][g]$, hence $\pi_1(X, x_0)$ is abelian. □

Corollary 3.21. *If G is a topological group, then $\pi_1(G, e)$ is abelian.*

The contrapositive of this last corollary is also interesting. If X is a space with $\pi_1(X, x_0)$ not abelian (eventually we shall see such X), then there is no way to define a multiplication on X making it a topological group. Indeed one cannot even equip such an X with the structure of an H-space.

We have seen that computing the fundamental group of a space yields useful information, but this computation, even for S^1, is not routine. In other chapters we shall develop techniques to facilitate this work.

CHAPTER 4

Singular Homology

Holes and Green's Theorem

For each $n \geq 0$, we now construct the homology functors H_n: **Top** $\to$ **Ab** that we used in Chapter 0 to prove Brouwer's fixed point theorem. The question we ask is whether a union of n-simplexes in a space X that "ought" to be the boundary of some union of $(n + 1)$-simplexes in X actually is such a boundary. Consider the case $n = 0$; a 0-simplex in X is a point. Given two points x_0, $x_1 \in X$, they "ought" to be the endpoints of a 1-simplex; that is, there ought to be a path in X from x_0 to x_1. Thus, $H_0(X)$ will bear on whether or not X is path connected. Consider the case $n = 1$. Let X be the punctured plane

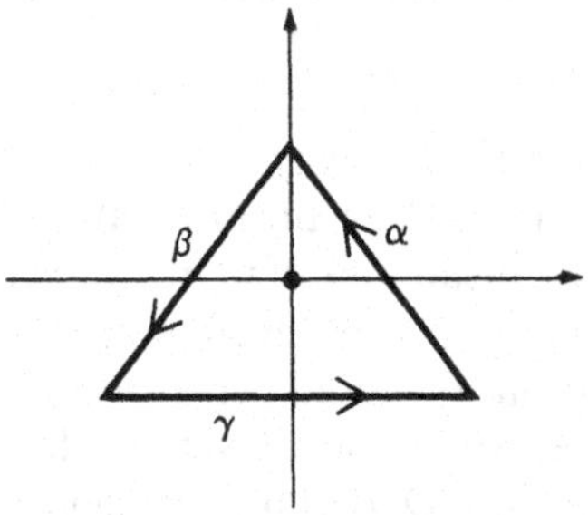

$\mathbf{R}^2 - \{0\}$, and let α, β, γ be the 1-simplexes as drawn; $\alpha \cup \beta \cup \gamma$ "ought" to bound the triangular 2-simplex, but the absence of the origin prevents this; loosely speaking, X has a "one-dimensional" hole in it. (Of course, $\alpha \cup \beta \cup \gamma$ would not bound the triangular 2-simplex if X were missing a small line segment through the origin, or even if X were missing a small neighborhood of the origin. When we say "one-dimensional" hole, we speak not of the size of the hole but of the size of the possible boundary. One must keep one's eye on the doughnut and not upon the hole!) $H_1(X)$ will describe the presence of such holes. We shall also see a close relation between $H_1(X)$ and $\pi_1(X, x_0)$; after all, the hole prevents one from deforming the closed path $\alpha * \beta * \gamma$ to a constant.

Better insight into homology is provided by Green's theorem from advanced calculus. Let D be an open disk in $\mathbf{R}^2$ with a finite number of points $z_1, z_2, \ldots, z_n$ deleted. Assume that there are closed curves $\gamma, \gamma_1, \ldots, \gamma_n$ in D as pictured below.

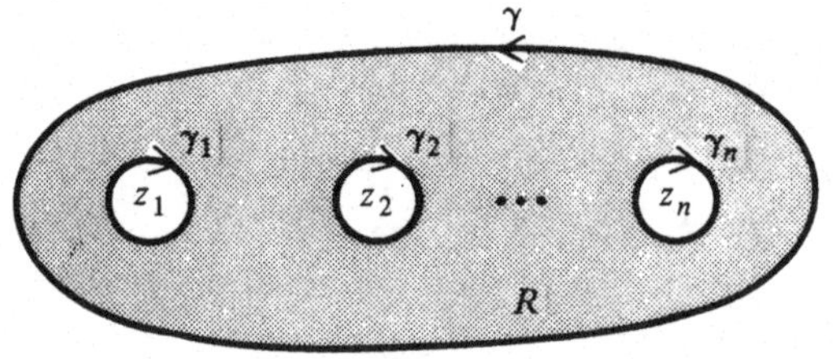

Here each γ_i is a simple closed curve (it does not intersect itself as does, say, a figure 8) having z_i inside and the other z's outside; all the γ_i are inside γ. If γ is oriented counterclockwise and each γ_i is oriented clockwise, then **Green's theorem** asserts, with certain differentiability hypotheses on these curves and on functions $P, Q: D \to \mathbf{R}$, that

$$\int_\gamma P\,dx + Q\,dy + \int_{\gamma_1} P\,dx + Q\,dy + \cdots + \int_{\gamma_n} P\,dx + Q\,dy$$
$$= \int_R \int \left(\frac{\partial Q}{\partial x} - \frac{\partial P}{\partial y} \right) dx\,dy,$$

where R is the shaded region in the picture. One is tempted to, and does, write the sum of the line integrals more concisely as

$$\int_{\gamma + \gamma_1 + \cdots + \gamma_n} P\,dx + Q\,dy.$$

Moreover, instead of describing how the orientations align, one could instead use signed coefficients to indicate this. If we no longer demand that the curves be simple and allow each γ_i to wind around z_i several times, we may even admit **Z**-linear combinations of closed curves in D.

Green's theorem arises when one considers whether, given two points a, $b \in D$, a line integral $\int_\beta P\,dx + Q\,dy$ is independent of the path β in D from a to b. If α is a second path in D from a to b, is $\int_\beta P\,dx + Q\,dy = \int_\alpha P\,dx + Q\,dy$?

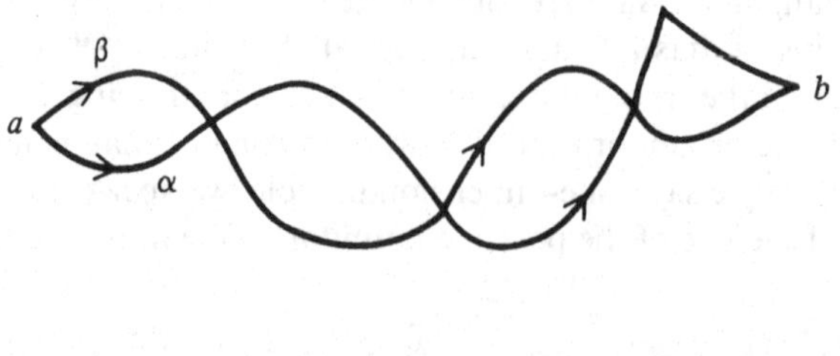

(Such paths α and β are examples of **chains.**) Plainly, for $\gamma = \beta - \alpha$ (proceed from a to b via β and then return to a backward via α, i.e., $\gamma = \beta * \alpha^{-1}$ in the multiplicative notation of Chapter 3), the two line integrals have the same value if and only if $\int_\gamma P\,dx + Q\,dy = 0$. We are thus led to closed paths (which are examples of **cycles**), and Green's theorem tells us to consider finite unions of oriented closed curves; algebraically, we consider formal **Z**-linear combinations of cycles. If we now restrict our attention to "exact" function pairs (P, Q) (there exists a function $F: D \to \mathbf{R}$ with $\partial F/\partial x = P$ and $\partial F/\partial y = Q$, hence $\partial Q/\partial x = \partial P/\partial y$), then the theorem asserts that the line integral vanishes if its oriented curves form the boundary of a two-dimensional region R in D.

One is thus led to consideration of oriented curves, closed oriented curves, and boundary curves (certain finite unions of oriented closed curves). The following equivalence relation on the set $S_1(D) = \{$all **Z**-linear combinations of oriented curves in $D\}$ is suggested: if $\alpha, \beta \in S_1(D)$, define $\alpha \sim \beta$ if $\int_\alpha P\,dx + Q\,dy = \int_\beta P\,dx + Q\,dy$ for *all* exact function pairs (P, Q). Such linear combinations α and β are called *homologous* (agreeing); equivalence classes of such linear combinations are called homology classes. It is known that two line integrals $\int_\alpha P\,dx + Q\,dy$ and $\int_\beta P\,dx + Q\,dy$ agree in value for every exact pair (P, Q) (i.e., α and β are homologous) precisely when $\alpha - \beta$ is a boundary. Thus integration is independent of paths lying in the same homology class. There are higher-dimensional analogues of this discussion: Stokes's and Gauss's theorems in two and three dimensions; more generally, there is a version for integration on differentiable manifolds.[1]

Free Abelian Groups

Let us proceed to the formal definitions: but first, some algebra.

Definition. Let B be a subset of an (additive) abelian group F. Then F is **free abelian** with **basis** B if the cyclic subgroup $\langle b \rangle$ is infinite cyclic for each $b \in B$ and $F = \sum_{b \in B} \langle b \rangle$ (direct sum).

A free abelian group is thus a direct sum of copies of **Z**. A typical element $x \in F$ has a unique expression

$$x = \sum m_b b,$$

where $m_b \in \mathbf{Z}$ and **almost all** m_b (all but a finite number of m_b) are zero.

Bases of free abelian groups behave as bases of vector spaces; one can construct a (unique) homomorphism if one knows its behavior on a basis; moreover, one can "do anything" to a basis.

[1] Further discussion of Green's theorem is in the first section of Chapter 12 on differential forms.

Theorem 4.1. *Let F be free abelian with basis B. If G is an abelian group and $\varphi: B \to G$ is a function, then there exists a unique homomorphism $\tilde{\varphi}: F \to G$ with $\tilde{\varphi}(b) = \varphi(b)$ for all $b \in B$.*

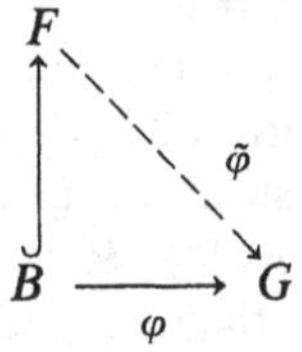

(ii) *Every abelian group G is isomorphic to a quotient group of the form F/R, where F is a free abelian group.*

Proof. (i) Each $x \in F$ may be written $x = \sum m_b b$; define $\tilde{\varphi}(x) = \sum m_b \varphi(b)$. Uniqueness of the expression for x shows that $\tilde{\varphi}$ is a well defined homomorphism. Finally, $\tilde{\varphi}$ is unique, because two homomorphisms agreeing on a set of generators—namely, B—must be equal.

(ii) For each $x \in G$, choose an infinite cyclic group $\mathbf{Z}_x$ having generator b_x, say. It follows that $F = \sum_{x \in G} \mathbf{Z}_x$ is a free abelian group with basis $B = \{b_x: x \in G\}$. Define a function $\varphi: B \to G$ by $\varphi(b_x) = x$. Since φ is surjective, it follows that the homomorphism $\tilde{\varphi}$ is surjective. By the first isomorphism theorem, $G \cong F/R$, where $R = \ker \tilde{\varphi}$. □

Definition. The construction of $\tilde{\varphi}$ from φ is called **extending by linearity**. Usually one abuses notation and denotes $\tilde{\varphi}$ by φ as well.

Part (ii) of the theorem suggests a way of describing abelian groups.

Definition. An abelian group G has **generators** $B = \{x_j: j \in J\}$ and **relations** $\Delta = \{r_k: k \in K\}$ if F is the free abelian group with basis B, if $\Delta \subset F$ (i.e., each r_k is a linear combination of the x_j with integer coefficients), and if $G \cong F/R$, where R is the subgroup of F generated by Δ. We say that $(B|\Delta)$ is a **presentation**[2] of the abelian group G.

Of course, an abelian group G has many presentations. The existence question for free abelian groups is essentially settled by the definition: one can exhibit a free abelian group with a basis of any cardinality merely by forming the direct sum of the desired number of copies of $\mathbf{Z}$. Here is a sharper version of the existence theorem.

Theorem 4.2. *Given a set T, there exists a free abelian group F having T as a basis.*

[2] Later we shall define presentations of groups that may not be abelian.

PROOF. If $T = \varnothing$, define $F = 0$. Otherwise, for each $t \in T$, define a group $\mathbf{Z}_t$ whose elements are all symbols mt with $m \in \mathbf{Z}$ and with addition defined by $mt + nt = (m + n)t$. It is easy to see that $\mathbf{Z}_t$ is infinite cyclic with generator t. The group $F = \sum_{t \in T} \mathbf{Z}_t$ is free abelian with basis the set of all $|T|$-tuples b_t, where b_t has all coordinates zero save for a 1 as its tth coordinate. The theorem is proved by first using a scissors to cut out all b_t's from F and then replacing each b_t by t itself. (One can be more fussy here if one wishes.) □

In our discussion of Green's theorem, we formed $\mathbf{Z}$-linear combinations of curves; Theorem 4.2 allows one to add and subtract curves without fear.

There is an analogue for free abelian groups of the dimension of a vector space.

Theorem 4.3. *Any two bases of a free abelian group F have the same cardinal.*

PROOF. Recall that any two bases of a vector space V (over any field) have the same cardinal. If V is finite-dimensional, this is standard linear algebra. If V is infinite-dimensional, one uses Zorn's lemma to prove that bases of V exist, and one then uses a set-theoretic fact (the family of all finite subsets of an infinite set A has the same cardinal as A) to prove invariance of the cardinal of a basis.

Now let A and B be bases of F. For a fixed prime p, it is easy to see that the quotient group F/pF is a vector space over $\mathbf{Z}/p\mathbf{Z}$ and that the cosets $\{a + pF: a \in A\}$ form a basis. Thus $\dim F/pF = \text{card } A$. Similarly, $\dim F/pF = \text{card } B$, hence $\text{card } A = \text{card } B$. □

Definition. If F is a free abelian group with basis B, then

$$\textbf{rank } F = \text{card } B.$$

Theorem 4.3 shows that rank F is well defined; that is, it does not depend on the choice of basis B. Exercise 4.2 below shows that the vector space analogy is a good one: free abelian groups are characterized by their rank as vector spaces are characterized by their dimension.

One can now define the rank of an arbitrary abelian group G.

Definition. An abelian group G has (possibly infinite) **rank** r if there exists a free abelian subgroup F of G with

(i) rank $F = r$;
(ii) G/F is torsion.

Such free abelian subgroups do exist. Define a subset B of G to be **independent** if $\sum m_i b_i = 0$ implies each $m_i = 0$ (where $m_i \in \mathbf{Z}$ and $b_i \in B$). It is easy to see that the subgroup generated by an independent subset B is free abelian with basis B. If F is the subgroup generated by a maximal independent subset (which exists, by Zorn's lemma), then F is free abelian and G/F is torsion. One can prove that the rank of F depends only on G (Exercise 9.32), so that the rank of G is indeed well defined.

EXERCISES

4.1. Let F be free abelian with basis B. If B is the disjoint union $B = \bigcup B_\lambda$, then $F = \sum F_\lambda$, where F_λ is free abelian with basis B_λ. Conclude that each $\gamma \in F$ has a unique expression $\gamma = \sum \gamma_\lambda$, where $\gamma_\lambda \in F_\lambda$ and almost all $\gamma_\lambda = 0$.

*4.2. Prove that two free abelian groups are isomorphic if and only if they have the same rank.

*4.3. For a given space X, define $S_1(X)$ to be the free abelian group with basis all paths $\sigma: \mathbf{I} \to X$, and let $S_0(X)$ be the free abelian group with basis X.
  (i) Show that there is a homomorphism $\partial_1: S_1(X) \to S_0(X)$ with $\partial_1 \sigma = \sigma(1) - \sigma(0)$ for every path σ in X.
  (ii) If $x_1, x_0 \in X$, show that $x_1 - x_0 \in \operatorname{im} \partial_1$ if and only if x_0, x_1 lie in the same path component of X.
  (iii) If σ is a path in X, then $\sigma \in \ker \partial_1$ if and only if σ is a closed path. Exhibit a nonzero element of $\ker \partial_1$ that is not a closed path.

The Singular Complex and Homology Functors

Exercise 4.3(ii) indicates that we are proceeding toward a definition that appears to capture the informal ideas discussed at the beginning of this chapter: $x_1 - x_0$ ought to be the boundary of a curve in X, but it may not be unless x_1, x_0 lie in the same path component of X. In preparation for the general definition, recall that Green's theorem suggests looking at *oriented* curves.

Definition. An **orientation** of $\Delta^n = [e_0, e_1, \ldots, e_n]$ is a linear ordering of its vertices.

An orientation thus gives a tour of the vertices. For example, the orientation $e_0 < e_1 < e_2$ of Δ^2 gives a counterclockwise tour.

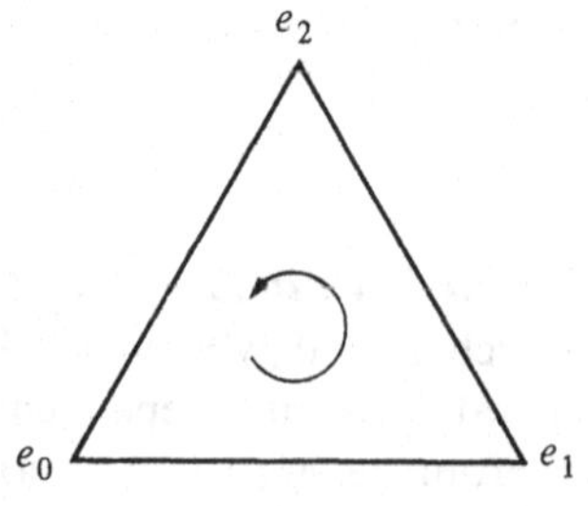

It is clear that two different orderings can give the same tour; thus $e_0 < e_1 < e_2$ and $e_1 < e_2 < e_0$ and $e_2 < e_0 < e_1$ all give the counterclockwise tour, while the other three orderings (orientations) give a clockwise tour.

If $n = 3$, the reader should see that there are essentially only two different tours, corresponding to the left-hand rule and right-hand rule, respectively.

Definition. Two orientations of Δ^n are the **same** if, as permutations of $\{e_0, e_1, \ldots, e_n\}$, they have the same parity (i.e., both are even or both are odd); otherwise the orientations are **opposite**.

Given an orientation of Δ^n, there is an **induced orientation** of its faces defined by orienting the ith face in the sense $(-1)^i[e_0, \ldots, \hat{e}_i, \ldots, e_n]$, where $-[e_0, \ldots, \hat{e}_i, \ldots, e_n]$ means the ith face (vertex e_i deleted) with orientation opposite to the one with the vertices ordered as displayed. For example, assume that Δ^2 is oriented counterclockwise.

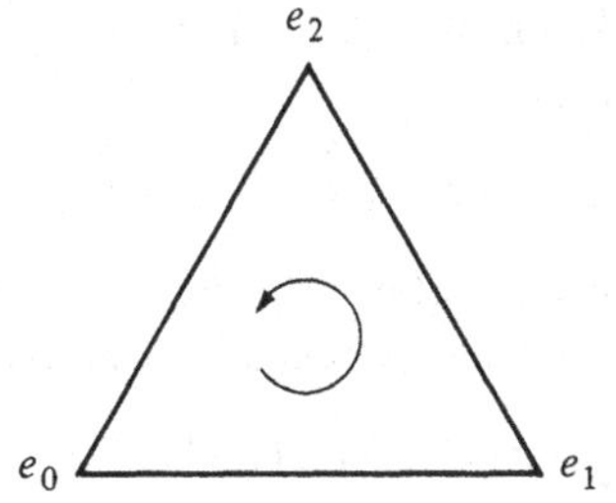

The 0th face of Δ^2 is $[\hat{e}_0, e_1, e_2] = [e_1, e_2]$, and it is oriented from e_1 to e_2; the first face $[e_0, \hat{e}_1, e_2] = [e_0, e_2]$ is oriented in the opposite direction: $-[e_0, e_2] = [e_2, e_0]$ is oriented from e_2 to e_0; the second face is $[e_0, e_1]$. It is plain that these orientations of the edges are "compatible" with the orientation of Δ^2.

The boundary of Δ^2 is

$$[e_1, e_2] \cup [e_0, e_2] \cup [e_0, e_1] = [\hat{e}_0, e_1, e_2] \cup [e_0, \hat{e}_1, e_2] \cup [e_0, e_1, \hat{e}_2].$$

The oriented boundary of Δ^2 is

$$[\hat{e}_0, e_1, e_2] \cup -[e_0, \hat{e}_1, e_2] \cup [e_0, e_1, \hat{e}_2] = [e_1, e_2] \cup [e_2, e_0] \cup [e_0, e_1].$$

More generally, the boundary of $\Delta^n = [e_0, \ldots, e_n]$ is $\bigcup_{i=0}^n [e_0, \ldots, \hat{e}_i, \ldots, e_n]$ and the oriented boundary of Δ^n is $\bigcup_{i=0}^n (-1)^i[e_0, \ldots, \hat{e}_i, \ldots, e_n]$.

For the moment, denote the ith face of Δ^2 by ε_i, where $i = 0, 1, 2$. Applying the homomorphism ∂_1 (see Exercise 4.3) to the oriented boundary, we see that

$$\partial_1(\varepsilon_0 - \varepsilon_1 + \varepsilon_2) = (e_2 - e_1) - (e_2 - e_0) + (e_1 - e_0) = 0,$$

and $\varepsilon_0 - \varepsilon_1 + \varepsilon_2 \in \ker \partial_1$; on the other hand, $\varepsilon_0 + \varepsilon_1 + \varepsilon_2 \notin \ker \partial_1$ (one thus sees that orientations are important). At last, here are the important definitions.

Definition. Let X be a topological space. A **(singular) n-simplex** in X is a continuous map $\sigma: \Delta^n \to X$, where Δ^n is the standard n-simplex.

Since Δ^1 is a closed interval ($\Delta^1 \approx \mathbf{I}$), a singular 1-simplex in X is essentially a path in X; since Δ^0 is a one-point set, a singular 0-simplex may be identified with a point in X.

Definition. Let X be a topological space. For each $n \geq 0$, define $S_n(X)$ as the free abelian group with basis all singular n-simplexes in X; define $S_{-1}(X) = 0$. The elements of $S_n(X)$ are called (singular) **n-chains** in X.

Of course, $S_n(X)$ essentially agrees with the construction in Exercise 4.3 when $n = 0$ and $n = 1$.

The oriented boundary of a singular n-simplex $\sigma: \Delta^n \to X$ ought to be $\sum_{i=0}^n (-1)^i(\sigma|[e_0, \ldots, \hat{e}_i, \ldots, e_n])$. A technical point arises: we prefer that this be a singular $(n-1)$-chain; it is not because the domain of $\sigma|[e_0, \ldots, \hat{e}_i, \ldots, e_n]$ is not the *standard* $(n-1)$-simplex Δ^{n-1}. This is easily remedied. For each n and i, define the ***i*th face map**

$$\varepsilon_i = \varepsilon_i^n: \Delta^{n-1} \to \Delta^n$$

to be the affine map taking the vertices $\{e_0, \ldots, e_{n-1}\}$ to the vertices $\{e_0, \ldots, \hat{e}_i, \ldots, e_n\}$ preserving the displayed orderings:

$$\varepsilon_0^n: (t_0, \ldots, t_{n-1}) \mapsto (0, t_0, \ldots, t_{n-1});$$

$$\varepsilon_i^n: (t_0, \ldots, t_{n-1}) \mapsto (t_0, \ldots, t_{i-1}, 0, t_i, \ldots, t_{n-1}) \quad \text{if } i \geq 1.$$

(The superscript n indicates that the target of ε_i^n is Δ^n.) For example, there are three face maps $\varepsilon_i^2: \Delta^1 \to \Delta^2$: $\varepsilon_0: [e_0, e_1] \to [e_1, e_2]$; $\varepsilon_1: [e_0, e_1] \to [e_0, e_2]$; $\varepsilon_2: [e_0, e_1] \to [e_0, e_1]$.

Definition. If $\sigma: \Delta^n \to X$ is continuous and $n > 0$, then its **boundary** is

$$\partial_n \sigma = \sum_{i=0}^n (-1)^i \sigma \varepsilon_i^n \in S_{n-1}(X);$$

if $n = 0$, define $\partial_0 \sigma = 0$.

Note that if $X = \Delta^n$ and $\delta: \Delta^n \to \Delta^n$ is the identity, then

$$\partial(\delta) = \sum_{i=0}^n (-1)^i \varepsilon_i^n.$$

Theorem 4.4. *For each $n \geq 0$, there is a unique homomorphism $\partial_n: S_n(X) \to S_{n-1}(X)$ with $\partial_n \sigma = \sum_{i=0}^n (-1)^i \sigma \varepsilon_i$ for every singular n-simplex σ in X.*

Proof. Use the formula for $\partial_n \sigma$ and extend by linearity. □

The homomorphisms ∂_n: $S_n(X) \to S_{n-1}(X)$ are called **boundary operators**. Strictly speaking, one ought to write ∂_n^X since these homomorphisms do depend on X; however, this is rarely done. We have constructed, for each X, a sequence of free abelian groups and homomorphisms

$$\cdots \longrightarrow S_n(X) \xrightarrow{\partial_n} S_{n-1}(X) \longrightarrow \cdots \longrightarrow S_1(X) \xrightarrow{\partial_1} S_0(X) \xrightarrow{\partial_0} 0,$$

called the **singular complex** of X; it is denoted by $(S_*(X), \partial)$ or, more simply, by $S_*(X)$.

Lemma 4.5. *If $k < j$, the face maps satisfy*

$$\varepsilon_j^{n+1}\varepsilon_k^n = \varepsilon_k^{n+1}\varepsilon_{j-1}^n \colon \Delta^{n-1} \to \Delta^{n+1}.$$

PROOF. Just evaluate these affine maps on every vertex e_i for $0 \le i \le n-1$. □

For example, $\varepsilon_2^4\varepsilon_0^3$ maps $e_0 \mapsto e_1 \mapsto e_1$; $e_1 \mapsto e_2 \mapsto e_3$; and $e_2 \mapsto e_3 \mapsto e_4$ (the image is thus the 2-face $[e_1, e_3, e_4]$ of Δ^4); $\varepsilon_0^4\varepsilon_1^3$: $e_0 \mapsto e_0 \mapsto e_1$; $e_1 \mapsto e_2 \mapsto e_3$; and $e_2 \mapsto e_3 \mapsto e_4$. If $k < j$, the image of $\varepsilon_j\varepsilon_k$ is the $(n-1)$-face of Δ^{n+1} obtained by deleting vertices e_j and e_k; when $k \ge j$, the image deletes vertices e_j and e_{k+1}.

Theorem 4.6. *For all $n \ge 0$, we have $\partial_n\partial_{n+1} = 0$.*

PROOF. Since $S_{n+1}(X)$ is generated by all $(n+1)$-simplexes σ, it suffices to show that $\partial\partial\sigma = 0$ for each such σ.

$$\begin{aligned}
\partial\partial\sigma &= \partial\left(\sum_j (-1)^j \sigma\varepsilon_j^{n+1}\right) \\
&= \sum_{j,k} (-1)^{j+k}\sigma\varepsilon_j^{n+1}\varepsilon_k^n \\
&= \sum_{j\le k} (-1)^{j+k}\sigma\varepsilon_j^{n+1}\varepsilon_k^n + \sum_{k<j} (-1)^{j+k}\sigma\varepsilon_j^{n+1}\varepsilon_k^n \\
&= \sum_{j\le k} (-1)^{j+k}\sigma\varepsilon_j^{n+1}\varepsilon_k^n + \sum_{k<j} (-1)^{j+k}\sigma\varepsilon_k^{n+1}\varepsilon_{j-1}^n, \quad \text{by Lemma 4.5.}
\end{aligned}$$

In the second sum, change variables: set $p = k$ and $q = j - 1$; it is now $\sum_{p\le q}(-1)^{p+q+1}\sigma\varepsilon_p^{n+1}\varepsilon_q^n$. Each term $\sigma\varepsilon_j^{n+1}\varepsilon_k^n$ occurs twice, once in the first sum with sign $(-1)^{j+k}$ and once in the second sum with (opposite) sign $(-1)^{j+k+1}$. Therefore terms cancel in pairs and $\partial\partial\sigma = 0$. □

Definition. The group of (singular) ***n*-cycles** in X, denoted by $Z_n(X)$,[3] is ker ∂_n; the group of (singular) ***n*-boundaries** in X, denoted by $B_n(X)$, is im ∂_{n+1}.

Clearly, $Z_n(X)$ and $B_n(X)$ are subgroups of $S_n(X)$ for all $n \ge 0$; but more is true.

---

[3] From the German *Zykel*.

Corollary 4.7. *For every space X and for every $n \geq 0$,*

$$B_n(X) \subset Z_n(X) \subset S_n(X).$$

PROOF. If $\beta \in B_n(X)$, then $\beta = \partial_{n+1}\alpha$ for some $\alpha \in S_{n+1}(X)$. But then $\partial_n(\beta) = \partial_n\partial_{n+1}\alpha = 0$, by Theorem 4.6, and $\beta \in Z_n(X)$. □

We have now made our earlier discussion precise: an n-cycle corresponds to those sum (unions) of oriented n-simplexes in X that ought to constitute the boundary of some union of $(n + 1)$-simplexes in X. Returning to the example of the punctured plane given at the beginning of this chapter, we see that $\alpha + \beta + \gamma$ is a 1-cycle in X. It is intuitively clear (but not so obvious to prove) that $\alpha + \beta + \gamma$ is not a 1-boundary (because the obvious candidate for the two-dimensional region it should bound is not a 2-simplex in X, lacking as it does the origin).

To detect "holes" in a space X, one should consider only cycles that are not boundaries; boundaries are "trivial" cycles. Indeed, Green's theorem also suggests this, for the line integral $\int_\gamma P\,dx + Q\,dy$ (where (P, Q) is an exact pair) is zero when γ is a union of oriented curves comprising the boundary of a region R in the space D. We are led to the following definition.

Definition. For each $n \geq 0$, the nth (singular) **homology group** of a space X is

$$H_n(X) = \frac{Z_n(X)}{B_n(X)} = \frac{\ker \partial_n}{\operatorname{im} \partial_{n+1}}.$$

The coset $z_n + B_n(X)$, where z_n is an n-cycle, is called the **homology class** of z_n, and it is denoted by cls z_n.

Our next aim is to show that each H_n is actually a functor **Top** → **Ab**. If $f: X \to Y$ is continuous and if $\sigma: \Delta^n \to X$ is an n-simplex in X, then $f \circ \sigma: \Delta^n \to Y$ is an n-simplex in Y. Extending by linearity gives a homomorphism $f_\#: S_n(X) \to S_n(Y)$, namely,

$$f_\#(\textstyle\sum m_\sigma \sigma) = \sum m_\sigma (f \circ \sigma), \quad \text{where } m_\sigma \in \mathbf{Z}.$$

This notation is careless, for $f_\#$ does depend on n. In fact there is one such $f_\#$ for every $n \geq 0$.

Lemma 4.8. *If $f: X \to Y$ is continuous, then $\partial_n f_\# = f_\# \partial_n$; that is, for every $n \geq 0$ there is a commutative diagram*

$$\begin{array}{ccc} S_n(X) & \xrightarrow{\partial_n} & S_{n-1}(X) \\ \downarrow{\scriptstyle f_\#} & & \downarrow{\scriptstyle f_\#} \\ S_n(Y) & \xrightarrow{\partial_n} & S_{n-1}(Y). \end{array}$$

Remark. Not content with omitting subscripts on the maps $f_\#$, we have omitted superscripts on the boundary maps ∂_n as well (these maps do depend on the spaces X and Y). This casual attitude is customary and necessary, for a jumble of indices, aside from being cumbersome, can mask a simple idea or a routine calculation. When the abbreviated symbol may cause confusion, however, we shall restore decorations as required.

PROOF. It suffices to evaluate each composite on a generator σ of $S_n(X)$. Now

$$\begin{aligned} f_\# \partial \sigma &= f_\#(\textstyle\sum (-1)^i \sigma \varepsilon_i) \\ &= \textstyle\sum (-1)^i f_\#(\sigma \varepsilon_i) = \sum (-1)^i f(\sigma \varepsilon_i). \end{aligned}$$

On the other hand,

$$\partial f_\# \sigma = \partial(f\sigma) = \textstyle\sum (-1)^i (f\sigma)\varepsilon_i. \qquad \square$$

Lemma 4.9. *If $f: X \to Y$ is continuous, then for every $n \geq 0$,*

$$f_\#(Z_n(X)) \subset Z_n(Y) \quad \text{and} \quad f_\#(B_n(X)) \subset B_n(Y).$$

PROOF. If $\alpha \in Z_n(X)$, then $\partial\alpha = 0$. Therefore $\partial f_\# \alpha = f_\# \partial \alpha = f_\#(0) = 0$, and $f_\# \alpha \in \ker \partial_n = Z_n(Y)$. If $\beta \in B_n(X)$, then $\beta = \partial\gamma$ for some $\gamma \in S_{n+1}(X)$, and $f_\# \beta = f_\# \partial \gamma = \partial f_\# \gamma \in \operatorname{im} \partial_{n+1} = B_n(Y)$. $\square$

Theorem 4.10. *For each $n \geq 0$, H_n:* **Top** $\to$ **Ab** *is a functor.*

PROOF. We have already defined H_n on objects X: $H_n(X) = Z_n(X)/B_n(X)$. If $f: X \to Y$ is continuous, define

$$H_n(f): H_n(X) \to H_n(Y)$$

by $z_n + B_n(X) \mapsto f_\#(z_n) + B_n(Y)$, where $z_n \in Z_n(X)$; that is,

$$H_n(f): \operatorname{cls} z_n \mapsto \operatorname{cls} f_\#(z_n).$$

There are some details to check. First, z_n being an n-cycle in X implies that $f_\# z_n$ is an n-cycle in Y, by Lemma 4.9. Second, this definition is independent of the choice of representative because $f_\#(B_n(X)) \subset B_n(Y)$: if $b_n \in B_n(X)$, then $f_\#(z_n + b_n) + B_n(Y) = f_\#(z_n) + f_\#(b_n) + B_n(Y) = f_\#(z_n) + B_n(Y)$. The remaining details—$H_n(f)$ is a homomorphism, $H_n(1_X)$ is the identity homomorphism, and $H_n(gf) = H_n(g)H_n(f)$—are all easy consequences of the definition of H_n. $\square$

Corollary 4.11. *If X and Y are homeomorphic, then $H_n(X) \cong H_n(Y)$ for all $n \geq 0$.*

PROOF. Theorem 0.5. $\square$

Each homology group $H_n(X)$ is thus an invariant of the space X; in particular, rank $H_n(X)$ is an invariant of X for each $n \geq 0$.

Definition. For each $n \geq 0$, rank $H_n(X)$ is called the nth **Betti number** of X.

If $H_n(X)$ is free abelian, then it is characterized by its rank; otherwise, there is more information contained in the homology group.

Dimension Axiom and Compact Supports

Before giving the first properties of the homology functors, we caution the reader. Many proofs, even of geometrically "obvious" facts, will seem too long (and too algebraic). One reason for this is our decision to define H_n as above, using **singular theory**. The advantages of this theory are the following: $H_n(X)$ is defined for every topological space X, that is, H_n is defined on all of **Top**; it is very easy to define induced maps and to prove that H_n is a functor. One disadvantage, as we have just said, is that some proofs appear too fussy and formal; another great disadvantage is that it is usually difficult to compute $H_n(X)$ for specific X. If we limit attention to spaces X that are polyhedra or CW complexes (these terms are defined later), then there are other definitions of H_n (the **simplicial theory** and the **cellular theory**) for which $H_n(X)$ is easier to calculate. The disadvantages of the other two theories are that they apply only to these special spaces and that induced maps are more complicated to define. These theories[4] will be presented along with a theorem of Eilenberg and Steenrod, which axiomatizes homology functors on the subcategory of (compact) polyhedra and which shows that the various theories agree on this subcategory. Once all this is known, the reader may then select the particular theory that is most convenient for a problem at hand. We have no such freedom of choice now, however, and so all our proofs are in singular style until Chapter 7. Thus warned, the reader should not be discouraged as we set forth the details of (singular) homology.

Theorem 4.12 (Dimension Axiom).[5] *If X is a one-point space, then $H_n(X) = 0$ for all $n > 0$.*

PROOF. For each $n \geq 0$, there is only one singular n-simplex $\sigma_n\colon \Delta^n \to X$, namely, the constant map. Therefore $S_n(X) = \langle \sigma_n \rangle$, the infinite cyclic group generated by σ_n. Let us now compute the boundary operators:

$$\partial_n \sigma_n = \sum_{i=0}^{n} (-1)^i \sigma_n \varepsilon_i = \left[\sum_{i=0}^{n} (-1)^i\right] \sigma_{n-1},$$

(for $\sigma_n \varepsilon_i$ is an $(n-1)$-simplex in X, and σ_{n-1} is the only such). It follows that

---

[4] There are homology theories other than the three we have mentioned here; these three are the most popular.

[5] The reason for this name will be explained in Chapter 9.

$$\partial_n \sigma_n = \begin{cases} 0 & \text{if } n \text{ is odd} \\ \sigma_{n-1} & \text{if } n \text{ is even and positive.} \end{cases}$$

Therefore $\partial_n = 0$ when n is odd, and ∂_n is an isomorphism when n is even and $n > 0$. Assume that $n > 0$, and consider the sequence

$$S_{n+1}(X) \xrightarrow{\partial_{n+1}} S_n(X) \xrightarrow{\partial_n} S_{n-1}(X).$$

If n is odd, then $\partial_n = 0$ implies that $S_n(X) = \ker \partial_n = Z_n(X)$; also ∂_{n+1} is an isomorphism ($n + 1$ is even), hence is surjective, and so $S_n(X) = \operatorname{im} \partial_{n+1} = B_n(X)$. Thus $H_n(X) = Z_n(X)/B_n(X) = 0$. If $n > 0$ is even, then ∂_n is an isomorphism, hence injective, and so $Z_n(X) = \ker \partial_n = 0$. It follows that $H_n(X) = Z_n(X)/B_n(X) = 0$ in this case as well. □

Definition. A space X is called **acyclic** if $H_n(X) = 0$ for all $n \geq 1$.

The dimension axiom shows that every one-point space is acyclic.

EXERCISES

*4.4. If $X = \varnothing$, then $H_n(X) = 0$ for all $n \geq 0$. (*Hint*: The free abelian group with empty basis is the trivial group $\{0\}$.)

4.5. If X is a one-point space, then $H_0(X) \cong \mathbf{Z}$.

*4.6. For each fixed $n \geq 0$, show that S_n: **Top** → **Ab** is a functor.

The next result will allow us to focus on path connected spaces.

Theorem 4.13. *If $\{X_\lambda: \lambda \in \Lambda\}$ is the set of path components of X, then, for every $n \geq 0$,*

$$H_n(X) \cong \sum_\lambda H_n(X_\lambda).$$

Remark. The elements of a direct sum $\sum G_\lambda$ are those "vectors" (g_λ) having only finitely many nonzero coordinates.

PROOF. If $\gamma = \sum m_i \sigma_i \in S_n(X)$, then Exercise 1.24 shows that each $\operatorname{im} \sigma_i$ is contained in a unique path component of X; we may thus write $\gamma = \sum \gamma_\lambda$, where γ_λ is the sum of those terms in γ involving a simplex σ_i for which $\operatorname{im} \sigma_i \subset X_\lambda$. It is easy to see that, for each n, the map $\gamma \mapsto (\gamma_\lambda)$ is an isomorphism $S_n(X) \to \sum_\lambda S_n(X_\lambda)$. Now γ is a cycle if and only if each γ_λ is a cycle: since $\partial\gamma_\lambda \in S_{n-1}(X_\lambda)$ (because $\operatorname{im} \sigma \subset X_\lambda$ implies $\operatorname{im} \sigma\varepsilon_i \subset X_\lambda$), the assumption $0 = \partial\gamma = \sum \partial\gamma_\lambda$ implies $\partial\gamma_\lambda = 0$ for all λ (because an element in the direct sum $\sum S_{n-1}(X_\lambda)$ is zero if and only if all its coordinates are zero). It follows that the map θ_n: $H_n(X) \to \sum H_n(X_\lambda)$, given by $\operatorname{cls} \gamma \mapsto (\operatorname{cls} \gamma_\lambda)$, is well defined. To see that θ_n is an isomorphism, we exhibit its inverse. Define Φ_n: $\sum H_n(X_\lambda) \to H_n(X)$ by $(\operatorname{cls} \gamma_\lambda) \mapsto \operatorname{cls}(\sum \gamma_\lambda)$; it is routine to check that both composites are identities. □

EXERCISES

4.7. Compute $H_n(S^0)$ for all $n \geq 0$.
4.8. Compute $H_n(X)$ for all $n \geq 0$, where X is the Cantor set.

Of course, the computation of $H_n(X)$, even when X is path connected, is usually difficult. However, one can always compute $H_0(X)$.

Theorem 4.14.

(i) *If X is a nonempty path connected space, then $H_0(X) \cong \mathbf{Z}$. Moreover, if $x_0, x_1 \in X$, then* cls x_0 = cls x_1 *is a generator of $H_0(X)$.*
(ii) *For any space X, the group $H_0(X)$ is free abelian of rank* = card Λ, *where $\{X_\lambda: \lambda \in \Lambda\}$ is the family of path components.*
(iii) *If X and Y are path connected spaces and $f: X \to Y$ is continuous, then $f_*: H_0(X) \to H_0(Y)$ takes a generator of $H_0(X)$ to a generator of $H_0(Y)$.*

PROOF. (i) Consider the end of the singular complex

$$S_1(X) \xrightarrow{\partial_1} S_0(X) \xrightarrow{\partial_0} 0.$$

As ∂_0 is zero, $Z_0(X) = \ker \partial_0 = S_0(X)$; therefore every 0-chain in X is a 0-cycle (in particular, cls $x \in H_0(X)$ for every $x \in X$). A typical 0-cycle is thus $\sum_{x \in X} m_x x$, where $m_x \in \mathbf{Z}$ and almost all $m_x = 0$. We claim that

$$B_0(X) = \{\textstyle\sum m_x x \in S_0(X): \sum m_x = 0\}.$$

If this claim is true, then define $\theta: Z_0(X) \to \mathbf{Z}$ by $\sum m_x x \mapsto \sum m_x$. It is clear that θ is a surjection with kernel $B_0(X)$, and so the first isomorphism theorem gives $H_0(X) \cong \mathbf{Z}$.

Let us prove the claim. Let $\gamma = \sum_{i=0}^{k} m_i x_i \in S_0(X)$, and assume that $\sum m_i = 0$. Choose a point $x \in X$ ($X \neq \varnothing$), and choose a path σ_i in X from x to x_i for each i (X is path connected). Note that $\partial_1 \sigma_i = \sigma_i(e_1) - \sigma_i(e_0) = x_i - x$ (we have identified $\mathbf{I} = [0, 1]$ with $\Delta^1 = [e_0, e_1]$). Now $\sum m_i \sigma_i \in S_1(X)$, and

$$\partial_1(\textstyle\sum m_i \sigma_i) = \sum m_i \partial_1(\sigma_i) = \sum m_i(x_i - x) = \sum m_i x_i - (\sum m_i)x = \gamma,$$

since $\sum m_i = 0$. Therefore $\gamma = \sum m_i x_i = \partial_1(\sum m_i \sigma_i) \in B_0(X)$. Conversely, if $\gamma \in B_0(X)$, then $\gamma = \partial_1(\sum n_j \tau_j)$, where $n_j \in \mathbf{Z}$ and τ_j is a 1-simplex in X. Hence

$$\gamma = \textstyle\sum n_j(\tau_j(e_1) - \tau_j(e_0)),$$

so that each coefficient n_j occurs twice and with opposite sign. Thus the sum of the coefficients is zero.

Let $x_0, x_1 \in X$. There is a path σ in X from x_0 to x_1, and $x_1 - x_0 = \partial_1 \sigma \in B_0(X)$; this says that $x_1 + B_0(X) = x_0 + B_0(X)$, that is, cls x_0 = cls x_1. Finally, if cls γ is a generator of $H_0(X)$, where $\gamma = \sum m_i x_i$, then $\theta(\gamma) = \sum m_i = \pm 1$. Replacing γ by $-\gamma$ if necessary, we may assume that $\sum m_i = 1$. If $x_0 \in X$, then $\gamma = x_0 + (\gamma - x_0)$; since $\gamma - x_0 \in B_0(X)$ (its coefficient sum is zero), we have cls γ = cls x_0, as desired.

(ii) Immediate from Theorem 4.13 and part (i) of this theorem.
(iii) Immediate from part (i). □

Compare the functors π_0 and $H_0 : \pi_0(X)$ is the set of path components of X; $H_0(X)$ carries exactly the same information and builds a free abelian group from it.

In Theorem 4.32, we shall give a geometric characterization of 1-cycles in a space X.

Lemma 4.15. *Let A be a subspace of X with inclusion $j: A \hookrightarrow X$. Then $j_\#: S_n(A) \to S_n(X)$ is an injection for every $n \geq 0$.*

PROOF. Let $\gamma = \sum m_i\sigma_i \in S_n(A)$; we may assume that all σ_i are distinct. If $\gamma \in \ker j_\#$, then $0 = j_\# \sum m_i\sigma_i = \sum m_i(j \circ \sigma_i)$. Since $j \circ \sigma_i$ differs from σ_i only in having its target enlarged from A to X, it follows that all $j \circ \sigma_i$ are distinct. But $S_n(X)$ is free abelian with basis all n-simplexes in X; it follows that every $m_i = 0$ and $\gamma = 0$. □

This lemma is invoked often, usually tacitly.

Definition. If $\zeta = \sum m_i\sigma_i \in S_n(X)$, with all $m_i \neq 0$ and all σ_i distinct, then the **support** of ζ, denoted by supp ζ, is $\bigcup \sigma_i(\Delta^n)$.

It is clear that supp ζ is a compact subset of X, since it is a finite union of compact subsets.

Theorem 4.16 (Compact Supports). *If* cls $\zeta \in H_n(X)$, *then there is a compact subspace A of X with* cls $\zeta \in \operatorname{im} j_*$, *where $j: A \hookrightarrow X$ is the inclusion.*

PROOF. Let $A = \operatorname{supp} \zeta$. If $\zeta = \sum m_i\sigma_i$, then for each i we may write $\sigma_i = j\sigma_i'$, where $\sigma_i': \Delta^n \to A$. Define $\gamma = \sum m_i\sigma_i' \in S_n(A)$. Now $j_\# \partial\gamma = \partial j_\# \gamma = \partial\zeta = 0$ (because ζ is an n-cycle in X); since $j_\#$ is an injection, it follows that $\partial\gamma = 0$, that is, γ is an n-cycle in A. Therefore cls $\gamma \in H_n(A)$ and j_* cls $\gamma =$ cls ζ. □

Corollary 4.17. *If X is a space for which there exists an integer $n \geq 0$ with $H_n(A) = 0$ for every compact subspace of X, then $H_n(X) = 0$.*

PROOF. If cls $\zeta \in H_n(X)$, then the theorem provides a compact subspace A of X (with inclusion $j: A \hookrightarrow X$) and an element cls $\gamma \in H_n(A)$ with j_* cls $\gamma =$ cls ζ. But $H_n(A) = 0$, by hypothesis, hence cls $\gamma = 0$, and hence cls $\zeta = 0$. □

The next technical result will be used in proving the Jordan curve theorem.

Theorem 4.18. *Let $X = \bigcup_{p=1}^{\infty} X^p$ with $X^p \subset X^{p+1}$ for all p (call the inclusion maps $\lambda^p: X^p \hookrightarrow X$ and $\varphi^p: X^p \hookrightarrow X^{p+1}$). If every compact subspace A of X is* *contained in some* X^p, *then* $\text{cls}\,\zeta \in H_n(X)$ *is zero if and only if there exist* p *and* $\text{cls}\,\zeta' \in H_n(X^p)$ *with*

$$\lambda_*^p \,\text{cls}\,\zeta' = \text{cls}\,\zeta \quad \textit{and} \quad \varphi_*^p \,\text{cls}\,\zeta' = 0.$$

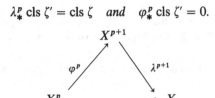

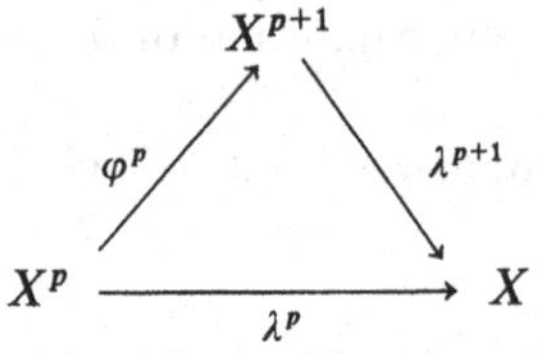

Proof. Sufficiency is obvious, for $\lambda^{p+1} \circ \varphi^p = \lambda^p$, hence $0 = \lambda_*^{p+1}(\varphi_*^p \,\text{cls}\,\zeta') = \lambda_*^p \,\text{cls}\,\zeta' = \text{cls}\,\zeta$.

Conversely, assume that $\text{cls}\,\zeta = 0$ in $H_n(X)$. Thus $\zeta = \sum m_i \sigma_i \in S_n(X)$, and there exists $\beta = \sum c_k \tau_k \in S_{n+1}(X)$ with $\partial\beta = \zeta$. Define $A = \text{supp}\,\zeta \cup \text{supp}\,\beta$,[6] and choose p with $A \subset X^p$. As in the proof of Theorem 4.16, there are n-simplexes $\sigma_i'\colon \Delta^n \to X^p$ and $(n+1)$-simplexes $\tau_k'\colon \Delta^{n+1} \to X^p$ for all i, k with $\sigma_i = \lambda^p \sigma_i'$ and $\tau_k = \lambda^p \tau_k'$; moreover, if $\zeta' = \sum m_i \sigma_i'$, then ζ' is an n-cycle in X^p and $\lambda_*^p \,\text{cls}\,\zeta' = \text{cls}\,\zeta$. On the other hand, if $\beta' = \sum c_k \tau_k'$, then $\partial \varphi_\#^p \beta' = \varphi_\#^p \partial \beta' = \varphi_\#^p \zeta'$; that is, $\varphi_*^p \,\text{cls}\,\zeta' = 0$ in $H_n(X^{p+1})$. □

Theorem 4.18 and Corollary 4.17 are instances of a more general result: each homology functor H_n preserves "direct limits" over a directed index set (see [Spanier, p. 162]).

The Homotopy Axiom

Our next goal is to show that $H_n(f) = H_n(g)$ for all n whenever f and g are homotopic. First, we present a preliminary result.

Theorem 4.19. *If* X *is a bounded convex subspace of euclidean space, then* $H_n(X) = 0$ *for all* $n \geq 1$. *In particular,* $H_n(D^k) = 0$ *for all* $n > 0$ *and all* k.

Remarks. (1) If $X \neq \varnothing$, then Theorem 4.14 shows that $H_0(X) = \mathbf{Z}$.

(2) This theorem will be used to prove a stronger result, Corollary 4.25, which replaces "convex subspace of euclidean space" by "contractible space".

Proof. Choose a point $b \in X$. For every n-simplex $\sigma\colon \Delta^n \to X$, consider the "cone over σ with vertex b" (recall that Exercise 2.10 shows that an affine simplex is the cone over any one of its faces with opposite vertex). Define an $(n+1)$-simplex $b.\sigma\colon \Delta^{n+1} \to X$ as follows:

[6] Actually, it is easy to see that $\text{supp}\,\zeta \subset \text{supp}\,\beta$, so that one may take $A = \text{supp}\,\beta$.

$$(b\,.\,\sigma)(t_0, t_1, \ldots, t_{n+1}) = \begin{cases} b & \text{if } t_0 = 1 \\ t_0 b + (1 - t_0)\sigma\left(\dfrac{t_1}{1-t_0}, \ldots, \dfrac{t_{n+1}}{1-t_0}\right) & \text{if } t_0 \neq 1 \end{cases}$$

(here $(t_0, t_1, \ldots, t_{n+1})$ are barycentric coordinates of points in Δ^{n+1}). Note that $t_0 = 1$ implies that $(t_0, \ldots, t_{n+1}) = (1, 0, \ldots, 0)$; moreover, $b\,.\,\sigma$ is well defined because $(1 - t_0)^{-1} \sum_{i=1}^{n+1} t_i = 1$ (hence the argument of σ lies in Δ^n) and X is convex. A routine argument shows that $b\,.\,\sigma$ is continuous.

Define $c_n\colon S_n(X) \to S_{n+1}(X)$ by setting $c_n(\sigma) = b\,.\,\sigma$ and extending by linearity. We claim that, for all $n \geq 1$ and every n-simplex σ in X,

$$\partial_{n+1} c_n(\sigma) = \sigma - c_{n-1}\partial_n(\sigma). \tag{$*$}$$

(If one ignores signs, formula $(*)$ says that the (oriented) boundary of the cone on σ is the union of σ with the cone on the boundary of σ. We illustrate this when σ is a 2-simplex.

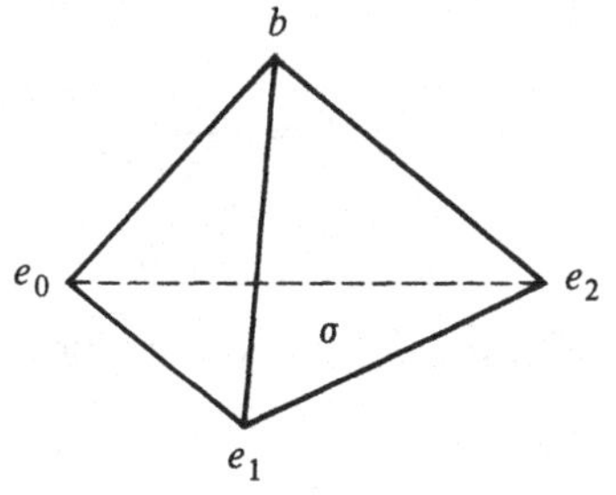

Here σ is represented by $[e_0, e_1, e_2]$; the cone $b\,.\,\sigma$ is the tetrahedron, and the boundary of this tetrahedron is σ together with the three faces $[b, e_0, e_1]$, $[b, e_0, e_2]$, and $[b, e_1, e_2]$, each of which is the cone on a face on σ.)

If formula $(*)$ holds, then the theorem follows easily. If $\gamma \in S_n(X)$, then extending by linearity gives

$$\gamma = \partial c\gamma + c\partial\gamma;$$

if γ is a cycle, that is, $\partial\gamma = 0$, then $\gamma = \partial c\gamma \in B_n(X)$. Hence $Z_n(X) = B_n(X)$, and $H_n(X) = 0$.

To verify $(*)$, let us first compute the faces of $c_n(\sigma) = b\,.\,\sigma$. If $n \geq 1$ and $i = 0$, then

$$((b\,.\,\sigma)\varepsilon_0^{n+1})(t_0, \ldots, t_n) = (b\,.\,\sigma)(0, t_0, \ldots, t_n) = \sigma(t_0, \ldots, t_n).$$

If $0 < i \leq n + 1$, then

$$((b\,.\,\sigma)\varepsilon_i^{n+1})(t_0, \ldots, t_n) = (b\,.\,\sigma)(t_0, \ldots, t_{i-1}, 0, t_i, \ldots, t_n).$$

If, in addition, $t_0 = 1$, then

$$(b\,.\,\sigma)(1, 0, \ldots, 0) = b;$$

if $t_0 \neq 1$, then the right side above is equal to

$$t_0 b + (1 - t_0)\sigma\left(\frac{t_1}{1 - t_0}, \ldots, \frac{t_{i-1}}{1 - t_0}, 0, \frac{t_i}{1 - t_0}, \ldots, \frac{t_n}{1 - t_0}\right)$$

$$= t_0 b + (1 - t_0)\sigma\varepsilon_{i-1}^n\left(\frac{t_1}{1 - t_0}, \ldots, \frac{t_n}{1 - t_0}\right)$$

$$= c_{n-1}(\sigma\varepsilon_{i-1}^n)(t_0, \ldots, t_n).$$

In conclusion, after evaluating each side on $(t_0, \ldots, t_n)$,

$$(c_n\sigma)\varepsilon_0^{n+1} = \sigma \quad \text{and} \quad (c_n\sigma)\varepsilon_i^{n+1} = c_{n-1}(\sigma\varepsilon_{i-1}^n) \quad \text{if } i > 0.$$

Taking alternating sums,

$$\partial_{n+1}c_n(\sigma) = \sum_{i=0}^{n+1} (-1)^i(c_n\sigma)\varepsilon_i = \sigma + \sum_{i=1}^{n+1} (-1)^i c_{n-1}(\sigma\varepsilon_{i-1})$$

$$= \sigma - \sum_{j=0}^{n} (-1)^j c_{n-1}(\sigma\varepsilon_j)$$

$$= \sigma - c_{n-1}\left(\sum_{j=0}^{n} (-1)^j \sigma\varepsilon_j\right)$$

$$= \sigma - c_{n-1}\partial_n\sigma.$$

□

Definition. The homomorphism c_n is called the **cone construction**.

Corollary 4.20.

(i) *Let X be convex and let $\gamma = \sum m_i\sigma_i \in S_n(X)$. If $b \in X$, then*

$$\partial(b\,.\,\gamma) = \begin{cases} \gamma - b\,.\,\partial\gamma & \text{if } n > 0 \\ (\sum m_i)b - \gamma & \text{if } n = 0. \end{cases}$$

(ii) *If γ is an n-cycle and $n > 0$, then*

$$\partial(b\,.\,\gamma) = \partial(c_n\gamma) = \gamma.$$

Remark. Part (ii) may be regarded as an **integration formula**.

Proof. (i) For $n > 0$, the formula has just been proved above. When $n = 0$, consider first a 0-simplex σ (which we identify with its image $x \in X$). The definition $b\,.\,\sigma\colon \Delta^1 \to X$ is

$$(b\,.\,\sigma)(t) = tb + (1 - t)x.$$

Therefore, if one identifies σ_i with its image x_i,

$$\partial(b\,.\,\gamma) = \partial(\sum m_i b\,.\,\sigma_i) = \sum m_i\partial(b\,.\,\sigma_i)$$

$$= \sum m_i(b - x_i) = (\sum m_i)b - \gamma.$$

(ii) Immediate from part (i). □

Lemma 4.21. *Assume that $f, g\colon X \to Y$ are continuous maps and that there are homomorphisms $P_n\colon S_n(X) \to S_{n+1}(Y)$ with*

$$f_\# - g_\# = \partial'_{n+1} P_n + P_{n-1}\partial_n.$$

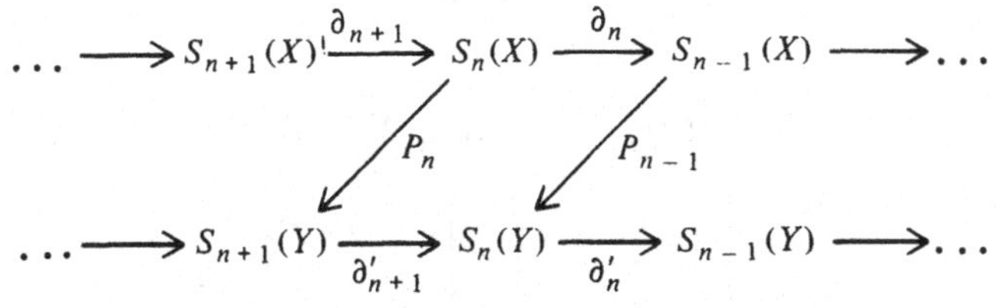

Then, for all $n \geq 0$, $H_n(f) = H_n(g)$.

Proof. By definition, $H_n(f)\colon z + B_n(X) \mapsto f_\#(z) + B_n(Y)$, where $\partial z = 0$. But

$$(f_\# - g_\#)z = (\partial' P + P\partial)z = \partial' P z \in B_n(Y),$$

and so $f_\#(z) + B_n(Y) = g_\#(z) + B_n(Y)$; that is, $H_n(f) = H_n(g)$. □

Remark. The equation in the statement of the lemma makes sense when $n = 0$, for $S_{-1}(X)$ was defined to be zero, hence $P_{-1}\colon S_{-1}(X) \to S_0(Y)$ must be the zero map.

Lemma 4.22. *Let X be a space and, for $i = 0, 1$, let $\lambda_i^X\colon X \to X \times \mathbf{I}$ be defined by $x \mapsto (x, i)$. If $H_n(\lambda_0^X) = H_n(\lambda_1^X)\colon H_n(X) \to H_n(X \times \mathbf{I})$, then $H_n(f) = H_n(g)$ whenever f and $g\colon X \to Y$ are homotopic.*

Proof. If $F\colon X \times \mathbf{I} \to Y$ is a homotopy $f \simeq g$, then

$$f = F\lambda_0^X \quad \text{and} \quad g = F\lambda_1^X.$$

Therefore

$$\begin{aligned} H_n(f) = H_n(F\lambda_0^X) &= H_n(F)H_n(\lambda_0^X) \\ &= H_n(F)H_n(\lambda_1^X) = H_n(F\lambda_1^X) = H_n(g). \end{aligned}$$ □

Theorem 4.23 (Homotopy Axiom). *If $f, g\colon X \to Y$ are homotopic, then*

$$H_n(f) = H_n(g) \textit{ for all } n \geq 0.$$

Proof. By Lemma 4.22, it suffices to prove that $H_n(\lambda_0^X) = H_n(\lambda_1^X)$ for all $n \geq 0$; by Lemma 4.21, it suffices to construct homomorphisms $P_n^X\colon S_n(X) \to S_{n+1}(X \times \mathbf{I})$ with

$$\lambda_{1\#}^X - \lambda_{0\#}^X = \partial_{n+1} P_n^X + P_{n-1}^X \partial_n. \tag{1}$$

We propose proving the existence of such homomorphisms P_n^X for all spaces

X by induction on $n \geq 0$. In order to prove the inductive step (and realizing that we must define P_n^X on a basis of the free abelian group $S_n(X)$), we strengthen the inductive hypothesis as follows. For all spaces X, there exist homomorphisms $P_n^X\colon S_n(X) \to S_{n+1}(X \times \mathbf{I})$ satisfying (1) and the following "naturality condition": the following diagram commutes for every simplex $\sigma\colon \Delta^n \to X$:

$$\begin{array}{ccc} S_n(\Delta^n) & \xrightarrow{P_n^{\Delta^n}} & S_{n+1}(\Delta^n \times \mathbf{I}) \\ {\scriptstyle \sigma_\#}\downarrow & & \downarrow{\scriptstyle (\sigma \times 1)_\#} \\ S_n(X) & \xrightarrow[P_n^X]{} & S_{n+1}(X \times \mathbf{I}); \end{array}$$

that is,

$$(\sigma \times 1)_\# P_n^{\Delta^n} = P_n^X \sigma_\#. \tag{2}$$

(Recall that $\sigma \times 1\colon \Delta^n \times \mathbf{I} \to X \times \mathbf{I}$ is defined by $(x, t) \mapsto (\sigma(x), t)$.)

Let $n = 0$. Begin by defining $P_{-1}^X = 0$ (there is no choice here because $S_{-1}(X) = 0$). Now $\Delta^0 = \{e_0\}$; given $\sigma\colon \Delta^0 \to X$, define $P_0^X(\sigma)\colon \Delta^1 \to X \times \mathbf{I}$ by $t \mapsto (\sigma(e_0), t)$, and then define $P_0^X\colon S_0(X) \to S_1(X \times \mathbf{I})$ by extending by linearity. To check Eq. (1), it suffices to evaluate on a typical basis element σ:

$$\partial_1 P_0^X \sigma = (\sigma(e_0), 1) - (\sigma(e_0), 0) = \lambda_1^X \circ \sigma - \lambda_0^X \circ \sigma = \lambda_{1\#}^X(\sigma) - \lambda_{0\#}^X(\sigma);$$

that is (since $P_{-1}^X = 0$),

$$\partial_1 P_0^X + P_{-1}^X \partial_0 = \lambda_{1\#}^X - \lambda_{0\#}^X.$$

To check the naturality condition (2), consider the diagram

$$\begin{array}{ccc} S_0(\Delta^0) & \xrightarrow{P_0^{\Delta^0}} & S_1(\Delta^0 \times \mathbf{I}) \\ {\scriptstyle \sigma_\#}\downarrow & & \downarrow{\scriptstyle (\sigma \times 1)_\#} \\ S_0(X) & \xrightarrow[P_0^X]{} & S_1(X \times \mathbf{I}). \end{array}$$

There is only one 0-simplex in Δ^0, namely, the identity function δ with $\delta(e_0) = e_0$. To check commutativity, it suffices to evaluate each composite on δ; note that each result is a map $\Delta^1 \to X \times \mathbf{I}$. Identify $(1 - t)e_0 + te_1 \in \Delta^1$ with t, and evaluate:

$$P_0^X \sigma_\#(\delta) = P_0^X(\sigma \circ \delta) = P_0^X(\sigma)\colon t \mapsto (\sigma(e_0), t);$$

$$(\sigma \times 1)_\# P_0^{\Delta^0}(\delta)\colon t \mapsto (\sigma \times 1)_\#(\delta(e_0), t) = (\sigma \times 1)_\#(e_0, t) = (\sigma(e_0), t),$$

as desired.

Assume that $n > 0$. We shall sometimes write Δ instead of Δ^n for the remainder of this proof. Were Eq. (1) true, then $(\lambda_{1\#}^{\Delta} - \lambda_{0\#}^{\Delta} - P_{n-1}^{\Delta}\partial_n)(\gamma)$ would be a cycle for every $\gamma \in S_n(X)$. This is indeed so.

$$\begin{aligned}\partial_n(\lambda_{1\#}^{\Delta} - \lambda_{0\#}^{\Delta} - P_{n-1}^{\Delta}\partial_n) &= \lambda_{1\#}^{\Delta}\partial_n - \lambda_{0\#}^{\Delta}\partial_n - \partial_n P_{n-1}^{\Delta}\partial_n \quad \text{(Lemma 4.8)}\\ &= \lambda_{1\#}^{\Delta}\partial_n - \lambda_{0\#}^{\Delta}\partial_n - (\lambda_{1\#}^{\Delta} - \lambda_{0\#}^{\Delta} - P_{n-2}^{\Delta}\partial_{n-1})\partial_n\\ &\qquad\qquad \text{(by induction)}\\ &= 0 \quad (\text{since } \partial\partial = 0).\end{aligned}$$

If $\delta\colon \Delta^n \to \Delta^n$ is the identity map, then $\delta \in S_n(\Delta^n)$; it follows that $(\lambda_{1\#}^{\Delta} - \lambda_{0\#}^{\Delta} - P_{n-1}^{\Delta}\partial_n)(\delta) \in Z_n(\Delta^n \times \mathbf{I})$. But $\Delta^n \times \mathbf{I}$ is convex, so that Theorem 4.19 gives $H_n(\Delta^n \times \mathbf{I}) = 0$ (because $n > 0$); therefore $Z_n(\Delta^n \times \mathbf{I}) = B_n(\Delta^n \times \mathbf{I})$, and there exists $\beta_{n+1} \in S_{n+1}(\Delta^n \times \mathbf{I})$ with

$$\partial_{n+1}\beta_{n+1} = (\lambda_{1\#}^{\Delta} - \lambda_{0\#}^{\Delta} - P_{n-1}^{\Delta}\partial_n)(\delta).$$

Define $P_n^X\colon S_n(X) \to S_{n+1}(X \times \mathbf{I})$, for any space X, by

$$P_n^X(\sigma) = (\sigma \times 1)_\#(\beta_{n+1})$$

(where σ is an n-simplex in X), and extend by linearity.

Before checking Eqs. (1) and (2), observe that, for $i = 0, 1$ and for $\sigma\colon \Delta^n \to X$ an n-simplex in X, we have

$$(\sigma \times 1)\lambda_i^{\Delta} = \lambda_i^X\sigma\colon \Delta^n \to X \times \mathbf{I} \tag{3}$$

[if $y \in \Delta^n$, then

$$(\sigma \times 1)\lambda_i^{\Delta}(y) = (\sigma \times 1)(y, i) = (\sigma(y), i) = \lambda_i^X(\sigma(y))].$$

To check Eq. (1), let $\sigma\colon \Delta^n \to X$ be an n-simplex in X.

$$\begin{aligned}\partial_{n+1}P_n^X(\sigma) &= \partial_{n+1}(\sigma \times 1)_\#(\beta_{n+1})\\ &= (\sigma \times 1)_\#\partial_{n+1}(\beta_{n+1}) \quad \text{(Lemma 4.8)}\\ &= (\sigma \times 1)_\#(\lambda_{1\#}^{\Delta} - \lambda_{0\#}^{\Delta} - P_{n-1}^{\Delta}\partial_n)(\delta) \quad (\text{definition of } \beta_{n+1})\\ &= (\sigma \times 1)\lambda_1^{\Delta} - (\sigma \times 1)\lambda_0^{\Delta} - (\sigma \times 1)_\# P_{n-1}^{\Delta}\partial_n(\delta) \quad (\text{since } \lambda_{i\#}^{\Delta}(\delta) = \lambda_i^{\Delta})\\ &= (\sigma \times 1)\lambda_1^{\Delta} - (\sigma \times 1)\lambda_0^{\Delta} - P_{n-1}^X\sigma_\#\partial_n(\delta) \quad (\text{Eq. (2) for } P_{n-1})\\ &= \lambda_1^X\sigma - \lambda_0^X\sigma - P_{n-1}^X\partial_n\sigma_\#(\delta) \quad \text{(Eq. (3) and Lemma 4.8)}\\ &= (\lambda_1^X - \lambda_0^X - P_{n-1}^X\partial_n)(\sigma) \quad (\text{since } \sigma_\#(\delta) = \sigma).\end{aligned}$$

To check the naturality equation (2), let $\tau\colon \Delta^n \to \Delta^n$ be an n-simplex in Δ^n. Then for every $\sigma\colon \Delta^n \to X$,

$$(\sigma \times 1)_\# P_n^{\Delta}(\tau) = (\sigma \times 1)_\#(\tau \times 1)_\#(\beta_{n+1}) = (\sigma\tau \times 1)_\#(\beta_{n+1}),$$

while

$$P_n^X\sigma_\#(\tau) = P_n^X(\sigma\tau) = (\sigma\tau \times 1)_\#(\beta_{n+1}),$$

as desired. □

Remarks. (1) If $X = \Delta^n$ and $\sigma = \delta$, the identity map in Δ^n, then one can give a geometric interpretation of β_{n+1}. Recall that $P_n^X(\sigma) = (\sigma \times 1)_\#(\beta_{n+1})$; in particular, $P_n^\Delta(\delta) = (\delta \times 1)_\#(\beta_{n+1}) = \beta_{n+1}$, since $\delta \times 1$ is the identity on $\Delta^n \times \mathbf{I}$. Now $\beta_{n+1} = P_n^\Delta(\delta)$ is, in no obvious way, a linear combination of simplexes because $\Delta^n \times \mathbf{I}$ is a prism (hence the letter P) which is not triangulated.

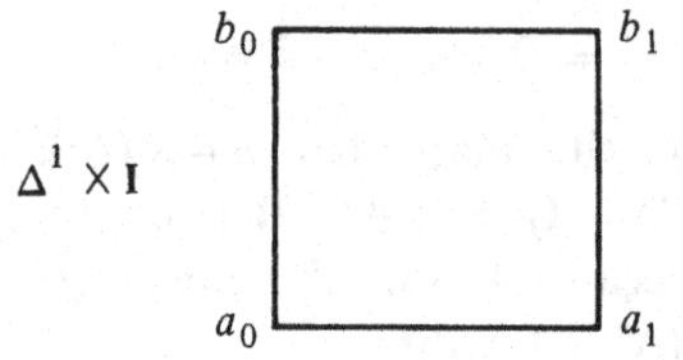

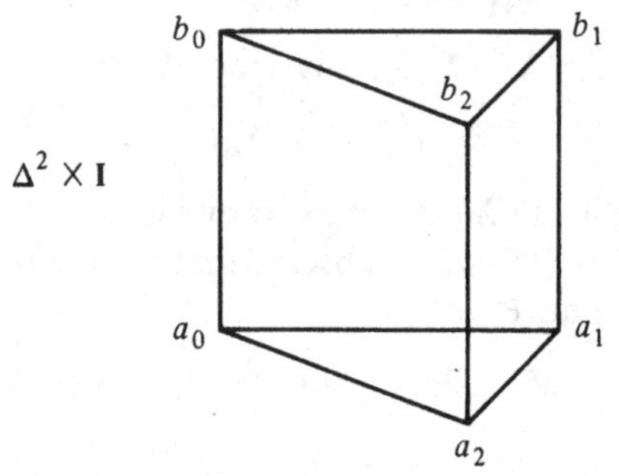

If $\Delta^n = [e_0, \ldots, e_n]$, define $a_i = (e_i, 0)$ and $b_i = (e_i, 1)$ for $0 \le i \le n$. A formula for β_{n+1} turns out to be

$$\beta_{n+1} = \sum_{i=0}^{n} (-1)^i [a_0, \ldots, a_i, b_i, b_{i+1}, \ldots, b_n], \tag{4}$$

where the brackets denote the affine map $\Delta^{n+1} \to \Delta^n \times \mathbf{I}$ taking the vertices $\{e_0, \ldots, e_{n+1}\}$ to the vertices $\{a_0, \ldots, a_i, b_i, \ldots, b_n\}$ preserving the displayed orderings. Aside from signs, formula (4) does triangulate the prisms. For example, $\Delta^1 \times \mathbf{I}$ is divided into two triangles $[a_0, b_0, b_1]$ and $[a_0, a_1, b_1]$.

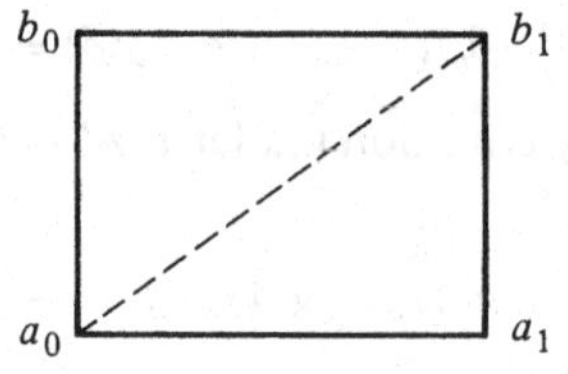

After drawing in dotted lines in $\Delta^2 \times \mathbf{I}$ pictured above, one sees three tetrahedra: $[a_0, b_0, b_1, b_2]$, $[a_0, a_1, b_1, b_2]$, and $[a_0, a_1, a_2, b_2]$. One thus views $P(\sigma)$ as the "triangulated prism over σ". The geometric interpretation of $\partial P + P\partial = \lambda_{1\#} - \lambda_{0\#}$ is: the (oriented) boundary ∂P of the solid prism without $P\partial$, the prism on the boundary, is the top and bottom (we ignore signs when being descriptive).

(2) One could prove Theorem 4.23 using the explicit formula (4) for β_{n+1}, but the proof is no shorter and one must always be alert that signs are correct.

(3) The construction of the sequence of homomorphisms P_n has been axiomatized (and will appear again); it is called the *method of acyclic models*, and we shall discuss it in Chapter 9.

We now draw the usual consequence of the homotopy axiom: the homology functors induce functors on the homotopy category; we may regard H_n as a functor **hTop** $\to$ **Ab**.

Corollary 4.24. *If X and Y have the same homotopy type, then $H_n(X) \cong H_n(Y)$ for all $n \geq 0$, where the isomorphism is induced by any homotopy equivalence.*

We now generalize Theorem 4.19.

Corollary 4.25. *If X is contractible, then $H_n(X) = 0$ for all $n > 0$.*

PROOF. X has the same homotopy type as a one-point space; apply Corollary 4.24 and the dimension axiom, Theorem 4.12. □

EXERCISES

4.9. (i) Using the explicit formula for β_{n+1}, show that

$$\partial_{n+1}\beta_{n+1} = (\lambda_{1\#}^{\Delta} - \lambda_{0\#}^{\Delta} - P_{n-1}^{\Delta}\partial_n)(\delta)$$

for $n = 0$ and $n = 1$.

(ii) Give an explicit formula for $P_1^X(\sigma)$, where $\sigma\colon \Delta^1 \to X$ is a 1-simplex.

*4.10. Prove that P_n is "natural": if $f\colon X \to Y$ is continuous, there is a commutative diagram

$$\begin{array}{ccc} S_n(X) & \xrightarrow{P_n^X} & S_{n+1}(X \times \mathbf{I}) \\ {\scriptstyle f_\#}\downarrow & & \downarrow{\scriptstyle (f \times 1)_\#} \\ S_n(Y) & \xrightarrow[P_n^Y]{} & S_{n+1}(Y \times \mathbf{I}). \end{array}$$

*4.11. If X is a deformation retract of Y, then $H_n(X) \cong H_n(Y)$ for all $n \geq 0$. In fact, if $i\colon X \to Y$ is the inclusion, then $H_n(i)$ is an isomorphism.

4.12. Compute the homology groups of the $\sin(1/x)$ space.

The Hurewicz Theorem

There is an intimate relation between π_1 and H_1.

Lemma 4.26. *Let $\eta\colon \Delta^1 \to \mathbf{I}$ be the homeomorphism $(1-t)e_0 + te_1 \mapsto t$. There is a well defined function*

$$\varphi\colon \pi_1(X, x_0) \to H_1(X)$$

given by

$$[f] \mapsto \operatorname{cls} f\eta,$$

where $f\colon \mathbf{I} \to X$ is a closed path in X at x_0.

Proof. It is plain that $f\eta$ is a 1-simplex in X, so that $f\eta \in S_1(X)$. Indeed, $f\eta \in Z_1(X)$, for $\partial_1(f\eta) = f\eta(e_1) - f\eta(e_0) = f(1) - f(0) = 0$, because f is a closed path; thus $\operatorname{cls} f\eta \in H_1(X)$. In particular, if $u\colon \mathbf{I} \to S^1$ is defined by $t \mapsto e^{2\pi i t}$, then $u\eta$ is a 1-cycle in S^1. We saw in Exercise 3.2 that there is a map $f'\colon S^1 \to X$ making the following diagram commute (f is a closed path in X):

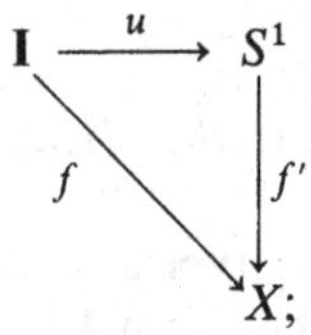

hence f' induces a homomorphism $f'_*\colon H_1(S^1) \to H_1(X)$, namely, $\operatorname{cls}(\sum m_i\sigma_i) \mapsto \operatorname{cls}(\sum m_i(f' \circ \sigma_i))$. It follows that

$$\operatorname{cls} f\eta = \operatorname{cls} f'u\eta = f'_* \operatorname{cls} u\eta \in H_1(X).$$

Now assume that g is a closed path in X at x_0 with $f \simeq g$ rel $\dot{\mathbf{I}}$; by Exercise 3.2, we have $f' \simeq g'$. The homotopy axiom (Theorem 4.23) thus gives

$$\operatorname{cls} f\eta = f'_* \operatorname{cls} u\eta = g'_* \operatorname{cls} u\eta = \operatorname{cls} g\eta.$$

Therefore φ is well defined. □

Lemma 4.26 may be paraphrased: homotopic closed curves in X must be homologous.

Definition. The function $\varphi\colon \pi_1(X, x_0) \to H_1(X)$ of Lemma 4.26 is called the **Hurewicz map.**

Theorem 4.27. *The Hurewicz map $\varphi\colon \pi_1(X, x_0) \to H_1(X)$ is a homomorphism.*

Proof. Let f and g be closed paths in X at x_0. Define a continuous map $\sigma\colon \Delta^2 \to X$ as indicated by the following picture.

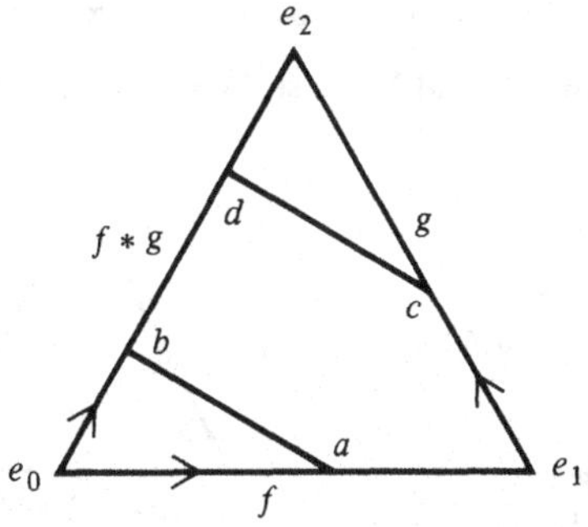

In more detail, first define σ on $\dot{\Delta}^2$: $\sigma(1-t, t, 0) = f(t)$; $\sigma(0, 1-t, t) = g(t)$; $\sigma(1-t, 0, t) = (f * g)(t)$. Now define σ on all of Δ^2 by setting it constant on the line segments with endpoints $a = a(t) = (1-t, t, 0)$ and $b = b(t) = ((2-t)/2, 0, t/2)$, and constant on the line segments with endpoints $c = c(t) = (0, 1-t, t)$ and $d = d(t) = ((1-t)/2, 0, (1+t)/2)$. It is easy to see that $\sigma: \Delta^2 \to X$ is continuous, that is, $\sigma \in S_2(X)$. Moreover, $\partial\sigma = \sigma\varepsilon_0 - \sigma\varepsilon_1 + \sigma\varepsilon_2$. But $\sigma\varepsilon_0(t) = \sigma(0, 1-t, t) = g(t)$, $\sigma\varepsilon_1 = f * g$, and $\sigma\varepsilon_2 = f$, so that $\partial\sigma = g - f * g + f$. Therefore

$$\varphi: [f][g] = [f * g] \mapsto \mathrm{cls}(f * g)\eta = \mathrm{cls}(f + g)\eta = \mathrm{cls}\, f\eta + \mathrm{cls}\, g\eta. \qquad \square$$

Exercises

*4.13. Prove that the Hurewicz map φ is "natural". If $h: (X, x_0) \to (Y, y_0)$ is a map of pointed spaces, then the following diagram commutes:

$$\begin{array}{ccc} \pi_1(X, x_0) & \xrightarrow{h_*} & \pi_1(Y, y_0) \\ \downarrow{\scriptstyle\varphi} & & \downarrow{\scriptstyle\varphi} \\ H_1(X) & \xrightarrow[h_*]{} & H_1(Y). \end{array}$$

*4.14. If f is a (not necessarily closed) path in X, prove that the 1-chain f is homologous to $-f^{-1}$. (*Hint*: Use Theorem 4.27 and Exercise 3.4 with the picture below.)

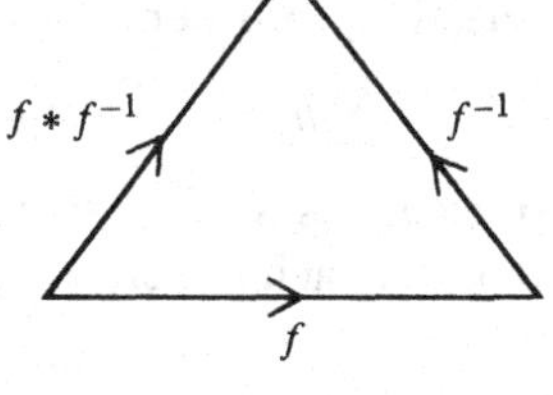

*4.15. Let X be a space and let α, β, γ be (not necessarily closed) paths in X such that $\alpha * \beta * \gamma$ is defined and is a closed path. Prove that, in $H_1(X)$,

$$\text{cls}(\alpha * \beta * \gamma) = \text{cls}(\alpha + \beta + \gamma) = \text{cls}\ \alpha + \text{cls}\ \beta + \text{cls}\ \gamma.$$

(*Hint*: Use Theorem 4.27 and Exercise 3.4 with maps $\Delta^2 \to X$ suggested by

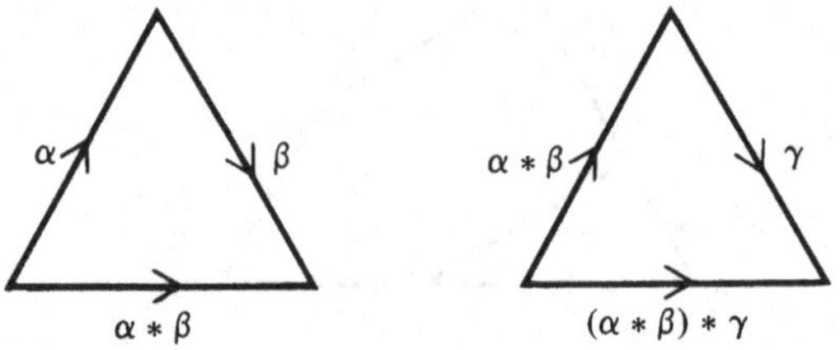

.)

Lemma 4.28 (Substitution Principle). *Let F be a free abelian group with basis B, let $x_0, x_1, \ldots, x_k$ be a list of elements of B, possibly with repetitions, and assume that*

$$m_0 x_0 = \sum_{i=1}^{k} m_i x_i, \quad \textit{where } m_i \in \mathbf{Z}.$$

If G is any abelian group, and if $y_0, y_1, \ldots, y_k \in G$ is a list such that $x_i = x_j$ implies $y_i = y_j$, then $m_0 y_0 = \sum_{i=1}^{k} m_i y_i$ in G.

Proof. Define a function $f: B \to G$ by $f(x_i) = y_i$ for $i = 0, 1, \ldots, k$, and $f(x) = 0$ otherwise (f is well defined by hypothesis). By Theorem 4.1, there is a homomorphism $\tilde{f}: F \to G$ extending f. But

$$0 = \tilde{f}(m_0 x_0 - \sum m_i x_i) = m_0 y_0 - \sum m_i y_i. \qquad \square$$

A key ingredient in the next proof is that if $\sigma: \Delta^2 \to X$ is a 2-simplex, then $\sigma|\dot{\Delta}^2$ is nullhomotopic (Theorem 1.6), and hence $\sigma|\dot{\Delta}^2 \simeq (\sigma\varepsilon_0) * (\sigma\varepsilon_1)^{-1} * (\sigma\varepsilon_2)$ is nullhomotopic (Exercise 3.4).

Theorem 4.29 (Hurewicz[7] Theorem). *If X is path connected, then the Hurewicz map $\varphi: \pi_1(X, x_0) \to H_1(X)$ is a surjection with kernel $\pi_1(X, x_0)'$, the commutator subgroup of $\pi_1(X, x_0)$. Hence*

$$\pi_1(X, x_0)/\pi_1(X, x_0)' \cong H_1(X).$$

Proof. To see that φ is a surjection, consider a 1-cycle $\zeta = \sum m_i \sigma_i$ in X; hence

$$0 = \partial_1(\zeta) = \sum m_i(\sigma_i(e_1) - \sigma_i(e_0)),$$

an equation among the basis elements X of the free abelian group $S_0(X)$. Now X path connected implies that, for each i, there are paths in X, say, γ_i from x_0 to $\sigma_i(e_1)$ and δ_i from x_0 to $\sigma_i(e_0)$.

[7] Although this result is due to Poincaré, there is a more general theorem of Hurewicz relating homotopy groups and homology groups.

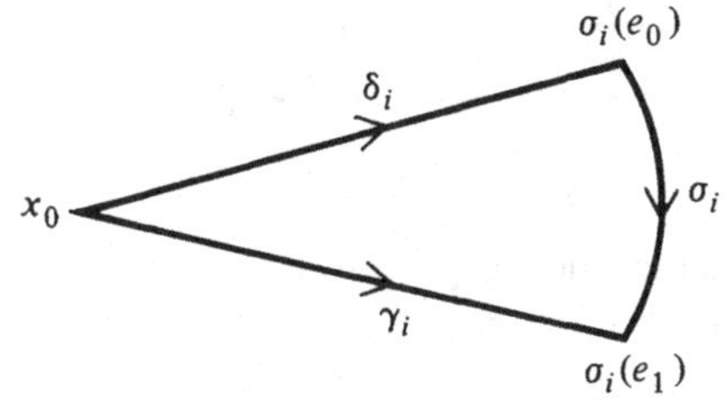

Choose $\gamma_i = \gamma_j$ if $\sigma_i(e_1) = \sigma_j(e_1)$; choose $\delta_i = \delta_j$ if $\sigma_i(e_0) = \sigma_j(e_0)$; choose $\gamma_i = \delta_j$ if $\sigma_i(e_1) = \sigma_j(e_0)$. The substitution principle (for the list $\sigma_1(e_1), \sigma_1(e_0), \sigma_2(e_1), \sigma_2(e_0), \ldots$ in the free abelian group $S_0(X)$ and for the list $\gamma_1\eta, \delta_1\eta, \gamma_2\eta, \delta_2\eta, \ldots$ in $S_1(X)$) gives the equation $0 = \sum m_i(\delta_i\eta - \gamma_i\eta)$ in $S_1(X)$. Hence

$$\sum m_i(\delta_i\eta + \sigma_i - \gamma_i\eta) = \sum m_i\sigma_i = \zeta. \tag{1}$$

But $\delta_i * \sigma_i\eta^{-1} * \gamma_i^{-1}$ is a closed path in X at x_0, so that Exercise 4.14 and Exercise 4.15 give

$$\begin{aligned}\varphi(\textstyle\prod [\delta_i * \sigma_i\eta^{-1} * \gamma_i^{-1}]^{m_i}) &= \textstyle\sum m_i\varphi[\delta_i * \sigma_i\eta^{-1} * \gamma_i^{-1}]\\ &= \textstyle\sum m_i \operatorname{cls}(\delta_i\eta + \sigma_i - \gamma_i\eta) = \operatorname{cls}\zeta.\end{aligned}$$

We now compute $\ker\varphi$. For the remainder of this proof, abbreviate $\pi_1(X, x_0)$ to π. Since $H_1(X)$ is abelian, $\pi' \subset \ker\varphi$. For the reverse inclusion, assume that γ is a closed path in X at x_0 with $[\gamma] \in \ker\varphi$; there are thus 2-simplexes $\tau_i\colon \Delta^2 \to X$ with $\gamma\eta = \partial_2(\sum n_i\tau_i)$ for $n_i \in \mathbf{Z}$. If $\tau_i\varepsilon_j$ is denoted by τ_{ij}, then $\partial_2(\tau_i) = \tau_{i0} - \tau_{i1} + \tau_{i2}$, and

$$\gamma\eta = \sum n_i(\tau_{i0} - \tau_{i1} + \tau_{i2}), \tag{2}$$

an equation among the basis elements of the free abelian group $S_1(X)$. It follows that $\gamma\eta = \tau_{pq}$ for some $p = i$ and $q \in \{0, 1, 2\}$ (because $\gamma\eta$ also is a basis element). As in the first part of the proof, we use path connectedness to construct auxiliary paths to make loops at x_0. For each i, choose paths λ_i, μ_i, ν_i from x_0 to $\tau_{i0}(e_0)$, $\tau_{i1}(e_1)$, $\tau_{i2}(e_0)$, respectively.

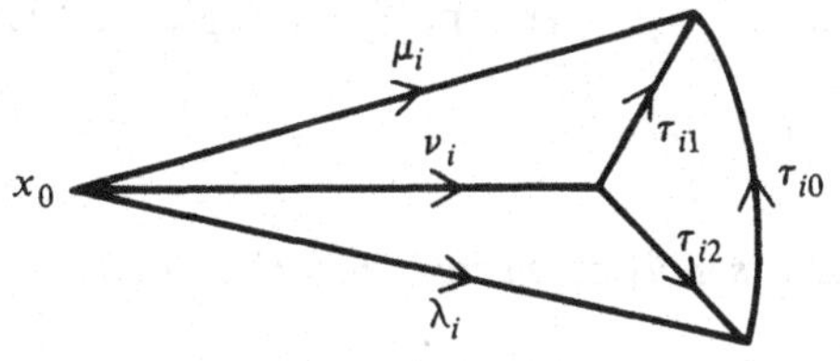

Should any of the ends $\tau_{i0}(e_0)$, $\tau_{i1}(e_1)$, $\tau_{i2}(e_0)$ be x_0, choose the corresponding λ, μ, ν to be the constant path at x_0; also, should $\tau_{i0}(e_0) = \tau_{j0}(e_0)$, choose $\lambda_i = \lambda_j$ (and similarly for μ, ν). Assemble paths to obtain elements of $\pi = \pi_1(X, x_0)$. Define

$$L_{i0} = [\lambda_i * \tau_{i0}\eta^{-1} * \mu_i^{-1}];$$
$$L_{i1} = [\nu_i * \tau_{i1}\eta^{-1} * \mu_i^{-1}];$$
$$L_{i2} = [\nu_i * \tau_{i2}\eta^{-1} * \lambda_i^{-1}].$$

The substitution principle when applied to Eq. (2) in $S_1(X)$ and the multiplicative abelian group π/π' gives an equation

$$\bar{L}_{pq} = \prod (\bar{L}_{i0}\bar{L}_{i1}^{-1}\bar{L}_{i2})^{n_i},$$

where bar denotes coset mod π'. Now $L_{pq} = [\alpha * \tau_{pq}\eta^{-1} * \beta]$, where α and β are appropriate λ, μ, ν. Since $\tau_{pq} = \gamma\eta$ is a closed path at x_0, the choice of auxiliary paths shows that α and β are constant paths at x_0; therefore $L_{pq} = [\tau_{pq}\eta^{-1}] = [\gamma]$. Finally, we have in π that

$$\begin{aligned} L_{i0}L_{i1}^{-1}L_{i2} &= [\lambda_i * \tau_{i0}\eta^{-1} * \mu_i^{-1} * \mu_i * (\tau_{i1}\eta^{-1})^{-1} * \nu_i^{-1} * \nu_i * \tau_{i2}\eta^{-1} * \lambda_i^{-1}] \\ &= [\lambda_i * \tau_{i0}\eta^{-1} * (\tau_{i1}\eta^{-1})^{-1} * \tau_{i2}\eta^{-1} * \lambda_i^{-1}] = 1, \end{aligned}$$

by Exercise 3.4(i). It follows that $\bar{L}_{pq} = \prod (\bar{L}_{i0}\bar{L}_{i1}^{-1}\bar{L}_{i2})^{n_i} = 1$ in π/π', hence $[\bar{\gamma}] = \bar{L}_{pq} = 1$ in π/π'; that is, $[\gamma] \in \pi'$. □

As we mentioned earlier, two homotopic closed curves in a space X are necessarily homologous (this is the statement that the Hurewicz map is well defined). One can show that the converse is not true by giving a space X whose fundamental group is not abelian (so that φ is not injective). An example of such a space X is the figure 8.

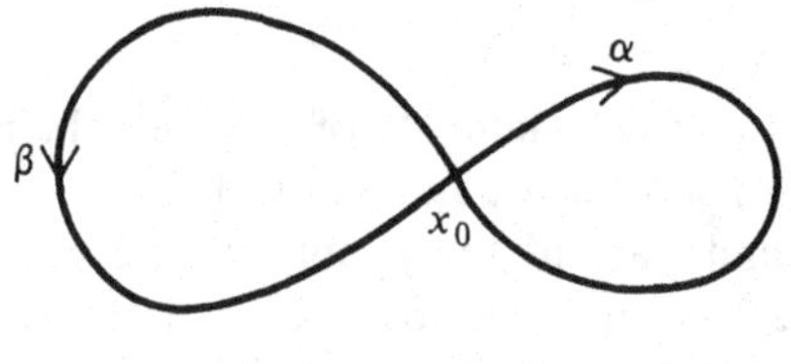

The closed paths $\alpha * \beta$ and $\beta * \alpha$ at x_0 are homologous, but they are not homotopic (i.e., $\alpha * \beta * \alpha^{-1} * \beta^{-1}$ is not nullhomotopic in X; see Corollary 7.42.

Corollary 4.30. $H_1(S^1) \cong \mathbf{Z}$.

Corollary 4.31. *If X is simply connected, then $H_1(X) = 0$.*

Exercise

4.16. If $f: S^1 \to S^1$ is continuous, define **degree** $f = m$ if the induced map $f_*: H_1(S^1) \to H_1(S^1)$ is multiplication by m. Show that this definition of degree coincides with the degree of a pointed map $(S^1, 1) \to (S^1, 1)$ defined in terms of $\pi_1(S^1, 1)$.

The last result in this chapter is a geometric characterization of $Z_1(X)$.

Definition. A **polygon** in a space X is a 1-chain $\pi = \sum_{i=0}^{k} \sigma_i$, where $\sigma_i(e_1) = \sigma_{i+1}(e_0)$ for all i (indices are read mod$(k+1)$).

Theorem 4.32. *Let X be a space. A 1-chain $\gamma = \sum m_i\sigma_i \in S_1(X)$ is a cycle if and only if γ is homologous to a linear combination of polygons.*

Proof. Sufficiency is clear, for every polygon π is a cycle:

$$\partial\pi = \partial(\sum \sigma_i) = \sum (\sigma_i(e_1) - \sigma_i(e_0)) = 0.$$

Conversely, let $\gamma = \sum m_i\sigma_i$ be a cycle. If some $m_i < 0$, then Exercise 4.14 says that $m_i\sigma_i$ is homologous to $(-m_i)\sigma_i^{-1}$. We may thus assume that each $m_i \geq 0$. The proof proceeds by induction on $\sum m_i \geq 0$; the induction does begin when $\sum m_i = 0$, for now $\gamma = 0$. For the inductive step, we may assume each $m_i > 0$. Define $E_i = \{\sigma_i(e_0), \sigma_i(e_1)\}$ and define $E = \bigcup E_i$. Since every closed path σ is itself a polygon, we may assume that no σ_i is closed (otherwise, apply induction to $\gamma - \sigma_i$). Denote $\sigma_1(e_0)$ by x_1 and $\sigma_1(e_1)$ by x_2, so that $\partial\sigma_1 = x_2 - x_1$. Since $\partial\gamma = 0$ and all $m_i > 0$, there must be some σ_i occurring in γ with $\sigma_i(e_0) = x_2$ [and so x_2 occurs with a negative sign in $\partial\sigma_i = \sigma_i(e_1) - \sigma_i(e_0)$]. Define $x_3 = \sigma_i(e_1)$. Iterate this procedure to obtain a sequence $x_1, x_2, x_3, \ldots$ of points in E. Because E is a finite set, there exists a "loop" $x_p, x_{p+1}, \ldots, x_n, x_{n+1} = x_p$; that is, there is a polygon $\pi = \sum_{j=p}^{n} \sigma_{i_j}$. Thus $\gamma - \pi$ is a 1-cycle to which the inductive hypothesis applies. Therefore $\gamma - \pi$ and hence γ is (homologous to) a linear combination of polygons. □

Just as one may regard $\pi_1(X, x_0)$ as (pointed) maps of S^1 into X, one can define higher homotopy groups $\pi_n(X, x_0)$ as pointed maps of S^n into X. There is a Hurewicz map $\pi_n(X) \to H_n(X)$, and the question whether there is an analog of Theorem 4.32 is related to the image of this map.

There are two more fundamental properties (axioms) of homology functors: the *long exact sequence* and *excision*. Once we know these, we shall be able to compute some homology groups and give interesting applications of this computation. These properties, along with properties we already know, serve to characterize the homology functors as well.

CHAPTER 5

Long Exact Sequences

The homology groups of a space X are defined in two stages: (1) construction of the singular complex $(S_*(X), \partial)$ and (2) formation of the groups $H_n(X) = \ker \partial_n/\text{im } \partial_{n+1}$. The first stage involves the topology of X in an essential way, for one needs to know the n-simplexes in X; the second stage is purely algebraic. Let us now acquaint ourselves with the algebraic half of the definition in order to establish the existence of certain long exact sequences; these are very useful for calculation because they display connections between the homology of a space and the homology of its subspaces.

The Category **Comp**

Definition. A **(chain) complex** is a sequence of abelian groups and homomorphisms

$$\cdots \longrightarrow S_{n+1} \xrightarrow{\partial_{n+1}} S_n \xrightarrow{\partial_n} S_{n-1} \longrightarrow \cdots, \qquad n \in \mathbf{Z},$$

such that $\partial_n \partial_{n+1} = 0$ for each $n \in \mathbf{Z}$. The homomorphism ∂_n is called the **differentiation** of **degree** n, and S_n is called the **term** of **degree** n.

The complex above is denoted by (S_*, ∂) or, more simply, by S_*. Observe that the condition $\partial_n \partial_{n+1} = 0$ is equivalent to

$$\text{im } \partial_{n+1} \subset \ker \partial_n.$$

Of course, the singular complex $(S_*(X), \partial)$ is an example of a complex (in which all terms with negative subscripts are zero). We shall see examples of complexes with negative subscripts in Chapter 12.

Definition. A sequence of two homomorphisms (of groups) $A \xrightarrow{f} B \xrightarrow{g} C$ is **exact** at B if $\text{im } f = \ker g$. A sequence of abelian groups and homomorphisms

$$\cdots \longrightarrow S_{n+1} \xrightarrow{\partial_{n+1}} S_n \xrightarrow{\partial_n} S_{n-1} \longrightarrow \cdots$$

is **exact** if it is exact at each S_n, that is, $\text{im } \partial_{n+1} = \ker \partial_n$ for all $n \in \mathbf{Z}$.

It is clear that every exact sequence is a complex: equality (im = ker) implies inclusion (im $\subset$ ker).

Exercises

*5.1. (i) If $0 \to A \xrightarrow{f} B$ is exact, then f is injective (there is no need to label the only possible homomorphism $0 \to A$).
(ii) If $B \xrightarrow{g} C \to 0$ is exact, then g is surjective (there is no need to label the only possible homomorphism $C \to 0$).
(iii) If $0 \to A \xrightarrow{f} B \to 0$ is exact, then f is an isomorphism.
(iv) If $0 \to A \to 0$ is exact, then $A = 0$.

*5.2. If $A \xrightarrow{f} B \xrightarrow{g} C \xrightarrow{h} D$ is exact, then f is surjective if and only if h is injective.

*5.3. A **short exact sequence** is an exact sequence of the form

$$0 \to A \xrightarrow{i} B \xrightarrow{p} C \to 0.$$

In this case, show that $iA \cong A$ and $B/iA \cong C$ via $b + iA \mapsto pb$.

*5.4. If $\cdots \longrightarrow C_{n+1} \longrightarrow A_n \xrightarrow{h_n} B_n \longrightarrow C_n \longrightarrow A_{n-1} \xrightarrow{h_{n-1}} B_{n-1} \longrightarrow C_{n-1} \longrightarrow \cdots$ is exact and every third arrow $h_n\colon A_n \to B_n$ is an isomorphism, then $C_n = 0$ for all n.

*5.5. (i) If $0 \to A \to B \to C \to 0$ is a short exact sequence of abelian groups, then $\text{rank } B = \text{rank } A + \text{rank } C$. (*Hint*: Extend a maximal independent subset of A to a maximal independent subset of B.)
(ii) If $0 \to A_n \to A_{n-1} \to \cdots \to A_1 \to A_0 \to 0$ is an exact sequence of (finitely generated) abelian groups, then $\sum_{i=0}^{n} (-1)^i \text{ rank } A_i = 0$.

Definition. If (S_*, ∂) is a complex, then $\ker \partial_n$ is called the group of ***n*-cycles** and is denoted by $Z_n(S_*, \partial)$; $\text{im } \partial_{n+1}$ is called the group of ***n*-boundaries** and is denoted by $B_n(S_*, \partial)$. The ***n*th homology group** of this complex is

$$H_n(S_*, \partial) = Z_n(S_*, \partial)/B_n(S_*, \partial).$$

Of course, we shall abbreviate this notation if no confusion ensues. If $z_n \in Z_n$, then $z_n + B_n \in H_n$ is called the **homology class** of z_n and it is denoted by $\text{cls } z_n$.

Theorem 5.1. *A complex (S_*, ∂) is an exact sequence if and only if $H_n(S_*, \partial) = 0$ for every n.*

PROOF. $Z_n = B_n$ if and only if $\ker \partial_n = \operatorname{im} \partial_{n+1}$. □

Thus the homology groups "measure" the deviation of a complex from being an exact sequence. Because of this theorem, an exact sequence is also called an **acyclic complex**.

Definition. If (S'_*, ∂') and (S_*, ∂) are complexes, a **chain map** $f: (S'_*, \partial') \to (S_*, \partial)$ is a sequence of homomorphisms $\{f_n: S'_n \to S_n\}$ such that the following diagram commutes:

$$\begin{array}{ccccccccc} \cdots & \longrightarrow & S'_{n+1} & \xrightarrow{\partial'_{n+1}} & S'_n & \xrightarrow{\partial'_n} & S'_{n-1} & \longrightarrow & \cdots \\ & & \downarrow{\scriptstyle f_{n+1}} & & \downarrow{\scriptstyle f_n} & & \downarrow{\scriptstyle f_{n-1}} & & \\ \cdots & \longrightarrow & S_{n+1} & \xrightarrow[\partial_{n+1}]{} & S_n & \xrightarrow[\partial_n]{} & S_{n-1} & \longrightarrow & \cdots, \end{array}$$

that is, $\partial_n f_n = f_{n-1}\partial'_n$ for all $n \in \mathbf{Z}$. If $f = \{f_n\}$, then one calls f_n the **term of degree** n.

If $f: X \to Y$ is continuous, then we saw in Lemma 4.8 that f induces a chain map $f_\#: S_*(X) \to S_*(Y)$.

Definition. All complexes and chain maps form a category, denoted by **Comp**, when one defines composition of chain maps coordinatewise: $\{g_n\} \circ \{f_n\} = \{g_n \circ f_n\}$.

The category **Comp** has the feature that, for every pair of complexes S'_* and S_*, $\operatorname{Hom}(S'_*, S_*)$ is an abelian group: if $f = \{f_n\}$ and $g = \{g_n\} \in \operatorname{Hom}(S'_*, S_*)$, then $f + g$ is the chain map whose term of degree n is $f_n + g_n$.

The reader may now show that there is a functor S_*: **Top** → **Comp** with $X \mapsto (S_*(X), \partial)$ and $f \mapsto f_\#$. Also for each $n \in \mathbf{Z}$, there is a functor H_n: **Comp** → **Ab** with $S_* \mapsto H_n(S_*) = Z_n(S_*)/B_n(S_*)$ and with $H_n(f)$: $\operatorname{cls} z_n \mapsto \operatorname{cls} f_n(z_n)$ for every chain map $f: S'_* \to S_*$ (one proves that $H_n(f)$ is well defined, as in Lemma 4.9, and one proves that H_n is a functor, as in Theorem 4.10). One usually writes f_* instead of $H_n(f)$, again omitting the subscript n unless it is needed for clarity. Obviously, each homology functor H_n: **Top** → **Ab** (for $n \geq 0$) is the composite of these functors **Top** → **Comp** → **Ab**; we have made precise the observation that our original construction of $H_n(X)$ involves a topological step followed by an algebraic one.

Theorem 5.2. *For each $n \in \mathbf{Z}$, the functor H_n:* **Comp** → **Ab** *is* **additive**; *that is, if $f, g \in \operatorname{Hom}(S'_*, S_*)$, then $H_n(f + g) = H_n(f) + H_n(g)$.*

PROOF. A routine exercise. □

It follows easily that $H_n(0) = 0$, where (the first) 0 denotes either the **zero complex** (all terms S_n zero) or the **zero chain map** (all terms f_n zero).

The category **Comp** strongly resembles the category **Ab** in the sense that one has analogues in **Comp** of the familiar notions of subgroup, quotient group, first isomorphism theorem, and so on. It is important that the reader feel as comfortable with a complex as with an abelian group. Here are the constructions.

Subcomplex. Define (S'_*, ∂') to be a **subcomplex** of (S_*, ∂) if each S'_n is a subgroup of S_n and if each $\partial'_n = \partial_n | S'_n$. Here are two other descriptions: (1) the following diagram commutes for all n:

$$\begin{array}{ccc} S'_n & \xrightarrow{\partial'_n} & S'_{n-1} \\ \downarrow{\scriptstyle i_n} & & \downarrow{\scriptstyle i_{n-1}} \\ S_n & \xrightarrow[\partial_n]{} & S_{n-1}, \end{array}$$

where $i_n: S'_n \hookrightarrow S_n$ is the inclusion map; (2) if $i = \{i_n\}$, then $i: S'_* \to S_*$ is a chain map. (That all three descriptions are equivalent is left as an exercise.)

Quotient. If (S'_*, ∂') is a subcomplex of (S_*, ∂), then the **quotient complex** is the complex

$$\cdots \longrightarrow S_n/S'_n \xrightarrow{\bar{\partial}_n} S_{n-1}/S'_{n-1} \longrightarrow \cdots,$$

where $\bar{\partial}_n: s_n + S'_n \mapsto \partial_n(s_n) + S'_{n-1}$ ($\bar{\partial}_n$ is well defined because $\partial_n(S'_n) \subset S'_{n-1}$).

Kernel and **Image.** If $f: (S_*, \partial) \to (S''_*, \partial'')$ is a chain map, then **ker** f is the subcomplex of S_*

$$\cdots \longrightarrow \ker f_n \xrightarrow{\partial'_n} \ker f_{n-1} \longrightarrow \cdots,$$

where ∂'_n is (necessarily) the restriction $\partial_n | \ker f_n$; **im** f is the subcomplex of S''_*

$$\cdots \longrightarrow \operatorname{im} f_n \xrightarrow{\Delta''_n} \operatorname{im} f_{n-1} \longrightarrow \cdots,$$

where Δ''_n is (necessarily) the restriction $\partial''_n | \operatorname{im} f_n$.

Exactness. A sequence of complexes and chain maps

$$\cdots \longrightarrow A_*^{q+1} \xrightarrow{f^{q+1}} A_*^q \xrightarrow{f^q} A_*^{q-1} \longrightarrow \cdots$$

is **exact** if $\operatorname{im} f^{q+1} = \ker f^q$ for every q. **A short exact sequence** of complexes is an exact sequence of the form

$$0 \to S'_* \xrightarrow{i} S_* \xrightarrow{p} S''_* \to 0,$$

where 0 denotes the zero complex.

Here is the picture of a short exact sequence of complexes in unabbreviated form.

$$\begin{array}{ccccccccc}
 & & \vdots & & \vdots & & \vdots & & \\
 & & \downarrow & & \downarrow & & \downarrow & & \\
0 & \longrightarrow & S'_{n+1} & \xrightarrow{i_{n+1}} & S_{n+1} & \xrightarrow{p_{n+1}} & S''_{n+1} & \longrightarrow & 0 \\
 & & \downarrow{\scriptstyle \partial'_{n+1}} & & \downarrow{\scriptstyle \partial_{n+1}} & & \downarrow{\scriptstyle \partial''_{n+1}} & & \\
0 & \longrightarrow & S'_n & \xrightarrow{i_n} & S_n & \xrightarrow{p_n} & S''_n & \longrightarrow & 0 \\
 & & \downarrow{\scriptstyle \partial'_n} & & \downarrow{\scriptstyle \partial_n} & & \downarrow{\scriptstyle \partial''_n} & & \\
0 & \longrightarrow & S'_{n-1} & \xrightarrow{i_{n-1}} & S_{n-1} & \xrightarrow{p_{n-1}} & S''_{n-1} & \longrightarrow & 0 \\
 & & \downarrow & & \downarrow & & \downarrow & & \\
 & & \vdots & & \vdots & & \vdots & &
\end{array}$$

This is a commutative diagram whose columns are complexes. By Exercise 5.8 below, its rows are short exact sequence of groups.

Intersection and Sum. Let S'_* and S''_* be subcomplexes of S_*. Then $S'_* \cap S''_*$ is the subcomplex of S_* whose nth term is $S'_n \cap S''_n$, and $S'_* + S''_*$ is the subcomplex of S_* whose nth term is $S'_n + S''_n$.

Direct Sum. Let $\{(S^\lambda_*, \partial^\lambda): \lambda \in \Lambda\}$ be a family of complexes, indexed by a set Λ. Their **direct sum** is the complex

$$\cdots \longrightarrow \textstyle\sum_\lambda S^\lambda_{n+1} \xrightarrow{\partial_{n+1}} \sum_\lambda S^\lambda_n \xrightarrow{\partial_n} \sum_\lambda S^\lambda_{n-1} \longrightarrow \cdots,$$

where $\partial_n = \sum_\lambda \partial^\lambda_n : \sum_\lambda s^\lambda_n \mapsto \sum_\lambda \partial^\lambda_n(s^\lambda_n)$ for $s^\lambda_n \in S^\lambda_n$. Note the special case $\Lambda = \{1, 2\}$.

An important example of a subcomplex arises from a subspace A of a space X. If $j: A \hookrightarrow X$ is the inclusion, we saw in Lemma 4.15 that $j_\#: S_n(A) \to S_n(X)$ is injective for every n. There is thus a short exact sequence of complexes

$$0 \to S_*(A) \to S_*(X) \to S_*(X)/S_*(A) \to 0$$

that will be very useful. It is convenient to regard $S_*(A)$ as being a subcomplex of $S_*(X)$ (instead of being isomorphic to im $j_\#$). This is accomplished by regarding every n-simplex $\sigma: \Delta^n \to A$ as an n-simplex in X whose image happens to be contained in A, that is, by identifying σ with $j\sigma$.

One cannot form the intersection of two arbitrary sets; one can only form the intersection of two subsets of a set. Let A_1 and A_2 be subspaces of a space X. As above, regard $S_*(A_1)$ and $S_*(A_2)$ as subcomplexes of $S_*(X)$. We claim that $S_*(A_1) \cap S_*(A_2) = S_*(A_1 \cap A_2)$. If $\sum m_i\sigma_i \in S_n(A_1) \cap S_n(A_2)$, then each σ_i is an n-simplex in X with im $\sigma_i \subset A_1$ and with im $\sigma_i \subset A_2$; hence each σ_i is an n-simplex in X with im $\sigma_i \subset A_1 \cap A_2$, that is, $\sum m_i\sigma_i \in S_n(A_1 \cap A_2)$. For the reverse inclusion, each n-simplex σ in X with im $\sigma_i \subset A_1 \cap A_2$ may be regarded as an n-simplex in either A_1 or A_2, and so $\sigma \in S_n(A_1) \cap S_n(A_2)$.

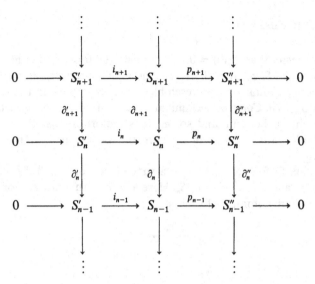

Our last example here involves the decomposition of a space X into the disjoint union of its path components: $X = \bigcup X_\lambda$. As above, each subspace X_λ gives a subcomplex $S_*(X_\lambda)$ of $S_*(X)$. Each n-simplex $\sigma: \Delta^n \to X$ actually takes values in some X_λ; therefore a linear combination of n-simplexes in X can be written, after collecting like terms, as a linear combination of n-simplexes in various X_λ. It follows (with routine details left to the reader) that $S_*(X) = \sum_\lambda S_*(X_\lambda)$.

EXERCISES

*5.6. If (S_*, ∂) is a complex with $\partial_n = 0$ for every $n \in \mathbf{Z}$, then $H_n(S_*) = S_n$ for every $n \in \mathbf{Z}$.

5.7. Prove that a chain map f is an equivalence in **Comp** if and only if each f_n is an isomorphism (one calls f an **isomorphism**).

*5.8. A sequence $S'_* \xrightarrow{f} S_* \xrightarrow{g} S''_*$ is exact in **Comp** if and only if $S'_n \xrightarrow{f_n} S_n \xrightarrow{g_n} S''_n$ is exact in **Ab** for every $n \in \mathbf{Z}$.

5.9. (i) Recall that the natural map $\nu: G \to G/K$ (in **Ab**) is defined by $\nu(g) = g + K$. If S'_* is a subcomplex of S_*, show that $\nu: S_* \to S_*/S'_*$, defined by $\nu = \{\nu_n: S_n \to S_n/S'_n: \nu_n$ is the natural map$\}$, is a chain map whose kernel is S'_* (ν is also called the **natural map**).

(ii) Prove that the **first isomorphism theorem** holds in **Comp**. If $f: S_* \to S''_*$ is a chain map, then there is an isomorphism

$$\theta: S_*/\ker f \xrightarrow{\sim} \operatorname{im} f$$

making the following diagram commute (ν is the natural map):

$$\begin{array}{ccccc} S_* & \xrightarrow{f} & \operatorname{im} f & \hookrightarrow & S''_* \\ & \searrow^{\nu} & \nearrow_{\theta} & & \\ & S_*/\ker f. & & & \end{array}$$

5.10. If S'_* and S''_* are subcomplexes of S_*, prove that the **second isomorphism theorem** holds in **Comp**:

$$S'_*/(S'_* \cap S''_*) \cong (S'_* + S''_*)/S''_*.$$

(*Hint*: Adapt the usual proof from group theory deriving the second isomorphism theorem from the first.)

*5.11. Prove that the **third isomorphism theorem** holds in **Comp**. If $U_* \subset T_* \subset S_*$ are subcomplexes, then there is a short exact sequence of complexes

$$0 \to T_*/U_* \xrightarrow{i} S_*/U_* \xrightarrow{p} S_*/T_* \to 0,$$

where $i_n: t_n + U_n \mapsto t_n + U_n$ (inclusion) and $p_n(s_n + U_n) = s_n + T_n$.

*5.12. For every n, $H_n(\sum_\lambda S^\lambda_*) \cong \sum_\lambda H_n(S^\lambda_*)$. (See the proof of Theorem 4.13.)

The next definition comes from Lemma 4.21.

Definition. If $f, g\colon (S'_*, \partial') \to (S_*, \partial)$ are chain maps, then f and g are **(chain) homotopic**, denoted by $f \simeq g$, if there is a sequence of homomorphisms $\{P_n\colon S'_n \to S_{n+1}\}$ such that, for all $n \in \mathbf{Z}$,

$$\partial_{n+1} P_n + P_{n-1} \partial'_n = f_n - g_n.$$

The sequence $P = \{P_n\}$ is called a **chain homotopy**.

A chain map $f\colon (S'_*, \partial') \to (S_*, \partial)$ is called a **chain equivalence** if there exists a chain map $g\colon (S_*, \partial) \to (S'_*, \partial')$ such that $g \circ f \simeq 1_{S'_*}$ and $f \circ g \simeq 1_{S_*}$. Two chain complexes are called **chain equivalent** if there exists a chain equivalence between them.

The relation of homotopy is an equivalence relation on the set of all chain maps $S'_* \to S_*$.

Theorem 5.3.

(i) *If $f, g\colon S'_* \to S_*$ are chain maps with $f \simeq g$, then, for all n,*

$$H_n(f) = H_n(g)\colon H_n(S'_*) \to H_n(S_*).$$

(ii)[1] *If $f\colon S'_* \to S_*$ is a chain equivalence, then, for all n,*

$$H_n(f)\colon H_n(S'_*) \to H_n(S_*)$$

is an isomorphism.

PROOF. (i) See the proof of Lemma 4.21.

(ii) An immediate consequence of part (i) and the definitions. □

The next definition recalls the cone construction of Theorem 4.19.

Definition. A **contracting homotopy** of a complex (S_*, ∂) is a sequence of homomorphisms $c = \{c_n\colon S_n \to S_{n+1}\}$ such that for all $n \in \mathbf{Z}$,

$$\partial_{n+1} c_n + c_{n-1} \partial_n = 1_{S_n}.$$

Plainly, a contracting homotopy is a chain homotopy between the identity map of S_* (namely, $\{1_{S_n}\}$) and the zero map on S_*.

Corollary 5.4.[2] *If a complex S_* has a contracting homotopy, then S_* is acyclic (i.e., $H_n(S_*) = 0$ for all n, i.e., S_* is an exact sequence).*

---

[1] The converse is almost true. In Theorem 9.8, we shall prove that if S'_* and S_* are chain complexes each of whose terms is *free* abelian and if $f\colon S'_* \to S_*$ is a chain map with every $H_n(f)$ an isomorphism, then f is a chain equivalence.

[2] The converse is true if each term S_n of S_* is free abelian (Theorem 9.4).

PROOF. If 1 denotes the identity on S_*, then Theorem 5.3 gives $H_n(1) = H_n(0) = 0$ for all n. Since H_n is a functor, $H_n(1)$ is the identity on $H_n(S_*)$; it follows that $H_n(S_*) = 0$. □

Indeed it is easy to see that a complex with a contracting homotopy is chain equivalent to the zero complex.

Exact Homology Sequences

A fundamental property of the homology functors H_n is that they are connected to one another. To see this, let us first see how H_n affects exactness.

Lemma 5.5. *If* $0 \to (S'_*, \partial') \xrightarrow{i} (S_*, \partial) \xrightarrow{p} (S''_*, \partial'') \to 0$ *is a short exact sequence of complexes, then for each n there is a homomorphism*

$$d_n\colon H_n(S''_*) \to H_{n-1}(S'_*)$$

given by

$$\text{cls } z''_n \mapsto \text{cls } i_{n-1}^{-1} \partial_n p_n^{-1} z''_n.$$

PROOF.[3] Because i and p are chain maps, the following diagram commutes; moreover, the rows are exact, by Exercise 5.8.

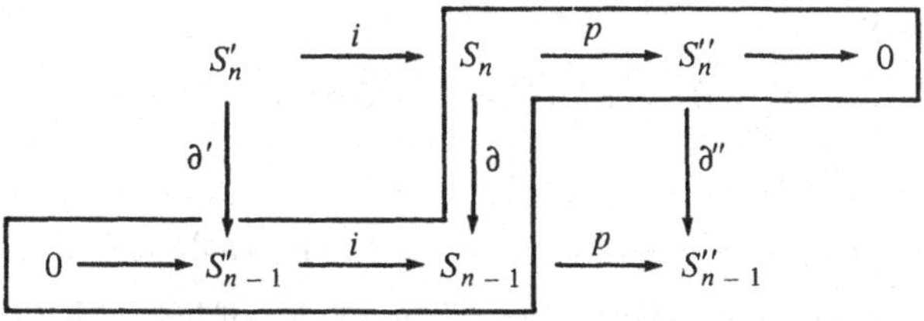

Suppose that $z'' \in Z''_n$ so $\partial'' z'' = 0$. Since p is surjective, we may lift z'' to $s_n \in S_n$ and then push down to $\partial s_n \in S_{n-1}$. By commutativity,

$$\partial s_n \in \ker(S_{n-1} \to S''_{n-1}) = \text{im } i.$$

It follows that $i^{-1}\partial s_n$ makes sense; that is, there is a unique (i is injective) $s'_{n-1} \in S'_{n-1}$ with $is'_{n-1} = \partial s_n$.

Suppose that we had lifted z'' to $\sigma_n \in S_n$. Then the construction above yields $\sigma'_{n-1} \in S'_{n-1}$ with $i\sigma'_{n-1} = \partial\sigma_n$. We also know that

[3] This method of proof is called **diagram chasing**. It is really a simple technique, for each step is essentially dictated, and so one proceeds without having to make any decisions.

$$s_n - \sigma_n \in \ker p = \operatorname{im}(S_n' \to S_n),$$

so there is $x_n' \in S_n'$ with $s_{n-1}' - \sigma_{n-1}' = \partial' x_n' \in B_{n-1}'$. There is thus a well defined homomorphism

$$Z_n'' \to S_{n-1}'/B_{n-1}'.$$

It is easy to see that this map sends B_n'' into 0 and that $s_{n-1}' = i^{-1}\partial p^{-1} z''$ is a cycle. Therefore the formula does give a map $H_n(S_*'') \to H_{n-1}(S_*')$, as desired. □

Definition. The maps d_n of Lemma 5.5 are called **connecting homomorphisms**.

Theorem 5.6 (Exact Triangle). *If $0 \to (S_*', \partial') \xrightarrow{i} (S_*, \partial) \xrightarrow{p} (S_*'', \partial'') \to 0$ is a short exact sequence of complexes, then there is an exact sequence*

$$\cdots \to H_n(S_*') \xrightarrow{i_*} H_n(S_*) \xrightarrow{p_*} H_n(S_*'') \xrightarrow{d} H_{n-1}(S_*') \xrightarrow{i_*} H_{n-1}(S_*) \xrightarrow{p_*} H_{n-1}(S_*'') \to \cdots.$$

PROOF. The argument is routine, but we give the details anyway. The notation below is self-explanatory and subscripts are omitted.

(1) $\operatorname{im} i_* \subset \ker p_*$.

This follows from $p_* i_* = (pi)_* = 0_* = 0$.

(2) $\ker p_* \subset \operatorname{im} i_*$.

If $p_*(z + B) = pz + B'' = B''$, then $pz = \partial'' s''$. But p surjective gives $s'' = ps$, so that $pz = \partial'' ps = p\partial s$ and $p(z - \partial s) = 0$. By exactness, there exists s' with $is' = z - \partial s$. Note that $s' \in Z'$, for $i\partial' s' = \partial i s' = \partial z - \partial\partial s = 0$ (z is a cycle). Since i is injective, $\partial' s' = 0$. Therefore

$$i_*(s' + B') = is' + B = z - \partial s + B = z + B.$$

(3) $\operatorname{im} p_* \subset \ker d$.

$$dp_*(z + B) = d(pz + B'') = i^{-1}\partial p^{-1}(pz) + B'.$$

As the definition of d is independent of the choice of lifting, we may choose $z = p^{-1}(pz)$, hence $i^{-1}\partial p^{-1}(pz) = i^{-1}\partial z = 0$.

(4) $\ker d \subset \operatorname{im} p_*$.

If $d(z'' + B'') = B'$, then $x' = i^{-1}\partial p^{-1} z'' \in B'$ and $x' = \partial' s'$. Now $ix' = i\partial' s' = \partial i s' = \partial p^{-1} z''$, so that $\partial(p^{-1} z'' - is') = 0$, and $p^{-1}z'' - is' \in Z$. Therefore

$$p_*(p^{-1}z'' - is' + B) = pp^{-1}z'' - pis' + B'' = z'' + B''.$$

(5) $\operatorname{im} d \subset \ker i_*$.

$$i_* d(z'' + B'') = i_*(i^{-1}\partial p^{-1} z'' + B') = \partial p^{-1} z'' + B = B.$$

(6) $\ker i_* \subset \operatorname{im} d$.

If $i_*(z' + B') = B$, then $iz' = \partial s$, and $\partial'' ps = p\partial s = piz' = 0$ and $ps \in Z''$. But $d(ps + B'') = i^{-1}\partial p^{-1} ps + B' = i^{-1}\partial s + B' = i^{-1} iz' + B = z' + B'$. □

Theorem 5.6 is called the Exact Triangle because of the mnemonic diagram,

$$\begin{array}{ccc} H(S'_*) & \xrightarrow{i_*} & H(S_*) \\ {\scriptstyle d}\nwarrow & & \swarrow{\scriptstyle p_*} \\ & H(S''_*) & \end{array}.$$

Theorem 5.7 (Naturality of the Connecting Homomorphism). *Assume that there is a commutative diagram of complexes with exact rows:*

$$\begin{array}{ccccccccc} 0 & \longrightarrow & S'_* & \xrightarrow{i} & S_* & \xrightarrow{p} & S''_* & \longrightarrow & 0 \\ & & \downarrow{\scriptstyle f'} & & \downarrow{\scriptstyle f} & & \downarrow{\scriptstyle f''} & & \\ 0 & \longrightarrow & T'_* & \xrightarrow[j]{} & T_* & \xrightarrow[q]{} & T''_* & \longrightarrow & 0. \end{array}$$

Then there is a commutative diagram of abelian groups with exact rows:

$$\begin{array}{ccccccccccc} \cdots & \longrightarrow & H_n(S'_*) & \xrightarrow{i_*} & H_n(S_*) & \xrightarrow{p_*} & H_n(S''_*) & \xrightarrow{d} & H_{n-1}(S'_*) & \longrightarrow & \cdots \\ & & \downarrow{\scriptstyle f'_*} & & \downarrow{\scriptstyle f_*} & & \downarrow{\scriptstyle f''_*} & & \downarrow{\scriptstyle f'_*} & & \\ \cdots & \longrightarrow & H_n(T'_*) & \xrightarrow[j_*]{} & H_n(T_*) & \xrightarrow[q_*]{} & H_n(T''_*) & \xrightarrow[d']{} & H_{n-1}(T'_*) & \longrightarrow & \cdots. \end{array}$$

PROOF. Exactness of the rows is Theorem 5.6. The first two squares commute because H_n is a functor (e.g., $fi = jf'$ implies that $f_* i_* = j_* f'_*$).

To see commutativity of the last square, we first set up notation: let $S_* = (S_*, \partial)$ and let $T_* = (T_*, \Delta)$. If cls $z'' \in H_n(S''_*)$, then p surjective implies that cls z'' = cls ps for some s. But now

$$\begin{aligned} f'_* d \text{ cls } z'' &= f'_* d \text{ cls } ps = f'_* \text{ cls } i^{-1}\partial s \\ &= \text{cls } f' i^{-1} \partial s = \text{cls } j^{-1} f \partial s \quad (\text{since } jf' = fi) \\ &= \text{cls } j^{-1} \Delta f s \quad (f \text{ is a chain map}) \\ &= d' \text{ cls } q f s \quad (\text{since } d' \text{ cls } \zeta'' = \text{cls } j^{-1}\Delta q^{-1} \zeta'') \\ &= d' \text{ cls } f'' p s = d' f''_* \text{ cls } ps = d' f''_* \text{ cls } z''. \end{aligned}$$ □

As we remarked earlier, a subspace A of a topological space X gives rise to a short exact sequence of complexes:

$$0 \to S_*(A) \to S_*(X) \to S_*(X)/S_*(A) \to 0.$$

We have already dubbed $H_n(S_*(A))$ and $H_n(S_*(X))$ as $H_n(A)$ and $H_n(X)$, respectively; we now give a name to the homology of the quotient complex.

Definition. If A is a subspace of X, the nth **relative homology group** $H_n(X, A)$ is defined to be $H_n(S_*(X)/S_*(A))$.

Theorem 5.8 (Exact Sequence of the Pair (X, A)). *If A is a subspace of X, there is an exact sequence*

$$\cdots \to H_n(A) \to H_n(X) \to H_n(X, A) \xrightarrow{d} H_{n-1}(A) \to \cdots.$$

Moreover, if $f: (X, A) \to (Y, B)$ (i.e., $f: X \to Y$ is continuous with $f(A) \subset B$), then there is a commutative diagram

$$\begin{array}{ccccccccc} \cdots \longrightarrow & H_n(A) & \longrightarrow & H_n(X) & \longrightarrow & H_n(X, A) & \longrightarrow & H_{n-1}(A) & \longrightarrow \cdots \\ & \downarrow & & \downarrow & & \downarrow & & \downarrow & \\ \cdots \longrightarrow & H_n(B) & \longrightarrow & H_n(Y) & \longrightarrow & H_n(Y, B) & \longrightarrow & H_{n-1}(B) & \longrightarrow \cdots, \end{array}$$

where the vertical maps are induced by f.

Proof. Immediate from Theorems 5.6 and 5.7. □

One now sees that the homology of a subspace A of X influences the homology of X, because Exercises 5.1–5.4 may be invoked when applicable.

A tower of subspaces gives a long exact sequence of relative homology groups.

Theorem 5.9 (Exact Sequence of the Triple (X, A, A')). *If $A' \subset A \subset X$ are subspaces, there is an exact sequence*

$$\cdots \to H_n(A, A') \to H_n(X, A') \to H_n(X, A) \xrightarrow{d} H_{n-1}(A, A') \to \cdots.$$

Moreover, if there is a commutative diagram of pairs of spaces

$$\begin{array}{ccccc} (A, A') & \longrightarrow & (X, A') & \longrightarrow & (X, A) \\ \downarrow & & \downarrow & & \downarrow \\ (B, B') & \longrightarrow & (Y, B') & \longrightarrow & (Y, B), \end{array}$$

then there is a commutative diagram with exact rows

$$\begin{array}{ccccccccc} \cdots \longrightarrow & H_n(A, A') & \longrightarrow & H_n(X, A') & \longrightarrow & H_n(X, A) & \longrightarrow & H_{n-1}(A, A') & \longrightarrow \cdots \\ & \downarrow & & \downarrow & & \downarrow & & \downarrow & \\ \cdots \longrightarrow & H_n(B, B') & \longrightarrow & H_n(Y, B') & \longrightarrow & H_n(Y, B) & \longrightarrow & H_{n-1}(B, B') & \longrightarrow \cdots. \end{array}$$

Proof. Apply Theorems 5.6 and 5.7 to the short exact sequences of complexes given by the third isomorphism theorem (Exercise 5.11):

$$0 \to S_*(A)/S_*(A') \to S_*(X)/S_*(A') \to S_*(X)/S_*(A) \to 0$$

and

$$0 \to S_*(B)/S_*(B') \to S_*(Y)/S_*(B') \to S_*(Y)/S_*(B) \to 0. \qquad \square$$

Remarks. (1) If $A = \varnothing$, then we saw in Exercise 4.4 that $S_*(A) = 0$. It follows that $H_n(X, \varnothing) = H_n(X)$; that is, **absolute** homology groups are particular relative homology groups. Thus Theorem 5.8 is a special case of Theorem 5.9.

(2) We claim that, except for connecting homomorphisms, all homomorphisms in Theorems 5.8 and 5.9 are induced by inclusions.

Recall that $\mathbf{Top}^2$ is the category whose objects are pairs (X, A) (where A is a subspace of X), whose morphisms $f: (X, A) \to (Y, B)$ are continuous functions $f: X \to Y$ with $f(A) \subset B$, and whose composition is ordinary composition of functions. Define a functor $S_*: \mathbf{Top}^2 \to \mathbf{Comp}$ as follows. On an object (X, A), define $S_*(X, A) = S_*(X)/S_*(A)$. To define S_* on a morphism $f: (X, A) \to (Y, B)$, note that the induced chain map $f_\#: S_*(X) \to S_*(Y)$ satisfies $f_\#(S_*(A)) \subset S_*(B)$. It follows that f induces a chain map $S_*(f): S_*(X)/S_*(A) \to S_*(Y)/S_*(B)$, namely,

$$\gamma_n + S_n(A) \mapsto f_\#(\gamma_n) + S_n(B),$$

where $\gamma_n \in S_n(X)$. One usually denotes $S_*(f)$ by $f_\#$. That S_* is a functor is routine.

In $\mathbf{Top}^2$, there are inclusions

$$(A, \varnothing) \overset{i}{\hookrightarrow} (X, \varnothing) \overset{j}{\hookrightarrow} (X, A);$$

there are thus chain maps $i_\#$ and $j_\#$ that give a short exact sequence of complexes ($j_\#$ is the natural map!):

$$0 \longrightarrow S_*(A, \varnothing) \xrightarrow{i_\#} S_*(X, \varnothing) \xrightarrow{j_\#} S_*(X, A) \longrightarrow 0.$$

Theorem 5.8 is the result of applying the exact triangle to this short exact sequence of complexes. In a similar way, using the third isomorphism theorem, one sees that Theorem 5.9 arises from the inclusions (in $\mathbf{Top}^2$)

$$(A, A') \hookrightarrow (X, A') \hookrightarrow (X, A).$$

(3) One can show that Theorem 5.8 implies Theorem 5.9. The proof is a long diagram chase using the following commutative diagram.

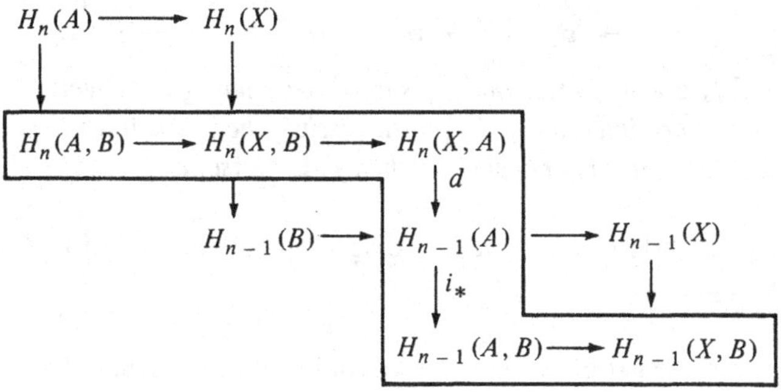

All maps are either connecting homomorphisms or are induced by inclusions; the map $H_n(X, A) \to H_{n-1}(A, B)$ is defined as the composite $i_* d\colon H_n(X, A) \to H_{n-1}(A) \to H_{n-1}(A, B)$. Full details can be found in [Eilenberg and Steenrod, pp. 25–28]. One should note that this proof applies to any sequence of functors $T_n\colon \mathbf{Top}^2 \to \mathbf{Ab}$ that satisfies Theorem 5.8; that is, there is a long exact sequence of a pair that has natural connecting homomorphisms.

(4) The following special case of Theorem 5.9 will be used in Chapter 8.

If (X, A, B) is a triple of topological spaces, then there is a commutative diagram

$$\begin{array}{ccc} H_n(X, A) & \xrightarrow{\ d\ } & H_{n-1}(A), \\ {\scriptstyle d'}\big\downarrow & \swarrow {\scriptstyle i_*} & \\ H_{n-1}(A, B) & & \end{array}$$

where $i\colon (A, \varnothing) \to (A, B)$ is the inclusion, where d is the connecting homomorphism of the pair (X, A), and where d' is the connecting homomorphism of the triple (X, A, B).

To see this, just apply Theorem 5.9 to the following commutative diagram of pairs and inclusions:

$$\begin{array}{ccccc} (A, \varnothing) & \longrightarrow & (X, \varnothing) & \longrightarrow & (X, A) \\ {\scriptstyle i}\big\downarrow & & \big\downarrow & & \big\downarrow{\scriptstyle 1} \\ (A, B) & \longrightarrow & (X, B) & \longrightarrow & (X, A). \end{array}$$

With Theorem 5.9 in mind, the reader can believe that the following theorem will be useful.

Theorem 5.10 (Five Lemma). *Consider the commutative diagram with exact rows*

$$\begin{array}{ccccccccc} A_1 & \longrightarrow & A_2 & \longrightarrow & A_3 & \longrightarrow & A_4 & \longrightarrow & A_5 \\ \big\downarrow{\scriptstyle f_1} & & \big\downarrow{\scriptstyle f_2} & & \big\downarrow{\scriptstyle f_3} & & \big\downarrow{\scriptstyle f_4} & & \big\downarrow{\scriptstyle f_5} \\ B_1 & \longrightarrow & B_2 & \longrightarrow & B_3 & \longrightarrow & B_4 & \longrightarrow & B_5. \end{array}$$

(i) *If f_2 and f_4 are surjective and f_5 is injective, then f_3 is surjective.*
(ii) *If f_2 and f_4 are injective and f_1 is surjective, then f_3 is injective.*
(iii) *If f_1, f_2, f_4, f_5 are isomorphisms, then f_3 is an isomorphism.*

Proof. Parts (i) and (ii) are proved by diagram chasing; part (iii) follows from the first two parts. □

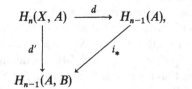

Having seen the proof of the exact triangle and having supplied the proof of the five lemma, the reader should now be comfortable with proofs by diagram chasing. Although such proofs may be long, they are not difficult; at each step, there is only one reasonable way to proceed, and so such proofs almost write themselves.

The definition of the relative homology group $H_n(X, A)$ as $H_n(S_*(X)/S_*(A))$ is perhaps too concise. Let us put this group in a more convenient form.

Recall the definition of the quotient complex

$$\cdots \longrightarrow \frac{S_{n+1}(X)}{S_{n+1}(A)} \xrightarrow{\bar{\partial}_{n+1}} \frac{S_n(X)}{S_n(A)} \xrightarrow{\bar{\partial}_n} \frac{S_{n-1}(X)}{S_{n-1}(A)} \longrightarrow \cdots,$$

where, for $\gamma \in S_n(X)$,

$$\bar{\partial}_n(\gamma + S_n(A)) = \partial_n \gamma + S_{n-1}(A).$$

Now

$$\ker \bar{\partial}_n = \{\gamma + S_n(A): \partial_n \gamma \in S_{n-1}(A)\}$$

and

$$\operatorname{im} \bar{\partial}_{n+1} = \{\gamma + S_n(A): \gamma \in \operatorname{im} \partial_{n+1} = B_n(X)\}.$$

Definition. The group of **relative n-cycles** mod A is

$$Z_n(X, A) = \{\gamma \in S_n(X): \partial_n \gamma \in S_{n-1}(A)\}.$$

The group of **relative n-boundaries** mod A is

$$\begin{aligned} B_n(X, A) &= \{\gamma \in S_n(X): \gamma - \gamma' \in B_n(X) \text{ for some } \gamma' \in S_n(A)\} \\ &= B_n(X) + S_n(A). \end{aligned}$$

It is easy to check that $S_n(A) \subset B_n(X, A) \subset Z_n(X, A) \subset S_n(X)$.

Theorem 5.11. *For all $n \geq 0$,*

$$H_n(X, A) \cong Z_n(X, A)/B_n(X, A).$$

PROOF. By definition,

$$H_n(X, A) = \ker \bar{\partial}_n / \operatorname{im} \bar{\partial}_{n+1}.$$

But it is easy to see from our remarks above that

$$\ker \bar{\partial}_n = Z_n(X, A)/S_n(A)$$

and

$$\operatorname{im} \bar{\partial}_{n+1} = B_n(X, A)/S_n(A).$$

The result now follows from the third isomorphism theorem (for groups). □

EXERCISES

*5.13. If A is a subspace of X, then for every $n \geq 0$, $S_n(X)/S_n(A)$ is a free abelian group with basis all (cosets of) n-simplexes σ in X for which $\operatorname{im} \sigma \not\subset A$.

*5.14. (i) Consider an exact sequence of abelian groups

$$\cdots \longrightarrow C_{n+1} \longrightarrow A_n \xrightarrow{i_n} B_n \xrightarrow{p_n} C_n \longrightarrow A_{n-1} \xrightarrow{i_{n-1}} B_{n-1} \xrightarrow{p_{n-1}} C_{n-1} \longrightarrow \cdots$$

in which every third map i_n is injective. Then

$$0 \longrightarrow A_n \xrightarrow{i_n} B_n \xrightarrow{p_n} C_n \longrightarrow 0$$

is exact for all n. (*Hint*: Exercise 5.2.)

(ii) If A is a retract of X, prove that for all $n \geq 0$,

$$H_n(X) \cong H_n(A) \oplus H_n(X, A).$$

(iii) If A is a deformation retract of X, then $H_n(X, A) = 0$ for all $n \geq 0$. (*Note*: $G \cong H \oplus K$ and $G \cong H$ do not imply that $K = 0$.)

5.15. Assume that $0 \to S'_* \to S_* \to S''_* \to 0$ is a short exact sequence of complexes. If two of the complexes are acyclic, then so is the third one.

5.16. If $f: (X, A) \to (X', A')$, then $f_\#: S_*(X) \to S_*(X')$ satisfies $f_\#(Z_n(X, A)) \subset Z_n(X', A')$ and $f_\#(B_n(X, A)) \subset B_n(X', A')$.

5.17. If $f: (X, A) \to (X', A')$, then the induced map $f_*: H_n(X, A) \to H_n(X', A')$ is given by

$$f_*: \gamma + B_n(X, A) \mapsto f_\#(\gamma) + B_n(X', A'),$$

where $\gamma \in Z_n(X, A)$. (The original definition of f_* is not in terms of relative cycles and relative boundaries.)

*5.18. If every face $\sigma\varepsilon_i$ of an n-simplex $\sigma: \Delta^n \to X$ has its values in $A \subset X$, then σ represents an element of $Z_n(X, A)$.

Exercise 5.18 gives a picture.

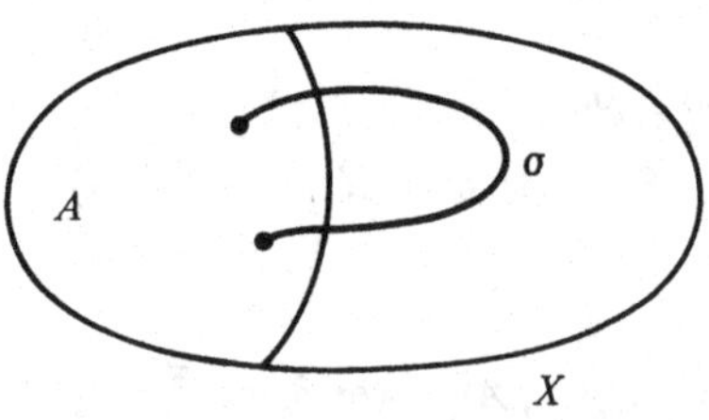

For example, a path σ in X is a 1-cycle if it is a closed path; it is a relative 1-cycle if it begins and ends in A. Observe, in this example, that if $A = \{x_0\}$, then "cycle" and "relative cycle" coincide. This is actually true (almost) always. First, we do a small computation.

Theorem 5.12. *If X is path connected and A is a nonempty subspace, then $H_0(X, A) = 0$.*

Proof. Choose $x_0 \in A$, and let $\gamma = \sum m_x x \in Z_0(X, A) = S_0(X)$. Since X is path connected, for each $x \in X$ there is a "path" $\sigma_x: \Delta^1 \to X$ with $\sigma_x(e_0) = x_0$ and

$\sigma_x(e_1) = x$. Then $\sum m_x \sigma_x \in S_1(X)$, and

$$\partial_1(\sum m_x \sigma_x) = \sum m_x x - (\sum m_x) x_0 = \gamma - (\sum m_x) x_0.$$

But γ' defined as $(\sum m_x) x_0$ lies in $S_0(A)$; hence $\gamma - \gamma' = \partial(\sum m_x \sigma_x) \in B_0(X)$, and so $\gamma \in B_0(X, A)$. Therefore $B_0(X, A) = Z_0(X, A)$ and $H_0(X, A) = 0$. □

Theorem 5.13. *If $\{X_\lambda : \lambda \in \Lambda\}$ is the family of path components of X, then, for each $n \geq 0$,*

$$H_n(X, A) \cong \sum_\lambda H_n(X_\lambda, A \cap X_\lambda).$$

Proof. Use Exercise 5.12 and Theorem 4.13. □

Corollary 5.14. *$H_0(X, A)$ is free abelian and*

$$\text{rank } H_0(X, A) = \text{card}\{\lambda \in \Lambda : A \cap X_\lambda = \varnothing\}$$

(where $\{X_\lambda : \lambda \in \Lambda\}$ is the family of path components of X).

Proof. By Theorem 5.13, $H_0(X, A) \cong \sum H_0(X_\lambda, A \cap X_\lambda)$. If $A \cap X_\lambda = \varnothing$, then $H_0(X_\lambda, A \cap X_\lambda) = H_0(X_\lambda) = \mathbf{Z}$ (by Theorem 4.14(*i*)). If, on the other hand, $A \cap X_\lambda \neq \varnothing$, then $H_0(X_\lambda, A \cap X_\lambda) = 0$ (by Theorem 5.12, for X_λ is path connected). □

Corollary 5.15. *If X is a space with basepoint x_0, then $H_0(X, x_0)$ is a free abelian group of (possibly infinite) rank r, where X has exactly $r + 1$ path components.*

Proof. Since path components are pairwise disjoint, the path component X_{λ_0} containing x_0 is unique, and so $\{x_0\} \cap X_\lambda = \varnothing$ for all $\lambda \neq \lambda_0$. Hence $H_0(X_\lambda, \{x_0\} \cap X_\lambda) \cong \mathbf{Z}$ for all $\lambda \neq \lambda_0$, while $H_0(X_{\lambda_0}, x_0) = 0$. □

Theorem 5.16. *Let X be a space with basepoint x_0. Then*

$$H_n(X, x_0) \cong H_n(X)$$

for all $n \geq 1$.

Proof. By Theorem 5.8, there is an exact sequence

$$\cdots \to H_n(\{x_0\}) \to H_n(X) \to H_n(X, x_0) \to H_{n-1}(\{x_0\}) \to \cdots.$$

If $n \geq 2$, then $n - 1 \geq 1$, and the dimension axiom (Theorem 4.12) gives $H_n(\{x_0\}) = 0 = H_{n-1}(\{x_0\})$; hence $H_n(X) \cong H_n(X, x_0)$ for all $n \geq 2$. To examine the remaining case $n = 1$, let us look at the tail of the exact sequence:

$$\cdots \to H_1(\{x_0\}) \to H_1(X) \xrightarrow{g} H_1(X, x_0) \to H_0(\{x_0\}) \xrightarrow{h} H_0(X) \xrightarrow{k} H_0(X, x_0) \to 0.$$

Since $H_1(\{x_0\}) = 0$, the map g is injective; by Exercise 5.2, g is surjective (hence is an isomorphism) if and only if h is injective. The map h has domain

$H_0(\{x_0\}) \cong \mathbf{Z}$ and target the free abelian group $H_0(X)$. If $h \neq 0$, then h must be injective (if $\ker h \neq 0$, then $H_0(X)$ would contain a nontrivial finite subgroup isomorphic to $\mathbf{Z}/\ker h$). Now $\operatorname{im} h = \ker k$, so that $\ker k \neq 0$ implies that $\operatorname{im} h \neq 0$, hence $h \neq 0$, as desired. But k, being induced by inclusion, is the map $S_0(X)/B_0(X) \to S_0(X)/B_0(X) + S_0(x_0)[S_0(X) = Z_0(X) = Z_0(X, x_0)]$ given by $\gamma + B_0(X) \mapsto \gamma + B_0(X) + S_0(x_0)$, and so $\ker k = (B_0(X) + S_0(x_0))/B_0(X)$. The proof of Theorem 4.14 describes $B_0(X)$ as all $\sum m_x x$ with $\sum m_x = 0$; hence $\ker k \neq 0$, and the proof is complete. □

For each $n \geq 1$, one may thus regard H_n as a functor with domain $\mathbf{Top}_*$, the category of pointed spaces.

Reduced Homology

The coming construction of reduced homology groups will allow us to avoid the fussy algebra at the end of the proof of Theorem 5.16.

Definition. Let $(S_*(X), \partial)$ be the singular complex of a space X. Define $\tilde{S}_{-1}(X)$ to be the infinite cyclic group with generator the symbol [ ], and define $\tilde{\partial}_0 \colon S_0(X) \to \tilde{S}_{-1}(X)$ by $\sum m_x x \mapsto (\sum m_x)[\ \ ]$. The **augmented singular complex** of X is

$$\tilde{S}_*(X)\colon \quad \cdots \to S_2(X) \xrightarrow{\partial_2} S_1(X) \xrightarrow{\partial_1} S_0(X) \xrightarrow{\tilde{\partial}_0} \tilde{S}_{-1}(X) \to 0.$$

It is a quick calculation that $\tilde{\partial}_0 \partial_1 = 0$, so that the augmented singular complex is in fact a complex (having $\tilde{S}_{-1}(X) \cong \mathbf{Z}$ as a nonzero term of negative degree).

There are several remarks to be made. First, the map $\tilde{\partial}_0$ has already appeared (in the proof of Theorem 4.14(i)). Second, suppose that one defines the empty set $\varnothing$ as the standard (-1)-simplex. For any space X, there is a unique (inclusion) function $\varnothing \to X$, and so $\tilde{S}_{-1}(X)$ as defined above is reasonable. Moreover, if one regards the boundary of a point $x \in X$ as empty, then $\tilde{\partial}_0$ is obtained from $\bar{\partial}_0 x = [\ \ ]$ by extending by linearity.

Definition. The **reduced homology groups** of X are

$$\tilde{H}_n(X) = H_n(\tilde{S}_*(X), \partial), \qquad \text{for all } n \geq 0.$$

Theorem 5.17. *For all* $n \geq 0$,

$$\tilde{H}_n(X) \cong H_n(X, x_0).$$

Proof. If $n \geq 1$, $\tilde{H}_n(X) = \ker \partial_n / \operatorname{im} \partial_{n+1} = H_n(X)$, so the result follows from Theorem 5.16. If $n = 0$, the end of $\tilde{S}_*(X)$ gives a short exact sequence

$$0 \to \ker \tilde{\partial}_0 \hookrightarrow S_0(X) \xrightarrow{\tilde{\partial}_0} \tilde{S}_{-1}(X) \to 0.$$

If $\alpha \in S_0(X)$ satisfies $\tilde{\partial}_0(\alpha) = 1$, then it is easy to see[4] that $S_0(X) = \ker \tilde{\partial}_0 \oplus \langle\alpha\rangle$ and $\langle\alpha\rangle \cong \mathbf{Z}$. But $\tilde{\partial}_0\partial_1 = 0$ implies that $B_0(X) = \operatorname{im} \partial_1 \subset \ker \tilde{\partial}_0$. Since $S_0(X) = Z_0(X)$, we have[5]

$$H_0(X) = S_0(X)/B_0(X) = (\ker \tilde{\partial}_0 \oplus \langle\alpha\rangle)/B_0(X)$$
$$\cong (\ker \tilde{\partial}_0/B_0(X)) \oplus \mathbf{Z} = \tilde{H}_0(X) \oplus \mathbf{Z}.$$

Since $H_0(X)$ is free abelian, the result follows from Corollary 5.15. □

One can squeeze a bit more from this proof to improve Theorem 5.17 by exhibiting a basis of $\tilde{H}_0(X)$.

Corollary 5.18. *Let $\{X_\lambda: \lambda \in \Lambda\}$ be the family of path components of X, and let $x_\lambda \in X_\lambda$ be a choice of points, one from each path component. If $x_0 \in X$ lies in X_{λ_0}, then $\tilde{H}_0(X)$ is free abelian with basis $\{\operatorname{cls}(x_\lambda - x_0): \lambda \neq \lambda_0\}$.*

Proof. We saw in the last proof that

$$S_0(X) = \ker \tilde{\partial}_0 \oplus \langle\alpha\rangle,$$

where α is any 0-chain with $\tilde{\partial}_0(\alpha) = 1$; let us choose $\alpha = x_0$. Since X is a basis of $S_0(X)$, we see that $\{x_0\} \cup Y$ is also a basis, where $Y = \{x - x_0: x \neq x_0\}$. We claim that Y is a basis of $\ker \tilde{\partial}_0$, for which it now suffices to prove that Y generates $\ker \tilde{\partial}_0$. As $\tilde{\partial}_0(x - x_0) = 0$, we see that $Y \subset \ker \tilde{\partial}_0$; furthermore, if $\sum m_i x_i \in S_0(X)$ and $\sum m_i = 0$, then

$$\sum m_i x_i = \sum m_i x_i - \left(\sum m_i\right)x_0 = \sum m_i(x_i - x_0)$$

(of course, we may delete $x_i - x_0$ from the sum if $x_i = x_0$).

$\tilde{H}_0(X) = \ker \tilde{\partial}_0/B_0(X)$ is a direct summand of $H_0(X) = S_0(X)/B_0(X) = (\ker \tilde{\partial}_0 + \langle x_0\rangle)/B_0(X)$. By Theorem 4.14, $\{\operatorname{cls} x_\lambda: \lambda \neq \lambda_0\} \cup \{\operatorname{cls} x_0\}$ is a basis of $H_0(X)$. As above, $\{\operatorname{cls}(x_\lambda - x_0): \lambda \neq \lambda_0\} \cup \{\operatorname{cls} x_0\}$ is also a basis of $H_0(X)$; since $\{\operatorname{cls}(x_\lambda - x_0): \lambda \neq \lambda_0\}$ generates $\tilde{H}_0(X)$, it is a basis. □

We shall see that reduced homology has other uses than allowing us to avoid algebraic arguments as in the proof of Theorem 5.16. For example, look at Theorem 6.5 and its proof.

---

[4] This is a special case of a more general result (Corollary 9.2): if $0 \to K \hookrightarrow G \to F \to 0$ is exact and F is free abelian, then $G = K \oplus F'$, where $F' \cong F$. Here we present a proof of this special case. If $x \in \ker \tilde{\partial}_0 \cap \langle\alpha\rangle$, then $x = m\alpha$ and $\tilde{\partial}_0(x) = 0 = m$, hence $x = 0$; if $\gamma \in S_0(X)$, then $\tilde{\partial}_0(\gamma) = k$, say, and so $\gamma = (\gamma - k\alpha) + k\alpha \in \ker \tilde{\partial}_0 + \langle\alpha\rangle$.

[5] If $B_i \subset A_i$ for $i = 1, 2$, then $(A_1 \oplus A_2)/(B_1 \oplus B_2) \cong (A_1/B_1) \oplus (A_2/B_2)$ (indeed the analogous statement for any index set is true): define a map $\theta: A_1 \oplus A_2 \to (A_1/B_1) \oplus (A_2/B_2)$ by $(a_1, a_2) \mapsto (a_1 + B_1, a_2 + B_2)$. Then θ is surjective and $\ker \theta = B_1 \oplus B_2$; now apply the first isomorphism theorem.

EXERCISES

*5.19. If $A \subset X$, then there is an exact sequence

$$\cdots \to \tilde{H}_n(A) \to \tilde{H}_n(X) \to H_n(X, A) \to \tilde{H}_{n-1}(A) \to \cdots,$$

which ends

$$\cdots \to \tilde{H}_0(A) \to \tilde{H}_0(X) \to H_0(X, A) \to 0.$$

(*Hint*: $\tilde{S}_*(X)/\tilde{S}_*(A) = S_*(X)/S_*(A)$.)

5.20. Show that $H_1(D^2, S^1) = 0$.

5.21. Assume that X has five path components. If CX is the cone on X, what is $H_1(CX, X)$?

5.22. What is $H_1(S^1, S^0)$?

5.23. Show that $H_n(X, X) = 0$ for all $n \geq 0$.

There is a geometric interpretation of relative homology groups other than Theorem 5.11. Recall that the quotient space X/A is obtained from X by collapsing A to a point. For a large class of pairs, for example, for A a "nice" subset of a polyhedron X, one can prove that $H_*(X, A) \cong \tilde{H}_*(X/A)$ (see Theorem 8.41). In this case, the exact sequence of Exercise 5.19 is

$$\cdots \to \tilde{H}_n(A) \to \tilde{H}_n(X) \to \tilde{H}_n(X/A) \to \tilde{H}_{n-1}(A) \to \cdots.$$

It turns out that the importance of relative homology groups is such that the category of pairs, $\mathbf{Top}^2$, is more convenient than $\mathbf{Top}$. Let us therefore give the obvious version of homotopy in $\mathbf{Top}^2$.

Definition. If $f, g: (X, A) \to (Y, B)$, then $f \simeq g \bmod A$ if there is a continuous $F: (X \times \mathbf{I}, A \times \mathbf{I}) \to (Y, B)$ with $F_0 = f$ and $F_1 = g$.

This notion of homotopy mod A is weaker than the previous notion of homotopy rel A, which requires that $f|A = g|A$ and also that $F(a, t)$ remain fixed for all $a \in A$ during the homotopy (i.e., for every time t). Now we require only that $F(a, t) \in B$ for all $a \in A$ and all $t \in \mathbf{I}$. Of course, the notions coincide when B is a one-point space.

Here is the appropriate version of the homotopy axiom in $\mathbf{Top}^2$.

Theorem 5.19 (Homotopy Axiom for Pairs). *If $f, g: (X, A) \to (Y, B)$ and $f \simeq g \bmod A$, then for all $n \geq 0$,*

$$H_n(f) = H_n(g): H_n(X, A) \to H_n(Y, B).$$

PROOF. If $j: A \hookrightarrow X$ is the inclusion, then Exercise 4.10 gives a commutative diagram

$$\begin{array}{ccc} S_n(A) & \xrightarrow{P_n^A} & S_{n+1}(A \times \mathbf{I}) \\ {\scriptstyle j_\#}\downarrow & & \downarrow{\scriptstyle (j \times 1)_\#} \\ S_n(X) & \xrightarrow[P_n^X]{} & S_{n+1}(X \times \mathbf{I}), \end{array}$$

where P_n is the nth term of the chain homotopy of Theorem 4.23. It follows that P_n induces a homomorphism $\bar{P}_n: S_n(X)/S_n(A) \to S_{n+1}(X \times \mathbf{I})/S_{n+1}(A \times \mathbf{I})$, that is, $\bar{P}_n: S_n(X, A) \to S_{n+1}(X \times \mathbf{I}, A \times \mathbf{I})$. The proof now proceeds exactly as that of Theorem 4.23; it is left to the reader to show that the maps $\bar{P}_n$ satisfy $\partial\bar{P} + \bar{P}\partial = \bar{\lambda}_{1\#} - \bar{\lambda}_{0\#}$ and hence comprise a chain homotopy. □

CHAPTER 6

Excision and Applications

Excision and Mayer–Vietoris

The last fundamental property (or axiom) of homology is *excision*. We state two versions. If A is a subspace of X, then $\bar{A}$ denotes its closure and A° denotes its interior.

Excision I. *Assume that $U \subset A \subset X$ are subspaces with $\bar{U} \subset A^\circ$. Then the inclusion $i: (X - U, A - U) \hookrightarrow (X, A)$ induces isomorphisms*

$$i_*: H_n(X - U, A - U) \xrightarrow{\approx} H_n(X, A)$$

for all n.

Stated in this way, we see that one may excise (cut out) U without changing relative homology groups.

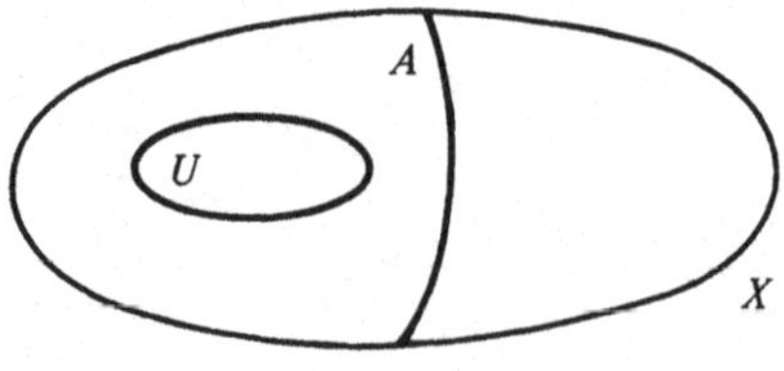

Excision II. *Let X_1 and X_2 be subspaces of X with $X = X_1^\circ \cup X_2^\circ$. Then the inclusion $j: (X_1, X_1 \cap X_2) \hookrightarrow (X_1 \cup X_2, X_2) = (X, X_2)$ induces isomorphisms*

$$j_*: H_n(X_1, X_1 \cap X_2) \xrightarrow{\approx} H_n(X, X_2)$$

for all n.

The second form is reminiscent of the second isomorphism theorem. Note that both forms involve two subspaces of X.

Theorem 6.1. *Excision I is equivalent to Excision II.*

PROOF. Assume Excision I, and let $X = X_1^\circ \cup X_2^\circ$. Define $A = X_2$ and $U = X - X_1$. First, we claim that $\bar{U} \subset A^\circ$: $X_1^\circ \subset X_1$ implies $X - X_1 \subset X - X_1^\circ$, hence $\bar{U} = \overline{(X - X_1)} \subset X - X_1^\circ$ (for the last set is closed); but $X - X_1^\circ = (X_2^\circ \cup X_1^\circ) - X_1^\circ = X_2^\circ - X_1^\circ \subset X_2^\circ = A^\circ$. Second, $X - U = X - (X - X_1) = X_1$ and $A - U = X_2 - (X - X_1) = X_2 \cap (X_1^c)^c$ (where $X_1^c = X - X_1$, the complement of X). Thus the pair $(X - U, A - U)$ is the pair $(X_1, X_1 \cap X_2)$. Finally, the pair (X, A) is the pair (X, X_2). The inclusions coincide and hence induce the same map in homology.

Assume Excision II, and let $\bar{U} \subset A^\circ$. Define $X_2 = A$ and $X_1 = X - U$. Now $U \subset \bar{U} \subset A^\circ$ implies $X - U \supset X - \bar{U} \supset X - A^\circ$. Since $X - \bar{U}$ is open, $X - \bar{U} = (X - \bar{U})^\circ \supset X - A^\circ$. Hence

$$X_1^\circ \cup X_2^\circ = (X - U)^\circ \cup A^\circ \supset (X - \bar{U})^\circ \cup A^\circ \supset (X - A^\circ) \cup A^\circ = X.$$

Finally, it is easy to see that $(X_1, X_1 \cap X_2) = (X - U, A - U)$ and $(X, X_2) = (X, A)$. □

Before we prove excision, let us see some of its consequences. We begin with a general diagram lemma.

Lemma 6.2 (Barratt-Whitehead). *Consider the commutative diagram with exact rows*

$$\begin{array}{ccccccccc} \cdots \longrightarrow & A_n & \xrightarrow{i_n} & B_n & \xrightarrow{p_n} & C_n & \xrightarrow{d_n} & A_{n-1} & \longrightarrow \cdots \\ & \downarrow{\scriptstyle f_n} & & \downarrow{\scriptstyle g_n} & & \downarrow{\scriptstyle h_n} & & \downarrow{\scriptstyle f_{n-1}} & \\ \cdots \longrightarrow & A'_n & \xrightarrow[j_n]{} & B'_n & \xrightarrow[q_n]{} & C'_n & \xrightarrow[\Delta_n]{} & A'_{n-1} & \longrightarrow \cdots \end{array}$$

in which every third vertical map h_n is an isomorphism. Then there is an exact sequence

$$\cdots \longrightarrow A_n \xrightarrow{(i_n, f_n)} B_n \oplus A'_n \xrightarrow{g_n - j_n} B'_n \xrightarrow{d_n h_n^{-1} q_n} A_{n-1} \longrightarrow \cdots.$$

PROOF. The map (i_n, f_n) is defined by $a_n \mapsto (i_n a_n, f_n a_n)$, and the map $g_n - j_n$ is defined by $(b_n, a'_n) \mapsto g_n b_n - j_n a'_n$. The proof of exactness is a diagram chase. □

Theorem 6.3 (Mayer–Vietoris). *If X_1, X_2 are subspaces of X with $X = X_1^\circ \cup X_2^\circ$, then there is an exact sequence*

$$\cdots \longrightarrow H_n(X_1 \cap X_2) \xrightarrow{(i_{1*}, i_{2*})} H_n(X_1) \oplus H_n(X_2) \xrightarrow{g_* - j_*} H_n(X) \xrightarrow{D} H_{n-1}(X_1 \cap X_2) \longrightarrow \cdots,$$

with i_1, i_2, g, j inclusions and $D = dh_*^{-1}q_*$, where h, q are inclusions and d is the connecting homomorphism of the pair $(X_1, X_1 \cap X_2)$.

PROOF. The following diagram of pairs of spaces commutes when all maps are inclusions:

$$\begin{array}{ccccc} (X_1 \cap X_2, \varnothing) & \xrightarrow{i_1} & (X_1, \varnothing) & \xrightarrow{p} & (X_1, X_1 \cap X_2) \\ \downarrow{\scriptstyle i_2} & & \downarrow{\scriptstyle g} & & \downarrow{\scriptstyle h} \\ (X_2, \varnothing) & \xrightarrow[j]{} & (X, \varnothing) & \xrightarrow[q]{} & (X, X_2). \end{array}$$

By Theorem 5.9, there is a commutative diagram with exact rows:

$$\begin{array}{ccccccccccc} \cdots \longrightarrow & H_n(X_1 \cap X_2) & \xrightarrow{i_{1*}} & H_n(X_1) & \xrightarrow{p_*} & H_n(X_1, X_1 \cap X_2) & \xrightarrow{d} & H_{n-1}(X_1 \cap X_2) & \longrightarrow \cdots \\ & \downarrow{\scriptstyle i_{2*}} & & \downarrow{\scriptstyle g_*} & & \downarrow{\scriptstyle h_*} & & \downarrow{\scriptstyle i_{2*}} & \\ \cdots \longrightarrow & H_n(X_2) & \xrightarrow[j_*]{} & H_n(X) & \xrightarrow[q_*]{} & H_n(X, X_2) & \xrightarrow[\Delta]{} & H_{n-1}(X_2) & \longrightarrow \cdots. \end{array}$$

Excision II asserts that each h_* is an isomorphism, so that Lemma 6.2 gives the result at once. □

EXAMPLE 6.1. Here is an example of a space $X = X_1 \cup X_2$, where X_1 and X_2 are (closed) subspaces of X (but where $X \neq X_1^\circ \cup X_2^\circ$) in which the Mayer–Vietoris theorem, and hence excision, fails.

Let X be the closed vertical strip in $\mathbf{R}^2$ lying between the y-axis and the line $x = 1/2\pi$. Define

$$X_1 = \{(0, y)\colon -1 \leq y\} \cup \{(x, y)\colon 0 < x \leq 1/2\pi \text{ and } \sin(1/x) \leq y\};$$

define

$$X_2 = \{(0, y)\colon y \leq 1\} \cup \{(x, y)\colon 0 < x \leq 1/2\pi \text{ and } \sin(1/x) \geq y\}.$$

Note that $X_1 \cup X_2 = X$ and that $X_1 \cap X_2$ is the $\sin(1/x)$ space. Were the Mayer–Vietoris theorem true here, there would be an exact sequence

$$H_1(X) \to H_0(X_1 \cap X_2) \to H_0(X_1) \oplus H_0(X_2) \to H_0(X) \to 0.$$

Since X, X_1, and X_2 are contractible, $H_1(X) = 0$ and $H_0(X) = \mathbf{Z} = H_0(X_i)$ for $i = 1, 2$. There is thus an exact sequence of the form

$$0 \to \mathbf{Z} \oplus \mathbf{Z} \to \mathbf{Z} \oplus \mathbf{Z} \to \mathbf{Z} \to 0,$$

and this contradicts Exercise 5.5.

Corollary 6.4 (Mayer–Vietoris Theorem for Reduced Homology). *If X_1, X_2 are subspaces of X with $X = X_1^\circ \cup X_2^\circ$ and $X_1 \cap X_2 \neq \varnothing$, then there is an exact* *sequence*

$$\cdots \to \tilde{H}_n(X_1 \cap X_2) \to \tilde{H}_n(X_1) \oplus \tilde{H}_n(X_2) \to \tilde{H}_n(X) \to \tilde{H}_{n-1}(X_1 \cap X_2) \to \cdots$$

with induced maps as in Theorem 6.3. This sequence ends

$$\cdots \to \tilde{H}_0(X_1) \oplus \tilde{H}_0(X_2) \to \tilde{H}_0(X) \to 0.$$

PROOF. If $x_0 \in X_1 \cap X_2$, proceed as in Theorem 6.3 from the commutative diagram of inclusions of pairs

$$\begin{array}{ccccc} (X_1 \cap X_2, x_0) & \longrightarrow & (X_1, x_0) & \longrightarrow & (X_1, X_1 \cap X_2) \\ \downarrow & & \downarrow & & \downarrow \\ (X_2, x_0) & \longrightarrow & (X, x_0) & \longrightarrow & (X, X_2). \end{array}$$ □

EXERCISES

*6.1. Assume that $X = A \cup B$ is a **disconnection** (A and B are nonempty open sets and $A \cap B = \varnothing$). Then $H_n(X) \cong H_n(A) \oplus H_n(B)$ for all $n \geq 0$. (*Hint*: The inclusion $A \hookrightarrow X$ is an excision here; or, use Theorem 4.13.)

6.2. If $X = A \cup B$ is a disconnection, then $H_n(X, A) \cong H_n(B)$ for all $n \geq 0$.

*6.3. Assume that $X = X_1^\circ \cup X_2^\circ$ and $Y = Y_1^\circ \cup Y_2^\circ$; assume further that $f: X \to Y$ is continuous with $f(X_i) \subset Y_i$ for $i = 1, 2$. Then the following diagram commutes:

$$\begin{array}{ccc} H_n(X) & \xrightarrow{D} & H_{n-1}(X_1 \cap X_2) \\ f_* \downarrow & & \downarrow g_* \\ H_n(Y) & \xrightarrow[D']{} & H_{n-1}(Y_1 \cap Y_2), \end{array}$$

where g is the restriction of f and D, D' are connecting homomorphisms of Mayer–Vietoris sequences.

*6.4. Assume that $X = X_1 \cup X_2 \cup X_3$, where each X_i is open. If all X_i, all three $X_i \cap X_j$, and $X_1 \cap X_2 \cap X_3$ are either contractible or empty, then $H_n(X) = 0$ for all $n \geq 2$. (*Hint*: Iterate Mayer–Vietoris.) (For a generalization to any open cover of X, see [K. S. Brown, p. 166]. Also, see Corollary 7.27.)

Homology of Spheres and Some Applications

Theorem 6.5. *Let S^n be the n-sphere, where $n \geq 0$. Then*

$$H_p(S^0) = \begin{cases} \mathbf{Z} \oplus \mathbf{Z} & \text{if } p = 0 \\ 0 & \text{if } p > 0; \end{cases}$$

if $n > 0$, then

$$H_p(S^n) = \begin{cases} \mathbf{Z} & \textit{if } p = 0 \textit{ or } p = n \\ 0 & \textit{otherwise.} \end{cases}$$

Remark. Using reduced homology, we can state these formulas more concisely:

$$\tilde{H}_p(S^n) = \begin{cases} \mathbf{Z} & \text{if } p = n \\ 0 & \text{if } p \neq n. \end{cases}$$

Proof. We do an induction on $n \geq 0$ that $\tilde{H}_p(S^n)$ is as claimed for all $p \geq 0$. The formula holds if $n = 0$, by the dimension axiom (Theorem 4.12) and Theorem 4.13; one can also use Exercise 6.1.

Assume that $n > 0$. Let a and b be the north and south poles of S^n, let $X_1 = S^n - \{a\}$, and let $X_2 = S^n - \{b\}$. Note that $S^n = X_1^\circ \cup X_2^\circ$ (because X_1 and X_2 are open), that X_1 and X_2 are contractible, and that $X_1 \cap X_2 = S^n - \{a, b\}$ has the same homotopy type as the equator S^{n-1} (by Exercise 1.31). Applying the Mayer–Vietoris sequence for reduced homology, we obtain an exact sequence

$$\tilde{H}_p(X_1) \oplus \tilde{H}_p(X_2) \to \tilde{H}_p(S^n) \to \tilde{H}_{p-1}(X_1 \cap X_2) \to \tilde{H}_{p-1}(X_1) \oplus \tilde{H}_{p-1}(X_2).$$

Contractibility of X_1 and X_2 shows that the flanking (direct sum) terms are both zero, and so

$$\tilde{H}_p(S^n) \cong \tilde{H}_{p-1}(X_1 \cap X_2) \cong \tilde{H}_{p-1}(S^{n-1}),$$

by Corollary 4.24 (note that we are using $n > 0$ as well). By induction, $\tilde{H}_{p-1}(S^{n-1}) = \mathbf{Z}$ if $p - 1 = n - 1$ and 0 otherwise; therefore $\tilde{H}_p(S^n) = \mathbf{Z}$ if $p = n$ and 0 otherwise. □

This theorem illustrates the value of reduced homology. Not only is the "reduced" statement better, but the proof is shorter. Without reduced homology, the inductive step would divide into two cases: $p - 1 > 0$ (which would proceed as above) and $p - 1 = 0$ (which would require an extra argument involving free abelian groups as in the proof of Theorem 5.16).

We may now draw some conclusions.

Theorem 6.6. *If $n \geq 0$, then S^n is not a retract of D^{n+1}.*

Proof. We have verified all the requirements for the proof of Lemma 0.2. □

Theorem 6.7 (Brouwer Fixed Point Theorem). *If $f\colon D^n \to D^n$ is continuous, then there is $x \in D^n$ with $f(x) = x$.*

Proof. Theorem 0.3. □

Theorem 6.8. *If $m \neq n$, then S^m and S^n are not homeomorphic. Indeed they do not have the same homotopy type.*

PROOF. If S^m and S^n had the same homotopy type, then $H_p(S^m) \cong H_p(S^n)$ for all p. □

Theorem 6.9. *If $m \neq n$, then $\mathbf{R}^m$ and $\mathbf{R}^n$ are not homeomorphic.*

PROOF. If there is a homeomorphism $f: \mathbf{R}^m \to \mathbf{R}^n$, choose $x_0 \in \mathbf{R}^m$ and obtain a homeomorphism $\mathbf{R}^m - \{x_0\} \approx \mathbf{R}^n - \{f(x_0)\}$. But $\mathbf{R}^m - \{x_0\}$ has the same homotopy type as S^{m-1} (Exercise 1.29), which leads to a contradiction of Theorem 6.8. □

Theorem 6.10. *If $n \geq 0$, then S^n is not contractible.*

PROOF. Otherwise S^n would have the same homology groups as a point. □

Using Exercise 3.21, we now have examples, namely, S^n for $n \geq 2$, of simply connected spaces that are not contractible.

Barycentric Subdivision and the Proof of Excision

The applications of Theorem 6.5 are not exhausted, but let us get on with the proof of excision (more precisely, of Excision II); we begin with an algebraic lemma.

If X_1 is a subspace of X, regard $S_*(X_1)$ as the subcomplex of $S_*(X)$ whose term of degree n is generated by all n-simplexes $\sigma: \Delta^n \to X$ for which $\sigma(\Delta^n) \subset X_1$.

Lemma 6.11. *Let X_1 and X_2 be subspaces of X. If the inclusion $S_*(X_1) + S_*(X_2) \hookrightarrow S_*(X)$ induces isomorphisms in homology, then excision holds for the subspaces X_1 and X_2 of X.*

PROOF. Applying the exact triangle to the short exact sequence

$$0 \to S_*(X_1) + S_*(X_2) \xrightarrow{i} S_*(X) \to S_*(X)/(S_*(X_1) + S_*(X_2)) \to 0,$$

we obtain a long exact sequence in which every third arrow $H_n(i)$ is an isomorphism (by hypothesis); it follows easily (Exercise 5.4) that $H_n(S_*(X)/(S_*(X_1) + S_*(X_2))) = 0$ for all n.

Now consider the short exact sequence of complexes

$$0 \to \frac{S_*(X_1) + S_*(X_2)}{S_*(X_2)} \xrightarrow{j} \frac{S_*(X)}{S_*(X_2)} \to \frac{S_*(X)}{S_*(X_1) + S_*(X_2)} \to 0.$$

The corresponding long exact sequence has every third term zero, so that $H_n(j)$ is an isomorphism for every n.

Finally, consider the commutative diagram of complexes

$$\begin{array}{ccc} \dfrac{S_*(X_1)}{S_*(X_1 \cap X_2)} & \xrightarrow{k} & \dfrac{S_*(X)}{S_*(X_2)}, \\ & {\scriptstyle \ell} \searrow \quad \nearrow {\scriptstyle j} & \\ & \dfrac{S_*(X_1) + S_*(X_2)}{S_*(X_2)} & \end{array}$$

where k is induced by the inclusion $(X_1, X_1 \cap X_2) \hookrightarrow (X, X_2)$ and ℓ is the isomorphism of the second isomorphism theorem (recall that $S_*(X_1 \cap X_2) = S_*(X_1) \cap S_*(X_2)$). Now $j\ell = k$ implies $H_n(j)H_n(\ell) = H_n(k)$. We have just seen that $H_n(j)$ is an isomorphism, while $H_n(\ell)$ is an isomorphism because ℓ is. It follows that $H_n(k)$ is an isomorphism for all n, which is the statement of Excision II. □

It thus remains to show that the inclusion $S_*(X_1) + S_*(X_2) \hookrightarrow S_*(X)$ induces isomorphisms in homology whenever $X = X_1^\circ \cup X_2^\circ$. This would appear reasonable if every n-cycle in X were a sum of chains in X_1 and chains in X_2. However, an n-simplex σ in X may have its image in neither X_1 nor X_2. The idea is to subdivide Δ^n into small pieces so that the restrictions of σ to these pieces do have images in either X_1 or X_2. The forthcoming construction, barycentric subdivision, is important in other contexts as well; let us therefore consider it leisurely.

We begin by examining (geometric) subdivisions of Δ^n for small n. With an understanding of these low-dimensional examples, we shall see how to define (inductively) subdivisions of every Δ^n; this definition will then be transferred to subdivisions of n-simplexes in an arbitrary space X.

Now Δ^0 is a one-point set; we admit it cannot be divided further and define Δ^0 to be its own subdivision. Consider the more interesting $\Delta^1 = [e_0, e_1]$. A reasonable way to subdivide Δ^1 is to cut it in half: let b be the midpoint of the interval $[e_0, e_1]$, that is, b is the barycenter of Δ^1. Define the barycentric subdivision of Δ^1 to be the 1-simplexes $[e_0, b]$ and $[b, e_1]$ and their faces. Let us now subdivide the standard 2-simplex Δ^2.

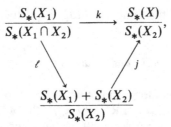

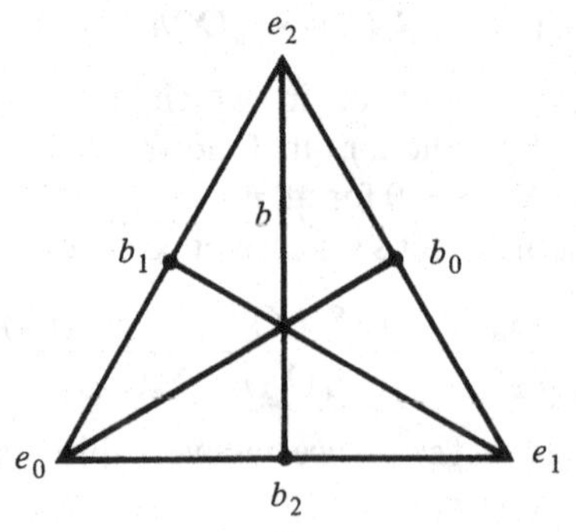

Subdivide the triangle Δ^2 as follows: first, subdivide each face (which is a 1-simplex) as above, using new vertices the barycenters b_0, b_1, b_2; second, let b be the barycenter of Δ^2; finally, draw the six new triangles illustrated above. Here is one way to view this construction. We have adjoined new vertices b_0, b_1, b_2, b to the original vertices. Which triangles do we form using these seven vertices? Note that each vertex is a barycenter of a face of Δ^2: the original vertices e_0, e_1, e_2 are barycenters of 0-faces (themselves); b_0, b_1, b_2 are barycenters of 1-faces; b is the barycenter of Δ^2 itself. Each vertex may thus be denoted by b^σ, where σ denotes a face of Δ^2, and $\{b^\tau, b^\sigma, b^\rho\}$ is a triangle precisely when $\tau < \sigma$ (τ is a proper face of σ) and $\sigma < \rho$. There are thus 3! triangles.

Definition. The **barycentric subdivision** of an affine n-simplex Σ^n, denoted by $\operatorname{Sd}\Sigma^n$, is a family of affine n-simplexes defined inductively for $n \geq 0$:

(i) $\operatorname{Sd}\Sigma^0 = \Sigma^0$;

(ii) if $\varphi_0, \varphi_1, \ldots, \varphi_{n+1}$ are the n-faces of Σ^{n+1} and if b is the barycenter of Σ^{n+1}, then $\operatorname{Sd}\Sigma^{n+1}$ consists of all the $(n+1)$-simplexes spanned by b and n-simplexes in $\operatorname{Sd}\varphi_i$, $i = 0, \ldots, n+1$.

It is plain that Σ^n is the union of the n-simplexes in $\operatorname{Sd}\Sigma^n$.

EXERCISES

6.5. Prove that $\operatorname{Sd}\Sigma^n$ consists of exactly $(n+1)!$ n-simplexes.

*6.6. (i) Every vertex b of $\operatorname{Sd}\Sigma^n$ is the barycenter of a unique face σ of Σ^n (denoted by $b = b^\sigma$).

(ii) Every n-simplex in $\operatorname{Sd}\Sigma^n$ has the form $[b^{\sigma_0}, b^{\sigma_1}, \ldots, b^{\sigma_n}]$, where each σ_i is an i-face of Σ^n and $\sigma_0 < \sigma_1 < \cdots < \sigma_n$.

Observe that even though an affine n-simplex may not be given with an orientation (i.e., an ordering of its vertices), Exercise 6.6(ii) shows that each n-simplex of $\operatorname{Sd}\Sigma^n$ comes equipped with an orientation.

Here is one last remark before we subdivide an arbitrary n-simplex $\sigma\colon \Delta^n \to X$. Recall that we saw in Exercise 2.10 that an (affine) n-simplex $[p_0, \ldots, p_n]$ is the cone over its ith face $[p_0, \ldots, \hat{p}_i, \ldots, p_n]$ with vertex p_i. This observation suggested the singular version, in a convex set, of the cone $b\,.\,\sigma$ over a singular n-simplex σ with vertex b (see Theorem 4.19).

Definition. Let E be a convex set. Then **barycentric subdivision** is a homomorphism $\operatorname{Sd}_n\colon S_n(E) \to S_n(E)$ defined inductively on generators $\tau\colon \Delta^n \to E$ as follows:

(i) If $n = 0$, then $\operatorname{Sd}_0(\tau) = \tau$;

(ii) if $n > 0$, then $\operatorname{Sd}_n(\tau) = \tau(b_n)\,.\,\operatorname{Sd}_{n-1}(\partial\tau)$, where b_n is the barycenter of Δ^n.

If X is any space, then the nth **barycentric subdivision**, for $n \geq 0$, is the homomorphism $\mathrm{Sd}_n \colon S_n(X) \to S_n(X)$ defined on generators $\sigma \colon \Delta^n \to X$ by

$$\mathrm{Sd}_n(\sigma) = \sigma_\# \, \mathrm{Sd}_n(\delta^n),$$

where $\delta^n \colon \Delta^n \to \Delta^n$ is the identity map.

Observe that $\sigma \colon \Delta^n \to X$ induces $\sigma_\# \colon S_n(\Delta^n) \to S_n(X)$, so that $\mathrm{Sd}_n(\sigma) = \sigma_\# \, \mathrm{Sd}_n(\delta^n)$ does make sense and does lie in $S_n(X)$. It is easy to see that both definitions of $\mathrm{Sd}_n(\sigma)$ agree when X is convex. If $\mathrm{Sd}_n(\delta^n) = \Sigma m_i \tau_i$, then $\mathrm{Sd}_n(\sigma) = \sigma_\#(\Sigma m_i \tau_i) = \Sigma m_i \sigma \tau_i$; thus one may view the n-simplexes τ_i as the smaller simplexes subdividing Δ^n, and one may view the n-simplexes $\sigma\tau_i$ as "restrictions" of σ to the τ_i that subdivide the image of σ.

Exercises

6.7. (i) Give explicit formulas for $\mathrm{Sd}_n(\delta^n)$ when $n = 1$ and $n = 2$.
(ii) Give explicit formulas for $\mathrm{Sd}_n(\sigma)$ whenever σ is an n-simplex in X and $n = 1$ and $n = 2$.

*6.8. If $f \colon X \to Y$ is continuous, prove that $\mathrm{Sd}\, f_\# = f_\# \, \mathrm{Sd}$, that is, the following diagram commutes for every $n \geq 0$:

$$\begin{array}{ccc} S_n(X) & \xrightarrow{\mathrm{Sd}} & S_n(X) \\ {\scriptstyle f_\#}\downarrow & & \downarrow{\scriptstyle f_\#} \\ S_n(Y) & \xrightarrow[\mathrm{Sd}]{} & S_n(Y). \end{array}$$

Lemma 6.12. $\mathrm{Sd} \colon S_*(X) \to S_*(X)$ *is a chain map.*

Proof. The proof is in two stages, according to the definition of Sd. Assume first that X is convex and that $\tau \colon \Delta^n \to X$ is an n-simplex. We prove, by induction on $n \geq 0$, that

$$\mathrm{Sd}_{n-1} \partial_n \tau = \partial_n \, \mathrm{Sd}_n \, \tau.$$

Since $\mathrm{Sd}_{-1} = 0$ (because $S_{-1}(X) = 0$) and $\partial_0 = 0$, the base step $n = 0$ is obvious. If $n > 0$, then

$$\begin{aligned} \partial_n \, \mathrm{Sd}_n \, \tau &= \partial_n(\tau(b_n) . \mathrm{Sd}_{n-1}(\partial_n \tau)) \quad \text{(definition of Sd)} \\ &= \mathrm{Sd}_{n-1} \partial_n \tau - \tau(b_n) . ((\partial_{n-1} \, \mathrm{Sd}_{n-1}) \partial_n \tau) \\ &\qquad \text{(Corollary 4.20(i): } \partial(b . \gamma) = \gamma - b . \partial\gamma) \\ &= \mathrm{Sd}_{n-1} \partial_n \tau - \tau(b_n) . ((\mathrm{Sd}_{n-2} \partial_{n-1}) \partial_n \tau) \quad \text{(induction)} \\ &= \mathrm{Sd}_{n-1} \partial_n \tau \quad (\partial\partial = 0). \end{aligned}$$

Now let X be any space, not necessarily convex. If $\sigma \colon \Delta^n \to X$ is an n-simplex, then

$$\begin{aligned}
\partial \operatorname{Sd}(\sigma) &= \partial\sigma_\# \operatorname{Sd}(\delta^n) && \text{(definition of Sd)}\\
&= \sigma_\# \partial \operatorname{Sd}(\delta^n) && (\sigma_\# \text{ is a chain map})\\
&= \sigma_\# \operatorname{Sd} \partial(\delta^n) && (\Delta^n \text{ is convex})\\
&= \operatorname{Sd} \sigma_\# \partial(\delta^n) && \text{(Exercise 6.8)}\\
&= \operatorname{Sd} \partial\sigma_\#(\delta^n) && (\sigma_\# \text{ is a chain map})\\
&= \operatorname{Sd} \partial\sigma && (\sigma_\#(\delta^n) = \sigma). \qquad \square
\end{aligned}$$

What is $z + B_n(X) \mapsto \operatorname{Sd} z + B_n(X)$, the homomorphism induced by Sd in homology?

Lemma 6.13. *For each $n \geq 0$, $H_n(\operatorname{Sd})\colon H_n(X) \to H_n(X)$ is the identity.*

Proof. It suffices (Theorem 5.3) to construct a chain homotopy between the chain maps Sd and 1, the identity on $S_*(X)$: we want homomorphisms $T_n\colon S_n(X) \to S_{n+1}(X)$ such that $\partial_{n+1} T_n + T_{n-1}\partial_n = 1_n - \operatorname{Sd}_n$.

Again the proof is in two steps. Assume first that X is convex; let us do an induction on n. If $n = 0$, define $T_0\colon S_0(X) \to S_1(X)$ as the zero map. If σ is a 0-simplex, then

$$0 = \partial T_0 \sigma \quad \text{and} \quad \sigma - \operatorname{Sd} \sigma = \sigma - \sigma = 0.$$

Assume that $n > 0$. If $\gamma \in S_n(X)$, then $T_n\gamma$ should satisfy

$$\partial T_n \gamma = \gamma - \operatorname{Sd} \gamma - T_{n-1}\partial\gamma.$$

Now the right-hand side is a cycle, because, using induction,

$$\partial(\gamma - \operatorname{Sd} \gamma - T_{n-1}\partial\gamma) = \partial\gamma - \partial \operatorname{Sd} \gamma - (1 - \operatorname{Sd} - T_{n-2}\partial)\partial\gamma = 0.$$

Since X is convex, the "integration formula", Corollary 4.20(ii), applies; define

$$T_n\gamma = b.(\gamma - \operatorname{Sd} \gamma - T_{n-1}\partial\gamma),$$

and note that $\partial T_n \gamma = \gamma - \operatorname{Sd} \gamma - T_{n-1}\partial\gamma$.

The remainder of the proof proceeds as the second stage of the preceding lemma. Let X be any space, not necessarily convex. If $\sigma\colon \Delta^n \to X$ is an n-simplex, define

$$T_n(\sigma) = \sigma_\# T_n(\delta^n) \in S_{n+1}(X),$$

where δ^n is the identity on Δ^n, and extend by linearity. We leave as an exercise that the T_n's constitute the desired chain homotopy. As a hint, one should show first that if $f\colon X \to Y$, then there is a commutative diagram

$$\begin{array}{ccc}
S_n(X) & \xrightarrow{f_\#} & S_n(Y)\\
{\scriptstyle T_n}\downarrow & & \downarrow{\scriptstyle T_n}\\
S_{n+1}(X) & \xrightarrow[f_\#]{} & S_{n+1}(Y).
\end{array} \qquad \square$$

Corollary 6.14. *If $q \geq 0$ is an integer and if $z \in Z_n(X)$, then*

$$\operatorname{cls} z = \operatorname{cls}(\operatorname{Sd}^q z).$$

PROOF. Since Sd induces the identity on $H_*(X)$, so does every composite $\operatorname{Sd}^q: z + B_n(X) \mapsto \operatorname{Sd}^q z + B_n(X)$. □

If E is a subspace of a euclidean space, then a continuous map $\sigma: \Delta^n \to E$ was called *affine* (in Chapter 2) if $\sigma(\sum t_i e_i) = \sum t_i \sigma(e_i)$, where $t_i \geq 0$ and $\sum t_i = 1$. Clearly, the identity $\delta^n: \Delta^n \to \Delta^n$ is affine.

Definition. If E is a subspace of euclidean space, then an n-chain $\gamma = \sum m_i \sigma_i \in S_n(E)$ is **affine** if each σ_i is affine.

EXERCISES

6.9. If σ is affine, then so is its ith face $\sigma\varepsilon_i$; moreover, the vertex set of $\sigma\varepsilon_i$, the set of all images of $e_0, e_1, \ldots, e_n$, is contained in the vertex set of σ. Conclude that $\partial\sigma$ is affine whenever σ is affine.

6.10. If E is convex and σ is affine, then so is the cone $b.\sigma$, where $b \in E$; moreover, the vertex set of $b.\sigma$ is the union of $\{b\}$ and the vertex set of σ. Conclude that Sd σ is affine whenever σ is affine.

Definition. If E is a subspace of some euclidean space, and if $\gamma = \sum m_j \sigma_j \in S_n(E)$, where all $m_j \neq 0$, then

$$\textbf{mesh } \gamma = \sup_j \{\operatorname{diam} \sigma_j(\Delta^n)\}$$

(note that $\sigma_j(\Delta^n)$ is compact (because Δ^n is) and hence has finite diameter).

Using Theorem 2.9, the reader may show that if E is a subspace of some euclidean space and γ is an affine n-chain in E, then $\operatorname{mesh}(\operatorname{Sd} \gamma) \leq [n/(n+1)] \operatorname{mesh} \gamma$. Iteration gives the next result.

Theorem 6.15. *If E is a subspace of some euclidean space and γ is an affine n-chain in E, then for all integers $q \geq 1$,*

$$\operatorname{mesh} \operatorname{Sd}^q \gamma \leq (n/n+1)^q \operatorname{mesh} \gamma.$$

This last theorem is fundamental; it says that the mesh of an affine chain, for example, $\delta^n: \Delta^n \to \Delta^n$, can be made arbitrarily small by repeated barycentric subdivision ($\lim_{q\to\infty} (n/n+1)^q = 0$ because $n/n+1 < 1$).

After this discussion of various features of barycentric subdivision, let us return to the proof of excision. Recall that we have only to show that the inclusion $S_*(X_1) + S_*(X_2) \hookrightarrow S_*(X)$ induces isomorphisms in homology whenever $X = X_1^\circ \cup X_2^\circ$.

Lemma 6.16. *If X_1 and X_2 are subspaces of X with $X = X_1^\circ \cup X_2^\circ$, and if σ is an n-simplex in X, then there exists an integer $q \geq 1$ with*

$$\mathrm{Sd}^q\, \sigma \in S_n(X_1) + S_n(X_2).$$

PROOF. Since $\sigma\colon \Delta^n \to X$ is continuous, $\{\sigma^{-1}(X_1^\circ), \sigma^{-1}(X_2^\circ)\}$ is an open cover of Δ^n. Since Δ^n is compact metric, this open cover has a Lebesgue number $\lambda > 0$: whenever $x, y \in \Delta^n$ satisfy $\|x - y\| < \lambda$, then there is an $i = 1, 2$ with $\sigma^{-1}(X_i^\circ)$ containing both x and y. Choose $q \geq 1$ with $(n/n + 1)^q$ diam $\Delta^n < \lambda$. Since the identity $\delta^n\colon \Delta^n \to \Delta^n$ is an affine n-simplex in Δ^n, Theorem 6.15 says that such a choice of q forces

$$\text{mesh } \mathrm{Sd}^q(\delta^n) < \lambda.$$

If $\mathrm{Sd}^q(\delta^n) = \sum m_j \tau_j$, then diam $\tau_j(\Delta^n) < \lambda$ for every j; hence $\tau_j(\Delta^n) \subset \sigma^{-1}(X_i^\circ)$ for some $i = i(\tau_j) \in \{1, 2\}$. Now $\mathrm{Sd}^q\, \sigma = \sigma_\# \mathrm{Sd}^q(\delta^n) = \sigma_\# \Sigma m_j \tau_j = \sum m_j \sigma \tau_j$; therefore $\sigma\tau_j(\Delta^n) \subset X_i^\circ \subset X_i$ (where $i = i(\tau_j)$) for every j. After collecting terms, $\mathrm{Sd}^q\, \sigma$ can be written $\gamma_1 + \gamma_2$, where $\gamma_i \in S_n(X_i)$. □

Theorem 6.17 (Excision). *If $X = X_1^\circ \cup X_2^\circ$, then the inclusion $j\colon (X_1, X_1 \cap X_2) \hookrightarrow (X, X_2)$ induces isomorphisms, for all $n \geq 0$,*

$$H_n(X_1, X_1 \cap X_2) \xrightarrow{\sim} H_n(X, X_2).$$

PROOF. By Lemma 6.11, it suffices to show that the maps

$$\theta_n\colon H_n(S_*(X_1) + S_*(X_2)) \to H_n(S_*(X)) = H_n(X)$$

induced by the inclusion $S_*(X_1) + S_*(X_2) \hookrightarrow S_*(X)$ are isomorphisms. If $\gamma_i \in S_n(X_i)$ for $i = 1, 2$, and if $\gamma_1 + \gamma_2$ is a cycle in the subcomplex (hence in $S_*(X)$), then

$$\theta\colon [\gamma_1 + \gamma_2] \mapsto \mathrm{cls}(\gamma_1 + \gamma_2),$$

where we denote the homology class in the subcomplex by [ ].

θ is surjective. Let cls $z \in H_n(X)$. By Lemma 6.16, there is an integer $q \geq 1$ with $\mathrm{Sd}^q\, z = \gamma_1 + \gamma_2$, where $\gamma_i \in S_n(X_i)$ for $i = 1, 2$. Since z is a cycle and Sd^q is a chain map, it follows that $\mathrm{Sd}^q\, z = \gamma_1 + \gamma_2$ is a cycle. Therefore $[\gamma_1 + \gamma_2]$ is an element of the nth homology group of the subcomplex, and $\theta([\gamma_1 + \gamma_2]) = \mathrm{cls}(\gamma_1 + \gamma_2) = \mathrm{cls}(\mathrm{Sd}^q\, z) = \mathrm{cls}\, z$, by Corollary 6.14.

θ is injective. Suppose that $[\gamma_1 + \gamma_2] \in \ker \theta$; then $\mathrm{cls}(\gamma_1 + \gamma_2) = 0$, so there is $\beta \in S_{n+1}(X)$ with $\partial\beta = \gamma_1 + \gamma_2$. By Lemma 6.16, there is an integer $q \geq 1$ with $\mathrm{Sd}^q\, \beta = \beta_1 + \beta_2$, where $\beta_i \in S_{n+1}(X_i)$ for $i = 1, 2$. Hence $\partial(\beta_1 + \beta_2) = \partial \mathrm{Sd}^q\, \beta = \mathrm{Sd}^q\, \partial\beta = \mathrm{Sd}^q(\gamma_1 + \gamma_2)$ (because Sd^q is a chain map). It follows that $[\mathrm{Sd}^q(\gamma_1 + \gamma_2)] = 0$. However, we know only that cls $\mathrm{Sd}^q(\gamma_1 + \gamma_2) = \mathrm{cls}(\gamma_1 + \gamma_2)$; we do not yet know that $[\mathrm{Sd}^q(\gamma_1 + \gamma_2)] = [\gamma_1 + \gamma_2]$.

By Exercise 6.8 applied to the inclusion map $X_i \hookrightarrow X$, one sees that $\mathrm{Sd}\colon S_*(X) \to S_*(X)$ carries $S_*(X_i)$ into $S_*(X_i)$ for $i = 1, 2$; hence Sd carries the subcomplex $S_*(X_1) + S_*(X_2)$ into itself. Moreover, the contracting homotopy

$\{T_n: S_n(X) \to S_{n+1}(X)\}$ of Lemma 6.13 restricts to contracting homotopies T' and T'' of $S_*(X_1)$ and $S_*(X_2)$, respectively (inspect the definition). Therefore $\gamma_1 - \mathrm{Sd}^q\,\gamma_1 = (T'\partial + \partial T')\gamma_1$ and $\gamma_2 - \mathrm{Sd}^q\,\gamma_2 = (T''\partial + \partial T'')\gamma_2$, hence

$$\gamma_1 + \gamma_2 - \mathrm{Sd}^q(\gamma_1 + \gamma_2) = T'\partial\gamma_1 + T''\partial\gamma_2 + \partial(T'\gamma_1 + T''\gamma_2).$$

$T'\partial\gamma_1 + T''\partial\gamma_2 = T\partial(\gamma_1 + \gamma_2)$, because T' and T'' are restrictions of T, and so it is 0, since $\gamma_1 + \gamma_2$ is a cycle. Because $\partial(T'\gamma_1 + T''\gamma_2) \in B_n(S_*(X_1) + S_*(X_2))$, it follows that $[\gamma_1 + \gamma_2] = [\mathrm{Sd}^q(\gamma_1 + \gamma_2)]$. But $[\mathrm{Sd}^q(\gamma_1 + \gamma_2)] = [\partial(\beta_1 + \beta_2)] = 0$. □

Having completed the proof of excision, we may now accept the Mayer–Vietoris theorem and the calculation of the homology groups of the spheres. We record two useful facts before giving more applications.

Lemma 6.18. *Let $X = X_1^\circ \cup X_2^\circ$, let $i_j: X_1 \cap X_2 \hookrightarrow X_j$ be inclusions for $j = 1, 2$, and let* cls $z \in H_n(X_1 \cap X_2)$. *If $H_{n+1}(X) = 0$, then* cls $z = 0$ *if and only if* i_{1*} cls $z = 0$ and i_{2*} cls $z = 0$.

Proof. Consider the portion of the Mayer–Vietoris sequence

$$\cdots \longrightarrow H_{n+1}(X) \longrightarrow H_n(X_1 \cap X_2) \xrightarrow{(i_{1*},\, i_{2*})} H_n(X_1) \oplus H_n(X_2).$$

Since $H_{n+1}(X) = 0$, the map (i_{1*}, i_{2*}) is injective. Thus cls $z = 0$ if and only if i_{1*} cls $z = 0$ in $H_n(X_1)$ and i_{2*} cls $z = 0$ in $H_n(X_2)$. □

Lemma 6.19. *Assume that $X = X_1^\circ \cup X_2^\circ$. Then each n-cycle z in X is homologous to a cycle of the form $\gamma_1 + \gamma_2$, where $\gamma_i \in S_n(X_i)$. Moreover, if $D: H_n(X) \to H_{n-1}(X_1 \cap X_2)$ is the connecting homomorphism in the Mayer–Vietoris sequence, then*

$$D(\mathrm{cls}\ z) = D(\mathrm{cls}(\gamma_1 + \gamma_2)) = \mathrm{cls}(\partial\gamma_1).$$

Remark. Of course, one may interchange X_1 and X_2.

Proof. That cls $z = \mathrm{cls}(\gamma_1 + \gamma_2)$ has already been proved (in Theorem 6.17). To see the last assertion, consider the diagram

$$\begin{array}{ccc} & & S_n(X_1)/S_n(X_1 \cap X_2) \\ & & \Big\downarrow h_\# \\ S_n(X) & \xrightarrow{q_\#} & S_n(X_1) + S_n(X_2)/S_n(X_2), \end{array}$$

where $q_\#$ sends $\gamma_1 + \gamma_2$ into its coset mod $S_n(X_2)$, and $h_\#$ is the isomorphism of the second isomorphism theorem; hence $h_\#^{-1}$ sends the coset $\gamma_1 + \gamma_2 + S_n(X_2)$ to $\gamma_1 + S_n(X_1 \cap X_2)$. The formula for D in Theorem 6.3 is $D = dh_\#^{-1}q_\#$, so that D cls$(\gamma_1 + \gamma_2)$ is $d(\gamma_1 + S_n(X_1 \cap X_2))$, where d is the connecting homomorphism from the exact sequence

$$0 \longrightarrow S_*(X_1 \cap X_2) \xrightarrow{i_\#} S_*(X_1) \xrightarrow{j_\#} S_*(X_1)/S_*(X_1 \cap X_2) \longrightarrow 0.$$

Now $d = i_{\#}^{-1}\partial j_{\#}^{-1}$ (Lemma 5.5); thus one lifts $\gamma_1 + S_n(X_1 \cap X_2)$ to γ_1, pushes down to $\partial\gamma_1$, and regards $\partial\gamma_1$ as being in $X_1 \cap X_2$. Hence $D\,\mathrm{cls}(\gamma_1 + \gamma_2) = \mathrm{cls}(\partial\gamma_1)$ in $H_{n-1}(X_1 \cap X_2)$. □

More Applications to Euclidean Space

Recall that if $h: \mathbf{Z} \to \mathbf{Z}$ is a homomorphism, then h is multiplication by some integer m: $h(n) = mn$ for all $n \in \mathbf{Z}$ (indeed $m = h(1)$).

Definition. A continuous map $f: S^n \to S^n$ (where $n > 0$) has **degree** m, denoted by $d(f) = m$, if $f_*: H_n(S^n) \to H_n(S^n)$ is multiplication by m.

Recall that we discussed a notion of degree for maps $f: S^1 \to S^1$ (denoted by $\deg f$) in terms of fundamental groups.

Theorem 6.20. *If $f: S^1 \to S^1$, then* $\deg(f) = d(f)$.

Proof. By Exercise 4.13, there is a commutative diagram

$$\begin{array}{ccc} \pi_1(S^1, 1) & \xrightarrow{f_*} & \pi_1(S^1, f(1)) \\ \downarrow \varphi & & \downarrow \varphi \\ \tilde{H}_1(S^1) & \xrightarrow[f_*]{} & \tilde{H}_1(S^1), \end{array}$$

where φ is the Hurewicz map. Since $\pi_1(S^1) \cong \mathbf{Z}$ is abelian, we know that φ is an isomorphism (Theorem 4.29). Finally, use Exercise 3.14, which says that one may view $f_*: \pi_1(S^1, 1) \to \pi_1(S^1, f(1))$ as multiplication by $\deg(f)$. □

Lemma 6.21. *Let $f, g: S^n \to S^n$ be continuous maps.*

(i) $d(g \circ f) = d(g)d(f)$.
(ii) $d(1_{S^n}) = 1$.
(iii) *If f is constant, then* $d(f) = 0$.
(iv) *If $f \simeq g$, then* $d(f) = d(g)$.[1]
(v) *If f is a homotopy equivalence, then* $d(f) = \pm 1$.

Proof. All parts follows from the fact that H_n is a functor defined on **hTop** (and that $H_n(S^n) = \mathbf{Z}$); in particular, (iii) follows from the existence of a com-

[1] The converse is also true, and it is a theorem of Brouwer (see [Spanier, p. 398]). We know the converse when $n = 1$ (Corollary 3.18). A theorem of Hopf (see [Hu (1959), p. 53]) generalizes this by classifying all homotopy classes of maps $X \to S^n$, where X is an n-dimensional polyhedron, in terms of the cohomology $H^n(X; \mathbf{Z})$.

mutative diagram

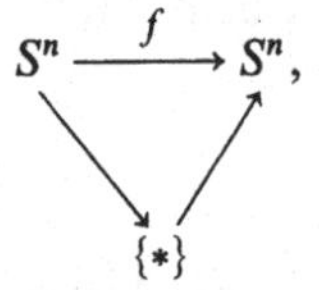

where $\{*\}$ is a one-point space. □

Using degrees, one may give another proof of Theorem 6.10: S^n is not contractible. Otherwise, $1_{S^n} \simeq c$, where $c: S^n \to S^n$ is some constant map, and these two maps would have the same degree, by Lemma 6.21(iv); but Lemma 6.21, parts (ii) and (iii), show that this is not so.

Computation of the degree of a map is facilitated if one has an explicit generator of $H_n(S^n)$. The next result exhibits a generator when $n = 1$.

Theorem 6.22. *Let $x = (-1, 0)$ and $y = (1, 0) \in S^1$, let σ be the (northerly) path in S^1 from y to x, and let τ be the (southerly) path in S^1 from x to y. Then $\sigma + \tau$ is a 1-cycle in S^1 whose homology class generates $H_1(S^1)$.*

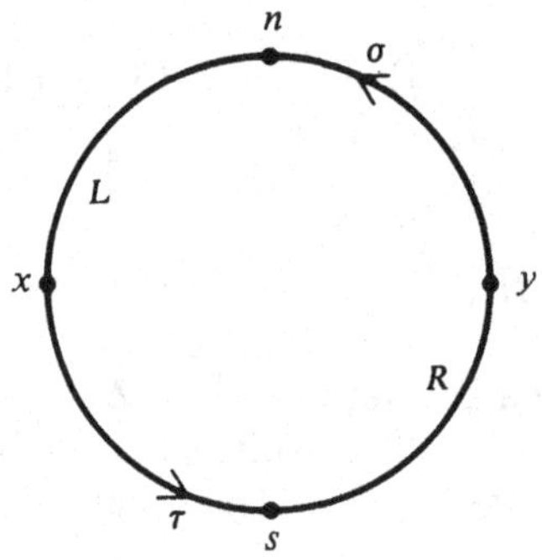

Proof. First, $\sigma + \tau$ is a 1-cycle, because

$$\partial(\sigma + \tau) = \partial\sigma + \partial\tau = (x - y) + (y - x) = 0.$$

Let $n = (0, 1)$ and $s = (0, -1)$ be the north and south poles; let $X_1 = S^1 - \{n\}$ and $X_2 = S^1 - \{s\}$. Note that $S^1 = X_1^\circ \cup X_2^\circ$, each X_i is contractible, and $X_1 \cap X_2 = S^1 - \{n, s\}$ consists of two disjoint open arcs L and R with $x \in L$ and $y \in R$. The Mayer–Vietoris theorem for reduced homology provides exactness of

$$\tilde{H}_1(X_1) \oplus \tilde{H}_1(X_2) \to \tilde{H}_1(S^1) \xrightarrow{D} \tilde{H}_0(X_1 \cap X_2) \to \tilde{H}_0(X_1) \oplus \tilde{H}_0(X_2).$$

Now D is an isomorphism, because contractibility of X_1 and X_2 forces both direct sums to be zero. Since $X_1 \cap X_2 = L \cup R$, Corollary 5.18 gives $\tilde{H}_0(X_1 \cap X_2)$ infinite cyclic with generator $\mathrm{cls}(x - y)$. But Lemma 6.19 shows that $D\,\mathrm{cls}(\sigma + \tau) = \mathrm{cls}\,\partial\sigma = \mathrm{cls}(x - y)$; it follows that $\mathrm{cls}(\sigma + \tau)$ generates $\tilde{H}_1(S^1) = H_1(S^1)$. □

Remark. One can show that a (simple) closed path generates $H_1(S^1)$, but we need Theorem 6.22 as stated.

Definition. If $x = (x_1, \ldots, x_{n+1}) \in S^n$, its **antipode** is $-x = (-x_1, \ldots, -x_{n+1})$. The **antipodal map** $a = a^n\colon S^n \to S^n$ is defined by $x \mapsto -x$.

Note that the distance from x to $-x$ is 2, the diameter of S^n, so that $-x$ is indeed antipodal to x.

Theorem 6.23. *If $n \geq 1$, then the antipodal map $a^n\colon S^n \to S^n$ has degree $(-1)^{n+1}$.*

Proof. As a preliminary step, we show, by induction on n, that the map $f\colon S^n \to S^n$ given by $f(x_1, \ldots, x_{n+1}) = (-x_1, x_2, \ldots, x_{n+1})$ has degree -1. Recall that the north pole of S^n is $(0, 0, \ldots, 0, 1)$ and that the south pole is $(0, 0, \ldots, -1)$.

Let $n = 1$. Set $X_1 = S^1 - \{\text{north pole}\}$ and $X_2 = S^1 - \{\text{south pole}\}$. By Exercise 6.3, there is a commutative diagram from Mayer–Vietoris

$$\begin{array}{ccc} H_1(S^1) & \xrightarrow{D} & H_0(X_1 \cap X_2) \\ {\scriptstyle f_*}\downarrow & & \downarrow{\scriptstyle g_*} \\ H_1(S^1) & \xrightarrow{D} & H_0(X_1 \cap X_2), \end{array}$$

where g is the restriction of f (note that $f(X_i) \subset X_i$ for $i = 1, 2$). Observe that D is injective, for the preceding term in the Mayer–Vietoris sequence is $H_1(X_1) \oplus H_1(X_2) = 0$. By Theorem 6.22 and Lemma 6.19, $\mathrm{cls}(\sigma + \tau)$ is a generator of $H_1(S^1)$, and $D(\mathrm{cls}(\sigma + \tau)) = \mathrm{cls}\,\partial\sigma = \mathrm{cls}(x - y)$. Hence commutativity of the diagram above gives

$$\begin{aligned} Df_*\,\mathrm{cls}(\sigma + \tau) &= g_* D\,\mathrm{cls}(\sigma + \tau) = g_*\,\mathrm{cls}(x - y) = \mathrm{cls}(g(x) - g(y)) \\ &= \mathrm{cls}(y - x), \end{aligned}$$

because f (and hence g) interchanges x and y. But

$$\mathrm{cls}(y - x) = -D\,\mathrm{cls}(\sigma + \tau) = D\,\mathrm{cls}(-(\sigma + \tau)).$$

Since D is injective, we have $f_*\,\mathrm{cls}(\sigma + \tau) = -\mathrm{cls}(\sigma + \tau)$, and $d(f) = -1$.

For the inductive step, we may assume that $n \geq 2$. Let $X_1 = S^n - \{\text{north pole}\}$, let $X_2 = S^n - \{\text{south pole}\}$, and let $i\colon S^{n-1} \hookrightarrow X_1 \cap X_2$ be the inclusion of the equator. Since S^{n-1} is a deformation retract of $X_1 \cap X_2$, we know that $i_*\colon H_{n-1}(S^{n-1}) \to H_{n-1}(X_1 \cap X_2)$ is an isomorphism. If f' is the restriction of f to S^{n-1}, there is a commutative diagram

$$\begin{array}{ccccc} H_n(S^n) & \xrightarrow{D} & H_{n-1}(X_1 \cap X_2) & \xleftarrow{i_*} & H_{n-1}(S^{n-1}) \\ \downarrow{f_*} & & \downarrow{g_*} & & \downarrow{f'_*} \\ H_n(S^n) & \xrightarrow{D} & H_{n-1}(X_1 \cap X_2) & \xleftarrow{i_*} & H_{n-1}(S^{n-1}). \end{array}$$

Since $n \geq 2$, we know that D is an isomorphism (for the flanking terms in the Mayer–Vietoris sequence are 0 because X_1 and X_2 are contractible). We thus have $f_* = D^{-1} i_* f'_* i_*^{-1} D$. By induction, $d(f') = -1$, so that f'_* is multiplication by -1; the other factors cancel each other and so f_* is also multiplication by -1, that is, $d(f) = -1$.

The next step shows that there is nothing magical about changing the sign of the first coordinate: if $f_i \colon S^n \to S^n$ is defined by $f_i(x_1, \ldots, x_{n+1}) = (x_1, \ldots, -x_i, \ldots, x_{n+1})$, we claim that $d(f_i) = -1$ also. If $h \colon S^n \to S^n$ is the homeomorphism of S^n interchanging the first and ith coordinates, then $f_i = hfh$. Using Lemma 6.21, we see that

$$d(f_i) = d(hfh) = (d(h))^2 d(f) = (d(h))^2(-1).$$

As $d(h) = \pm 1$ (since h is a homeomorphism), we have $d(f_i) = -1$.

Finally, observe that the antipodal map a^n is the composite

$$a^n = ff_2 f_3 \cdots f_{n+1},$$

so that $d(a^n) = (-1)^{n+1}$, as desired. □

Another proof of this theorem is given as Corollary 9.24.

Theorem 6.24.

(i) *If $f \colon S^n \to S^n$ has no fixed points, then f is homotopic to the antipodal map $a = a^n$.*

(ii) *If $g \colon S^n \to S^n$ is nullhomotopic, then g has a fixed point.*

PROOF. (i) We can give a homotopy explicitly. Define $F \colon S^n \times \mathbf{I} \to S^n$ by

$$F(x, t) = \frac{(1 - t)a(x) + tf(x)}{\|(1 - t)a(x) + tf(x)\|}.$$

The right-hand side is a unit vector (hence lies in S^n) as long as $(1 - t)a(x) + tf(x) \neq 0$. Were this zero, then we would have

$$f(x) = (-(1 - t)/t)a(x).$$

Taking the norm of each side, noting that $\|f(x)\| = 1 = \|a(x)\|$, we see that $(1 - t)/t = 1$; therefore $f(x) = -a(x)$. But, by definition, $a(x) = -x$, so that $f(x) = x$, a contradiction.

(ii) If g has no fixed points, then $g \simeq a$, by part (i), and so $d(g) = d(a) = \pm 1$ (Lemma 6.21). But g nullhomotopic implies that $d(g) = 0$, a contradiction. □

Theorem 6.25. *If $f: S^{2n} \to S^{2n}$, then either f has a fixed point or some point is sent to its antipode.*

Proof. Assume that f has no fixed points. By Theorem 6.24, $f \simeq a^{2n}$; by Theorem 6.23, $d(f) = (-1)^{2n+1} = -1$. Suppose that $f(x) \neq -x$ for all $x \in S^{2n}$. If we define $g(x) = -f(x)$, we see that g has no fixed points, and so $-f = g \simeq a^{2n}$. It follows that $f \simeq -a^{2n} = 1_{S^{2n}}$ and $d(f) = 1$, a contradiction. □

This result is false for odd-dimensional spheres; for example, rotation $\rho: S^1 \to S^1$ about the origin through almost any angle has neither fixed points nor points sent into antipodes. More generally, regard a point $x \in S^{2n-1}$ as an n-tuple $(z_1, \ldots, z_n)$ of complex numbers, and define $f: S^{2n-1} \to S^{2n-1}$ by

$$(z_1, \ldots, z_n) \mapsto (\rho z_1, \rho z_2, \ldots, \rho z_n),$$

where $\rho: S^1 \to S^1$ is a suitable rotation.

Theorem 6.26. *There is no continuous $f: S^{2n} \to S^{2n}$ such that x and $f(x)$ are orthogonal for every x.*

Proof. If $f(x_0) = -x_0$ for some x_0, then their inner product $(x_0, f(x_0)) = -1$, contradicting the hypothesis that $(x_0, f(x_0)) = 0$. Hence f sends no point to its antipode, so that f must have a fixed point, say, x_1. But then x_1 is orthogonal to itself, contradicting $\|x_1\| = 1$. □

Theorem 6.26 is false for S^1; indeed it is false for every odd-dimensional sphere. If $x \in S^{2n-1}$, then $x = (x_1, x_2, \ldots, x_{2n-1}, x_{2n})$; define $f: S^{2n-1} \to S^{2n-1}$ by

$$f: x \mapsto (-x_2, x_1, -x_4, x_3, \ldots, -x_{2n}, x_{2n-1}).$$

Definition. A **vector field** on S^m is a continuous map $f: S^m \to \mathbf{R}^{m+1}$ with $f(x)$ tangent to S^m at x for every $x \in S^m$; one says that f is **nonzero** if $f(x) \neq 0$ for all x.

Corollary 6.27 (Hairy Ball Theorem). *There exists no nonzero vector field on S^{2n}.*

Proof. If $f: S^{2n} \to \mathbf{R}^{2n+1}$ is a nonzero vector field, then $g: S^{2n} \to S^{2n}$ defined by $x \mapsto f(x)/\|f\|$ is a continuous map with $g(x)$ tangent to S^{2n} at x for every x. □

A function $f: S^m \to \mathbf{R}^{m+1}$ may be viewed as a family of vectors with $f(x)$ attached to S^m at x (thus S^m is a "hairy ball"). If we say that a hair is "combed" if it lies flat, that is, if it is tangent to the sphere, then Theorem 6.27 can be interpreted as saying that one cannot comb the hair on an even-dimensional sphere.

Definition. A continuous map $g: S^m \to S^n$ is called **antipodal** if $ga^m = a^n g$, that is, if $g(-x) = -g(x)$ for all $x \in S^m$.

An antipodal map g thus carries the antipode of x to the antipode of $g(x)$. Hence, if $y = -x$, then $g(y) = -g(-y)$, that is, g maps antipodal pairs into antipodal pairs. Note that the antipodal map $a^n: S^n \to S^n$ is antipodal.

EXERCISES

*6.11. If $\gamma: \Delta^1 \to S^n$ is a "path" with $\gamma(e_1) = -\gamma(e_0)$, then $(1 + a^n_\#)\gamma$ is a 1-cycle in S^n. (*Note*: By "path" we mean that σ has domain $\Delta^1 = [-1, 1]$ instead of **I**.)

*6.12. If γ is a 1-chain in S^n, then

$$(1 + a^n_\#)(1 - a^n_\#)\gamma = 0.$$

*6.13. If β is a 1-chain in S^n, then

$$(1 + a^n_\#)(1 + a^n_\#)\beta = 2(1 + a^n_\#)\beta.$$

*6.14. If σ is the northerly path in S^1 from $y = (1, 0)$ to $a^1(y) = (-1, 0)$, then $(1 + a^1_\#)\sigma$ is a 1-cycle whose homology class generates $H_1(S^1)$. (*Hint*: Theorem 6.22.)

Theorem 6.28. *If $m > 1$, there exists no antipodal map $g: S^m \to S^1$.*

PROOF (after J. W. Walker). Assume that such a map g exists. Let $y = (1, 0) \in S^1$ and let σ be the northerly path in S^1 from y to $a^1(y) = (-1, 0)$. Choose a point $x_0 \in S^m$, and let λ be a path in S^m from x_0 to $-x_0$. Finally, choose a path f in S^1 from $g(x_0)$ to y. Now

$$\sigma - g_\#\lambda + f - af$$

is a 1-cycle in S^1, for its boundary is

$$(a^1 y - y) - (g(-x_0) - g(x_0)) + (y - g(x_0)) - (a^1 y - a^1 g(x_0)) = 0$$

(because g is antipodal). Let $\theta = 1 + a^1_\#$. Since cls $\theta\sigma$ is a generator of $H_1(S^1)$, by Exercise 6.14, there is some integer m with

$$\text{cls}(\sigma - g_\#\lambda + f - a^1 f) = m \text{ cls } \theta\sigma.$$

On the other hand, applying θ to this equation gives

$$\text{cls}(\theta\sigma - \theta g_\#\lambda) = 2m \text{ cls } \theta\sigma$$

(using Exercises 6.12 and 6.13). Therefore

$$\overline{\text{cls}(\theta\sigma)} = \overline{\text{cls}(\theta g_\#\lambda)} \quad \text{in } H_1(S^1)/2H_1(S^1)$$

(where bar denotes coset mod $2H_1(S^1)$). As cls$(\theta\sigma)$ is a generator of $H_1(S^1)$, it follows that $\overline{\text{cls}(\theta\sigma)}$ and $\overline{\text{cls}(\theta g_\#\lambda)}$ are nonzero in $H_1(S^1)/2H_1(S^1)$; therefore cls$(\theta g_\#\lambda) \neq 0$ in $H_1(S^1)$. But g is antipodal, so that $\theta g_\#\lambda = (1 + a^1_\#)g_\#\lambda = g_\#(1 + a^m_\#)\lambda$. Since $(1 + a^m_\#)\lambda$ is a 1-cycle in S^m,

Exercise 6.11, we must have $\text{cls}((1 + a^1_{\#})\lambda) \neq 0$ in $H_1(S^m)$. As $m > 1$, this contradicts $H_1(S^m) = 0$. □

Corollary 6.29 (Borsuk–Ulam). *Given a continuous* $f: S^2 \to \mathbf{R}^2$, *there exists* $x \in S^2$ *with* $f(x) = f(-x)$.

PROOF. If no such x exists, define $g: S^2 \to S^1$ by

$$g(x) = (f(x) - f(-x))/\|f(x) - f(-x)\|.$$

Clearly, g is an antipodal map, and this contradicts the theorem. □

Remark. In Chapter 12, we shall prove the more general version of Theorem 6.28 and its corollary: if $m > n$, then there is no antipodal map $S^m \to S^n$; if $f: S^m \to \mathbf{R}^m$, then there exists $x \in S^m$ with $f(x) = f(-x)$.

EXERCISES

6.15. Prove directly that if $f: S^1 \to \mathbf{R}$ is continuous, then there exists $x \in S^1$ with $f(x) = f(-x)$.

6.16. Prove that there is no homeomorphic copy of S^2 in the plane $\mathbf{R}^2$. This result says that a map of the earth cannot be drawn (homeomorphically) on a page of an atlas. (*Remark*: This result remains true if "2" is replaced by "n"; it follows from the general Borsuk–Ulam theorem.)

Corollary 6.30 (Lusternik–Schnirelmann). *If* $S^2 = F_1 \cup F_2 \cup F_3$, *where each* F_i *is closed, then some* F_i *contains a pair of antipodal points.*

PROOF. If $a^2: S^2 \to S^2$ is the antipodal map $x \mapsto -x$, then we may assume that $a^2(F_1) \cap F_1 = \varnothing = a^2(F_2) \cap F_2$, or we are done. By the Urysohn lemma, there are continuous maps $g_i: S^2 \to \mathbf{I}$, for $i = 1, 2$, with $g_i(F_i) = 0$ and $g_i(a^2 F_i) = 1$. Define $f: S^2 \to \mathbf{R}^2$ by

$$f(x) = (g_1(x), g_2(x)).$$

By Corollary 6.29, there exists $x_0 \in S^2$ with $f(x_0) = f(-x_0)$, that is, $g_i(x_0) = g_i(-x_0)$ for $i = 1, 2$. It follows that $x_0 \notin F_i$ for $i = 1, 2$, because $x \in F_i$ implies that $g_i(x_0) = 0$ and $g_i(-x_0) = 1$ (for $-x_0 = a^2(x_0) \in a^2 F_i$). Since $S^2 = F_1 \cup F_2 \cup F_3$, we must have $x_0 \in F_3$. A similar argument shows that $-x_0 \notin F_1 \cup F_2$, hence $-x_0 \in F_3$, as desired. □

EXERCISES

6.17. If $f: S^2 \to \mathbf{R}^2$ satisfies $f(-x) = -f(x)$ for all x, then there exists $x_0 \in S^2$ with $f(x_0) = 0$.

6.18. Assume that there is no antipodal map $S^m \to S^n$ for $m > n$. Prove that if $f: S^n \to \mathbf{R}^n$, then there exists $x_0 \in S^n$ with $f(x_0) = f(-x_0)$.

6.19. Assume that there is no antipodal map $S^m \to S^n$ for $m > n$. Prove that if S^n is the union of $n + 1$ closed subsets $F_1, F_2, \ldots, F_{n+1}$, then at least one F_i contains a pair of antipodal points. Prove that this conclusion is false if one replaces $n+1$ by $n+2$.

We now prepare for the Jordan–Brouwer separation theorem.

Definition. If $r \geq 0$, a **(closed) r-cell** e_r is a homeomorphic copy of $\mathbf{I}^r$, the cartesian product of r copies of $\mathbf{I}$. In particular, e_0 is a point.

Theorem 6.31. *If S^n contains an r-cell e_r, then $S^n - e_r$ is acyclic:*

$$\tilde{H}_q(S^n - e_r) = 0 \quad \textit{for all } q.$$

Proof. We prove the theorem by induction on $r \geq 0$. If $r = 0$, then e_0 is a point, $S^n - e_0 \approx \mathbf{R}^n$ (stereographic projection), and so $S^n - e_0$ is contractible; the result follows.

Suppose that $r > 0$, let $B = \mathbf{I}^{r-1}$, and let $h\colon \mathbf{I}^r = B \times \mathbf{I} \to e_r$ be a homeomorphism. Define $e' = h(B \times [0, \frac{1}{2}])$ and $e'' = h(B \times [\frac{1}{2}, 1])$. Then $e_r = e' \cup e''$ while $e' \cap e'' = h(B \times \{\frac{1}{2}\})$ is an $(r-1)$-cell. By induction, $(S^n - e') \cup (S^n - e'') = S^n - (e' \cap e'')$ is acyclic. Since $S^n - e'$ and $S^n - e''$ are open subsets, Mayer–Vietoris for reduced homology gives exactness of

$$\tilde{H}_{q+1}(S^n - (e' \cap e'')) \to \tilde{H}_q(S^n - (e' \cup e'')) \to \tilde{H}_q(S^n - e') \oplus \tilde{H}_q(S^n - e'') \to \tilde{H}_q(S^n - (e' \cap e'')).$$

The outside terms being zero and $S^n - (e' \cup e'') = S^n - e_r$ give an isomorphism

$$\tilde{H}_q(S^n - e_r) \cong \tilde{H}_q(S^n - e') \oplus \tilde{H}_q(S^n - e'').$$

Assume that $\operatorname{cls} \zeta \in \tilde{H}_q(S^n - e_r)$ and $\operatorname{cls} \zeta \neq 0$; we shall reach a contradiction. Now Lemma 6.18 gives either $i'_* \operatorname{cls} \zeta \neq 0$ or $i''_* \operatorname{cls} \zeta \neq 0$, where $i'\colon S^n - e_r \hookrightarrow S^n - e'$ and $i''\colon S^n - e_r \hookrightarrow S^n - e''$ are inclusions. Assume that $i'_* \operatorname{cls} \zeta \neq 0$, and define $E^1 = e'$. We have thus constructed an r-cell $E^1 \subset e_r$ such that the inclusion $i\colon S^n - e_r \hookrightarrow S^n - E^1$ satisfies $i_* \operatorname{cls} \zeta \neq 0$. Repeat this construction with $B \times \mathbf{I}$ replaced by $B \times [0, \frac{1}{2}]$ and with $[0, \frac{1}{2}]$ bisected. Iterating, we see that there is a descending sequence of r-cells

$$e_r \supset E^1 \supset E^2 \supset \cdots \supset E^p \supset E^{p+1} \supset \cdots,$$

with $E^p = h(B \times J^p)$ (where $J^p \subset J^{p-1}$ is a subinterval of length 2^{-p}), with $i^p_* \operatorname{cls} \zeta \neq 0$ (where $i^p\colon S^n - e_r \hookrightarrow S^n - E^p$ is inclusion), and with $\bigcap E^p$ an $(r-1)$-cell, namely, $h(B \times \{\text{point}\})$.

We are going to apply Theorem 4.18. There is a commutative diagram with all arrows inclusions:

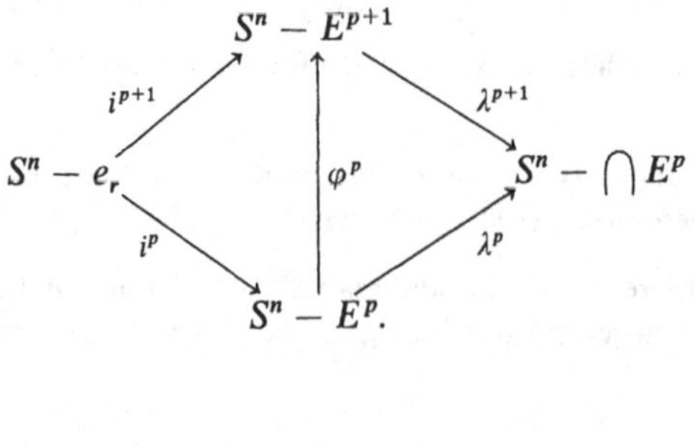

Since $i_*^p \text{ cls } \zeta \neq 0$ for all p, it follows that $\varphi_*^p(i_*^p \text{ cls } \zeta) \neq 0$ for all p. Now assume that A is a compact subset of $S^n - \bigcap E^p = \bigcup(S^n - E^p)$, an ascending union of open sets in S^n. This open cover of A has a finite subcover, that is, $A \subset S^n - E^p$ for some p. Theorem 4.18 now applies to give $\lambda_*^k i_*^k \text{ cls } \zeta \neq 0$ in $\tilde{H}_q(S^n - \bigcap E^p)$ for all k. But $\tilde{H}_q(S^n - \bigcap E^p) = 0$, by induction, because $\bigcap E^p$ is an $(r-1)$-cell, and we have reached a contradiction. Therefore $\text{cls } \zeta = 0$, that is, $\tilde{H}_q(S^n - e_r) = 0$, and $S^n - e_r$ is acyclic. □

Corollary 6.32. *If e_r is a closed r-cell in S^n, then $S^n - e_r$ is path connected.*

PROOF. $\tilde{H}_0(S^n - e_r) = 0$, and so Corollary 5.15 applies. □

Theorem 6.33. *Let s_r be a homeomorphic copy of S^r in S^n, where $n > 0$. Then*

$$\tilde{H}_q(S^n - s_r) = \begin{cases} \mathbf{Z} & \text{if } q = n - r - 1 \\ 0 & \text{otherwise.} \end{cases}$$

PROOF. We do an induction on r. If $r = 0$, then s_0 consists of two points and $S^n - s_0$ has the same homotopy type as S^{n-1} (think of s_0 as the north and south poles, and deform $S^n - s_0$ to the equator). Hence

$$\tilde{H}_q(S^n - s_0) \cong \tilde{H}_q(S^{n-1}),$$

and this is 0 for $q \neq n - 1$, and $\mathbf{Z}$ for $q = n - 1$, as desired.

Assume that $r > 0$, and let $\varphi: S^r \to s_r$ be a homeomorphism. Write $S^r = E^+ \cup E^-$, where E^+ is the closed northern hemisphere and E^- is the closed southern hemisphere. Note that $E^+ \cap E^- = S^{r-1}$, the equator. If $e' = \varphi(E^+)$ and $e'' = \varphi(E^-)$, then it is an easy exercise to show that e' and e'' are closed r-cells in S^n.

Define $X_1 = S^n - e'$ and $X_2 = S^n - e''$; then X_1 and X_2 are open subsets of S^n, hence $X_1 \cup X_2 = X_1^\circ \cup X_2^\circ$. Furthermore,

$$X_1 \cup X_2 = (S^n - e') \cup (S^n - e'') = S^n - (e' \cap e'') = S^n - s_{r-1}.$$

We also have

$$X_1 \cap X_2 = (S^n - e') \cap (S^n - e'') = S^n - (e' \cup e'') = S^n - s_r.$$

There is an exact Mayer–Vietoris sequence

$$\tilde{H}_{q+1}(S^n - e') \oplus \tilde{H}_{q+1}(S^n - e'') \to \tilde{H}_{q+1}(S^n - s_{r-1}) \to \tilde{H}_q(S^n - s_r) \to \tilde{H}_q(S^n - e') \oplus \tilde{H}_q(S^n - e'').$$

By Theorem 6.31, the flanking (direct sum) terms are 0, so that

$$\tilde{H}_{q+1}(S^n - s_{r-1}) \cong \tilde{H}_q(S^n - s_r).$$

By induction,

$$\tilde{H}_{q+1}(S^n - s_{r-1}) = \begin{cases} \mathbf{Z} & \text{if } q + 1 = n - (r-1) - 1 \\ 0 & \text{otherwise,} \end{cases}$$

and this gives

$$\tilde{H}_q(S^n - s_r) = \begin{cases} \mathbf{Z} & \text{if } q = n - r - 1 \\ 0 & \text{otherwise.} \end{cases}$$ □

Corollary 6.34. *Let s_r be contained in S^n. If $r \neq n - 1$, then $S^n - s_r$ is path connected; if $r = n - 1$, then $S^n - s_r$ has exactly two path components.*

Proof. Compute $\tilde{H}_0(S^n - s_r)$ by the theorem, and apply Corollary 5.15. □

We have shown that one cannot disconnect S^n by removing a homeomorph of a sphere of dimension $\leq n - 2$. Since S^n is the one-point compactification of $\mathbf{R}^n$, it follows that one cannot disconnect $\mathbf{R}^n$ by removing s_r with $r \leq n - 2$.

Definition. Let U be a subspace of a space X. An element $x \in X$ is a **boundary point** of U if every neighborhood of x meets both U and $X - U$. The **boundary** of U (or the **frontier** of U), denoted by $\dot{U}$, is the set of all boundary points of U.

Clearly, $\dot{U}$ depends on the ambient space X; moreover, the closure $\bar{U}$ of U is just $U \cup \dot{U}$, while U open implies that $\dot{U} = \bar{U} - U$.

Theorem 6.35 (Jordan–Brouwer[2] Separation Theorem). *If s_{n-1} is a subspace of S^n that is homeomorphic to S^{n-1}, then $S^n - s_{n-1}$ has exactly two components, and s_{n-1} is their common boundary.*

Proof. Denote s_{n-1} by s. By Corollary 6.34, $S^n - s$ has exactly two path components, say, U and V. By Exercise 1.28, $S^n - s$ is locally path connected, and so Corollary 1.20 shows that U and V are components; by Theorem 1.18, U and V are open sets of $S^n - s$ and hence are open in S^n.

Since V is open in S^n, $S^n - V = U \cup s$ is a closed set containing U; hence $\bar{U} \subset U \cup s$, and so $\dot{U} = \bar{U} - U \subset s$. A similar argument shows that $\dot{V} \subset s$. For the reverse inclusion, let $x \in s$ and let N be an open neighborhood of x. Clearly, N meets $S^n - U = V \cup s$; to show that $x \in \dot{U}$, it remains to prove that N meets U. Now every nonempty open subset of S^{n-1} contains an (open) subset D whose complement is an $(n-1)$-cell (because every open set contains a homeomorph of $\mathbf{R}^{n-1}$); since $s \approx S^{n-1}$, there exists a subset A of $N \cap s$ with $s - A$ a closed $(n-1)$-cell. By Theorem 6.31, $\tilde{H}_0(S^n - (s - A)) = 0$, hence $S^n - (s - A)$ is path connected. If $u \in U$ and $v \in V$, there exists a path f in $S^n - (s - A)$ from u to v. Since u and v lie in distinct path components of $S^n - s$, we must have $f(\mathbf{I}) \cap A \neq \varnothing$. But $f(\mathbf{I}) \cap A = f(\mathbf{I}) \cap s$: $f(\mathbf{I}) \cap A \subset f(\mathbf{I}) \cap s$ because $A \subset s$; for the reverse inclusion,

[2] The special case $n = 2$ is called the Jordan curve theorem; it was conjectured by Jordan but proved by Veblen; Theorem 6.35 was later proved by Brouwer.

$$f(\mathbf{I}) \cap s \subset f(\mathbf{I}) \cap (S^n - (s - A)) \cap s \subset f(\mathbf{I}) \cap A.$$

Hence, if $t_0 = \inf\{t \in \mathbf{I}: f(t) \in s\}$, then $t_0 = \inf\{t \in \mathbf{I}: f(t) \in A\}$. Thus $f(t_0) \in f(\mathbf{I}) \cap A \subset N$. If $J = [0, t_0)$, then $f(J)$ is a connected set containing $u = f(0)$ and with $f(J) \subset f(\mathbf{I}) \cap S^n - s = f(\mathbf{I}) \cap (U \cup V)$; it follows that $f(J) \subset U$. Therefore any open neighborhood of $f(t_0)$ in N meets U, as desired. A similar argument using $t_1 = \sup\{t \in \mathbf{I}: f(t) \in s\}$ shows that N meets V as well. □

If we regard $S^n = \mathbf{R}^n \cup \{\infty\}$, and if $\infty \notin s_{n-1}$, then that component of $S^n - s_{n-1}$ containing ∞ is called the **outside** of s_{n-1}, and the other component is called the **inside**. Is the inside of s_{n-1} homeomorphic to an open ball (i.e., the interior of D^n)? When $n = 2$, then s_1 is called a **Jordan curve**, and the **Schoenflies theorem** gives an affirmative answer: the inside of a Jordan curve is homeomorphic to the interior of D^2. However, for $n = 3$, Alexander gave an example (the "horned sphere") showing that the inside need not be homeomorphic to an open ball (the interior of D^3) (see [Hocking and Young, p. 176]). Alexander's example can be modified to show the same phenomenon of bad insides can occur for all $n \geq 3$.

Let us mention a famous example (the **lakes of Wada**), which comes very close to the Jordan curve theorem. There exists a compact connected subset K of $\mathbf{R}^2$ whose complement $\mathbf{R}^2 - K$ has three components U_1, U_2, U_3 and $K = \dot{U}_i$ for $i = 1, 2, 3$. Of course, K is not a Jordan curve, otherwise its complement would have two components. (See [Kosniowski, p. 100] for details of this example.)

Here is another important theorem of Brouwer.

Theorem 6.36 (Invariance of Domain). *Let U and V be subsets of S^n having a homeomorphism $h: U \to V$. If U is open, then V is open.*

PROOF. Let $y \in V$ and let $h(x) = y$. Take a closed neighborhood N of x in U with $N \approx \mathbf{I}^n$ and $\dot{N} \approx S^{n-1}$; of course, $h(N) \subset V$. Now N and $h(N)$ are closed n-cells, so that Theorem 6.31 says that $S^n - h(N)$ is connected. On the other hand, $S^n - h(\dot{N})$ has two components, by Theorem 6.35. Since

$$S^n - h(\dot{N}) = (S^n - h(N)) \cup (h(N) - h(\dot{N}))$$

and the two terms on the right are disjoint, nonempty, and connected, they must be the components of $S^n - h(\dot{N})$. It follows that each is open in $S^n - h(\dot{N})$; in particular, $h(N) - h(\dot{N})$ is open in $S^n - h(\dot{N})$ and hence is open in S^n. But $y \in h(N) - h(\dot{N}) \subset V$; since $h(\dot{N})$ is the boundary of each component, it follows that y is an interior point of V. Therefore, V is open in S^n. □

For more applications to euclidean space, we recommend [Eilenberg and Steenrod, Chap. XI].

EXERCISES

6.20. Show that S^n is not homeomorphic to any proper subspace of itself. (*Hint*: Use compactness of S^n and invariance of domain.)

*6.21. Prove invariance of domain if the ambient space S^n is replaced by $\mathbf{R}^n$.

*6.22. If invariance of domain holds with ambient space X, then show it holds with any ambient space homeomorphic to X.

6.23. Show that invariance of domain does not hold with ambient space D^n.

CHAPTER 7

Simplicial Complexes

Definitions

We have been studying arbitrary spaces X using fundamental groups and homology groups, and we have been rewarded with interesting applications in the few cases in which we could compute these groups. At this point, however, we would have difficulty computing the homology groups of a space as simple as the torus $T = S^1 \times S^1$; indeed $S_n(T)$ is uncountable for every $n \geq 0$, so it is conceivable that $H_n(T)$ is uncountable for every n (we shall soon see that this is not so). Many interesting spaces, as the torus, can be "triangulated", and we shall see that this (strong) condition greatly facilitates calculation of homology groups. Moreover, we shall also be able to give a presentation of the fundamental groups of such spaces.

In contrast to the singular theory, a q-simplex will once again be an honest space (and not a continuous map with domain Δ^q). Recall that if $\{v_0, \ldots, v_q\}$ is an affine independent subset of some euclidean space, then it spans the q-simplex $s = [v_0, \ldots, v_q]$ consisting of all convex combinations of these vertices.

Definition. If $s = [v_0, \ldots, v_q]$ is a q-simplex, then we denote its **vertex set** by $\text{Vert}(s) = \{v_0, \ldots, v_q\}$.

Definition. If s is a simplex, then a **face** of s is a simplex s' with $\text{Vert}(s') \subset \text{Vert}(s)$; one writes $s' \leq s$. If $s' < s$ (i.e., $\text{Vert}(s') \subsetneqq \text{Vert}(s)$), then s' is called a **proper face** of s.

Definition. A finite **simplicial complex** K is a finite collection of simplexes in some euclidean space such that:

(i) if $s \in K$, then every face of s also belongs to K;
(ii) if $s, t \in K$, then $s \cap t$ is either empty or a common face of s and of t.

We write Vert(K) to denote the **vertex set** of K, namely, the set of all 0-simplexes in K.

Definition. If K is a simplicial complex, its **underlying space** $|K|$ is the subspace (of the ambient euclidean space)

$$|K| = \bigcup_{s \in K} s,$$

the union of all the simplexes in K.

Clearly, $|K|$ is a compact subspace of some euclidean space. Note that if s is a simplex in K, then $|s| = s$.

Definition. A topological space X is a **polyhedron** if there exists a simplicial complex K and a homemorphism $h: |K| \to X$. The ordered pair (K, h) is called a **triangulation** of X.

EXAMPLE 7.1. The standard 2-simplex Δ^2 is contained in euclidean space $\mathbf{R}^3$. Define K to be the family of all vertices and 1-simplexes of Δ^2 (i.e., K is the family of all proper faces of Δ^2). Then K is a simplicial complex whose underlying space $|K|$ is the perimeter of the triangle Δ^2 in $\mathbf{R}^3$. If $X = S^1$, choose distinct points $a_0, a_1, a_2 \in S^1$, and define a homemorphism $h: |K| \to S^1$ with $h(e_i) = a_i$ for $i = 0, 1, 2$, and with h taking each 1-simplex $[e_i, e_{i+1}]$ (read indices mod 3) onto the arc joining a_i to a_{i+1}. Then (K, h) is a triangulation of S^1, and so S^1 is a polyhedron.

EXAMPLE 7.2. If K is the family of all proper faces of an n-simplex s, then there is a triangulation (K, h) of S^{n-1}. Denote this simplicial complex K by $\dot{s}$. (Note that $|K|$ is the boundary $\dot{s} \approx S^{n-1}$, so that our two dot notations are compatible.)

EXAMPLE 7.3. It is easy to give examples of finite collections of simplexes satisfying condition (i) of the definition of simplicial complexes but not condition (ii).

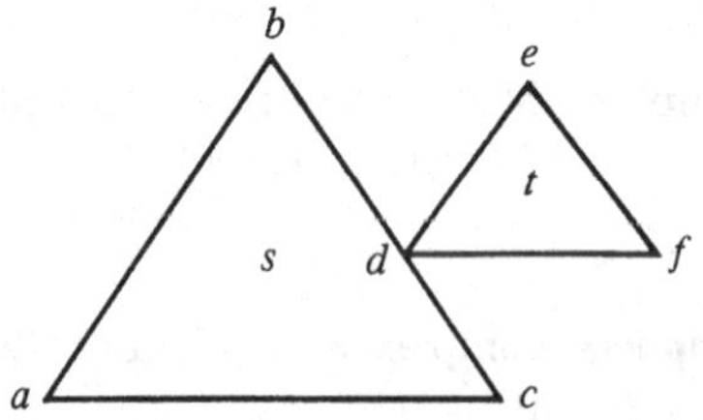

The simplexes $s = [a, b, c]$ and $t = [d, e, f]$ (and all their faces) do not comprise a simplicial complex because $s \cap t$, though a face of t, is not a face of s. The space $X = |s| \cup |t|$ is a polyhedron, but one needs another simplicial complex to triangulate it; for example, $K = \{[a, b, d], [a, d, c], [d, e, f]$, and all their faces$\}$ will serve.

EXAMPLE 7.4. Every q-simplex s determines a simplicial complex K, namely, the family of all (not necessarily proper) faces of s. Clearly, $|K| = s$. If $h: |K| \to s$ is the identity map, then s is a polyhedron (as it ought to be!).

EXAMPLE 7.5. Consider the square $\mathbf{I} \times \mathbf{I}$ with sides identified as indicated.

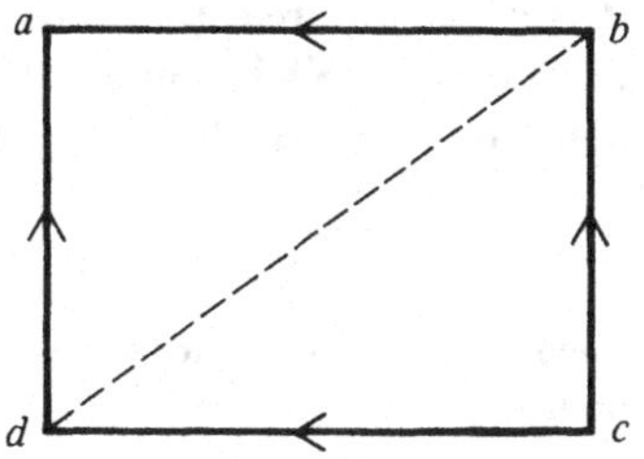

In detail, $(t, 0)$ is identified with $(t, 1)$ for each $t \in \mathbf{I}$, giving a cylinder, and $(0, s)$ is identified with $(1, s)$ for each $s \in \mathbf{I}$, giving a torus. A triangulation of $\mathbf{I} \times \mathbf{I}$ (e.g., insertion of the diagonal bd) may not give a triangulation of the torus because, after the identification, the two distinct triangles (2-simplexes) abd and bcd have the same vertex set. The following triangulation of $\mathbf{I} \times \mathbf{I}$ does lead to a triangulation of the torus, hence the torus is a polyhedron.

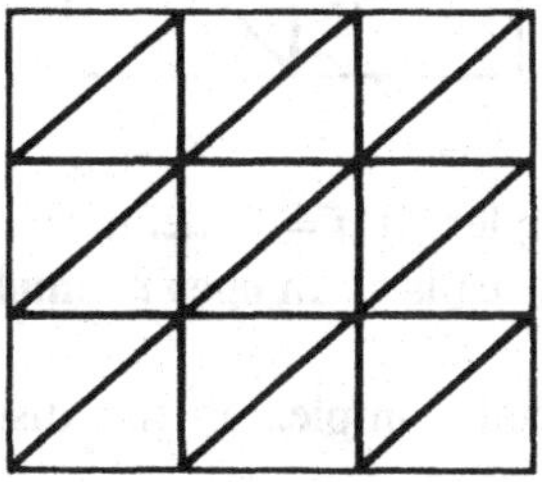

This triangulation of the torus has 18 triangles (2-simplexes), 27 edges (1-simplexes), and 9 vertices; it is known that the minimum number of triangles in a triangulation of the torus is 14 (see [Massey (1967), p. 34, Exercise 2] for an inequality implying this result).

EXAMPLE 7.6.[1] Other identifications of the boundary points of $\mathbf{I} \times \mathbf{I}$ lead to other polyhedra.

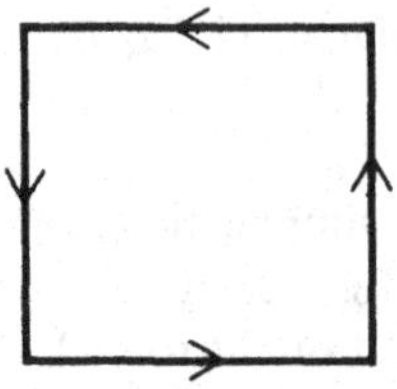
Figure (i)

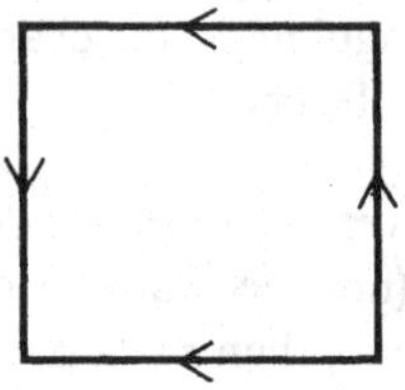
Figure (ii)

The space obtained from Fig. (i) by identifying $(t, 0)$ with $(1 - t, 1)$ and $(0, t)$ with $(1, 1 - t)$ for all $t \in \mathbf{I}$ is called the **real projective plane** $\mathbf{R}P^2$; the space obtained from Fig. (ii) is called the **Klein bottle**. If one identifies all the boundary to a common point, one obtains the sphere S^2.

EXERCISES

7.1. Show that $\mathbf{R}P^2$ is homemorphic to the quotient space of the disk D^2 after identifying antipodal points.

7.2. Exhibit a compact connected subset of $\mathbf{R}^2$ that is not a polyhedron.

7.3. Why does the following triangulation of $\mathbf{I} \times \mathbf{I}$ not give a triangulation of the torus?

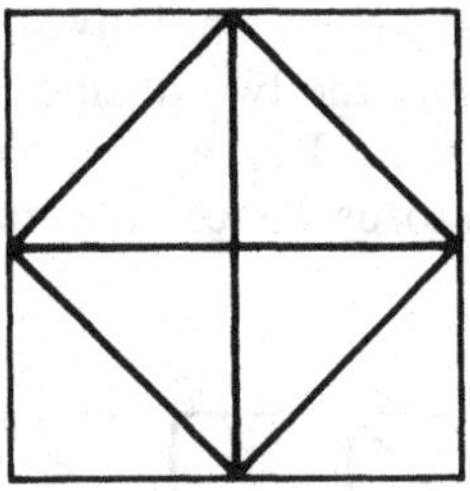

Definition. Let s be a q-simplex. If $q = 0$, define $s^\circ = s$; if $q > 0$, define $s^\circ = s - \dot{s}$ (see Example 7.2). One calls s° an **open q-simplex**.

Observe that a simplicial complex is the disjoint union of its open simplexes.

It is plain that an open q-simplex s° is an open subset of s (it is its interior), but if s lies in a simplicial complex K, then s° may not be an open subset of

[1] It will be shown that the homology groups obtained from $\mathbf{I} \times \mathbf{I}$ by "twisting", for example, $\mathbf{R}P^2$ and the Klein bottle, have homology groups with elements of finite order. This is probably the reason that torsion groups are so called. An etymology of twisting also appears in the discussion of lens spaces in [Seifert and Threlfall, p. 220].

$|K|$. For example, an open 1-simplex s° is an open interval; its endpoints are open 0-simplexes that are not open subsets of the (closed) 1-simplex spanned by the endpoints. Similarly, an open 1-simplex is not an open subset of a 2-simplex.

Definition. Let K be a simplicial complex and let $p \in \operatorname{Vert}(K)$. Then the **star** of p, denoted by $\operatorname{st}(p)$, is defined by

$$\operatorname{st}(p) = \bigcup_{\substack{s \in K \\ p \in \operatorname{Vert}(s)}} s^\circ \subset |K|.$$

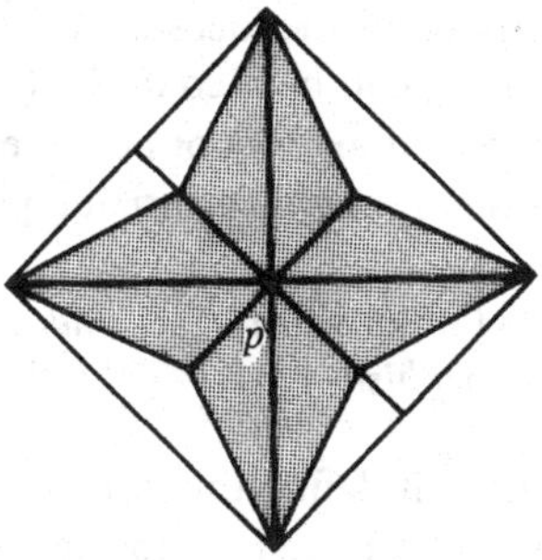

In the figure, $\operatorname{st}(p)$ consists of the open shaded region. One sees that $\operatorname{st}(p)$ consists of all the open simplexes of which p is a neighbor.

EXERCISES

*7.4. (i) If K is a simplicial complex and F is a subset of $|K|$, then F is closed if and only if $F \cap s$ is closed in s for every $s \in K$.
(ii) If s is a simplex in K of largest dimension, then $s^\circ = s - \dot{s}$ is an open subset of $|K|$.

7.5. If K is a simplicial complex, then $|K|$ is the disjoint union of all the open simplexes s° with $s \in K$. Conclude that each $x \in |K|$ lies in a unique open simplex.

7.6. Let K be a simplicial complex, let $x \in |K|$, and let s° be the (unique) open simplex with $x \in s^\circ$. If $\operatorname{Vert}(s) = \{p_0, \ldots, p_q\}$, then $x \in \operatorname{st}(p)$ if and only if $p = p_i$ for some $i = 0, 1, \ldots, q$.

*7.7. (i) For each vertex $p \in \operatorname{Vert}(K)$, prove that $\operatorname{st}(p)$ is an open subset of $|K|$ and that the family of all such stars is an open cover of $|K|$.
(ii) If $x \in \operatorname{st}(p)$, then the line segment with endpoints x and p is contained in $\operatorname{st}(p)$.

*7.8. Let $p_0, p_1, \ldots, p_n \in \operatorname{Vert}(K)$. Prove that $\{p_0, \ldots, p_n\}$ spans a simplex of K if and only if $\bigcap_{i=0}^{n} \operatorname{st}(p_i) \neq \varnothing$.

Definition. If K is a simplicial complex, define its **dimension**, denoted by $\dim K$, to be

$$\dim K = \sup_{s \in K} \{\dim s\}$$

(of course, a q-simplex has dimension q).

The construction of polyhedra as quotient spaces of $\mathbf{I} \times \mathbf{I}$ raises an interesting question. It is plain that there exists a simplicial complex—indeed a collection of triangles in $\mathbf{R}^3$—which can be assembled to form a space homeomorphic to the torus. It is less obvious, though not difficult, that such a simplicial complex with fourteen 2-simplexes exists in $\mathbf{R}^3$. Now the Klein bottle exists in $\mathbf{R}^4$, but it cannot be imbedded in $\mathbf{R}^3$. There is a general existence theorem (see [Hurewicz and Wallman, p. 56]) that a finite simplicial complex of dimension d always triangulates a subspace of $\mathbf{R}^{2d+1}$; moreover, this result is best possible: if K consists of all the faces of Δ^{2d+2} having dimension $\leq d$, then $\dim K = d$ and K cannot be imbedded in $\mathbf{R}^{2d}$ (see [Flores]) (when $d = 1$, this says that the complete pentagon cannot be imbedded in the plane, as one knows from Kuratowski's theorem characterizing planar graphs).

Theorem 7.1 (Invariance of Dimension). *If K and L are simplicial complexes and if there exists a homeomorphism $f: |K| \to |L|$, then* $\dim K = \dim L$.

Remark. It follows that one can define the dimension of a polyhedron X as the common dimension of the simplicial complexes involved in triangulations of X.

PROOF. Assume, on the contrary, that $m = \dim K > \dim L = n$ (replacing f by f^{-1} handles the reverse inequality). Take an m-simplex σ in K, and let $\sigma^\circ = \sigma - \dot{\sigma}$ be its interior. Now σ° is an open set in $|K|$, by Exercise 7.4(ii). Since f is a homeomorphism, $f(\sigma^\circ)$ is open in $|L|$. There thus exists some p-simplex τ in L (of course, $p \leq n < m$) with $f(\sigma^\circ) \cap \tau^\circ = W$, a nonempty open set in $|L|$ (for the stars of vertices form an open cover of $|L|$, by Exercise 7.7(i)). Choose a homeomorphism $\varphi: \Delta^m \to \sigma$ with $\varphi(\dot{\Delta}^m) = \dot{\sigma}$; then U, defined by $U = \varphi^{-1}f^{-1}(W)$, is an open subset of $(\Delta^m)^\circ$. Since $p < m$, there exists an imbedding $g: \Delta^p \to (\Delta^m)^\circ$ such that im g contains no nonempty open subsets of $(\Delta^m)^\circ$. Both U and $g(W)$ are homeomorphic subsets of $(\Delta^m)^\circ$; as U is open and $g(W)$ is not, this contradicts invariance of domain (Theorem 6.36) as modified by Exercises 6.21 and 6.22. □

Simplicial Approximation

If we want a category whose objects are simplicial complexes (and we do), what are the morphisms?

Definition. Let K and L be simplicial complexes. A **simplicial map** $\varphi: K \to L$ is a function φ: Vert(K) $\to$ Vert(L) such that whenever $\{p_0, p_1, \ldots, p_q\}$ spans a simplex of K, then $\{\varphi(p_0), \varphi(p_1), \ldots, \varphi(p_q)\}$ spans a simplex of L.

Of course, repetitions among $\varphi(p_0), \ldots, \varphi(p_q)$ are allowed.

Theorem 7.2. *If $\mathscr{K}$ consists of all simplicial complexes and all simplicial maps (with usual composition), then $\mathscr{K}$ is a category, and underlying defines a functor* $|\ \ |: \mathscr{K} \to \mathbf{Top}$.

Proof. It is routine to check that $\mathscr{K}$ is a category; let us construct $|\ \ |$. On objects, assign the space $|K|$ to K. If $\varphi: K \to L$ is a simplicial map, define $|\varphi|: |K| \to |L|$ as follows. For each $s \in K$, define $f_s: s \to |L|$ as the affine map determined by $\varphi|\text{Vert}(s)$ (Theorem 2.10). Condition (ii) of the definition of simplicial complex implies that the functions f_s agree on overlaps, so that the gluing lemma 1.1 assembles them into a unique continuous function $|K| \to |L|$, denoted by $|\varphi|$. That we have defined a functor is left as an exercise. □

Definition. A map of the form $|\varphi|: |K| \to |L|$, where $\varphi: K \to L$ is a simplicial map, is called **piecewise linear**.

There is no obvious functor $\mathbf{Top} \to \mathscr{K}$, even if we confine our attention to the subcategory of polyhedra. Given a continuous $f: |K| \to |L|$, it may not be true that $f = |\varphi|$ for some simplicial map φ: after all, there are only finitely many φ's. But we are flexible. Is it true that $f \simeq |\varphi|$ for some φ? The answer is still "no": if $K = L = \{\text{all proper faces of } [p_0, p_1, p_2]\}$, then $|K| \approx S^1 \approx |L|$. Since $\pi_1(S^1) \cong \mathbf{Z}$, there are infinitely many nonhomotopic maps $f: S^1 \to S^1$, while there are still only finitely many simplicial maps $\varphi: K \to L$. We shall subdivide K (the same process as in the proof of excision) to obtain more (and better) approximations by simplicial maps.

Definition. Let K and L be simplicial complexes, let $\varphi: K \to L$ be a simplicial map, and let $f: |K| \to |L|$ be continuous. Then φ is a **simplicial approximation** to f if, for every vertex p of K,

$$f(\text{st}(p)) \subset \text{st}(\varphi(p)).$$

It is easy to see that $|\varphi|(\text{st}(p)) \subset \text{st}(\varphi(p))$. Thus we are saying that f behaves like $|\varphi|$ in that it carries neighboring simplexes of p inside the union of the simplexes near $\varphi(p)$.

Exercises

7.9. Let K and K' be simplicial complexes, and let $\varphi: \text{Vert}(K) \to \text{Vert}(K')$ be a function. Prove that φ is a simplicial map if and only if, whenever $\bigcap \text{st}(p_i) \neq \varnothing$, then $\bigcap \text{st}(\varphi p_i) \neq \varnothing$. (*Hint*: Use Exercise 7.8.)

*7.10. Prove that a simplicial map $\varphi: K \to L$ is a simplicial approximation to $f: |K| \to |L|$ if and only if, whenever $x \in |K|$ and $f(x) \in s^\circ$ (where s is a simplex of L), then $|\varphi|(x) \in s$.

*7.11. If $\varphi: K \to L$ is a simplicial approximation to $f: |K| \to |L|$, then $|\varphi| \simeq f$.

Our remarks about $\pi_1(S^1)$ show that, for a given triangulation, a continuous map need not have a simplicial approximation. Let us therefore change the triangulation.

Definition. If s is a simplex, let b^s denote its barycenter. If K is a simplicial complex, define Sd K, the **barycentric subdivision** of K, to be the simplicial complex with

$$\mathrm{Vert}(\mathrm{Sd}\, K) = \{b^s: s \in K\}$$

and with simplexes $[b^{s_0}, b^{s_1}, \ldots, b^{s_q}]$, where the s_i are simplexes in K with $s_0 < s_1 < \cdots < s_q$.

Recall that if s is a vertex of K, that is, s is a 0-simplex, then $b^s = s$; therefore $\mathrm{Vert}(K) \subset \mathrm{Vert}(\mathrm{Sd}\, K)$. It is easy to check axioms (i) and (ii) in the definition of simplicial complex; using Exercise 7.13 below, one shows that $[b^{s_0}, \ldots, b^{s_q}]$ is a q-simplex.

EXAMPLE 7.7. If $\sigma = [p_0, p_1, p_2]$, then $\mathrm{Vert}(\mathrm{Sd}\, \sigma) = \{p_0, p_1, p_2, b_0, b_1, b_2, b^\sigma\}$.

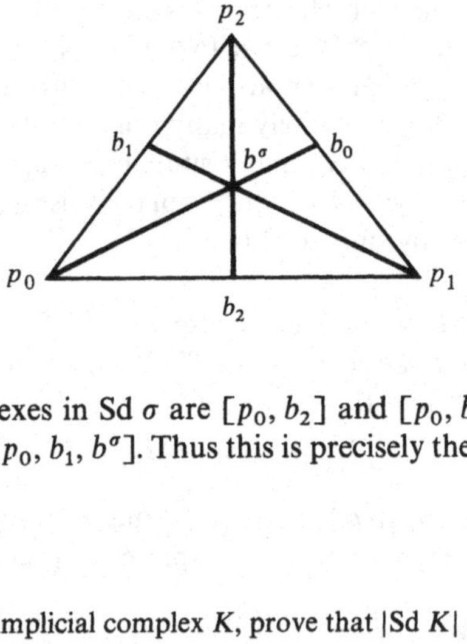

Examples of 1-simplexes in Sd σ are $[p_0, b_2]$ and $[p_0, b^\sigma]$; an example of a 2-simplex in Sd σ is $[p_0, b_1, b^\sigma]$. Thus this is precisely the earlier construction of Chapter 6.

EXERCISES

*7.12. (i) For every simplicial complex K, prove that $|\mathrm{Sd}\, K| = |K|$.
(ii) Prove that there exists a simplicial map φ: Sd $K \to K$ that is a simplicial approximation to the identity $|\mathrm{Sd}\, K| \to |K|$. (*Hint*: Define φ: $\mathrm{Vert}(\mathrm{Sd}\, K) \to \mathrm{Vert}(K)$ so that $\varphi(b^s) \in \mathrm{Vert}(s)$.)
(iii) If X is a polyhedron and $x \in X$, there exists a triangulation (K, h) of X with $x = h(v)$ for some vertex v of K.

*7.13. If $s_0 < s_1 < \cdots < s_q$ are simplexes in some euclidean space, then $\{b^{s_0}, b^{s_1}, \ldots, b^{s_q}\}$ is affine independent.

7.14. Every open simplex of Sd K is contained in a unique open simplex of K.

Definition. If K is a simplicial complex, then

$$\textbf{mesh } K = \sup_{s \in K} \{\text{diam}(s)\},$$

where diam(s) denotes the diameter of s.

EXERCISES

*7.15. If mesh $K = \mu$ and $p \in \text{Vert}(K)$, then $\text{diam}(\text{st}(p)) \le 2\mu$.

*7.16. If dim $K = n$, then

$$\text{mesh Sd } K \le (n/n + 1) \text{ mesh } K.$$

(*Hint*: Theorem 2.9.) Conclude that, for $q \ge 1$,

$$\text{mesh Sd}^q\, K \le (n/n + 1)^q \text{ mesh } K.$$

Theorem 7.3 (Simplicial Approximation Theorem). *If K and L are simplicial complexes and if $f\colon |K| \to |L|$ is continuous, then there is an integer $q \ge 1$ and a simplicial approximation $\varphi\colon \text{Sd}^q\, K \to L$ to f.*

PROOF. Let $\text{Vert}(L) = \{w_j : j \in J\}$ and let $\{\text{st}(w_j)\}$ be the open cover of $|L|$ by its stars. Since f is continuous, $\{f^{-1}\, st(w_j)\}$ is an open cover of $|K|$. Since $|K|$ is compact metric, this cover has a Lebesgue number $\lambda > 0$. By Exercise 7.16, we can choose q large enough so that mesh $\text{Sd}^q\, K < \frac{1}{2}\lambda$; it follows from Exercise 7.15 that $\text{diam}(\text{st}(p)) < \lambda$ for every $p \in \text{Vert}(\text{Sd}^q\, K)$.

Define $\varphi\colon \text{Vert}(\text{Sd}^q\, K) \to \text{Vert}(L)$ by $\varphi(p) = w_j$, where w_j is some vertex with $\text{st}(p) \subset f^{-1}(\text{st}(w_j))$ (w_j exists, by definition of Lebesgue number; if there are several choices, pick any one). It follows that $f(\text{st}(p)) \subset \text{st}(w_j) = \text{st}(\varphi(p))$, so that we are done if we can show that φ is a simplicial map: if $\{p_0, \ldots, p_m\}$ spans a simplex in $\text{Sd}^q\, K$, does $\{\varphi(p_0), \ldots, \varphi(p_m)\}$ span a simplex in L? Now Exercise 7.8 gives $\bigcap_{i=0}^m \text{st}(p_i) \ne \varnothing$, so that

$$\varnothing \ne f\left(\bigcap \text{st}(p_i)\right) \subset \bigcap f(\text{st}(p_i)) \subset \bigcap \text{st}(\varphi(p_i)).$$

Exercise 7.8 thus shows that $\{\varphi(p_0), \ldots, \varphi(p_m)\}$ spans a simplex of L. □

Corollary 7.4. *Let K and L be simplicial complexes, and let $f\colon |K| \to |L|$ be continuous. Assume that K' is a simplicial complex such that*

(i) $|K'| = |K|$;
(ii) $\text{Vert}(K) \subset \text{Vert}(K')$;
(iii) mesh K' *is "small".*

Then there exists a simplicial approximation $\varphi\colon K' \to L$ to f.

PROOF. The listed properties are the only properties of $\text{Sd}^q\, K$ used in the proof of the theorem. □

Definition. A **subcomplex** L of a simplicial complex K is a simplicial complex contained in K (i.e., $s \in L$ implies that $s \in K$) with $\text{Vert}(L) \subset \text{Vert}(K)$.

Note that Sd K is not a subcomplex of K (nor is K a subcomplex of Sd K).

Definition. For any $q \geq -1$, the ***q*-skeleton** of K, denoted by $K^{(q)}$, is the subcomplex of K consisting of all simplexes $s \in K$ with $\dim(s) \leq q$.

EXERCISES

*7.17. If $\varphi: K \to L$ is a simplicial map, then $\varphi(K^{(q)}) \subset L^{(q)}$ for every q. Conclude that $\dim K = n$ implies that $\text{im}|\varphi| \subset |L^{(n)}|$.

7.18. If K is the n-skeleton of an $(n + 1)$-simplex, then $|K| \approx S^n$.

Theorem 7.5. *If $m < n$, then every continuous map $f: S^m \to S^n$ is nullhomotopic.*

PROOF. Let K be the m-skeleton of an $(m + 1)$-simplex, and let L be the n-skeleton of an $(n + 1)$-simplex; we may regard f as a continuous map from $|K|$ into $|L|$. Let $\varphi: \text{Sd}^q K \to L$ be a simplicial approximation to f. Since $\dim \text{Sd}^q K = \dim K = m$, Exercise 7.17 gives $\text{im}|\varphi| \subset |L^{(m)}|$, and so $|\varphi|$ is not surjective. Hence $\text{im}|\varphi| \subset |L| - \{\text{point}\}$, which is contractible, and so $|\varphi|$ is nullhomotopic. But $|\varphi| \simeq f$, hence f is nullhomotopic. □

Corollary 7.6. *If $n \geq 2$, then S^n is simply connected.*

PROOF. The theorem shows that every continuous map $f: S^1 \to S^n$ is nullhomotopic, and so the result follows from Theorem 1.6. □

We have already sketched a proof of this corollary in Exercises 3.20 and 3.21. The result does not follow, however, from the Hurewicz theorem and the fact that $H_1(S^n) = 0$ for $n \geq 2$ (one can conclude from these data only that $\pi_1(S^n, 1)$ is its own commutator subgroup, and such groups do exist; e.g., every nonabelian simple group is such a group).

Let us mention a famous problem. The **Poincaré conjecture** asks whether a simply connected compact n-manifold having the same homology groups as S^n is homeomorphic to S^n. It is not too difficult to show that the conjecture is true when $n = 2$; for $n \geq 5$, the conjecture was solved affirmatively by Smale in the 1960s; the case $n = 4$ was solved (affirmatively) in the 1980s by Freedman. The familiar dimension 3 is thus the only open case.

Abstract Simplicial Complexes

We are going to define homology groups of a simplicial complex K (which will turn out to be isomorphic to the homology groups of the space $X = |K|$ as defined in Chapter 4). This construction works in a simpler setting, which we now describe.

Definition.[2] Let V be a finite set. An **abstract simplicial complex** K is a family of nonempty subsets of V, called **simplexes**, such that

(i) if $v \in V$, then $\{v\} \in K$;
(ii) if $s \in K$ and $s' \subset s$, then $s' \in K$.

One calls V the **vertex set** of K and denotes it by $\mathrm{Vert}(K)$; a simplex $s \in K$ having $q + 1$ distinct vertices is called a ***q*-simplex**.

Definition. If K and L are abstract simplicial complexes, then a **simplicial map** $\varphi: K \to L$ is a function $\varphi: \mathrm{Vert}(K) \to \mathrm{Vert}(L)$ such that whenever $\{v_0, \ldots, v_q\}$ is a simplex in K, then $\{\varphi v_0, \ldots, \varphi v_q\}$ is a simplex in L (of course, it is possible that the latter list of vertices has repetitions).

Theorem 7.7. *All abstract simplicial complexes and simplicial maps determine a category, denoted by* $\mathscr{K}^{\mathrm{a}}$.

Proof. A routine check. □

Equivalences in the category $\mathscr{K}^{\mathrm{a}}$ and in the category $\mathscr{K}$ are called **isomorphisms**.

Example 7.8. Every simplicial complex K determines an abstract simplicial complex K^{a} with the same vertex set: let each simplex $s \in K$ determine its vertex set $\mathrm{Vert}(s) \subset \mathrm{Vert}(K)$.

Example 7.9. Let X be a topological space, and let $\mathscr{U}$ be a finite open cover of X. Define an abstract simplicial complex having vertices the open sets in $\mathscr{U}$ and declare that open sets $U_0, U_1, \ldots, U_q$ in $\mathscr{U}$ form a simplex if $\bigcap_{i=0}^{q} U_i \neq \varnothing$. This simplicial complex is called the **nerve** of the open cover $\mathscr{U}$ and is denoted by $N(\mathscr{U})$.

Example 7.10. Let G be a finite group. Define an abstract simplicial complex $Q_p(G)$ whose vertex set consists of all nontrivial p-subgroups (for some fixed prime divisor p of $|G|$) and with subgroups $P_0, P_1, \ldots, P_q$ forming a simplex if $\bigcap_{i=0}^{q} P_i \neq \{1\}$.

Example 7.11. If K is an abstract simplicial complex, we construct its **barycentric subdivision** $\mathrm{Sd}\, K$ as follows (here $\mathrm{Sd}\, K$ is also an abstract simplicial complex): define $\mathrm{Vert}(\mathrm{Sd}\, K) = \{\text{simplexes } \sigma: \sigma \in K\}$; define a simplex in $\mathrm{Sd}(K)$ to be a set $\{\sigma_0, \sigma_1, \ldots, \sigma_q\}$ with $\sigma_0 < \sigma_1 < \cdots < \sigma_q$ (where $\sigma < \sigma'$ means $\sigma \subsetneqq \sigma'$).

---

[2] Here is the definition of a possibly infinite abstract simplicial complex K; let V be any set and define K as a family of finite nonempty subsets of V satisfying properties (i) and (ii).

The construction $K \mapsto K^a$ in Example 7.8 defines a functor $\mathscr{K} \to \mathscr{K}^a$. The next theorem says that there is a good way to reverse the procedure, obtaining a simplicial complex from an abstract one.

Theorem 7.8. *There is a functor* $u: \mathscr{K}^a \to \mathscr{K}$ *such that* $K \cong u(K^a)$ *for all* $K \in \operatorname{obj} \mathscr{K}$ *and* $L \cong (uL)^a$ *for every* $L \in \operatorname{obj} \mathscr{K}^a$.

Proof. Let L be an abstract simplicial complex, and let $V = \operatorname{Vert}(L) = \{v_0, v_1, \ldots, v_n\}$. Recall that the standard n-simplex Δ^n has vertices $\{e_0, e_1, \ldots, e_n\}$. If $s = \{v_{i_0}, \ldots, v_{i_q}\}$ is a q-simplex in L, define $|s| = [e_{i_0}, \ldots, e_{i_q}]$, the q-simplex in Δ^n spanned by the displayed vertices. Finally, define $u(L)$ as the family of all $|s|$ for $s \in L$. It is plain that $u(L)$ is a simplicial complex; indeed $u(L)$ is a subcomplex of Δ^n.

It is easy to see that a simplicial map $\varphi: L \to L'$ in $\mathscr{K}^a$ (which is a certain function $\varphi: \operatorname{Vert}(L) \to \operatorname{Vert}(L')$) corresponds to the obvious simplicial map $u(\varphi): u(L) \to u(L')$ (which is a certain function $\{e_0, \ldots, e_n\} \to \{e_0, \ldots, e_m\}$). Moreover, one verifies quickly that $u: \mathscr{K}^a \to \mathscr{K}$ is a functor and that the isomorphisms mentioned in the statement do exist. □

Definition. If L is an abstract simplicial complex, then a **geometric realization** of L is a space homeomorphic to $|u(L)|$.[3]

Corollary 7.9. *Isomorphic abstract simplicial complexes have homeomorphic geometric realizations.*

Proof. Every functor (in particular, the composite $\mathscr{K}^a \to \mathscr{K} \to$ **Top**) preserves equivalences. □

As a result of Theorem 7.8, one usually does not emphasize the distinction between simplicial complexes and abstract simplicial complexes. Henceforth, we drop the adjective "abstract", although we shall usually be thinking of the simpler notion of abstract simplicial complex. We shall also not distinguish between the categories $\mathscr{K}$ and $\mathscr{K}^a$; either will be denoted by $\mathscr{K}$. Indeed some authors do not bother to distinguish simplicial complexes from polyhedra!

Simplicial Homology

Definition. An **oriented** simplicial complex K is a simplicial complex and a partial order on $\operatorname{Vert}(K)$ whose restriction to the vertices of any simplex in K is a linear order.

---

[3] The geometric realization of an infinite abstract simplicial complex can also be defined (see Example 8.11); in general, it does not lie in any (finite-dimensional) euclidean space.

Every linear ordering of Vert(K) makes K into an oriented simplicial complex. For every simplicial complex K, the barycentric subdivision Sd K is always oriented (see Exercise 6.6 (ii)).

We shall define homology groups of oriented simplicial complexes K; eventually, we shall see that they coincide with the homology groups of $|K|$ (hence are independent of the partial order on Vert(K)).

Definition. If K is an oriented simplicial complex and $q \geq 0$, let $C_q(K)$ be the abelian group having the following presentation.

Generators: all $(q+1)$-tuples $(p_0, \ldots, p_q)$ with $p_i \in \text{Vert}(K)$ such that $\{p_0, \ldots, p_q\}$ spans a simplex in K.

Relations: (i) $(p_0, \ldots, p_q) = 0$ if some vertex is repeated;
(ii) $(p_0, \ldots, p_q) = (\text{sgn } \pi)(p_{\pi 0}, p_{\pi 1}, \ldots, p_{\pi q})$, where π is a permutation of $\{0, 1, \ldots, q\}$.

Denote the element of $C_q(K)$ corresponding to $(p_0, \ldots, p_q)$ by $\langle p_0, \ldots, p_q \rangle$. Of course, sgn $\pi = \pm 1$ (depending on the parity of π).

Lemma 7.10. *Let K be an oriented simplicial complex of dimension m.*

(i) *$C_q(K)$ is a free abelian group with basis all symbols $\langle p_0, \ldots, p_q \rangle$, where $\{p_0, \ldots, p_q\}$ spans a q-simplex in K and $p_0 < p_1 < \cdots < p_q$. Moreover, $\langle p_{\pi 0}, \ldots, p_{\pi q} \rangle = (\text{sgn } \pi)\langle p_0, \ldots, p_q \rangle$.*
(ii) *$C_q(K) = 0$ for all $q > m$.*

PROOF. (i) Define F_q to be the free abelian group with basis all $(q+1)$-tuples $(p_0, \ldots, p_q)$ of vertices of K such that $\{p_0, \ldots, p_q\}$ spans a simplex in K. If R_q is the subgroup of relations (as in the definition above), then $F_q/R_q = C_q(K)$. But it is easy to see that there is a new basis of F_q of the form $B_1 \cup B_2 \cup B_3$, where B_1 consists of all $(q+1)$-tuples in F_q with a repeated vertex, B_2 consists of all $(p_0, \ldots, p_q)$ with $p_0 < p_1 < \cdots < p_q$, and B_3 consists of all terms of the form $(p_0, \ldots, p_q) - (\text{sgn } \pi)(p_{\pi 0}, \ldots, p_{\pi q})$, where π is a nonidentity permutation of $\{0, 1, \ldots, q\}$. It is now clear that R_q (with basis $B_1 \cup B_3$) is a direct summand of F_q. Therefore $C_q(K) = F_q/R_q$ is free abelian as claimed.

(ii) If $q > m$, then every $(q+1)$-tuple $(p_0, \ldots, p_q)$ of vertices, which spans a simplex of K, must have a repeated vertex; hence $\langle p_0, \ldots, p_q \rangle = 0$ in $C_q(K)$. □

The reason for not defining $C_q(K)$ as described in the lemma will soon be clear.

Definition. Define $\partial_q \colon C_q(K) \to C_{q-1}(K)$ by setting

$$\partial_q(\langle p_0, \ldots, p_q \rangle) = \sum_{i=0}^{q} (-1)^i \langle p_0, \ldots, \hat{p}_i, \ldots, p_q \rangle$$

(where $\hat{p}_i$ means delete the vertex p_i) and extending by linearity.

Theorem 7.11. *If K is an oriented simplicial complex of dimension m, then*

$$C_*(K) = 0 \to C_m(K) \xrightarrow{\partial} \cdots \to C_1(K) \xrightarrow{\partial} C_0(K) \to 0$$

is a chain complex.

PROOF. The argument of Theorem 4.6 can be used here to show that $\partial\partial = 0$. □

Definition.[4] If K is an oriented simplicial complex, then

$Z_q(K) = \ker \partial_q$, the group of **simplicial q-cycles,**
$B_q(K) = \operatorname{im} \partial_{q+1}$, the group of **simplicial q-boundaries,**
and
$H_q(K) = Z_q(K)/B_q(K)$, the **qth simplicial homology group.**

We now associate an induced homomorphism to every simplicial map.

Definition. Let K and L be oriented simplicial complexes. If $\varphi: K \to L$ is a simplicial map, define $\varphi_\#: C_q(K) \to C_q(L)$, for each $q \geq 0$, by

$$\varphi_\#(\langle p_0, \ldots, p_q \rangle) = \langle \varphi(p_0), \ldots, \varphi(p_q) \rangle.$$

Of course, if some vertex $\varphi(p_i)$ is repeated, then the term on the right is zero. Furthermore, the ordering of the vertices on the right side may not be compatible with the orientation of L; our fussy definition of $C_q(K)$ (and $C_q(L)$) thus allows $\varphi_\#$ to be defined. Better, it allows the next result to be proved.

Lemma 7.12. *If $\varphi: K \to L$ is a simplicial map, then $\varphi_\#: C_*(K) \to C_*(L)$ is a chain map; that is, $\varphi_\# \partial = \partial \varphi_\#$.*

PROOF. The usual calculation, as in Lemma 4.8. □

Theorem 7.13. *For each $q \geq 0$, $H_q: \mathscr{K} \to$ **Ab** is a functor.*

PROOF. $H_q(K)$ has already been defined on objects K. On morphisms $\varphi: K \to L$, that is, on simplicial maps, define $\varphi_*: H_q(K) \to H_q(L)$ by

$$\varphi_*: z + B_q(K) \mapsto \varphi_\#(z) + B_q(L).$$

That H_q is a functor is routine. □

One wants to promote the definition of simplicial homology functors to the subcategory of **Top** of polyhedra. One problem is the definition of f_* when f is a continuous map. Plainly, the simplicial approximation theorem will be

[4] This definition also makes sense for infinite oriented simplicial complexes.

useful, and this will force comparison of $H_q(K)$ and $H_q(\text{Sd } K)$. This complication is one reason that we presented the singular theory first.

Theorem 7.14. *Let K be a (finite) oriented simplicial complex of dimension m.*

(i) $H_q(K)$ *is* f.g. *(finitely generated) for every* $q \geq 0$.
(ii) $H_q(K) = 0$ *for all* $q > m$.
(iii) $H_m(K)$ *is free abelian, possibly zero.*

PROOF. (i) $C_q(K)$ is f.g., hence its subgroup $Z_q(K)$ is f.g. (Theorem 9.3), and, finally, its quotient $H_q(K)$ is f.g.

(ii) Immediate from Lemma 7.10(ii).

(iii) Since $C_{m+1}(K) = 0$, we have $B_m(K) = 0$ and so $H_m(K) = Z_m(K)$. But a subgroup of a free abelian group is also free abelian (Theorem 9.3). □

Remark. If dim $K = m$, we do not assert that $H_m(K) \neq 0$ (this may be false). Moreover, if there are α_q q-simplexes in K, then $H_q(K)$ needs at most α_q generators.

We have just defined "absolute" simplicial homology groups. If K is an oriented simplicial complex and L is a subcomplex, then L is also oriented in the **induced orientation**, namely, the partial order on Vert(L) inherited from that on Vert K. It is easy to see that each $C_q(L)$ is a subgroup of $C_q(K)$ and that $C_*(L)$ is a subcomplex of $C_*(K)$.

Definition. If L is a subcomplex of an oriented simplicial complex K, then the qth **relative simplicial homology group** is

$$H_q(K, L) = H_q(C_*(K)/C_*(L)).$$

Let (K, f) be any triangulation of S^2; let V be the number of vertices, let E be the number of edges (1-simplexes), and let F be the number of faces (2-simplexes) in K. Euler's famous formula is

$$V - E + F = 2;$$

this formula is a key ingredient in showing that the five Platonic solids (tetrahedron, cube, octahedron, dodecahedron, and icosahedron) are the only regular solids in $\mathbf{R}^3$. Let us now generalize Euler's formula.

Definition. Let K be a simplicial complex of dimension m, and for each $q \geq 0$, let α_q be the number of q-simplexes in K. The **Euler–Poincaré characteristic** of K, denoted by $\chi(K)$, is defined by

$$\chi(K) = \sum_{q=0}^{m} (-1)^q \alpha_q.$$

Theorem 7.15. *If K is an oriented simplicial complex of dimension m, then*

$$\chi(K) = \sum_{q=0}^{m} (-1)^q \operatorname{rank} H_q(K).$$

Remark. The Euler–Poincaré characteristic is the alternating sum of the Betti numbers (once we show that $H_q(K) \cong H_q(|K|)$ for all q).

PROOF. Consider the chain complex $C_*(K)$:

$$0 \longrightarrow C_m(K) \xrightarrow{\partial_m} C_{m-1}(K) \longrightarrow \cdots \longrightarrow C_1(K) \xrightarrow{\partial_1} C_0(K) \longrightarrow 0.$$

Each $C_q(K)$ is a (free abelian) group of rank α_q. Of course, $H_q(K) = Z_q(K)/B_q(K) = \ker \partial_q/\operatorname{im} \partial_{q+1}$; Exercise 5.5 thus gives

$$\operatorname{rank} H_q(K) = \operatorname{rank} Z_q(K) - \operatorname{rank} B_q(K).$$

Note that $\operatorname{rank} B_m(K) = 0$ (in fact $B_m(K) = 0$). For each $q \geq 0$, there is an exact sequence

$$0 \longrightarrow Z_q(K) \longrightarrow C_q(K) \xrightarrow{\partial_q} B_{q-1}(K) \longrightarrow 0;$$

again Exercise 5.5 applies, and

$$\alpha_q = \operatorname{rank} C_q(K) = \operatorname{rank} Z_q(K) + \operatorname{rank} B_{q-1}(K).$$

Hence

$$\begin{aligned}\chi(K) = \sum_{q=0}^{m} (-1)^q \alpha_q &= \sum_{q=0}^{m} (-1)^q (\operatorname{rank} Z_q(K) + \operatorname{rank} B_{q-1}(K)) \\ &= \sum_{q=0}^{m} (-1)^q \operatorname{rank} Z_q(K) + \sum_{q=0}^{m} (-1)^q \operatorname{rank} B_{q-1}(K).\end{aligned}$$

Changing index of summation in the last sum and using the fact that $\operatorname{rank} B_{-1}(K) = 0 = \operatorname{rank} B_m(K)$, we have

$$\begin{aligned}\chi(K) &= \sum_{q=0}^{m} (-1)^q \operatorname{rank} Z_q(K) + \sum_{q=0}^{m} (-1)^{q+1} \operatorname{rank} B_q(K) \\ &= \sum_{q=0}^{m} (-1)^q (\operatorname{rank} Z_q(K) - \operatorname{rank} B_q(K)) \\ &= \sum_{q=0}^{m} (-1)^q \operatorname{rank} H_q(K).\end{aligned}$$

□

Remark. We have actually proved a more general result. If

$$C_* = 0 \longrightarrow C_n \xrightarrow{d_n} \cdots \longrightarrow C_1 \xrightarrow{d_0} C_0 \longrightarrow 0$$

is a chain complex in which each C_i is a f.g. free abelian group of rank α_i, then

$$\sum_{i=0}^{n} (-1)^i \alpha_i = \sum_{i=0}^{n} (-1)^i \operatorname{rank} H_i(C_*).$$

EXERCISES

7.19. Prove that

$$\chi(S^n) = \begin{cases} 2 & \text{if } n \text{ is even} \\ 0 & \text{if } n \text{ is odd.} \end{cases}$$

7.20. Compute $\chi(T)$, where T is the torus (see Example 7.5 above).

*7.21. If B is a set, let $F(B)$ denote the free abelian group with basis B. If $B = B_1 \cup B_2$, show there is an exact sequence

$$0 \to F(B_1 \cap B_2) \xrightarrow{i} F(B_1) \oplus F(B_2) \xrightarrow{p} F(B) \to 0,$$

where i is the "diagonal" map $x \mapsto (x, x)$ and p is the "subtraction" map $p\colon (x, y) \mapsto x - y$.

Theorem 7.16 (Excision). *If K_1 and K_2 are subcomplexes of a simplicial complex K with $K_1 \cup K_2 = K$, then the inclusion $(K_1, K_1 \cap K_2) \hookrightarrow (K, K_2)$ induces isomorphisms, for all $q \geq 0$,*

$$H_q(K_1, K_1 \cap K_2) \xrightarrow{\sim} H_q(K, K_2).$$

PROOF. By Lemma 6.11, it suffices to show that the inclusion $C_*(K_1) + C_*(K_2) \hookrightarrow C_*(K)$ induces isomorphisms in homology. But this map is the identity: $C_*(K_1) + C_*(K_2) = C_*(K)$. If $\sum m_i\sigma_i \in C_q(K)$, where σ_i denotes a q-simplex in K, then $\sigma_i \in K = K_1 \cup K_2$; that is, $\sigma_i \in K_1$ or $\sigma_i \in K_2$. One may thus collect terms and write $\sum m_i\sigma_i = \gamma_1 + \gamma_2$, where γ_1 is the sum of all those terms involving σ_i in K_1, and γ_2 is the sum of the other terms involving σ_i necessarily in K_2. □

Corollary 7.17 (Mayer–Vietoris). *If K_1 and K_2 are subcomplexes of a simplicial complex K with $K_1 \cup K_2 = K$, then there is an exact sequence*

$$\cdots \to H_{q+1}(K) \to H_q(K_1 \cap K_2) \to H_q(K_1) \oplus H_q(K_2) \to H_q(K) \to H_{q-1}(K_1 \cap K_2) \to \cdots.$$

PROOF. Use the proof of Theorem 6.3; even the induced maps are the same. □

Comparison with Singular Homology

We are now going to compare $H_*(K)$ with $H_*(|K|)$. To facilitate our work, we introduce reduced simplicial homology groups by augmenting $C_*(K)$, because it is more convenient to compare $\tilde{H}_*(K)$ (defined below) with $\tilde{H}_*(|K|)$.

Definition. If K is an oriented simplicial complex, define $C_{-1}(K)$ to be the infinite cyclic group generated by the symbol $\langle\ \rangle$, define $\tilde{\partial}_0\colon C_0(K) \to C_{-1}(K)$ by $\sum m_p\langle p\rangle \mapsto (\sum m_p)\langle\ \rangle$, and define the **augmented complex**

$$\tilde{C}_*(K) = 0 \to C_m(K) \xrightarrow{\partial_m} C_{m-1}(K) \to \cdots \to C_1(K) \xrightarrow{\partial_1} C_0(K) \xrightarrow{\tilde{\partial}_0} C_{-1}(K) \to 0.$$

Finally, define **reduced simplicial homology groups** by

$$\tilde{H}_q(K) = H_q(\tilde{C}_*(K)).$$

Essentially, reduced simplicial homology differs from ordinary simplicial homology in that it recognizes $\varnothing$ as the (unique) (-1)-simplex.

EXERCISES

7.22. For all $q \geq 1$, $\tilde{H}_q(K) \cong H_q(K)$; $H_0(K) \cong \tilde{H}_0(K) \oplus \mathbf{Z}$. (*Hint*: See Theorem 5.17.)

7.23. Show that $\tilde{H}_{-1}(\tilde{C}_*(K)) = 0$.

Corollary 7.18. *Let K be the simplicial complex consisting of all the faces of an n-simplex whose vertex set is linearly ordered (so that $|K| \approx \Delta^n$). Then*

$$\tilde{H}_q(K) = 0 \quad \textit{for all } q \geq 0.$$

PROOF. The statement is that the augmented complex $\tilde{C}_*(K)$ is an exact sequence; we prove this by appealing to Corollary 5.4. Thus it suffices to exhibit a contracting homotopy,

$$\{h_q: C_q(K) \to C_{q+1}(K), \text{ all } q \geq -1\}$$

so that

$$\partial_{q+1} h_q + h_{q-1} \partial_q = 1_q, \quad \text{the identity on } C_q(K). \tag{$*$}$$

The construction of h is patterned after the cone construction in Theorem 4.19.

Let v_0 be the smallest vertex in the orientation. For $q = -1$, define $h_{-1}: C_{-1}(K) \to C_0(K)$ by $\langle\ \rangle \mapsto \langle v_0 \rangle$ and extending by linearity. For $q \geq 0$, define $h_q: C_q(K) \to C_{q+1}(K)$ by $\langle p_0, \ldots, p_q \rangle \mapsto \langle v_0, p_0, \ldots, p_q \rangle$ and extending by linearity. Note that the last value is 0 if $v_0 = p_0$. It remains to verify Eq. $(*)$.

If $q = -1$, the desired formula is $\tilde{\partial}_0 h_{-1} = 1$; this is clear because $\tilde{\partial}_0 h_{-1}(\langle\ \rangle) = \tilde{\partial}_0(\langle v_0 \rangle) = \langle\ \rangle$. If $q \geq 0$,

$$\begin{aligned} h_{q-1}\partial_q \langle p_0, \ldots, p_q \rangle &= h_{q-1} \sum_{i=0}^{q} (-1)^i \langle p_0, \ldots, \hat{p}_i, \ldots, p_q \rangle \\ &= \sum_{i=0}^{q} (-1)^i \langle v_0, p_0, \ldots, \hat{p}_i, \ldots, p_q \rangle. \end{aligned}$$

On the other hand,

$$\begin{aligned} \partial_{q+1} h_q \langle p_0, \ldots, p_q \rangle &= \partial_{q+1} \langle v_0, p_0, \ldots, p_q \rangle \\ &= \langle p_0, \ldots, p_q \rangle - \sum_{i=0}^{q} (-1)^i \langle v_0, p_0, \ldots, \hat{p}_i, \ldots, p_q \rangle \end{aligned}$$

Therefore $(h\partial + \partial h)(\langle p_0, \ldots, p_q \rangle) = \langle p_0, \ldots, p_q \rangle$. □

Remarks. (1) If K is any oriented simplicial complex and if s is a simplex in K, then the induced orientation on s is a linear ordering of $\mathrm{Vert}(s)$.

(2) Note how much simpler this proof is than the corresponding result for singular homology. The next result is also simpler than its singular version, and so we present it; however, we do not use it in the proof of Theorem 7.22.

Corollary 7.19. *Let K consist of all the faces of an oriented n-simplex (so that $\mathrm{Vert}(K)$ is linearly ordered), and let L be the subcomplex of all the proper faces (so that $|L| \approx S^{n-1}$). Then*

$$\tilde{H}_q(L) = \begin{cases} 0 & \text{if } q \neq n-1 \\ \mathbf{Z} & \text{if } q = n-1. \end{cases}$$

PROOF. Since L is a subcomplex of K, $\tilde{C}_*(L)$ is a subcomplex of $\tilde{C}_*(K)$, and so there is a commutative diagram

$$\begin{array}{ccccccccccccc} 0 & \longrightarrow & 0 & \longrightarrow & C_{n-1}(L) & \longrightarrow & \cdots & \longrightarrow & C_0(L) & \longrightarrow & C_{-1}(L) & \longrightarrow & 0 \\ & & \downarrow & & \downarrow & & & & \downarrow & & \downarrow & & \\ 0 & \longrightarrow & C_n(K) & \underset{\partial_n}{\longrightarrow} & C_{n-1}(K) & \underset{\partial_{n-1}}{\longrightarrow} & \cdots & \longrightarrow & C_0(K) & \longrightarrow & C_{-1}(K) & \longrightarrow & 0, \end{array}$$

where the vertical maps (for $-1 \leq q \leq n-1$) are identities. Now the bottom row is an exact sequence, by Corollary 7.18. It follows easily that $\tilde{H}_q(L) = \tilde{H}_q(K) = 0$ for all $q < n-1$; moreover, $\tilde{H}_q(L) = 0$ for all $q > n-1$, by Theorem 7.14(ii).

Note that $C_n(K) \cong \mathbf{Z}$ because K has only one n-simplex. Exactness of the bottom row of the diagram thus gives

$$\mathbf{Z} \cong C_n(K) \cong \operatorname{im} \partial_n = \ker \partial_{n-1}.$$

On the other hand, $\tilde{H}_{n-1}(L) = \ker \partial_{n-1}$ (because $C_n(L) = 0$ implies that $B_{n-1}(L) = \operatorname{im} \partial_n = 0$). We conclude that $\tilde{H}_{n-1}(L) \cong \mathbf{Z}$, as desired. □

The reader can readily construct an example of a simplicial complex K having subcomplexes K_1 and K_2 with $K = K_1 \cup K_2$ such that $|K| \neq |K_1|^\circ \cup |K_2|^\circ$. Nevertheless, excision and Mayer–Vietoris do hold for $|K|$, $|K_1|$, and $|K_2|$; this will follows from Theorem 7.16 and Corollary 7.18 once we prove that $H_q(K, K_1) \cong H_q(|K|, |K_1|)$. The next lemma is a special case of this extended (singular) Mayer–Vietoris theorem that will be used to establish the general case.

Lemma 7.20. *Let K be a finite simplicial complex, and let s be a simplex of highest dimension; define $K_1 = K - \{s\}$ and $K_2 = \{s$ and all its proper faces$\}$. Then there is an exact Mayer–Vietoris sequence in singular homology*

$$\cdots \to H_q(|K_1| \cap |K_2|) \to H_q(|K_1|) \oplus H_q(|K_2|) \to H_q(|K|) \to H_{q-1}(|K_1| \cap |K_2|) \to \cdots$$

Proof. It suffices to prove excision here. Define $V = s - \{x\}$, where x is an interior point of s. Then V is an open subset of $|K_2|$ (because s has highest dimension), and $|K_1 \cap K_2| = |K_1| \cap |K_2| = |\dot{s}|$ is a deformation retract of V (deform along radii from x). There is a commutative diagram with exact rows and with vertical arrows induced by inclusions:

$$\begin{array}{ccccccccc} \cdots \longrightarrow & H_q(|K_1 \cap K_2|) & \longrightarrow & H_q(|K_2|) & \longrightarrow & H_q(|K_2|, |K_1 \cap K_2|) & \longrightarrow \cdots \\ & \downarrow & & \downarrow 1 & & \downarrow & \\ \cdots \longrightarrow & H_q(V) & \longrightarrow & H_q(|K_2|) & \longrightarrow & H_q(|K_2|, V) & \longrightarrow \cdots \end{array}$$

Since $|K_1 \cap K_2|$ is a deformation retract of V, the inclusion is a homotopy equivalence, hence it induces isomorphisms for all q. The five lemma now shows that inclusion induces an isomorphism for all q

$$H_q(|K_2|, |K_1| \cap |K_2|) \cong H_q(|K_2|, V).$$

Let $X_1 = |K_1| \cup V$. Note that $X_1 \cap |K_2| = (|K_1| \cup V) \cap |K_2| = (|K_1| \cap |K_2|) \cup (V \cap |K_2|) = V$ because $|K_1| \cap |K_2| = |K_1 \cap K_2| \subset V \subset |K_2|$. Furthermore, $|K_1| \subset X_1^\circ$ and, since $|K_2| - |K_1|$ is an open subset of $|K_2|$, it follows that $|K_2| - |K_1| \subset |K_2|^\circ$. Therefore $X_1^\circ \cup |K_2|^\circ = |K|$ and (singular) excision holds: inclusion induces isomorphisms for all q

$$H_q(|K_2|, V) \cong H_q(|K|, |K_1|).$$

Composing with the earlier isomorphisms gives the desired isomorphisms

$$H_q(|K_2|, |K_1| \cap |K_2|) \cong H_q(|K|, |K_1|) \quad \text{for all } q. \qquad \square$$

Lemma 7.21. *For each oriented simplicial complex K, there is a chain map $j = j^K\colon \tilde{C}_*(K) \to \tilde{S}_*(|K|)$ with each j_q an injection. For every simplicial map $\varphi\colon K \to K'$, there is a commutative diagram*

$$\begin{array}{ccc} \tilde{C}(K) & \xrightarrow{\varphi_*} & \tilde{C}(K') \\ \downarrow j & & \downarrow j \\ \tilde{S}(|K|) & \xrightarrow{|\varphi|_*} & \tilde{S}(|K'|). \end{array}$$

Moreover, if K_1 and K_2 are subcomplexes of K as in Lemma 7.20, then there is a commutative diagram

$$\begin{array}{ccc} H_q(K) & \longrightarrow & H_{q-1}(K_1 \cap K_2) \\ \downarrow & & \downarrow \\ H_q(|K|) & \longrightarrow & H_{q-1}(|K_1 \cap K_2|), \end{array}$$

where the horizontal maps are connecting homomorphisms of Mayer–Vietoris sequences.

PROOF. Define $j_{-1}: C_{-1}(K) \to S_{-1}(|K|)$ by $\langle\ \ \rangle \mapsto [\ \ ]$ and extending by linearity. If $q \geq 0$, define $j_q: C_q(K) \to S_q(|K|)$ by

$$j_q(\langle p_0, \ldots, p_q \rangle) = \sigma,$$

where $\sigma: \Delta^q \to |K|$ is the affine map $\sum t_i e_i \mapsto \sum t_i p_i$. The routine verifications of the stated properties of j are left to the reader. □

Theorem 7.22. *For every oriented finite simplicial complex K, the chain map* $j: \tilde{C}_*(K) \to \tilde{S}_*(|K|)$ *(of Lemma 7.21) induces isomorphisms,*[5] *for all* $q \geq 0$,

$$\tilde{H}_q(K) \cong \tilde{H}_q(|K|).$$

PROOF. We do an induction on the number N of simplexes in K. If $N = 1$, then $K = \varnothing$ and $|K| = \varnothing$ (reduced homology recognizes $\varnothing$ as a simplex), and $\tilde{H}_*(K) = 0 = \tilde{H}_*(|K|)$ in this case.

Assume that $N > 1$ and choose a simplex $s \in K$ of highest dimension. As in Lemma 7.20, define

$$K_1 = K - \{s\} \quad \text{and} \quad K_2 = \{s \text{ and all of its proper faces}\}.$$

Thus $K_1 \cup K_2 = K$ and $K_1 \cap K_2 = \{\text{all proper faces of } s\}$. Note that the vertex sets of K_2 and of $K_1 \cap K_2$ are each linearly ordered in the induced orientation. Since each of K_1 and $K_1 \cap K_2$ have fewer than N simplexes (the alternative is that $K_1 = K - \{s\} = \varnothing$ and $K = s$, which must now be a 0-simplex), the inductive hypothesis says that the respective chain maps j induce isomorphisms for each $q \geq 0$:

$$\tilde{H}_q(K_1) \xrightarrow{\cong} \tilde{H}_q(|K_1|) \quad \text{and} \quad \tilde{H}_q(K_1 \cap K_2) \xrightarrow{\cong} \tilde{H}_q(|K_1 \cap K_2|).$$

There are two Mayer–Vietoris sequences, from Corollary 7.17 and Lemma 7.20, and the maps j between them give a commutative diagram with exact rows, by Lemma 7.21.

$$\begin{array}{ccccccccc}
\tilde{H}_q(K_1 \cap K_2) & \to & \tilde{H}_q(K_1) \oplus \tilde{H}_q(K_2) & \to & \tilde{H}_q(K) & \to & \tilde{H}_{q-1}(K_1 \cap K_2) & \to & \tilde{H}_{q-1}(K_1) \oplus \tilde{H}_{q-1}(K_2) \\
\downarrow & & \downarrow & & \downarrow & & \downarrow & & \downarrow \\
\tilde{H}_q(|K_1 \cap K_2|) & \to & \tilde{H}_q(|K_1|) \oplus \tilde{H}_q(|K_2|) & \to & \tilde{H}_q(|K|) & \to & \tilde{H}_{q-1}(|K_1 \cap K_2|) & \to & \tilde{H}_{q-1}(|K_1|) \oplus \tilde{H}_{q-1}(|K_2|).
\end{array}$$

By Corollary 7.18

$$\tilde{H}_q(K_1) \oplus \tilde{H}_q(K_2) = \tilde{H}_q(K_1);$$

similarly, since $|K_2| = s$ is contractible,

$$\tilde{H}_q(|K_1|) \oplus \tilde{H}_q(|K_2|) = \tilde{H}_q(|K_1|).$$

Since all vertical maps are now induced by j, all save the middle one are known to be are isomorphisms. But the five lemma (Theorem 5.10) applies to show that the middle map $j_*: \tilde{H}_q(K) \to \tilde{H}_q(|K|)$ is also an isomorphism. □

[5] j is actually a chain equivalence; this follows from Theorem 9.8.

Corollary 7.23 (Alexander–Veblen). *Let X be a polyhedron having triangulations (K, h) and (K', h'). Then $H_q(K) \cong H_q(K')$ for all $q \geq 0$.*

Proof. By hypothesis, $|K| \approx |K'|$. But $H_q(K) \cong H_q(|K|)$ and $H_q(K') \cong H_q(|K'|)$, by Theorem 7.22. □

Corollary 7.24. *If X is a polyhedron of dimension m, then*

(i) $H_q(X)$ *is* f.g. *for every* $q \geq 0$;
(ii) $H_q(X) = 0$ *for all* $q > m$;
(iii) $H_m(X)$ *is free abelian.*

Proof. Immediate from Theorems 7.22 and 7.14. □

Corollary 7.25. (i) *If K is an oriented simplicial complex, then $H_q(K)$ is independent of the orientation.*

(ii) *If X is a polyhedron with triangulation (K, h), then the Euler–Poincaré characteristic is independent of the triangulation.*

Proof. (i) $H_q(|K|)$ is independent of any ordering of Vert(K).
(ii) Combine Theorems 7.22 and 7.15. □

One can now define $\chi(X)$, the **Euler–Poincaré characteristic** of a polyhedron X as $\chi(K)$, where there is a triangulation (K, h) of X.

One last comment before proceeding. First attempts to prove Corollary 7.23 were aimed at the polyhedron itself. For many years, one tried to prove the **Hauptvermutung** (principle conjecture): if (K, h) and (L, g) are triangulations of a polyhedron X, then there are subdivisions (not necessarily barycentric) K' of K and L' of L with $K' \cong L'$. Were this true, there would be an easy proof of the topological invariance of $H_*(K)$. The Hauptvermutung was proved for $n = 3$ by Moise (in the 1950s), but in 1961 Milnor constructed counterexamples to it for every $n \geq 6$.

The following notion is a substitute for homotopy in $\mathscr{K}$.

Definition. Let $\varphi, \psi\colon K \to L$ be simplicial maps. Then φ is **contiguous** to ψ, denoted by $\varphi \sim \psi$, if, for each simplex $s = \{p_0, \ldots, p_q\}$ of K, there exists a simplex s' of L with both $\{\varphi p_0, \ldots, \varphi p_q\}$ and $\{\psi p_0, \ldots, \psi p_q\}$ faces of s'.

Exercises

7.24. Let $\varphi, \psi\colon K \to L$ be contiguous.
(i) Prove that $|\varphi| \simeq |\psi|$.
(ii) $\varphi_* = \psi_*\colon H_q(K) \to H_q(L)$ for all $q \geq 0$ (*Hint*: Theorem 7.22.)

7.25. Give an example showing that contiguity may not be a transitive relation.

Definition. Let (X, x_0) and (Y, y_0) be pointed spaces. Their **wedge** $X \vee Y$ is the quotient space of their disjoint union, $X \coprod Y$ in which the basepoints are identified.

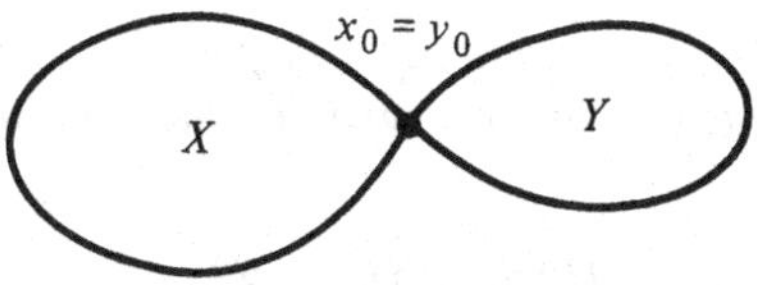

It is easy to see that if X and Y are polyhedra, then so is $X \vee Y$.

EXERCISES

*7.26. (i) Show that wedge is associative; that is, $(X \vee Y) \vee Z$ and $X \vee (Y \vee Z)$ are homeomorphic pointed spaces.

(ii) If K_1 and K_2 are polyhedra, then for all $n \geq 1$,

$$H_n(K_1 \vee K_2) \cong H_n(K_1) \oplus H_n(K_2).$$

(*Hint*: Mayer–Vietoris, Corollary 7.17.)

(iii) Let $m_1, \ldots, m_n$ be a sequence of nonnegative integers. Prove that there exists a connected polyhedron X of dimension n with $H_q(X)$ free abelian of rank m_q for every $q = 1, \ldots, n$.

7.27. (i) If L is a subcomplex of K, prove that for all $q \geq 0$,

$$H_q(K, L) \cong H_q(|K|, |L|).$$

(*Hint*: Five lemma and Theorem 7.22.)

(ii) There is an exact Mayer–Vietoris sequence in singular homology corresponding to any pair of subcomplexes K_1, K_2 of a simplicial complex K for which $K = K_1 \cup K_2$, namely,

$$\cdots \to H_q(|K_1| \cap |K_2|) \to H_q(|K_1|) \oplus H_q(|K_2|) \to H_q(|K|) \to H_{q-1}(|K_1| \cap |K_2|) \to \cdots.$$

*7.28. Let K be a simplicial complex and let $p \in \operatorname{Vert}(K)$. Define the **closed star** of p to be the subcomplex of K consisting of all the faces of those σ in the star $\operatorname{st}(p)$. Prove that the closed star of p is contractible. (*Hint*: Exercise 7.7(ii).).

The next result considers the question, generalizing the Mayer–Vietoris theorem, of relating the homology of K to the homology of subcomplexes whose union is K (also see [K. S. Brown, p. 166]).

Definition. A **cover** of a simplicial complex K is a family of subcomplexes $\mathscr{L} = \{L_\alpha : \alpha \in A\}$ with $K = \bigcup L_\alpha$.

Definition. Let $\mathscr{L} = \{L_\alpha : \alpha \in A\}$ be a cover of a simplicial complex K. The **nerve** of $\mathscr{L}$, denoted by $N(\mathscr{L})$, is the simplicial complex having vertices $\text{Vert}(N(\mathscr{L})) = A$ and with $\{\alpha_0, \ldots, \alpha_q\}$ a simplex if $\bigcap L_{\alpha_i} \neq \varnothing$ (see Example 7.9.)

Theorem 7.26 (Leray).[6] *Assume that $\mathscr{L} = \{L_\alpha : \alpha \in A\}$ is a cover of a simplicial complex K such that each L_α and every finite intersection $L_{\alpha_1} \cap L_{\alpha_2} \cap \cdots \cap L_{\alpha_q}$ is acyclic. Then*

$$H_q(K) \cong H_q(N(\mathscr{L}))$$

for all $q \geq 0$.

Proof. It suffices to construct a simplicial map $f: \text{Sd } K \to N(\mathscr{L})$ that induces isomorphisms $H_q(\text{Sd } K) \to H_q(N(\mathscr{L}))$ (because $H_q(K) \cong H_q(\text{Sd } K)$ since $|K| = |\text{Sd } K|$). We view Sd K as an abstract simplicial complex, as in Example 7.11. Linearly order the index set A, say, $A = \{\alpha_1, \ldots, \alpha_n\}$. Define $f: \text{Vert}(\text{Sd } K) \to \text{Vert}(N(\mathscr{L})) = A$ as follows: for each simplex $\sigma \in K$, there exists L_α with $\sigma \in L_\alpha$ (for $\mathscr{L}$ is a cover of K); define $f(\sigma) = \alpha_i$, where α_i is the first α for which $\sigma \in L_\alpha$. We claim that f is a simplicial map. If $\{\sigma_0, \ldots, \sigma_q\}$ is a simplex in Sd K, then $\sigma_0 < \sigma_1 < \cdots < \sigma_q$; thus $\sigma_0 \leq \sigma_i$ for every $i \leq q$, hence $\sigma_0 \in L_{\alpha_i}$ for every $i \leq q$. Therefore $\{f\sigma_0, \ldots, f\sigma_q\}$ is a simplex in $N(\mathscr{L})$, for $\sigma_0 \in L_{\alpha_0} \cap \cdots \cap L_{\alpha_q}$, and so this intersection is nonempty.

The proof that f induces isomorphisms in homology is by induction on $n = |A|$. If $n = 1$, then $N(\mathscr{L})$ is a point and $K = L_\alpha$ has the homology of a point: $H_q(L_\alpha) = 0$ for all $q > 0$ (K is acyclic because, by hypothesis, every L_α is acyclic). The result is thus obvious in this case.

Assume that $A = \{\alpha_1, \ldots, \alpha_{n+1}\}$. Define

$$K_1 = L_{\alpha_1} \cup \cdots \cup L_{\alpha_n} \quad \text{and} \quad K_2 = L_{\alpha_{n+1}}$$

(thus $\mathscr{L}_1 = \{L_{\alpha_1}, \ldots, L_{\alpha_n}\}$ is a cover of K_1); define

$$N_1 = N(\mathscr{L}_1).$$

Note that N_1 is a subcomplex of $N(\mathscr{L})$, as is N_2 defined by

$$N_2 = \text{closed star of } \alpha_{n+1}.$$

The construction of f shows that if $\sigma \in K_1$, then $f(\sigma) \in \{\alpha_1, \ldots, \alpha_n\}$. It follows by induction that $f|\text{Sd } K_1$ induces isomorphisms $H_q(\text{Sd } K_1) \xrightarrow{\sim} H_q(N_1)$. Furthermore $f|\text{Sd } L_{\alpha_{n+1}}$ induces isomorphisms $H_q(\text{Sd } K_2) \xrightarrow{\sim} H_q(N_2)$ because $K_2 = L_{\alpha_{n+1}}$, hence Sd K_2, is acyclic (by hypothesis) and N_2 is acyclic (by Exercise 7.28).

There is an obvious cover of $K_1 \cap K_2$, namely, $\mathscr{M} = \{M_{\alpha_1}, \ldots, M_{\alpha_n}\}$, where M_{α_i} is defined by $M_{\alpha_i} = L_{\alpha_i} \cap L_{\alpha_{n+1}}$. Note that $\mathscr{M}$ has the property that each

[6] It is proved in [Bott and Tu, p. 148] that if every finite intersection is contractible, then $\pi_1(|K|) \cong \pi_1(N(\mathscr{L}))$.

M_{α_i} and every finite intersection of them is acyclic. Since $\operatorname{Sd}(K_1 \cap K_2) = \operatorname{Sd} K_1 \cap \operatorname{Sd} K_2$, it follows by induction that $f|\operatorname{Sd}(K_1 \cap K_2)$ induces isomorphisms $H_q(\operatorname{Sd}(K_1 \cap K_2)) \xrightarrow{\sim} H_q(N(\mathscr{M}))$. But it is easy to see that $N(\mathscr{M}) = N_1 \cap N_2$, hence $H_q(N(\mathscr{M})) = H_q(N_1 \cap N_2)$. Of course, $\operatorname{Sd} K = \operatorname{Sd} K_1 \cup \operatorname{Sd} K_2$ and $N_1 \cup N_2 = N(\mathscr{L})$, so that we may apply Mayer–Vietoris (Corollary 7.17) to assert exactness of the rows in the following commutative diagram.

$$\begin{array}{ccccccccc}
H_q(\beta_1 \cap \beta_2) & \to & H_q(\beta_1) \oplus H_q(\beta_2) & \to & H_q(\operatorname{Sd} K) & \to & H_{q-1}(\beta_1 \cap \beta_2) & \to & H_{q-1}(\beta_1) \oplus H_{q-1}(\beta_2) \\
\downarrow & & \downarrow & & \downarrow & & \downarrow & & \downarrow \\
H_q(N_1 \cap N_2) & \to & H_q(N_1) \oplus H_q(N_2) & \to & H_q(N(\mathscr{L})) & \to & H_{q-1}(N_1 \cap N_2) & \to & H_{q-1}(N_1) \oplus H_{q-1}(N_2);
\end{array}$$

here β_i denotes $\operatorname{Sd} K_i$ for $i = 1, 2$, and the vertical maps are induced by restrictions of f. We have already seen that the four outside vertical maps are isomorphisms, and so the five lemma gives $f_*\colon H_q(\operatorname{Sd} K) \to H_q(N(\mathscr{L}))$ an isomorphism for all q. □

Definition. An **acyclic cover** of a simplicial complex is a cover satisfying the hypotheses of Theorem 7.26.

Corollary 7.27. *If $\mathscr{L}$ is an acyclic cover of a simplicial complex K, then $H_q(K) = 0$ for all $q >$ dim $N(\mathscr{L})$.*

Remark. Compare Exercise 6.4.

EXERCISES

7.29. In the proof of Theorem 7.26, suppose that we define g: Sd $K \to N(\mathscr{L})$ as follows: $g(\sigma) = \alpha$, where $\sigma \in L_\alpha$ (but α may not be the first such index in the ordering of A). Show that g and f are contiguous.

7.30. Let $\{M, L_1, \ldots, L_n\}$ be a cover of a simplicial complex K such that (i) each L_i is acyclic, (ii) $M \cap L_i$ is acyclic for each i, and (iii) $L_i \cap L_j \subset M$ for all $i \neq j$. Prove that $H_*(K) \cong H_*(M)$.

Calculations

The significance of Theorem 7.22 is that one can compute homology groups of polyhedra using simplicial homology. That this is valuable is clear from Corollary 7.24, for we now know that $H_q(|K|)$ is always f.g., and this is important because such groups are completely classified.

Fundamental Theorem. *Let G be a* f.g. *abelian group.*

(i) *$G = F \oplus T$, where F is free abelian of finite rank $r \geq 0$ and T is finite.*
(ii) *T is a direct sum of cyclic groups, $T = C_1 \oplus \cdots \oplus C_k$, with order $C_i = b_i$, say, and with $b_1 | b_2 | \cdots | b_k$ ($b_1 | b_2$ means "b_1 divides b_2"). The numbers $b_1, \ldots, b_k$ are called the* **torsion coefficients** *of G.*
(iii) rank *F and the torsion coefficients are invariants of G, and two* f.g. *abelian groups are isomorphic if and only if they have the same rank and the same torsion coefficients.*

Let F and F' be free abelian groups with finite bases $\{x_1, \dots, x_n\}$ and $\{x'_1, \dots, x'_m\}$, respectively. If $h: F \to F'$ is a homomorphism, then $h(x_i) = \sum d_{ji} x'_j$ with $d_{ji} \in \mathbf{Z}$. Thus, given ordered bases of F and F', h gives an $m \times n$ matrix $D = [d_{ji}]$ over $\mathbf{Z}$, where the ith column consists of the coefficients of $h(x_i)$ in terms of the x'_j. In light of Theorem 4.1(i), h is completely determined by this matrix.

Definition. A **normal form** is an $m \times n$ matrix N over $\mathbf{Z}$ such that

$$N = \begin{bmatrix} \Delta & 0 \\ 0 & 0 \end{bmatrix},$$

where Δ is a diagonal matrix, say, $\Delta = \operatorname{diag}\{b_1, b_2, \dots, b_k\}$, with $b_1 | b_2 | \cdots | b_k$. (Zero rows or columns bordering Δ need not be present.)

There is an analogue of Gaussian elimination for matrices over $\mathbf{Z}$. Define three types of **elementary row operation**: (i) interchange two rows, (ii) multiply a row by ± 1, and (iii) replace a row by that row plus an integer multiple of another row; there are three similar **elementary column operations**.

Theorem (Smith Normal Form). *Every rectangular matrix D over $\mathbf{Z}$ can be transformed, using elementary row and column operations, into a normal form; moreover, this normal form is independent of the elementary operations and is thus uniquely determined by D.*

The proof of this theorem uses nothing more sophisticated than the division algorithm in $\mathbf{Z}$; indeed the usual proof is itself an algorithm (e.g., see [Jacobson, p. 176]).

Theorem. *For any oriented simplicial complex K, there is an algorithm to compute $H_q(K)$ for all $q \geq 0$.*

For a proof, see [Munkres (1984), p. 60].

Here is the algorithm. Each $C_q(K)$ is a free abelian group equipped with a (finite) basis of oriented q-simplexes. As above, each $\partial_q: C_q(K) \to C_{q-1}(K)$ determines a matrix D_q over $\mathbf{Z}$ (with entries 0, 1, -1). Let N_q be the Smith normal form of D_q, let $\Delta_q = \operatorname{diag}\{b_1^q, \dots, b_{k_q}^q\}$ be the diagonal block of N_q, let $c_q \geq 0$ be the number of zero columns of N_q, and let $r_q \geq 0$ be the number of nonzero rows of N_q. Then the qth Betti number of K is $c_q - r_{q+1}$, and the torsion coefficients of $H_q(K)$ are those $b_1^q, \dots, b_{k_q}^q$ if any that are distinct from 1.

Here is the reason that the algorithm computes Betti numbers. Regard each integer matrix D_q as a matrix of rational numbers. Then the rank of the matrix D_q is the rank of the abelian group $B_{q-1}(K)$ and the nullity of D_q is the rank of $Z_q(K)$. Therefore

$$\operatorname{rank} H_q(K) = \operatorname{rank} Z_q(K) - \operatorname{rank} B_q(K) = \operatorname{nullity} D_q - \operatorname{rank} D_{q+1}.$$

In spite of this algorithm, one cannot in practice compute $H_q(X)$ with its use because of the large number of q-simplexes in a triangulation (K, h) of X. In short, the matrices D_q are too big and the calculations are too long. One is led to modify the definition of triangulation to obtain *cellular decompositions* of a space, which are more useful (we shall soon give an algebraic method of reducing the number of simplexes).

We illustrate these remarks by trying to compute $H_*(X)$ via simplicial homology when X is either the torus or the real projective plane.

EXAMPLE 7.12. In Example 7.5, we gave an explicit triangulation (K, h) of the torus X with $\dim K = 2$; moreover, if α_q denotes the number of q-simplexes in K, then $\alpha_2 = 18$, $\alpha_1 = 27$, and $\alpha_0 = 9$. It follows from Corollary 7.24 that $H_q(X) = 0$ for all $q > 2$ and that $H_2(X)$ is free abelian. Since X is path connected, $H_0(X) = \mathbf{Z}$. Now $\chi(K) = 9 - 27 + 18 = 0$, so that $1 + \operatorname{rank} H_2(X) = \operatorname{rank} H_1(X)$. To complete the computation using the algorithm, we must examine the matrices of ∂_2 and ∂_1; the first is 18×27 and the second is 27×9. These matrices are too big! Even a minimal triangulation having 14 triangles is not a significant improvement. These matrices will be shrunk in Example 7.14.

Let us instead use a Mayer–Vietoris sequence to compute $H_*(X)$. Take two circles a and b on the torus. Choose two overlapping open cylinders X_1 and

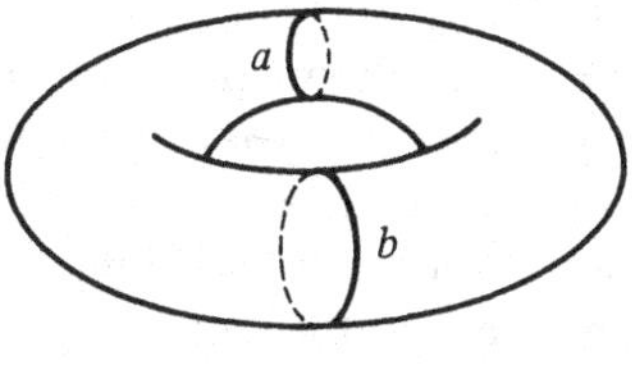

X_2, each containing a and b, with $X_1 \cup X_2 = X$ and such that $X_1 \cap X_2 = U_a \cup U_b$, a disjoint union of two open cylinders with $a \subset U_a$ and $b \subset U_b$. Note that X_1, X_2, U_a, and U_b each have the homotopy type of a circle S^1. There is thus an exact sequence of reduced homology groups:

$$0 \to \tilde{H}_2(X) \to \tilde{H}_1(U_a \cup U_b) \xrightarrow{f} \tilde{H}_1(X_1) \oplus \tilde{H}_1(X_2) \to \tilde{H}_1(X) \to \tilde{H}_0(U_a \cup U_b) \to 0.$$

Since we know generators of $H_1(S^1)$, we can abuse notation and write $\tilde{H}_1(U_a) = \langle a \rangle$ and $\tilde{H}_1(U_b) = \langle b \rangle$. Recall that if $i_1\colon U_a \cup U_b \hookrightarrow X_1$ and $i_2\colon U_a \cup U_b \hookrightarrow X_2$ are inclusions, then the map f in the sequence is given by $\operatorname{cls} z \mapsto (i_{1*} \operatorname{cls} z, i_{2*} \operatorname{cls} z)$. In particular, $i_{1*}a$ and $i_{1*}b$ are generators of $\tilde{H}_1(X_1)$ and $i_{2*}a$ and $i_{2*}b$ are generators of $\tilde{H}_1(X_2)$. It follows easily that f cannot be injective; therefore $\tilde{H}_2(X) \neq 0$. Furthermore, since $\tilde{H}_1(U_a \cup U_b) \cong \mathbf{Z} \oplus \mathbf{Z}$, we must have $\tilde{H}_2(X) \cong \mathbf{Z}$ (otherwise, $\operatorname{im} f$ is a subgroup of $\tilde{H}_1(X_1) \oplus \tilde{H}_1(X_2) = \mathbf{Z} \oplus \mathbf{Z}$ of rank 0, i.e., $\operatorname{im} f = 0$, and this is not so). One can show that $(\mathbf{Z} \oplus \mathbf{Z})/\operatorname{im} f \cong \mathbf{Z}$. Since $\tilde{H}_1(X) \cong \mathbf{Z} \oplus (\mathbf{Z} \oplus \mathbf{Z})/\operatorname{im} f$ (see footnote on page 103), it follows that $\tilde{H}_1(X) \cong \mathbf{Z} \oplus \mathbf{Z}$. (A more sophisticated argument showing that $H_1(X) = \mathbf{Z} \oplus \mathbf{Z}$ uses the Hurewicz theorem (Theorem

4.29) since we know that

$$\pi_1(X) = \pi_1(S^1 \times S^1) \cong \pi_1(S^1) \times \pi_1(S^1) = \mathbf{Z} \oplus \mathbf{Z}.)$$

We agree that this proof is not satisfying, because it really does not use a triangulation of X. See Example 7.14 for a better version.

EXAMPLE 7.13. Let X be the real projective plane $\mathbf{R}P^2$ regarded as the quotient space of D^2 by identifying antipodal points.

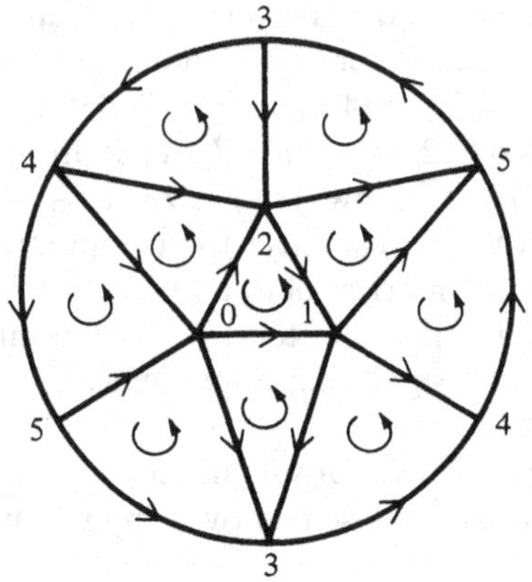

Note that dim $K = 2$, $\alpha_2 = 10$, $\alpha_1 = 15$, and $\alpha_0 = 6$. Again, $H_0(X) = \mathbf{Z}$, but now $\chi(X) = 6 - 15 + 10 = 1$. It follows that rank $H_2(X) =$ rank $H_1(X)$. If $\zeta = \langle 4, 5\rangle - \langle 3, 5\rangle + \langle 3, 4\rangle$, then it is easy to see that ζ is a cycle, that is, $\partial\zeta = 0$. To see that cls $\zeta \neq 0$, assume that $\zeta = \partial\xi$, where $\xi = \sum_{i=1}^{10} m_i\sigma_i$. Computing $\partial\sigma_i$ explicitly for each of the ten σ_i and comparing coefficients, one sees that all the m_i are equal; this leads to the contradiction that the coefficient of, say, $\langle 4, 5\rangle$ in the expression $\partial\xi = \zeta$ is even. If β is the 2-chain which is the sum of all the 2-simplexes in K with signs chosen according to the orientations above, then $\partial\beta = 2\zeta$ (every edge inside D^2 occurs exactly twice as a face of a 2-simplex and with opposite orientations; hence only the edges on the circumference survive). It turns out that $H_1(X) = \mathbf{Z}/2\mathbf{Z}$ (we have shown only that it has an element, namely, cls ζ, of order 2), hence rank $H_1(X) = 0$. It follows that $H_2(X) = 0$, since it is free abelian of rank 0. This example thus shows that the top homology group may be zero and also that there may be torsion coefficients. We shall complete this calculation in Example 7.15.

We have seen in Example 7.12 that the algorithm for computing homology is impractical for a space as simple as the torus. The following technique is more practical.

Definition. A subcomplex E'_* of a chain complex E_* is **adequate** if, for all $q \geq 0$:

(i) if $z \in Z_q$, then there exists $z' \in Z'_q$ with $z - z' \in B_q$;
(ii) if $z' \in Z'_q$ and $z' = \partial c$ for some $c \in E_{q+1}$, then there exists $c' \in E'_{q+1}$ with $z' = \partial c'$.

(Of course, Z_q and B_q are cycles and boundaries of E_*, and Z'_q and B'_q are cycles and boundaries of E'_*.)

Lemma 7.28. *If E'_* is an adequate subcomplex of E_*, then, for every q, the map $z' + B'_q \mapsto z' + B_q$ is an isomorphism*

$$H_q(E'_*) \xrightarrow{\sim} H_q(E_*).$$

Proof. Let $\theta\colon z' + B'_q \mapsto z' + B_q$ be the "enlargement of coset" map; it is well defined because $B'_q \subset B_q$. Now $\ker \theta$ is $(Z'_q \cap B_q)/B'_q$, and this is zero because axiom (ii) says that $Z'_q \cap B_q \subset B'_q$. Finally, $\operatorname{im} \theta = (Z'_q + B_q)/B_q = Z_q/B_q$, because axiom (i) says that $Z'_q + B_q = Z_q$. □

Definition. A chain complex E_* is **finitely based** if each term E_q is a f.g. free abelian group with a specified basis; the elements of the specified basis of E_q are called (algebraic) ***q*-cells**.

If K is a finite oriented simplicial complex, then $C_*(K)$ is finitely based with q-cells all symbols $\langle p_0, \ldots, p_q \rangle$, where $p_0 < p_1 < \cdots < p_q$ and $\{p_0, \ldots, p_q\}$ spans a simplex in K.

Lemma 7.29. *Let E_* be a finitely based chain complex, and let σ be a q-cell such that $\sigma = \partial\tau$ for some $(q + 1)$-cell τ. If σ is not involved in $\partial\tau'$ for any $(q + 1)$-cell $\tau' \neq \tau$, then one may remove σ and τ leaving an adequate subcomplex E'_*.*

Proof. Let E'_{q+1} be the free abelian group with basis all $(q + 1)$-cells $\tau' \neq \tau$, let E'_q be the free abelian group with basis all q-cells $\sigma' \neq \sigma$, and let $E'_p = E_p$ for all $p \neq q, q + 1$. It is easy to see that E'_* is a subcomplex of E_* if we show that $\operatorname{im} \partial_{q+2} \subset E'_{q+1}$. It suffices to see that there are no $(q + 2)$-cells c with $\partial c = \pm\tau + \gamma$, where τ is not involved in γ. If such a c exists, then $0 = \pm\sigma + \partial\gamma$ and σ is involved in $\partial\gamma$, contrary to the hypothesis. It remains to check axioms (i) and (ii) in each dimension.

Dimension $q + 1$. To check (i), let $z \in Z_{q+1}$. Is there $z' \in Z'_{q+1}$ with $z - z' \in B_q$? Now $z = m\tau + \alpha$, where $m \in \mathbf{Z}$ and τ is not involved in α. Since z is a cycle, $0 = \partial z = m\sigma + \partial\alpha$, where (by hypothesis) σ is not involved in $\partial\alpha$. But $0 = m\sigma + \partial\alpha$ is an equation relating basis elements, hence $m = 0 = \partial\alpha$. Thus $\alpha \in Z'_{q+1}$ and $z - \alpha = 0 \in B_{q+1}$. To check (ii), assume that $z' \in Z'_{q+1}$ and $z' = \partial c$ for some $c \in E_{q+2}$. Since $E_{q+2} = E'_{q+2}$, we have $c \in E'_{q+2}$.

Dimension q. To check (i), let $z \in Z_q$, and write $z = n\sigma + \beta$, where $n \in \mathbf{Z}$ and σ is not involved in β. Then $0 = \partial z = n\partial\sigma + \partial\beta = \partial\beta$ (because $\sigma = \partial\tau$). Therefore $\beta \in Z'_q$ and $z - \beta = n\sigma = n\partial\tau \in B_q$. To check (ii), take $z' \in Z'_q$ with $z' = \partial c$ for some $c \in E_{q+1}$; thus $c = m\tau + c'$ for $m \in \mathbf{Z}$ and $c' \in E'_{q+1}$. Hence

$$z' = \partial c = m\partial\tau + \partial c' = m\sigma + \partial c'.$$

Since σ is not involved in either $\partial c'$ or z', it follows that $m = 0$. □

Theorem 7.30 (Reduction). *Let E_* be a finitely based chain complex. Let σ be a q-cell involved in the boundary of precisely two $(q + 1)$-cells τ_1 and τ_2 and with opposite sign; that is,*

$$\partial\tau_1 = \sigma + c_1 \quad \textit{and} \quad \partial\tau_2 = -\sigma + c_2,$$

where σ is involved in neither c_1 nor c_2.

Then replacing τ_1 and τ_2 by $\tau_1 + \tau_2$ and deleting σ yields an adequate subcomplex.

PROOF. Change the basis of E_{q+1} by replacing τ_1 by $\tau_1 + \tau_2$ and τ_2 by τ_1; change the basis of E_q by replacing σ by $\sigma' = \sigma + c_1$. We claim that this new finitely based chain complex satisfies the hypotheses of Lemma 7.29. Note that $\sigma' = \partial\tau_1$. Let τ be a $(q + 1)$-cell with $\tau \neq \tau_1$. Either $\tau = \tau_1 + \tau_2$ or τ is an original $(q + 1)$-cell. In the first case, $\partial\tau = \partial(\tau_1 + \tau_2) = c_1 + c_2$, and this does not involve σ' because it does not involve σ. In the second case, the hypothesis says that $\partial\tau$ does not involve σ, and hence it does not involve $\sigma' = \sigma + c_1$. It follows from Lemma 7.29 that removal of τ_1 and σ' leaves an adequate subcomplex (note that τ_2 was removed at the outset, being replaced by τ_1). Finally, rewrite the basis of E_q in terms of the original basis. □

In Examples 7.5 and 7.6, certain spaces were constructed from the square $\mathbf{I} \times \mathbf{I}$ by identifying various edges; the following discussion will compute the homology groups of these spaces.

Let P be a polygon in the plane having m sides, with vertices ordered counterclockwise, and let X be the quotient space of P that identifies certain edges. The following triangulation of P induces a triangulation of X. Let A,

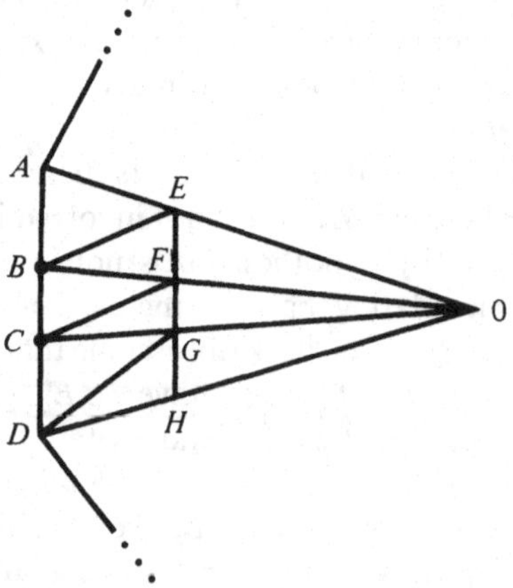

D be consecutive vertices of P. Insert an interior vertex 0 and new (boundary) vertices B, C as illustrated, and draw the edges $0A$, $0B$, $0C$, $0D$. Insert new

vertices E on $0A$ and H on $0D$. Finally, draw EH, label the new vertices F and G, and insert edges BE, CF, and DG. This triangulation should be repeated for each sector of P. Now orient every triangle counterclockwise. Note that the triangle $0AD$ has been subdivided into nine triangles, so that P is subdivided into $9m$ triangles.

Let K denote this triangulation of X, and let $C_*(K)$ be the simplicial chain complex of K. Then $C_*(K)$ has nonzero q-cells only for $q \leq 2$. We shall use reduction (Theorem 7.30) to replace $C_*(K)$ by an adequate subcomplex having fewer cells.

Apply reduction to remove successively the 1-cells corresponding to the edges $0F$, $0G$, BE, BF, CF, CG, DG (each lies on the boundary of exactly two 2-cells); the picture is now

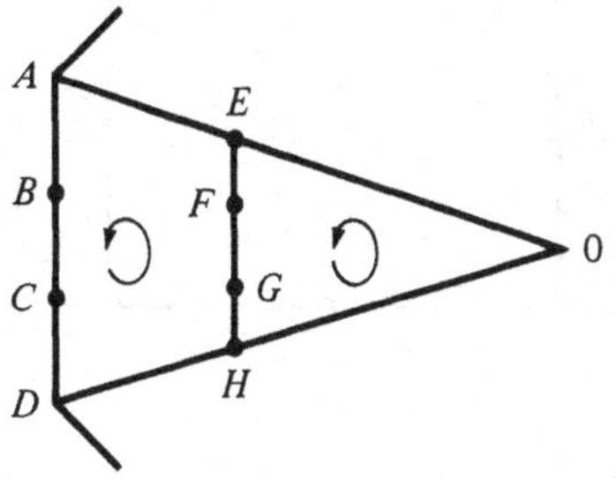

Reduce by removing the 0-cells B, C, F, G; now remove the 1-cell EH, and then the 0-cells E and H. What remains is

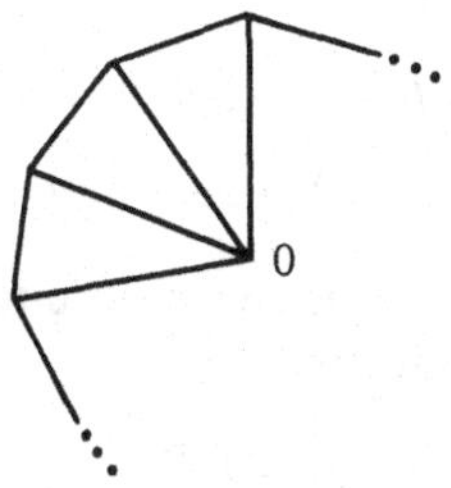

Now successively remove all but two of the radii, leaving

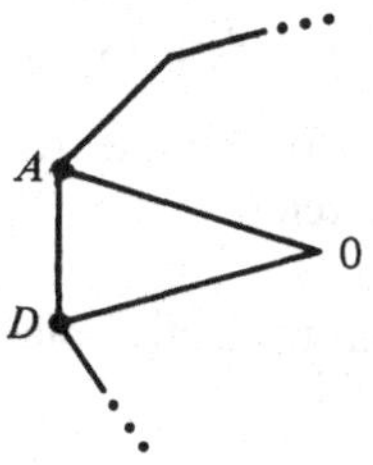

Finally, remove the 0-cell 0 (which lies on exactly two 1-cells), making the broken line $A0 + 0D$ into a new 1-cell; reduce once again to eliminate this 1-cell (which lies on exactly two 2-cells).

In sum, we have arrived at an adequate subcomplex of $C_*(K)$ having at most m 0-cells (Example 7.14 shows that there can be fewer than m), at most m 1-cells, and one 2-cell (the polygon itself).

EXAMPLE 7.14. Let X be the torus arising from identifying opposite edges of a square P as follows.

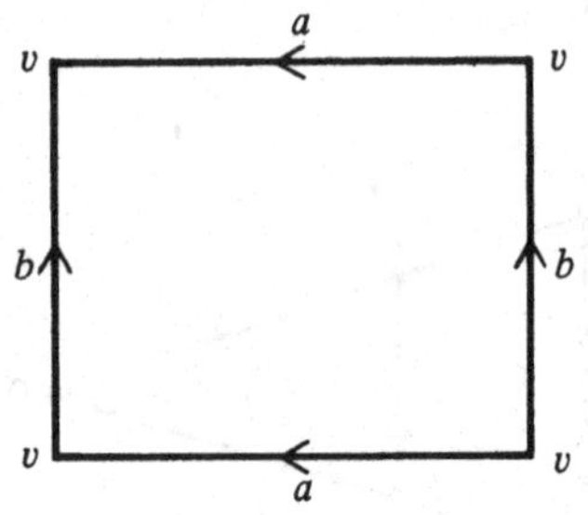

Note that, in this case, all the vertices (corners) of the square are identified to a common vertex. The adequate subcomplex obtained above has chains

$$E_2 = \langle P \rangle, \quad E_1 = \langle a \rangle \oplus \langle b \rangle, \quad E_0 = \langle v \rangle,$$

and differentiations

$$\partial P = a - b - a + b = 0,$$

$$\partial a = v - v = 0 = \partial b, \quad \text{and} \quad \partial v = 0.$$

(The differentiations in a subcomplex are restrictions of the differentiations in the original subcomplex.) We see that

$$Z_2 = \langle P \rangle, \quad Z_1 = \langle a \rangle \oplus \langle b \rangle, \quad Z_0 = \langle v \rangle,$$

$$B_2 = 0, \qquad B_1 = 0, \qquad B_0 = 0,$$

and we conclude that

$$H_2 = \mathbf{Z}, \qquad H_1 = \mathbf{Z} \oplus \mathbf{Z}, \qquad H_0 = \mathbf{Z}.$$

Of course, this result agrees with Example 7.12. Note that a basis of $H_1(X)$ consists of the two "obvious" circles.

EXAMPLE 7.15. Let X be the real projective plane $\mathbf{R}P^2$. Here are two pictures ($m = 1$ and $m = 2$):

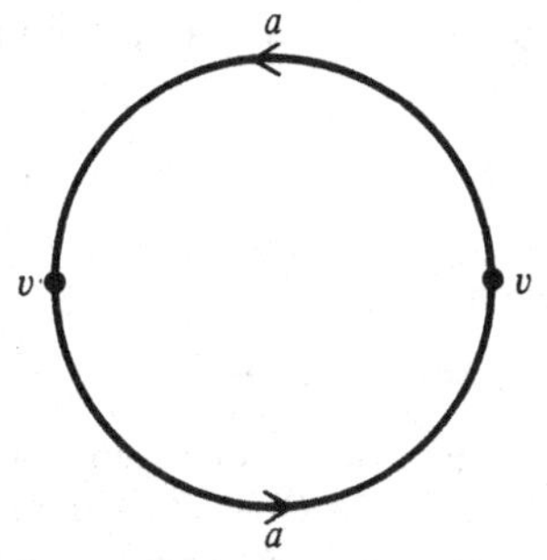

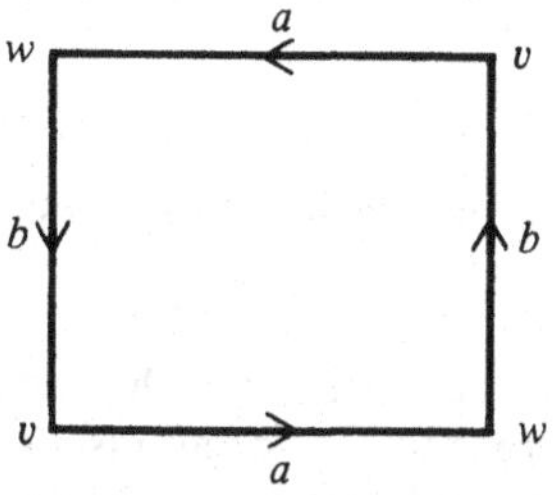

For the first picture, the adequate subcomplex has chains

$$E_2 = \langle P \rangle, \quad E_1 = \langle a \rangle, \quad E_0 = \langle v \rangle,$$

and differentiations

$$\partial P = a + a = 2a, \quad \partial a = v - v = 0, \quad \text{and} \quad \partial v = 0.$$

We conclude that

$$Z_2 = 0, \quad Z_1 = \langle a \rangle, \quad Z_0 = \langle v \rangle,$$
$$B_2 = 0, \quad B_1 = \langle 2a \rangle, \quad B_0 = 0,$$

hence

$$H_2 = 0, \quad H_1 = \mathbf{Z}/2\mathbf{Z}, \quad H_0 = \mathbf{Z}.$$

If we compute using the second picture, the adequate subcomplex has chains

$$E_2 = \langle P \rangle, \quad E_1 = \langle a \rangle \oplus \langle b \rangle, \quad E_0 = \langle v \rangle \oplus \langle w \rangle$$

and differentiations

$$\partial P = 2(a + b), \quad \partial a = w - v, \quad \partial b = v - w, \quad \partial v = 0 = \partial w.$$

We conclude that

$$Z_2 = 0, \quad Z_1 = \langle a + b \rangle, \quad Z_0 = \langle v \rangle \oplus \langle w \rangle,$$
$$B_2 = 0, \quad B_1 = \langle 2(a + b) \rangle, \quad B_0 = \langle w - v \rangle,$$

and again

$$H_2 = 0, \quad H_1 = \mathbf{Z}/2\mathbf{Z}, \quad H_0 = \mathbf{Z}$$

(one needs a little algebra to see that H_0 is infinite cyclic: the homomorphism $Z_0 \to \mathbf{Z}$ defined by $mv + nw \mapsto m + n$ is a surjection with kernel B_0, and so the first isomorphism theorem gives $H_0 = Z_0/B_0 \cong \mathbf{Z}$).

Remark. It is known (see [Massey (1967), Chap. 1]) that every compact connected 2-manifold can be obtained by identifying edges of an even-sided polygon. The method of adequate subcomplexes is thus strong enough to compute their homology groups.

EXERCISES

7.31. Show that the homology groups of the Klein bottle are

$$H_p = 0 \text{ for } p \geq 2, \quad H_1 = \mathbf{Z} \oplus (\mathbf{Z}/2\mathbf{Z}), \quad \text{and} \quad H_0 = \mathbf{Z}.$$

7.32. Let P be a polygon with k vertices $v_0, v_1, \ldots, v_{k-1}$ (where we assume that v_i is adjacent to v_{i-1} for all i (subscripts are read mod k)). Orient the edges in the direction from v_{i-1} to v_i, and now identify all edges with one another. The quotient space is called the k-fold **dunce cap** (when $k = 2$, the dunce cap is the real projective plane).

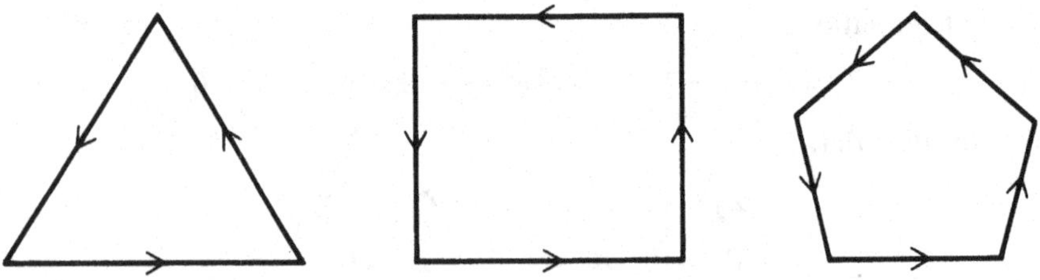

Prove that the homology groups of the k-fold dunce cap are:

$$H_p = 0 \text{ for } p \geq 2, \quad H_1 = \mathbf{Z}/k\mathbf{Z}, \quad H_0 = \mathbf{Z}.$$

Fundamental Groups of Polyhedra

Let us turn our attention from the homology groups of a polyhedron to its fundamental group. We begin by mimicking, in an atopological setting, our earlier discussion of multiplication of paths.

Definition. An **edge** $e = (p, q)$ in a simplicial complex K is an ordered pair of (not necessarily distinct) vertices lying in a simplex of K; p is called the **origin** of e and q is called the **end** of e.

Definition. An **edge path** α (of **length** n) in K is a finite sequence of edges,

$$\alpha = e_1 e_2 \cdots e_n,$$

where end e_i = origin e_{i+1} for all $i = 1, 2, \ldots, n - 1$. We call origin e_1 the **origin** of α, denoting it by $o(\alpha)$, and we call end e_n the **end** of α, denoting it by $\mathrm{e}(\alpha)$. An edge path α is **closed** if $o(\alpha) = \mathrm{e}(\alpha)$. If $\alpha = e_1 \cdots e_n$ and $\alpha' = e'_1 \cdots e'_m$ are edge paths with $\mathrm{e}(\alpha) = o(\alpha')$ then their **product** is

$$\alpha\alpha' = e_1 \cdots e_n e'_1 \cdots e'_m.$$

Clearly, the product of edge paths, when defined, is associative.

Notation. If $e = (p, q)$ is an edge, then $e^{-1} = (q, p)$ (which is also an edge). If $\alpha = e_1 \cdots e_n$ is an edge path, then its **inverse** is $\alpha^{-1} = e_n^{-1} \cdots e_1^{-1}$. If $e = (p, p)$, then e is called a **constant path** and is denoted by i_p.

In order to force a group structure on edge paths, we must (as with fundamental groups) impose an equivalence relation on them.

Definition. Two edge paths α and α' in K are **homotopic**, denoted by $\alpha \simeq \alpha'$, if one can be obtained from the other by a finite number of **elementary moves** consisting of replacing one side of the following equation by the other:

$$\beta(p, q)(q, r)\gamma = \beta(p, r)\gamma,$$

where $\{p, q, r\}$ lie in a simplex of K, and β, γ are (possibly empty) edge paths in K.

EXAMPLE 7.16. If K is the 2-simplex $[p_0, p_1, p_2]$, then the edge paths $\alpha = (p_0, p_1)(p_1, p_2)$ and $\alpha' = (p_0, p_2)$ are homotopic; if $K^{(1)}$ is the 1-skeleton of K, then these edge paths are not homotopic in $K^{(1)}$.

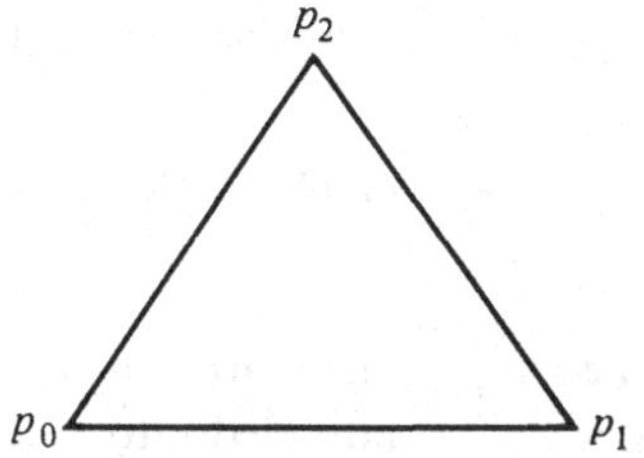

EXERCISES

7.33. Homotopy is an equivalence relation on the set of all edge paths in K; the equivalence class of α is denoted by $[\alpha]$ and is called a **path class**.

7.34. (i) If $\alpha \simeq \alpha'$, then $o(\alpha) = o(\alpha')$ and $e(\alpha) = e(\alpha')$. Conclude that $o[\alpha]$ and $e[\alpha]$ are well defined.
(ii) If $\alpha \simeq \alpha'$, $\beta \simeq \beta'$ and $e(\alpha) = o(\beta)$, then $\alpha\beta \simeq \alpha'\beta'$. Conclude that $[\alpha][\beta] = [\alpha\beta]$ is well defined when $e[\alpha] = o[\beta]$.

Let $\pi(K)$ denote the set of all path classes in K.

Theorem 7.31. *$\pi(K)$ is a groupoid, that is, it is an algebraic system satisfying the following axioms:*

(i) *each path class* $[\alpha]$ *has an origin* p *and an end* q, *where* $p, q \in \text{Vert}(K)$, *and*

$$[i_p][\alpha] = [\alpha] = [\alpha][i_q];$$

(ii) *associativity holds when defined;*
(iii) $[\alpha][\alpha^{-1}] = [i_p]$ *and* $[\alpha^{-1}][\alpha] = [i_q]$.

PROOF. Straightforward (much simpler than the analogous Theorem 3.2). □

Definition. Fix a vertex $p \in \text{Vert}(K)$ and call it a **basepoint**. The **edge path group** is

$$\pi(K, p) = \{[\alpha] \in \pi(K): o[\alpha] = p = e[\alpha]\}.$$

Theorem 7.32. *The edge path group* $\pi(K, p)$ *is a group.*

PROOF. Immediate from Theorem 7.31. □

Definition. A simplicial complex K is **connected** if, for every pair of vertices p, $q \in \text{Vert}(K)$, there exists an edge path in K from p to q.

EXERCISES

7.35. Show that the following are equivalent: K is connected; the 1-skeleton $K^{(1)}$ is connected; $|K|$ is connected; $|K|$ is path connected.

7.36. If K is connected and $p_0, p_1 \in \text{Vert}(K)$, then $\pi(K, p_0) \cong \pi(K, p_1)$. (*Hint*: See the proof of Theorem 3.6.)

7.37. If K is a connected simplicial complex with 2-skeleton $K^{(2)}$, then $\pi(K, p) \cong \pi(K^{(2)}, p)$.

Let $\alpha = e_1 \cdots e_m$ be an edge path in K from p_0 to p_m, where $e_i = (p_{i-1}, p_i)$ for $i = 1, \ldots, m$. Let $\mathbf{I}_m$ denote $\mathbf{I}$ subdivided into m intervals of equal length; more precisely, let $\mathbf{I}_m$ be the simplicial complex with $\text{Vert}(\mathbf{I}_m) = \{v_0, v_1, \ldots, v_m\}$ (so $v_i = i/m$) and 1-simplexes $\{v_{i-1}, v_i\}$ for $i = 1, \ldots, m$. An edge path α of length m defines a simplicial map $\alpha^\circ: \mathbf{I}_m \to K$ by $\alpha^\circ(v_i) = p_i$. Of course, $|\alpha^\circ|: \mathbf{I} \to |K|$ is an honest path in $|K|$, where $|\alpha^\circ|$ is the piecewise linear map determined by α°.

EXERCISES

7.38. Define a relation R on $\text{Vert}(K)$ by vRw if there exists an edge path in K from v to w.
(i) Show that R is an equivalence relation on $\text{Vert}(K)$.
(ii) For each $x \in \text{Vert}(K)$, define the **component** of K containing x as the family of all simplexes $s \in K$ with $\text{Vert}(s)$ contained in the R-equivalence class of x. Show that each component of K is a connected subcomplex and that K is their disjoint union.
(iii) If $x \in \text{Vert}(K)$ and L is the component of K containing x, then

$$\pi(K, x) \cong \pi(L, x).$$

*7.39. (i) If α and β are edge paths in K of lengths m and n, respectively, and if $e(\alpha) = o(\beta)$, then there are simplicial maps $\alpha^\circ: \mathbf{I}_m \to K$ and $\beta^\circ: \mathbf{I}_n \to K$ (as above). Define a simplicial map $\gamma: \mathbf{I}_{m+n} \to K$ by $\gamma(v_i) = \alpha^\circ(v_i)$ for $0 \le i \le m$ and $\gamma(v_{m+j}) = \beta^\circ(v_j)$ for $0 \le j \le n$. Show that $(\alpha\beta)^\circ = \gamma$.

(ii) If α and β are edge paths in K with $\alpha \simeq \beta$, then $|\alpha^\circ| \simeq |\beta^\circ|$ rel $\dot{\mathbf{I}}$.

In the sequel, we drop the distinction between α and α°, and we shall regard an edge path as a simplicial map when convenient.

Definition. An edge path $\alpha = e_1 \cdots e_n$ is **reduced** if no e_j is a constant i_p and if $e_j \neq e_{j+1}^{-1}$ for all $j = 1, 2, \ldots, n-1$; a **circuit** is a reduced closed edge path.

Definition. A **tree** is a connected simplicial complex T with $\dim T \le 1$ and which contains no circuits.

A tree of dimension 0 must have only one vertex.

EXERCISES

*7.40. If $\alpha = e_1 \cdots e_n$ is a closed edge path in K with $o(\alpha) = p = e(\alpha)$, and if there is a tree T in K containing every edge e_j, then $[\alpha] = 1$ in $\pi(K, p)$. (Intuitively, trees are contractible, and every path in a contractible space is nullhomotopic.)

*7.41. Let T_1 and T_2 be trees in a simplicial complex K. If $T_1 \cap T_2$ is connected, then $T_1 \cup T_2$ is a tree.

Lemma 7.33. *Every tree T' in a connected simplicial complex K is contained in a tree T with* $\mathrm{Vert}(T) = \mathrm{Vert}(K)$; *moreover, a tree T in K is maximal if and only if* $\mathrm{Vert}(T) = \mathrm{Vert}(K)$.

PROOF. Suppose there is a vertex $q \in \mathrm{Vert}(K)$ with $q \notin \mathrm{Vert}(T')$. Choose $p \in \mathrm{Vert}(T')$. Since K is connected, there is a (reduced) edge path α in K from $p = p_0$ to q, say, $\alpha = (p, p_1)(p_1, p_2)\cdots(p_n, q)$. Since $p \in \mathrm{Vert}(T')$ and $q \notin \mathrm{Vert}(T')$, there is a smallest index i with $p_i \in \mathrm{Vert}(T')$ and $p_{i+1} \notin \mathrm{Vert}(T')$. Define a subcomplex T'' of K with vertices $\mathrm{Vert}(T') \cup \{p_{i+1}\}$ and one additional 1-simplex $\{p_i, p_{i+1}\}$. Clearly, $T'' \supsetneqq T'$. But T'' is a tree, for any circuit in T'' must pass through p_{i+1} (since T' is a tree), and such an edge path cannot be reduced. This procedure may be iterated as long as the tree obtained has vertex set smaller than $\mathrm{Vert}(K)$. We conclude that a maximal tree T containing T' exists, and that $\mathrm{Vert}(T) = \mathrm{Vert}(K)$. The proof of the second statement is left as an (easy) exercise. □

It follows from Lemma 7.33 that maximal trees in finite simplicial complexes always exist; one can prove their existence in general by Zorn's lemma.

Some maximal trees of a "figure 8" are indicated below.

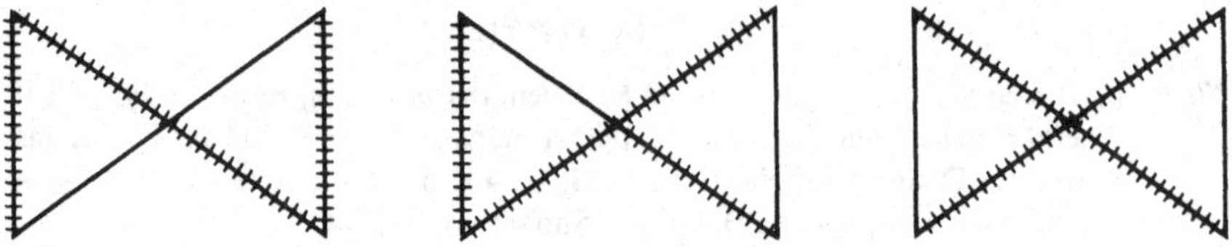

We now introduce free groups so that we can describe (not necessarily abelian) groups by generators and relations.

Definition. Let F be a group containing a subset X. Then F is **free** with **basis** X if for every group G and every function $\varphi: X \to G$, there exists a unique homomorphism $\tilde{\varphi}: F \to G$ with $\tilde{\varphi}(x) = \varphi(x)$ for all $x \in X$.

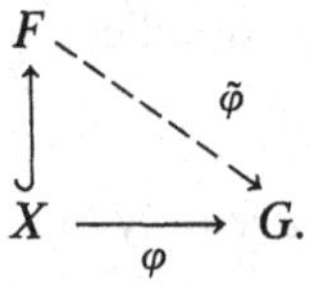

The reader should compare this definition with the corresponding property of free abelian groups in Theorem 4.1(i) (we emphasize that the groups G here may not be abelian). Assuming that free groups exist, one can prove, as in Theorem 4.1(ii), that every group is isomorphic to a quotient group of a free group. The positive answer to the existence question for free groups is given by the following construction.

Let X' be a set disjoint from X and let $x \mapsto x^{-1}$ be a bijection $X \to X'$. Let X'' be a set disjoint from $X \cup X'$ that contains one element we denote by "1". Call $X \cup X' \cup X''$ the *alphabet*, and call its elements *letters*. Let S be the set of all sequence of letters $(\alpha_1, \alpha_2, \ldots)$; that is, each $\alpha_k = 1$ or $x^{\pm 1}$ for some $x \in X$ (we agree that x^1 may denote x). A **word** on X is a sequence $(\alpha_1, \alpha_2, \ldots) \in S$ such that all coordinates are 1 from some point on; that is, there is an integer n such that $\alpha_k = 1$ for all $k \geq n$. In particular, the constant sequence

$$(1, 1, 1, \ldots)$$

is a word; it is called the **empty word** and is also denoted by 1. A **reduced word** on X is a word on X that satisfies the extra conditions:

(i) x and x^{-1} are never adjacent;
(ii) if $\alpha_m = 1$ for some m, then $\alpha_k = 1$ for all $k > m$.

In particular, the empty word is a reduced word. Since words contain only a finite number of letters before they become constant, we use the more economical (and suggestive) notation

$$w = x_1^{\varepsilon_1} x_2^{\varepsilon_2} \cdots x_n^{\varepsilon_n},$$

where $\varepsilon_i = \pm 1$. Observe that this spelling of a reduced word is unique, for this is just the definition of equality of sequences.

The idea of the construction of the free group F is just this: the elements of F are the reduced words and the binary operation is essentially juxtaposition. Unfortunately, the juxtaposition of two reduced words need not be reduced, and one defines the product of two reduced words as the reduced word obtained from their juxtaposition after performing all possible cancellations. One verifies that this product is well defined and associative. Moreover, given a function $\varphi: X \to G$, one defines

$$\tilde{\varphi}(x_1^{\varepsilon_1} x_2^{\varepsilon_2} \cdots x_n^{\varepsilon_n}) = \varphi(x_1)^{\varepsilon_1} \varphi(x_2)^{\varepsilon_2} \cdots \varphi(x_n)^{\varepsilon_n}.$$

Definition. If F is free with basis X, then **rank** $F = \text{card } X$.

EXERCISES

*7.42. If F is free with basis X and if F' is the commutator subgroup of F, then F/F' is free abelian with basis all cosets xF' with $x \in X$.

7.43. The rank of a free group does not depend on the choice of basis. (*Hint*: Exercise 7.42 and Theorem 4.3.)

Definition. A group G is defined by **generators** $X = \{x_k: k \in K\}$ and **relations** $\Delta = \{r_j = 1: j \in J\}$ if $G \cong F/R$, where F is free on X and R is the *normal* subgroup of F generated by $\{r_j: j \in J\}$. The ordered pair $(X|\Delta)$ is called a **presentation** of G.

There are two reasons forcing us to use the normal subgroup R generated by $\{r_j: j \in J\}$: we wish to form the quotient group F/R; if $r_j = 1$ in G and $w \in F$, then $wr_jw^{-1} = 1$ in G.

Definition. Let K be a connected simplicial complex and let T be a maximal tree in K. Define a group $G = G_{K,T}$ having the presentation:

Generators: all edges (p, q) in K;
Relations: (i) $(p, q) = 1$ if (p, q) is an edge in T;
(ii) $(p, q)(q, r) = (p, r)$ if $\{p, q, r\}$ is a simplex in K.

Theorem 7.34. *If K is a connected simplicial complex with basepoint p, then* $\pi(K, p) \cong G_{K,T}$.

Remark. We are describing $\pi(K, p)$ by generators and relations.

PROOF. Let F be the free group with basis all edges (u, v) in K, and let R be the normal subgroup of F generated by all relations of types (i) and (ii) above (so that $G_{K,T} = F/R$).

Let $v \in \text{Vert}(K)$. If $v = p$, define $\alpha_p = (p, p)$. If $v \neq p$, there is a reduced edge path α_v in T from p to v (for T is connected and $\text{Vert}(T) = \text{Vert}(K)$, by Lemma 7.33). Note that α_v is the unique such path lest T contain a circuit.

Define $\varphi: F \to \pi(K, p)$ by $\varphi(u, v) = [\alpha_u(u, v)\alpha_v^{-1}]$. To see that φ defines a homomorphism on $F/R = G_{K,T}$, it suffices to show that all the relators (hence R) lie in $\ker \varphi$.

Type (i). If $(u, v) \in T$, then every edge in $\alpha_u(u, v)\alpha_v^{-1}$ lies in T, and Exercise 7.40 shows that $\varphi(u, v) = 1$.

Type (ii). If $\{u, v, w\}$ is a simplex of K, then

$$[\alpha_u(u, v)\alpha_v^{-1}][\alpha_v(v, w)\alpha_w^{-1}] = [\alpha_u(u, v)\alpha_v^{-1}\alpha_v(v, w)\alpha_w^{-1}]$$
$$= [\alpha_u(u, v)(v, w)\alpha_w^{-1}] = [\alpha_u(u, w)\alpha_w^{-1}],$$

the last equation being the definition of the homotopy relation. Hence φ induces a homomorphism $\bar{\varphi}: G_{K,T} \to \pi(K, p)$, namely,

$$(u, v)R \mapsto \varphi(u, v) = [\alpha_u(u, v)\alpha_v^{-1}].$$

We prove that $\bar{\varphi}$ is an isomorphism by constructing its inverse. If $\alpha = e_1 \cdots e_n$ is a closed edge path in K at p, define

$$\theta(\alpha) = e_1 \cdots e_n R \in G_{K,T}.$$

If α' is a second such edge path with $\alpha' \simeq \alpha$, the relations of type (ii) show that $\theta(\alpha) = \theta(\alpha')$. Therefore θ induces a homomorphism $\bar{\theta}: \pi(K, p) \to G_{K,T}$ by

$$\bar{\theta}: [e_1 \cdots e_n] = [\alpha] \mapsto \theta(\alpha) = e_1 \cdots e_n R.$$

Let us compute composites. If $[\alpha] \in \pi(K, p)$ and $\alpha = e_1 \cdots e_n$, then

$$\begin{aligned}\bar{\varphi}\bar{\theta}[\alpha] &= \bar{\varphi}(\theta(\alpha)) = \bar{\varphi}(e_1 \cdots e_n R)\\ &= \varphi(e_1) \cdots \varphi(e_n) \quad \text{(since } \varphi \text{ is a homomorphism killing } R)\\ &= [\alpha_p e_1 \cdots e_n \alpha_p^{-1}]\\ &= [\alpha], \quad \text{since } [\alpha_p] = 1.\end{aligned}$$

Finally, assume that (u, v) is a generator of $G_{K,T}$.

$$\bar{\theta}\bar{\varphi}((u, v)R) = \bar{\theta}(\varphi(u, v)) = \bar{\theta}[\alpha_u(u, v)\alpha_v^{-1}]$$
$$= \alpha_u(u, v)\alpha_v^{-1}R.$$

Now α_u and α_v^{-1} lie in R (since their edges do), so that normality of R gives $\alpha_u(u, v)\alpha_v^{-1}R = \alpha_u(u, v)R = (u, v)R$.

Therefore both composites are identities, and $\bar{\varphi}$ is an isomorphism. □

Corollary 7.35. *If K is a graph, that is, a connected 1-complex, then $\pi(K, p)$ is a free group. Moreover, it has a basis in bijective correspondence with $\{$1-simplexes $s \in K: s \notin T\}$, where T is some chosen maximal tree in K.*

Proof. By relations of type (i), $\pi(K, p) \cong G_{K,T}$ is generated by all edges (u, v) that are not in T. Examining relations of type (ii), we see that $(u, v)(v, u) = 1$, so that, for each of the edges just mentioned, picking just one of (u, v) or (v, u) still leaves a set of generators. Next, if $\{u, v, w\}$ is a simplex of K, then at least one vertex is repeated, for $\dim K \leq 1$. The relations of type (ii) are thus all trivial $[(u, u)(u, v) = (u, v), (u, v)(v, v) = (u, v)$, and $(u, v)(v, u) = (u, u)]$. □

EXAMPLE 7.17. (i) If K is $\dot{s}$, the 1-skeleton of a 2-simplex s, then $\pi(K, p) \cong \mathbf{Z}$. (ii) If K is $\dot{s} \vee \dot{s}$, that is, a "figure 8", then $\pi(K, p)$ is a free group having two generators. In particular, $\pi(K, p)$ is not abelian.

Theorem 7.36. *If K is a connected simplicial complex with basepoint p, then*

$$\pi(K, p) \cong \pi_1(|K|, p).$$

Remark. A presentation for $\pi(K, p)$ is given in Theorem 7.34.

PROOF. If we regard an edge path α of length m as a simplicial map $\mathbf{I}_m \to K$, then Exercise 7.39(i) shows that there is a homomorphism $\tau: \pi(K, p) \to \pi_1(|K|, p)$ given by $[\alpha] \mapsto [|\alpha|]$, where $|\alpha|: |\mathbf{I}_m| = \mathbf{I} \to |K|$.

To see that τ is surjective, let $f: \mathbf{I} \to |K|$ be a closed path in $|K|$ at p. By Corollary 7.4, there is an integer m and a simplicial approximation $\varphi: \mathbf{I}_m \to K$ to f. Of course, we may regard φ as a closed edge path in K at p; moreover, $|\varphi| \simeq f$ rel $\dot{\mathbf{I}}$ [Exercise 7.39(ii)], and so $\tau: [\varphi] \mapsto [f]$.

To see that τ is injective, assume that α is a closed edge path in K at p with $|\alpha|$ nullhomotopic in $|K|$; we must show that $\alpha \simeq i_p$. Let $F: \mathbf{I} \times \mathbf{I} \to |K|$ be a (relative) homotopy $F: |\alpha| \simeq c$ rel $\dot{\mathbf{I}}$, where c is the constant path at p. Assume that α has length m, that is, $\alpha: \mathbf{I}_m \to K$, where $\alpha(0) = \alpha(m) = p$ and $\alpha(i) = p_i \in \text{Vert}(K)$ for $0 < i < m$. Subdivide $\mathbf{I} \times \mathbf{I}$ by a rectangular grid of vertical and horizontal lines, which contains a vertical line passing through each point $(v_i, 0)$, $0 \leq i \leq m$ (recall that $\text{Vert}(\mathbf{I}_m) = \{v_0, \ldots, v_m\}$). Further subdivide by bisecting each little square in the grid into two triangles, using one of its diagonals. Clearly, such subdivisions can be made with arbitrarily small mesh, and so Corollary 7.4 gives a simplicial approximation $\Phi: L \to K$ to F, where L is a suitable subdivision of $\mathbf{I} \times \mathbf{I}$.

Let $\mathbf{I}^*$ be the bottom edge of $\mathbf{I} \times \mathbf{I}$ as subdivided by L. Clearly, $\Phi|\mathbf{I}^*: \mathbf{I}^* \to K$ is a simplicial approximation to $F|\mathbf{I} \times \{0\} = |\alpha|$. Suppose that

$$\text{Vert}(\mathbf{I}^*) = \{\ldots, v_i, a_1, a_2, \ldots, a_k, v_{i+1}, \ldots\}.$$

Since $\Phi|\mathbf{I}^*$ is a simplicial map, $\Phi(a_j) \in \text{Vert}(K)$, $\Phi(v_i) = p_i$, and $\Phi(v_{i+1}) = p_{i+1}$. Furthermore, that $\Phi|\mathbf{I}^*$ is a simplicial approximation to $|\alpha|$ gives $\Phi(a_j) \in \{p_i, p_{i+1}\}$, that is, $\Phi(a_j) = p_i$ or p_{i+1}. The edge path $\alpha' = \Phi|\mathbf{I}^*$ is thus obtained from α by insertion of edges of the form (p_i, p_{i+1}), (p_{i+1}, p_i), (p_{i+1}, p_{i+1}), or (p_i, p_i); it follows that $\alpha \simeq \alpha'$. A similar investigation of the top edge of $\mathbf{I} \times \mathbf{I}$ (as well as the left and right sides) shows that each is just a product of i_p's (p is the basepoint of K), which is obviously homotopic to i_p. The bottom row of $\mathbf{I} \times \mathbf{I}$ has the following form.

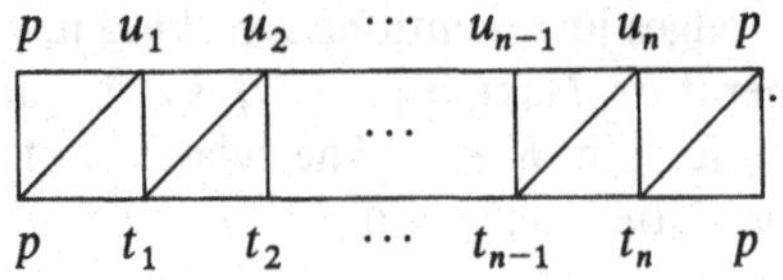

Now $\alpha \simeq \alpha' = (p, t_1)(t_1, t_2)\cdots(t_{n-1}, t_n)(t_n, p)$

$$\simeq (p, u_1)(u_1, t_1)(t_1, u_2)(u_2, t_2)\cdots(t_{n-1}, u_n)(u_n, t_n)(t_n, p)(p, p)$$

$$\simeq (p, p)(p, u_1)(u_1, u_2)\cdots(u_{n-1}, u_n)(u_n, p)(p, p)$$

$$\simeq (p, u_1)(u_1, u_2)\cdots(u_{n-1}, u_n)(u_n, p).$$

Thus, α is homotopic to the edge path on the top of the row. An induction on the number of rows gives $\alpha \simeq i_p$, as desired. □

Corollary 7.37 (Tietze). *If X is a connected polyhedron, then $\pi_1(X, x_0)$ is* **finitely presented**, *that is, $\pi_1(X, x_0)$ has a presentation with only finitely many generators and finitely many relations. Indeed, if (K, h) is a triangulation of X and T is a maximal tree in K, then*

$$\pi_1(X, x_0) \simeq G_{K,T}.$$

Proof. Theorems 7.36 and 7.34. □

We remark that it is a stronger condition on a group that it be finitely presented than that it be finitely generated; in fact there are uncountably many nonisomorphic f.g. groups while there are only countably many finitely presented groups. It is easy to prove, however, that every f.g. abelian group is finitely presented.

To see the power of Corollary 7.37, recall our earlier labor in proving that $\pi_1(S^1, 1) \cong \mathbf{Z}$. This is immediate from the corollary and Corollary 7.35. Example 7.17 also gives us our first example of a nonabelian fundamental group. On the other hand, there is a limit to the power of Corollary 7.37, which shows that fundamental groups are inherently more difficult than homology groups. We have already mentioned that there exists an algorithm to compute the homology groups of a polyhedron. In contrast, it is known (see [Rotman (1984), p. 395]) that there is no algorithm that can decide of an arbitrary finite presentation whether or not the presented group has order 1. In our context, there is no algorithm using Corollary 7.37 which can always decide whether or not a polyhedron is simply connected!

Exercises

7.44. (i) Using Example 7.13, prove that $\pi_1(\mathbf{R}P^2, x_0) \cong \mathbf{Z}/2\mathbf{Z}$.

(ii) Using part (i) and the Hurewicz theorem, show (again) that $H_1(\mathbf{R}P^2) \cong \mathbf{Z}/2\mathbf{Z}$.

7.45. Let X be a polyhedron.
  (i) Show that X is connected if $X^{(0)} = \text{Vert}(X)$ is contractible in X, i.e., there is $F: X \times \mathbf{I} \to X$ with $F(v, 0) = v$ for all $v \in X^{(0)}$ and $F(\ \ , 1)$ a constant function.
  (ii) Show that X is simply connected if $X^{(1)}$ is contractible in X.

7.46. Let K be a connected one-dimensional simplicial complex having m edges and n vertices. Prove that $\pi_1(|K|, x_0)$ is a free group of rank $m - n + 1$.

7.47. If X is a connected polyhedron of dimension 1, show that:

$H_0(X) = \mathbf{Z}$;
$H_1(X)$ is free abelian of rank $(1 - \chi(X))$;
$H_q(X) = 0$ for all $q \geq 2$.

7.48. Use Theorem 7.36 to prove (again) that S^m is simply connected for all $m \geq 2$.

7.49. A subcomplex L of a simplicial complex K is called **full** if, whenever $\sigma \in K$ and $\text{Vert}(\sigma) \subset \text{Vert}(L)$, then $\sigma \in L$.
  (i) The q-skeleton $K^{(q)}$ is not full for $q < \dim K$.
  (ii) If $A \subset \text{Vert}(K)$, then there is a unique full subcomplex L of K with $\text{Vert}(L) = A$; moreover, if L' is a subcomplex of K with $\text{Vert}(L') = A$, then $L' \subset L$.

7.50. Let K be a connected simplicial complex, let L be a full connected subcomplex of K, and let $v_0 \in \text{Vert}(L)$. If every closed edge path in K at v_0 is homotopic to a closed edge path in L at v_0, then the inclusion $L \hookrightarrow K$ induces a surjection $\pi(L, v_0) \twoheadrightarrow \pi(K, v_0)$. Show that this map need not be an isomorphism (take K simply connected).

The Seifert–van Kampen Theorem

Definition. Let A and B be (not necessarily abelian) groups. Their **free product**, denoted by $A * B$, is a group satisfying the following condition:

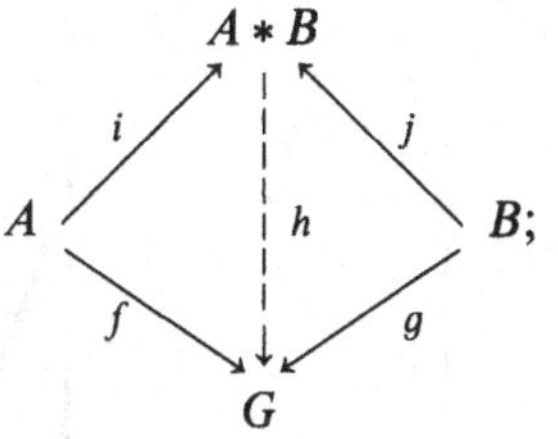

there are homomorphisms i and j such that, for every pair of homomorphisms $f: A \to G$ and $g: B \to G$ for any group G, there exists a unique homomorphism $h: A * B \to G$ making the diagram commute.

In categorical language, $A * B$ is the coproduct in **Groups** and hence is unique to isomorphism if it exists. Existence is proved by showing that there is a group, each of whose nonidentity elements has a unique factorization of the form

$$a_1 b_1 a_2 b_2 \cdots a_n b_n,$$

where $a_i \in A$, $b_i \in B$, and only a_1 or b_n is allowed to be 1.[7] An alternative description of $A * B$ can be given via presentations. Let $A = (X|R)$ and $B = (Y|S)$ be presentations in which the sets X and Y of generators (and hence the relations R and S) are disjoint; then a presentation for $A * B$ is $(X \cup Y | R \cup S)$.

EXAMPLE 7.18. $\mathbf{Z} * \mathbf{Z}$ is a free group (of rank 2).

Definition. Let B, A_1, A_2 be objects in a category $\mathscr{C}$, and let f_1, f_2 be morphisms:

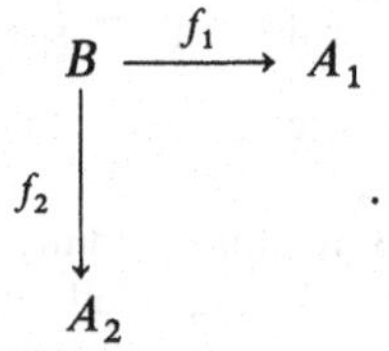

$$
\begin{array}{ccc}
B & \xrightarrow{\ f_1\ } & A_1 \\
{\scriptstyle f_2}\downarrow & & \\
A_2 & &
\end{array}
\qquad (\delta)
$$

A **solution** of the diagram (δ) is an object C and morphisms g_1, g_2 such that the following diagram commutes:

$$
\begin{array}{ccc}
B & \xrightarrow{\ f_1\ } & A_1 \\
{\scriptstyle f_2}\downarrow & & \downarrow{\scriptstyle g_1} \\
A_2 & \xrightarrow{\ g_2\ } & C.
\end{array}
$$

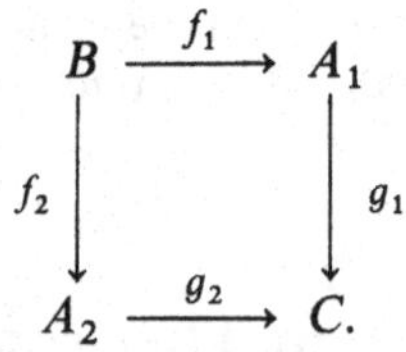

A **pushout** of the diagram (δ) is a solution (C, g_1, g_2) such that, for any other solution (D, h_1, h_2), there exists a unique morphism $\varphi \colon C \to D$ making the following diagram commute:

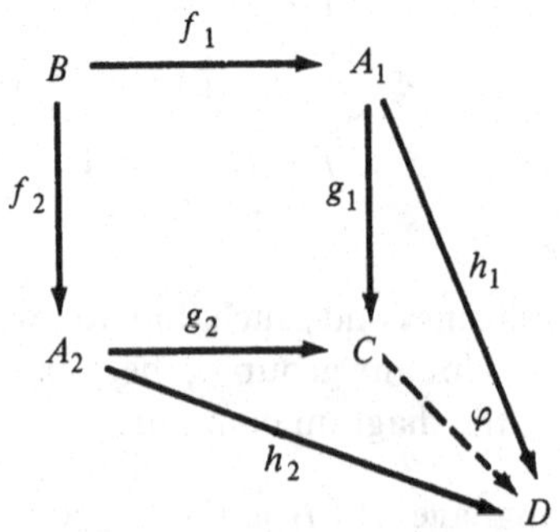

$$(\delta^*)$$

----

[7] Multiplication is essentially juxtaposition: moreover, in the definition of $A * B$, one defines $h(a_1 b_1 \cdots a_n b_n) = f(a_1) g(b_1) \cdots f(a_n) g(b_n)$.

One proves quickly that pushouts, when they exist, are unique to equivalence: if (C, g_1, g_2) and (D, h_1, h_2) are both pushouts, then the morphism $\varphi: C \to D$ is an equivalence.

Theorem 7.38. *A pushout exists for the diagram* (δ) *in* **Groups**. *Moreover, if for* $i = 1, 2$, A_i *has presentation* $(X_i | \Delta_i)$, *then the pushout has the presentation*

$$C = (X_1 \cup X_2 | \Delta_1 \cup \Delta_2 \cup \{f_1(b)f_2(b^{-1}): b \in B\}).$$

Proof. Let N be the normal subgroup of $A_1 * A_2$ generated by $\{f_1(b)f_2(b^{-1}): b \in B\}$. Define $C = (A_1 * A_2)/N$ and define $g_i: A_i \to C$ by $g_i(a_i) = a_i N$ for $i = 1, 2$. It is easy to verify that (C, g_1, g_2) is a solution of (δ).

Suppose that (D, h_1, h_2) is a second solution of (δ). The definition of free product provides a unique homomorphism $\psi: A_1 * A_2 \to D$ with $\psi | A_i = h_i$ for $i = 1, 2$. Since $h_2 f_2 = h_1 f_1$, it follows that $N \subset \ker \psi$ and ψ induces a homomorphism $\varphi: C \to D$. One shows easily that the diagram (δ^*) commutes and that φ is unique. Finally, it is plain from the construction that C has a presentation as described in the statement. □

Corollary 7.39. *If* $A_2 = \{1\}$ *in diagram* (δ), *then the pushout* C *is* A_1/N, *where* N *is the normal subgroup generated by* $f_1(B)$.

An observation is needed. If G is an infinite cyclic group with generator x, we know that $G * G$ is a free group of rank 2 and that a presentation of $G * G$ is $(x, y | \varnothing)$. It is necessary to write y for the second generator to avoid confusing it with x. More generally, if groups A_i have presentations $(X_i | \Delta_i)$ for $i = 1, 2$, then $A_1 * A_2$ has presentation $(X_1 \cup X_2 | \Delta_1 \cup \Delta_2)$ if X_1 and X_2 are disjoint; if X_1 and X_2 are not disjoint, new notation must be introduced to make them disjoint. We have tacitly done this in Theorem 7.38; we shall be more explicit in the next proof.

The next theorem shows that pushouts occur quite naturally.

Theorem 7.40 (Seifert–van Kampen).[8,9] *Let* K *be a simplicial complex having connected subcomplexes* L_1 *and* L_2 *such that* $L_1 \cup L_2 = K$ *and* $L_1 \cap L_2$ *is connected. If* $v_0 \in \operatorname{Vert}(L_1 \cap L_2)$ *(so that* $L_1 \cap L_2 \neq \varnothing$*), then* $\pi(K, v_0)$ *is the pushout of the diagram*

$$\begin{array}{ccc} \pi(L_1 \cap L_2, v_0) & \longrightarrow & \pi(L_1, v_0) \\ \big\downarrow & & \\ \pi(L_2, v_0) & & \end{array},$$

where the arrows are induced by the inclusion maps $j_i: L_1 \cap L_2 \hookrightarrow L_i$ *for* $i = 1, 2$.

---

[8] Many authors call this van Kampen's theorem.

[9] In light of Theorem 7.36, this theorem may be rephrased so that "simplicial complex" may everywhere be replaced by "polyhedron".

Remark. The hypothesis implies that K is connected.

PROOF. Denote $L_1 \cap L_2$ by L_0. Choose a maximal tree T_0 in L_0 and for each $i = 1, 2$ choose a maximal tree T_i in L_i containing T_0. By Exercise 7.41, $T_1 \cup T_2$ is a tree in K; moreover, $T_1 \cup T_2$ is a maximal tree because $\text{Vert}(T_1 \cup T_2) = \text{Vert}(T_1) \cup \text{Vert}(T_2) = \text{Vert}(L_1) \cup \text{Vert}(L_2) = \text{Vert}(K)$. Theorem 7.34 says that $\pi(K, v_0)$ has a presentation $(E|\Delta' \cup \Delta'')$, where E is the set of edges (u, v) in K, $\Delta' = E \cap (T_1 \cup T_2)$, and

$$\Delta'' = \{(u, v)(v, w)(u, w)^{-1}: \{u, v, w\} = s \in K\}.$$

There are similar presentations for $\pi(L_i, v_0)$, namely, $(E_i|\Delta_i' \cup \Delta_i'')$, where E_i is the set of edges in L_i.

Denote the set of edges in $L_0 = L_1 \cap L_2$ by E_0. We make E_1 and E_2 disjoint by affixing the symbols j_1 and j_2 (which designate the inclusions). Theorem 7.38 thus gives the presentation for the pushout

$$(j_iE_1 \cup j_2E_2|j_1\Delta_1' \cup j_1\Delta_1'' \cup j_2\Delta_2' \cup j_2\Delta_2'' \cup \{(j_1e)(j_2e)^{-1}: e \in E_0\}).$$

The generators may be rewritten as

$$j_1E_0 \cup j_1(E_1 - E_0) \cup j_2E_0 \cup j_2(E_2 - E_0).$$

The relations include $j_1E_0 = j_2E_0$ (so that one of these subsets is superfluous). Next, $\Delta_i' = E_i \cap T_i = (E_i \cap T_0) \cup (E_i \cap (T_i - T_0))$, and this gives a decomposition of $j_1\Delta_1' \cup j_2\Delta_2'$ into four subsets, one of which is superfluous. Furthermore, $\Delta'' = \Delta_1'' \cup \Delta_2''$, for if $(u, v)(v, w)(u, w)^{-1} \in \Delta$, then $\{u, v, w\} \in K = L_1 \cup L_2$ and $\{u, v, w\} \in L_i$ for some i. Transform this presentation as follows: (1) isolate those generators and relations involving L_0; (2) delete superfluous generators and relations involving L_0 (say, delete such having symbol j_2); (3) erase the now unnecessary symbols j_1 and j_2. It is now apparent that the pushout and $\pi(K, v_0)$ have the same presentation and hence are isomorphic. □

Corollary 7.41. *With the hypothesis and notation of the previous theorem, a presentation for* $\pi(K, v_0)$ *is*

$$(j_1E_1 \cup j_2E_2|j_1\Delta_1' \cup j_1\Delta_1'' \cup j_2\Delta_2' \cup j_2\Delta_2'' \cup \{(j_1e)(j_2e)^{-1}: e \in E_0\}).$$

Corollary 7.42. *If K is a simplicial complex having connected subcomplexes L_1 and L_2 such that $L_1 \cup L_2 = K$ and $L_1 \cap L_2$ is simply connected, then for* $v_0 \in \text{Vert}(L_1 \cap L_2)$,

$$\pi(K, v_0) \cong \pi(L_1, v_0) * \pi(L_2, v_0).$$

Remark. There is a version of the Seifert–van Kampen theorem for spaces other than polyhedra, but the analogue of Corollary 7.42 $[\pi_1(X_1 \vee X_2, x_0) \cong \pi_1(X_1, x_0) * \pi_1(X_2, x_0)]$ may be false (see [Olum]).

Note that a "figure 8" is $S^1 \vee S^1$, so that Corollary 7.42 gives another proof of Example 7.17.

Corollary 7.43. *Let K be a simplicial complex having connected subcomplexes L_1 and L_2 such that $L_1 \cup L_2 = K$ and $L_1 \cap L_2$ is connected. If $v_0 \in \text{Vert}(L_1 \cap L_2)$ and if L_2 is simply connected, then*

$$\pi(K, v_0) \cong \pi(L_1, v_0)/N,$$

where N is the normal subgroup generated by the image of $\pi(L_1 \cap L_2, v_0)$. Moreover, in the notation of the theorem, $\pi(K, v_0)$ has the presentation

$$(E_1 | \Delta_1' \cup \Delta_1'' \cup j_1 E_0).$$

Proof. Since $\pi(L_2, v_0) = \{1\}$, the first statement is immediate from the Seifert–van Kampen theorem and Corollary 7.39; the second statement is immediate from Corollary 7.41. □

We now exploit Corollary 7.43. Let K be a connected 2-complex with basepoint v_0 and let α be a closed edge path in K at v_0, say,

$$\alpha = e_1 \cdots e_n = (v_0, v_1)(v_1, v_2) \cdots (v_{n-1}, v_0).$$

Define a **triangulated polygon** $D(\alpha)$ as the 2-complex with vertices $\text{Vert}(D(\alpha)) = \{p_0, \ldots, p_{n-1}, q_0, \ldots, q_{n-1}, r\}$ and 2-simplexes $\{r, q_i, q_{i+1}\}, \{q_i, q_{i+1}, p_{i+1}\}$, and $\{q_i, p_i, p_{i+1}\}$, where $0 \le i \le n-1$ and subscripts are read modulo n.

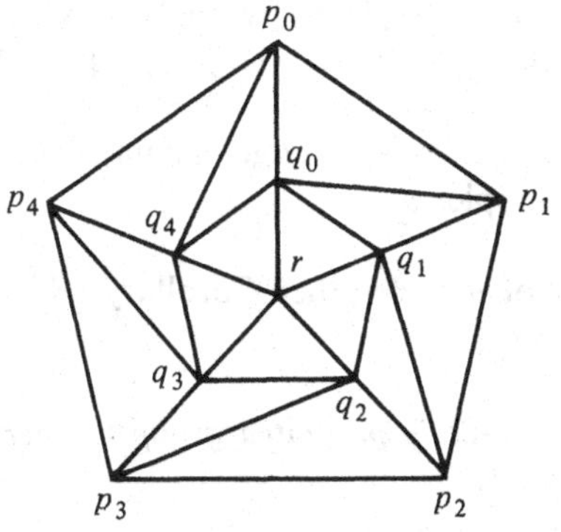

Let $\partial D(\alpha)$ denote the boundary of $D(\alpha)$, that is, $\partial D(\alpha)$ is the full subcomplex with vertices $\{p_0, \ldots, p_{n-1}\}$. Define the **attaching map** $\varphi_\alpha \colon \partial D(\alpha) \to K$ by $\varphi_\alpha(p_i) = v_i$ for $0 \le i \le n-1$. Clearly, φ_α carries the boundary edge path $(p_0, p_1) \cdots (p_{n-1}, p_0)$ onto the edge path α.

Definition. Let K be a simplicial complex and let $\sim$ be an equivalence relation on $\text{Vert}(K)$. The **quotient complex** $K/\sim$ is the simplicial complex with vertex set all equivalence classes $[v]$ for $v \in \text{Vert}(K)$ and with simplexes $\{[v_0], \ldots, [v_q]\}$ if there exists a simplex $\{u_0, \ldots, u_q\} \in K$ with $u_i \sim v_i$ for $i = 0, \ldots, q$.

One verifies quickly that $K/\sim$ is in fact a simplicial complex.

Definition. Let α be a closed edge path in K at v_0, let $D(\alpha)$ be the corresponding triangulated polygon, and let $\varphi_\alpha: \partial D(\alpha) \to K$ be the attaching map. The quotient complex $K_\alpha = (K \coprod D(\alpha))/\sim$, where $\sim$ identifies each p_i with $\varphi_\alpha(p_i)$, is called the simplicial complex obtained from K by **attaching a 2-cell along α.**

Theorem 7.44. *Let α be a closed edge path in K at v_0 and let K_α be obtained by attaching a 2-cell along α. Then*

$$\pi(K_\alpha, v_0) \cong \pi(K, v_0)/N,$$

where N is the normal subgroup generated by $[\alpha]$.

PROOF. Define L_1 to be the full subcomplex of K_α with vertices $\text{Vert}(K) \cup \{q_0, \ldots, q_{n-1}\}$, and define L_2 to be the full subcomplex of K_α with vertices $\{r, v_0, q_0, q_1, \ldots, q_{n-1}\}$. Note that $L_1 \cup L_2 = K_\alpha$ and $L_1 \cap L_2$ is the edge (v_0, q_0) and the loop $\{q_0, \ldots, q_{n-1}\}$; it follows that $\pi(L_1 \cap L_2, v_0) \cong \mathbf{Z}$. Now L_2 (isomorphic to the full subcomplex of $D(\alpha)$ with vertices $\{r, q_0, \ldots, q_{n-1}\}$) is simply connected. The inclusion $j: K \hookrightarrow L_1$ induces an isomorphism $\pi(K, v_0) \xrightarrow{\simeq} \pi(L_1, v_0)$. Define a function $\psi: \text{Vert}(L_1) \to \text{Vert}(K)$ by $\psi(v) = v$ for all $v \in K$ and $\psi(q_i) = \varphi_\alpha(p_i)$ for all i. It is easy to see that ψ is a simplicial map and $j\psi: L_1 \to L_1$ is homotopic to the identity; hence the induced map ψ_* is the inverse of j_*. The proof is completed by applying Corollary 7.43, since the image of the infinite cyclic group $\pi(L_1 \cap L_2, v_0)$ is generated by $[\alpha]$. □

Definition. A **bouquet of circles** is a wedge of complexes $\bigvee K_i$, where each K_i has the form $\dot{s}$ for a 2-simplex s.

If $\bigvee K_i$ is a bouquet of m circles, then Corollary 7.35 shows that $\pi(\bigvee K_i, b)$ is a free group of rank m.

Theorem 7.45. *Given a finitely presented group G, there exists a connected 2-complex K with $G \cong \pi(K, v_0)$.*

Remark. If one uses infinite simplicial complexes K, one can prove that for any (not necessarily finitely presented) group G, there exists a topological space $X \approx K$ with $G \cong \pi_1(X, x_0)$.

PROOF. Let $(X|\Delta)$ be a presentation of G and let B be a bouquet of $|X|$ circles: $\text{Vert}(B) = \{v_0, u_1^x, v_1^x: x \in X\}$. If we identify the closed edge path $(v_0, u_1^x)(u_1^x, v_1^x)(v_1^x, v_0)$ with x, then each word $w \in \Delta$ may be regarded as a closed edge path in B at v_0. Let $D(w)$ be the triangulated polygon of w and let $\varphi_w: \partial D(w) \to B$ be the attaching map; let D be the wedge $\bigvee_{w \in \Delta} D(w)$ and let $\varphi: \bigvee \partial D(w) \to B$ satisfy $\varphi | \partial D(w) = \varphi_w$. Finally, define K as the quotient complex of $B \cup D$ in which we identify each p^w [in $D(w)$] with $\varphi(p^w) = \varphi_w(p^w)$ [vertices of $D(w)$ are $r^w, p_0^w, p_1^w, \ldots, q_0^w, q_1^w, \ldots$].

Let T be the tree in K with vertices $\{v_0, u_1^x: x \in X\}$. Define L_1 to be the full subcomplex of K with vertices

$$\operatorname{Vert}(B) \cup \left(\bigcup_{w \in \Delta} \{q_0^w, q_1^w, \ldots\} \right),$$

and define L_2 to be the full subcomplex of K with vertices

$$\operatorname{Vert}(T) \cup \left(\bigcup_{w \in \Delta} \{r^w, q_0^w, q_1^w, \ldots\} \right).$$

Note that $L_1 \cup L_2 = K$ and $L_1 \cap L_2$ is the union of T with loops $\{q_0^w, q_1^w, \ldots\}$; it follows that $\pi(L_1 \cap L_2, v_0)$ is free on these loops. Now L_2, being a wedge of simply connected complexes, is simply connected and $\pi(L_1, v_0) \cong \pi(B, v_0)$, as in the proof of Theorem 7.44. This proof is completed by applying Corollary 7.43, for the image of the free group $\pi(L_1 \cap L_2, v_0)$ is generated by Δ. □

Corollary 7.46.

(i) *Let K be a bouquet of $2g$ circles, and let K_α be obtained from K by attaching a 2-cell along α, where $\alpha = a_1 b_1 a_1^{-1} b_1^{-1} \cdots a_g b_g a_g^{-1} b_g^{-1}$. Then*

$$\pi(K_\alpha, v_0) = \left(a_1, b_1, \ldots, a_g, b_g \,\middle|\, \prod_{i=1}^{g} a_i b_i a_i^{-1} b_i^{-1} \right).$$

(ii) *Let K be a bouquet of g circles, and let K_α be obtained from K by attaching a 2-cell along α, where $\alpha = c_1^2 c_2^2 \cdots c_g^2$. Then*

$$\pi(K_\alpha, v_0) = \left(c_1, \ldots, c_g \,\middle|\, \prod_{i=1}^{g} c_i^2 \right).$$

PROOF. Theorem 7.44. □

Definition. The one-relator groups occurring in Corollary 7.46 are called **surface groups**.

Surface groups are the fundamental groups of surfaces (compact connected 2-manifolds); see Exercise 8.18 and the subsequent discussion.

Corollary 7.47. *A group G is finitely presented if and only if there exists a polyhedron X with $G \cong \pi_1(X, x_0)$.*

PROOF. Necessity follows from the theorem; sufficiency follows from Corollary 7.37. □

The quotient group in the statement of the Seifert–van Kampen theorem can be complicated. In the special case when the maps j_{i*} induced by the inclusions (for $i = 1, 2$) are injections, the resulting group is called an **amalgam** (or a **free product with amalgamated subgroups**). Such groups have been studied extensively.

CHAPTER 8

CW Complexes

We return to homology, seeking to compute homology groups more effectively. The spaces for which this search is successful, the so-called CW complexes introduced by J. H. C. Whitehead, generalize simplicial complexes; they have also proved to be of fundamental importance in homotopy theory.

The basic idea is quite simple. Recall that a simplicial complex is a union of simplexes (homeomorphs of standard simplexes) that fit together nicely: any two of its simplexes that intersect do so in a common face. Standard simplexes as building blocks of nice spaces is too good an idea to abandon. On the other hand, we can replace simplexes by spaces that are "almost" homeomorphic to standard simplexes in the sense that boundary points may be identified. Think for a moment of the interesting spaces obtained from the square $\mathbf{I} \times \mathbf{I}$ (which is homeomorphic to Δ^2) by identifying points on its boundary: torus; real projective plane; Klein bottle; 2-sphere; there are others, of course. After constructing these spaces, one must take care in actually triangulating them; moreover, triangulations are wanted only because simplicial homology requires them. The idea now is to consider spaces built from generalized simplexes ("almost" homeomorphic to standard simplexes) that are glued together along their boundaries (more details later). We shall see that the homology groups of these spaces arise from chain groups having smaller ranks than the chain groups appearing in simplicial theory.

Hausdorff Quotient Spaces

In Chapter 1 we considered quotient spaces $X/\sim$, where $\sim$ is an equivalence relation on a space X; the points of $X/\sim$ are the equivalence classes $[x]$ for $x \in X$. An important item in this context is the natural map $\nu: X \to X/\sim$, defined by $x \mapsto [x]$; it is a continuous surjection; indeed v is an identification. Recall that if A is a subset of X, then X/A denotes the quotient space corresponding to the equivalence relation that identifies every pair of elements of A and no others.

The following examples show that a quotient space of a (compact) Hausdorff space may not be Hausdorff and that the natural map need be neither an open map nor a closed map.

EXAMPLE 8.1. Consider the quotient space X/A, where $X = \mathbf{I}$ and A is the (open) subset $A = [0, 1)$; let $v: X \to X/A$ be the natural map. Then the point $[0] \in X/A$ is open (because $v^{-1}([0]) = A$ is open), but the other point $[1] \in X/A$ is not open (because $v^{-1}([1]) = \{1\}$ is not open). Therefore X/A is Sierpinski space, which is not Hausdorff.

EXAMPLE 8.2. (i) Let $v: X \to X/A$ be the natural map of Example 8.1. Then $v([0, 1/2]) = \{[0]\}$ is not closed in X/A, and so v is not a closed map.

(ii) Let X be the $\sin(1/x)$ space, and define $f: X \to \mathbf{I}$ as the vertical projection $(x, y) \mapsto x$; it is easy to see (using the definition of an identification) that the continuous surjection f is an identification. If $U = V \cap X$, where V is the open disk with center $(0, \frac{1}{2})$ and radius $\frac{1}{4}$, then $f(U)$ is not open in $\mathbf{I}$, and so f is not an open map. (Note that the target $\mathbf{I}$ is Hausdorff.)

We seek sufficient conditions guaranteeing that $X/\sim$ be Hausdorff when X is Hausdorff. Recall an elementary fact. The **diagonal** of a space Y is the subset D of $Y \times Y$ (product topology) defined by

$$D = \{(y, y) \in Y \times Y: y \in Y\};$$

a space Y is Hausdorff if and only if its diagonal D is closed in $Y \times Y$.

Definition. If $\sim$ is a binary relation on a space X, then its **graph** G is the subset of $X \times X$ defined by

$$G = \{(x_1, x_2) \in X \times X: x_1 \sim x_2\};$$

we say that $\sim$ is **closed** if its graph G is a closed subset of $X \times X$.

The identity relation on X is closed if and only if X is Hausdorff.

If $\sim$ is an equivalence relation on a space X, then its graph G is equal to $(v \times v)^{-1}(D)$, where $v: X \to X/\sim$ is the natural map and D is the diagonal of $X/\sim$. When $X/\sim$ is Hausdorff, the diagonal $D \subset (X/\sim) \times (X/\sim)$ is closed, hence G is closed in $X \times X$ (because v and hence $v \times v$ are continuous) and $\sim$ is closed. We give a partial converse after a general lemma.

Lemma 8.1. *Let $v: W \to Z$ be a closed map, let S be a subset of Z, and let U be an open subset of W containing $v^{-1}(S)$. Then there exists an open set V in Z with*

$$S \subset V \quad \text{and} \quad v^{-1}(V) \subset U.$$

Proof. Define $V = \nu(U^c)^c$, where ^{c} means complement. Since U is open, U^c is closed; since ν is a closed map, $\nu(U^c)$ is closed and so $V = \nu(U^c)^c$ is open. Now $\nu^{-1}(S) \subset U$ gives $S \cap \nu(U^c) = \emptyset$, so that $S \subset \nu(U^c)^c = V$. Finally, $\nu^{-1}(V) = W - \nu^{-1}\nu(W - U) \subset W - (W - U) = U$, as claimed. □

Theorem 8.2. *If $\sim$ is a closed equivalence relation on a compact Hausdorff space X, then the quotient space $X/\sim$ is also (compact) Hausdorff.*

Proof. Let $\nu: X \to X/\sim$ be the natural map, and let $G \subset X \times X$ be the graph of $\sim$; for $i = 1, 2$, let $p_i: X \times X \to X$ be the projection $(x_1, x_2) \mapsto x_i$. We claim that ν is a closed map. If C is any subset of X, then

$$p_2(p_1^{-1}(C) \cap G) = \{y \in X: y \sim x \text{ for some } x \in C\} = \nu^{-1}\nu(C).$$

If C is closed, then so are $p_1^{-1}(C)$ and $p_1^{-1}(C) \cap G$; since X is compact Hausdorff, p_2 is a closed map, and so $\nu^{-1}\nu(C)$ is closed. But ν is an identification, and so $\nu(C)$ is closed, as claimed.

It follows that every point of $X/\sim$ is closed (being the image of a (necessarily closed) singleton in X). If $[x], [y]$ are distinct points of $X/\sim$, then $\nu^{-1}([x])$ and $\nu^{-1}([y])$ are disjoint closed subsets of X. Since X is compact and Hausdorff, it is normal; there thus exist disjoint open sets $U_{[x]}$ and $U_{[y]}$ in X with $\nu^{-1}([x]) \subset U_{[x]}$ and $\nu^{-1}([y]) \subset U_{[y]}$. By the lemma, there are open sets $V_{[x]}$ and $V_{[y]}$ in $X/\sim$ with $[x] \in V_{[x]}$, $[y] \in V_{[y]}$, $\nu^{-1}(V_{[x]}) \subset U_{[x]}$, and $\nu^{-1}(V_{[y]}) \subset U_{[y]}$. It follows that $V_{[x]} \cap V_{[y]} = \emptyset$, and this shows that $X/\sim$ is Hausdorff. □

Corollary 8.3. *If X is a compact Hausdorff space and A is a closed subset, then X/A is (compact) Hausdorff.*

Proof. The graph of the appropriate equivalence relation is $(A \times A) \cup D$, where D is the diagonal of X, and this is a closed subset in $X \times X$ because A is closed in X. □

Here is an important class of examples.

Definition. Let F be a division ring and let $n \geq 0$. Define an equivalence relation on $F^{n+1} - \{0\}$ (where F^{n+1} is the (left) vector space over F consisting of all $(n + 1)$-tuples $x = (x_0, x_1, \ldots, x_n)$ with coordinates x_i in F) by $x \sim y$ if there exists $\lambda \in F - \{0\}$ with $x = \lambda y$. The quotient set $(F^{n+1} - \{0\})/\sim$, that is, the set of all equivalence classes, is called ***F*-projective *n*-space** and is denoted by FP^n. The class of $x = (x_0, \ldots, x_n)$ is denoted by $[x] = [x_0, \ldots, x_n] \in FP^n$.

Note, for each $n \geq 0$, that there is an imbedding $FP^n \hookrightarrow FP^{n+1}$ given by $[x_0, \ldots, x_n] \mapsto [x_0, \ldots, x_n, 0]$. One calls the union[1] $\bigcup_{n \geq 0} FP^n$ **infinite-dimensional *F*-projective space** and denotes it by FP^∞.

---

[1] Actually, one can only take the union of a family of subsets of a given set; the notion of *direct limit* is needed to make this definition precise.

There are three division rings in which we are interested: the reals **R**, the complexes **C**, and the quaternions **H**. Of course, the reader is familiar with **C** as a two-dimensional vector space over **R** with basis $\{1, i\}$. Let us recall that **H** is a four-dimensional vector space over **R** with basis $\{1, i, j, k\}$. The ring structure on **H** is determined by the distributivity laws and the rules: $i^2 = j^2 = k^2 = -1$; $ij = -ji = k$; $jk = -kj = i$; $ki = -ki = j$. Each of **R**, **C**, and **H** has a **norm** $|\ \ |$ with values nonnegative real numbers: in **R**, $|x| = \sqrt{x^2}$ is absolute value; in **C**, $|z| = |a + bi| = \sqrt{a^2 + b^2}$; in **H**, $|w| = |a + bi + cj + dk| = \sqrt{a^2 + b^2 + c^2 + d^2}$. One verifies that, in each case, the norm is a continuous multiplicative map: $|xy| = |x||y|$ (this calculation in **H** is tedious).

In **C**, we know that if $z = a + bi$, then $\bar{z}$ defined as $a - bi$ satisfies $z\bar{z} = |z|^2$; hence, if $z \neq 0$, then $z^{-1} = \bar{z}/z\bar{z} = \bar{z}/|z|^2$. Similarly, if $w = a + bi + cj + dk \in$ **H**, then $\bar{w}$ defined as $\bar{w} = a - bi - cj - dk$ satisfies $w\bar{w} = |w|^2$. If $w \neq 0$, define $w^{-1} = \bar{w}/|w|^2$; it is now straightforward to check that **H** is a division ring.

For each of the three division rings $F = \mathbf{R}, \mathbf{C}$, and **H**, we see that $F^{n+1} - \{0\}$ is a topological space, and so the corresponding projective spaces FP^n are also topological spaces when given the quotient topology.

Notation. For each $n \geq 0$, **real projective *n*-space** is denoted by $\mathbf{R}P^n$, **complex projective *n*-space** is denoted by $\mathbf{C}P^n$, and **quaternionic projective *n*-space** is denoted by $\mathbf{H}P^n$.

EXERCISES

8.1. For every division ring F, show that FP^0 is a point.

*8.2. Show that $\mathbf{R}P^1 \approx S^1$, $\mathbf{C}P^1 \approx S^2$, and $\mathbf{H}P^1 \approx S^4$.

8.3. Define $U(F) = \{x \in F: |x| = 1\}$, where $F = \mathbf{R}$, **C**, or **H**. Show that $U(\mathbf{R}) \approx S^0$, $U(\mathbf{C}) \approx S^1$, and $U(\mathbf{H}) \approx S^3$.

8.4. Show that $\mathbf{R}P^2$ is homeomorphic to the real projective plane (defined earlier as a certain quotient space of $\mathbf{I} \times \mathbf{I}$).

*8.5. For each $n \geq 0$, define an equivalence relation on S^n by $x \sim y$ if $x = \pm y$ (identify antipodal points). Prove that $S^n/\!\sim\ \approx \mathbf{R}P^n$.

*8.6. For each $n \geq 0$, define an equivalence relation on S^{2n+1} by $x \sim y$ if $x = \lambda y$ for some complex λ with $|\lambda| = 1$. Prove that $S^{2n+1}/\!\sim\ \approx \mathbf{C}P^n$. (*Hint*: Write $x = (x_1, x_2, \ldots, x_{2n+2}) \in S^{2n+1}$ as an $(n + 1)$-tuple of complex numbers: $x = (z_1, \ldots, z_{n+1})$; then $x = \lambda y$ implies $|x| = |\lambda y| = |\lambda||y|$ and $|\lambda| = 1$.)

*8.7. For each $n \geq 0$, define an equivalence relation on S^{4n+3} by $x \sim y$ if $x = \lambda y$ for some quaternion λ with $|\lambda| = 1$. Prove that $S^{4n+3}/\!\sim\ \approx \mathbf{H}P^n$. (*Hint*: If $x, y \in S^{4n+3} \subset \mathbf{H}^{n+1} - \{0\}$, then $x = \lambda y$ implies $|\lambda| = 1$.)

Theorem 8.4. *For every $n \geq 0$, the projective spaces $\mathbf{R}P^n$, $\mathbf{C}P^n$, and $\mathbf{H}P^n$ are compact Hausdorff.*

PROOF. In light of Exercise 8.5, we may regard $\mathbf{R}P^n$ as a quotient space of S^n. Moreover, it is easy to see that the graph of the equivalence relation (in that exercise) is $D \cup D^*$, where D is the diagonal in $S^n \times S^n$ and $D^* = \{(x, -x) \in S^n \times S^n : x \in S^n\}$. This graph is closed, and so Theorem 8.2 applies.

As in Exercise 8.6, we may regard $\mathbf{C}P^n$ as a quotient space of S^{2n+1}. Moreover, it is easy to see that the graph of the equivalence relation is

$$G = \{(x, \lambda x) \in S^{2n+1} \times S^{2n+1} : x \in S^{2n+1} \text{ and } \lambda \in S^1\}.$$

Thus G is the image of $S^{2n+1} \times S^1$ under the continuous map of $S^{2n+1} \times S^{2n+1}$ to itself given by $(x, y) \mapsto (x, yx)$; therefore G is compact, hence closed.

A similar argument, using Exercise 8.7, applies to $\mathbf{H}P^n$ regarded as a quotient of S^{4n+3}; the graph is a continuous image of $S^{4n+3} \times S^3$. □

Attaching Cells

We now prepare for an important construction of examples of quotient spaces. Recall the definition of the **coproduct** $X_1 \coprod X_2$ of two spaces X_1 and X_2 (the disjoint union in which each X_i is an open subset). If $f_i: X_i \to Y$ is continuous, for $i = 1, 2$, then the (continuous) map $f_1 \coprod f_2: X_1 \coprod X_2 \to Y$ is defined by $(f_1 \coprod f_2)(x) = f_i(x)$, where $x \in X_i$.

Definition. Let X and Y be spaces, let A be a closed subset of X, and let $f: A \to Y$ be continuous. The space obtained from Y by **attaching X via f** is $(X \coprod Y)/\sim$, where $\sim$ is the equivalence relation on $X \coprod Y$ generated[2] by $\{(a, f(a)) \in (X \coprod Y) \times (X \coprod Y): a \in A\}$, This space is denoted by $X \coprod_f Y$; the map f is called the **attaching map**.

The mapping cylinder shows that every continuous map can be viewed as an attaching map.

EXERCISES

8.8. Let $v: X \coprod Y \to X \coprod_f Y$ be the natural map. Let Z be a space, and let $\alpha: X \to Z$ and $\beta: Y \to Z$ be continuous maps such that

$$\alpha(a) = \beta(f(a)) \quad \text{for all } a \in A.$$

Then $(\alpha \coprod \beta) \circ v^{-1}$ is a well defined continuous map $X \coprod_f Y \to Z$. (*Hint*: Use Corollary 1.9.)

*8.9. If B is a subset of $Z \times Z$ for some set Z, then define

$$B^{-1} = \{(v, u) \in Z \times Z: (u, v) \in B\}.$$

---

[2] If R is a binary relation on a set X, then define a new binary relation R' on X by $(x, y) \in R'$ if there exists $n \geq 1$ and elements $x_0, x_1, \ldots, x_n \in X$ with $x_0 = x$ and $x_n = y$ such that, for all $0 \leq i \leq n - 1$, either $x_i = x_{i+1}$, $(x_i, x_{i+1}) \in R$, or $(x_{i+1}, x_i) \in R$. It is easy to see that R' is an equivalence relation on X; it is called the equivalence relation **generated** by R.

(i) Let X and Y be sets, let A be a subset of X, and let $f: A \to Y$ be a function. If $B = \{(a, f(a)) \in (X \coprod Y) \times (X \coprod Y): a \in A\}$, then the equivalence relation on $X \coprod Y$ generated by B is

$$D \cup B \cup B^{-1} \cup K,$$

where D is the diagonal of $X \coprod Y$ and $K = \ker f = \{(a, a') \in A \times A: f(a) = f(a')\}$ (regard $A \times A$ as a subset of $(X \coprod Y) \times (X \coprod Y)$).

(ii) Let Δ be the diagonal of $Y \times Y$. Show that

$$K = (f \times f)^{-1}(\Delta \cap \operatorname{im}(f \times f)).$$

8.10. Show that the diagram

$$\begin{array}{ccc} A & \xrightarrow{f} & Y \\ {\scriptstyle i}\downarrow & & \downarrow \\ X & \longrightarrow & X \coprod_f Y \end{array}$$

(where $i: A \hookrightarrow X$ is inclusion) is a pushout in **Top**.

8.11. If X and Y are path connected, then $X \coprod_f Y$ is path connected.

*8.12. Let X and Y be spaces, let A be a nonempty closed subset of X, let $f: A \to Y$ be continuous, and let $v: X \coprod Y \to X \coprod_f Y$ be the natural map.

(i) Assume that $C \subset X \coprod Y$ is such that $C \cap X$ is closed in X. Show that $v(C)$ is closed in $X \coprod_f Y$ if and only if $(C \cap Y) \cup f(C \cap A)$ is closed in Y. (*Hint*: For any $C \subset X \coprod Y$, show that

$$v^{-1}v(C) = C \cup f(C \cap A) \cup f^{-1}(f(C \cap A)) \cup f^{-1}(C \cap Y).)$$

(ii) Show that the composite

$$Y \hookrightarrow X \coprod Y \to X \coprod_f Y$$

is a homeomorphism from Y to a subspace of $X \coprod_f Y$. (One usually identifies Y with its image under this map.)

(iii) Show that the composite

$$\Phi: X \hookrightarrow X \coprod Y \to X \coprod_f Y$$

maps $X - A$ homeomorphically onto an open subset of $X \coprod_f Y$.

(iv) Under the identification in (ii), show that one may regard $\Phi|A$ as the attaching map f.

*8.13. Suppose that, in Exercise 8.12, A is compact and both X and Y are Hausdorff.

(i) Show that the natural map $v: X \coprod Y \to X \coprod_f Y$ is a closed map.

(ii) If $z \in X \coprod_f Y$, show that its fiber $v^{-1}(z)$ is a nonempty compact subset of $X \coprod Y$.

Definition. The map $\Phi: X \to X \coprod_f Y$ (which is the composite $X \hookrightarrow X \coprod Y \to X \coprod_f Y$) is called the **characteristic map**. (See the remark after Theorem 8.7 as well as the definition of CW complex.)

Theorem 8.5. *Let X and Y be Hausdorff, let A be a compact subset of X, and let $f: A \to Y$ be continuous; then $X \coprod_f Y$ is Hausdorff.*

Proof. Let z_1 and z_2 be distinct points of $X \coprod_f Y$. The fibers $v^{-1}(z_1)$ and $v^{-1}(z_2)$ are disjoint compact subsets of $X \coprod Y$, by Exercise 8.13(ii). Since $X \coprod Y$ is Hausdorff, a standard subcover argument provides disjoint open sets U_1 and U_2 in $X \coprod Y$ with $v^{-1}(z_i) \subset U_i$ for $i = 1, 2$. Since v is a closed map (Exercise 8.13(i)), Lemma 8.1 gives open subsets V_i in $X \coprod_f Y$ with $z_i \in V_i$ and $v^{-1}(V_i) \subset U_i$, $i = 1, 2$. But V_1 and V_2 must be disjoint (because U_1 and U_2 are disjoint); hence $X \coprod_f Y$ is Hausdorff. □

Remark. Theorem 8.2 cannot be used to prove Theorem 8.5 because we are not assuming that X and Y are compact.

The next (technical) result enables one to recognize when a given space is homeomorphic to a space obtained from another space via an attaching map. This situation is analogous to that in group theory when one passes from the description of an external direct product in terms of ordered pairs to a description of an internal direct product whose elements need not be ordered pairs.

Lemma 8.6. *Let X and Y be compact Hausdorff, let A be a closed subset of X, let $f: A \to Y$ be continuous, and let $X \coprod_f Y = (X \coprod Y)/\sim$. Assume that W is a compact Hausdorff space for which there exists a continuous surjection $h: X \coprod Y \to W$ such that, for $u, v \in X \coprod Y$, one has $u \sim v$ if and only if $h(u) = h(v)$. Then $[u] \mapsto h(u)$ is a homeomorphism $X \coprod_f Y \to W$.*

Proof. Consider the diagram

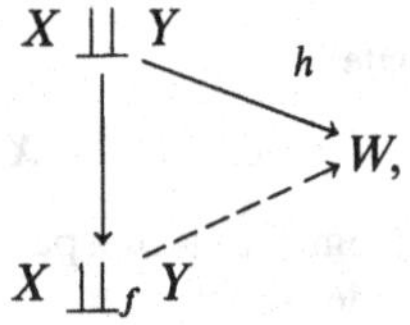

where the vertical arrow is the natural map $u \mapsto [u]$. The hypothesis says that the equivalence relation $\sim$ on $X \coprod Y$ coincides with $\ker h$, hence $X \coprod_f Y = (X \coprod Y)/\ker h$. Since all spaces are compact Hausdorff, Corollary 1.10 applies at once to show that the dashed arrow $[u] \mapsto h(u)$ is a homeomorphism. □

Definition. An ***n*-cell** e^n (or simply e) is a homeomorphic copy of the open n-disk $D^n - S^{n-1}$.

Of course, $D^n - S^{n-1}$ is homeomorphic to $\mathbf{R}^n$ (hence has dimension n). This definition of n-cell is compatible with our earlier notion of "closed n-cell" (a homeomorphic copy of D^n) as the following exercise shows.

Exercise

*8.14. Assume that Y is Hausdorff and that E is a closed n-cell in Y, where $n > 0$. If $\Phi: D^n \to E$ is a homeomorphism, then $\Phi(D^n - S^{n-1})$ is an n-cell whose closure in Y is $E = \Phi(D^n)$.

Definition.[3] Let Y be a Hausdorff space and let $f: S^{n-1} \to Y$ be continuous. Then $D^n \coprod_f Y$ is called the space obtained from Y by **attaching an n-cell via** f, and it is denoted by Y_f.

The elements of Y_f have the form $[x]$ or $[y]$, where $x \in D^n$ and $y \in Y$. Exercise 8.9(i) says that the only identifications are $[x] = [x']$ (when $x, x' \in S^{n-1}$ and $f(x) = f(x')$) and $[x] = [y]$ (when $y = f(x)$). The characteristic map Φ in this case is the composite

$$D^n \hookrightarrow D^n \coprod Y \to D^n \coprod_f Y = Y_f,$$

so that $\Phi: (D^n, S^{n-1}) \to (Y_f, Y)$ is a function of pairs (as in Exercise 8.12(ii), we have identified Y with its homeomorphic copy in Y_f via $y \mapsto [y]$). Our previous discussion gives the following: (i) Y_f is a Hausdorff space (Theorem 8.5), which is compact when Y is: (ii) $\Phi|S^{n-1}$ is the attaching map f (Exercise 8.12(iv)); (iii) $\Phi(D^n - S^{n-1})$ is an n-cell, which is an open subset of Y_f (Exercise 8.12(iii)).

The next definition isolates a property of characteristic maps.

Definition. A continuous map $g: (X, A) \to (Y, B)$ is a **relative homeomorphism** if $g|(X - A): X - A \to Y - B$ is a homeomorphism.

EXAMPLE 8.3. If Y is a Hausdorff space, then a characteristic map $\Phi: (D^n, S^{n-1}) \to (Y_f, Y)$ is a relative homeomorphism.

EXAMPLE 8.4. If $U \subset A \subset X$ with $\bar{U} \subset A^\circ$, then the excision map (namely, the inclusion $(X - U, A - U) \hookrightarrow (X, A)$) is a relative homeomorphism.

EXAMPLE 8.5. If X is a compact Hausdorff space and A is a closed subset, then the natural map $v: (X, A) \to (X/A, *)$, where $*$ denotes the equivalence class comprised of all the points of A, is a relative homeomorphism.

In contrast to the constructive approach we have been giving, the following result shows that spaces with attached cells exist in nature.

Theorem 8.7. *Let Z be a compact Hausdorff space, let Y be a closed subset of Z, and let e be an n-cell in Z with $e \cap Y = \varnothing$. If there is a relative homeomorphism $\Phi: (D^n, S^{n-1}) \to (e \cup Y, Y)$, then the "obvious" map $Y_f = D^n \coprod_f Y \to e \cup Y$ (where $f = \Phi|S^{n-1}$), defined by $[u] \mapsto (\Phi \coprod 1_Y)(u)$, is a homeomorphism.*

PROOF. We are going to use Lemma 8.6. The map $h: D^n \coprod Y \to e \cup Y$ defined by $h = \Phi \coprod 1_Y$ is a continuous surjection, hence $e \cup Y$ is compact. Assume that $u, v \in D^n \coprod Y$. We must show that $u \sim v$ (where $\sim$ is the defining equivalence relation for attaching via f) if and only if $h(u) = h(v)$. If $u \sim v$, then Exercise 8.9(i) allows us to assume that $u \in S^{n-1}$ and that either (1) $v \in Y$ and $v = \Phi(u)$, or (2) $v \in S^{n-1}$ and $\Phi(v) = f(v) = f(u) = \Phi(u)$. In either case, one sees

[3] Compare the simplicial version of this construction at the end of Chapter 7.

that $h(u) = h(v)$. Conversely, if $h(u) = h(v)$, then either u, v are both in D^n, both in Y, or one in each. The only nontrivial case is u, v in D^n; here the hypothesis that $\Phi|D^n - S^{n-1}$ is a homeomorphism from $D^n - S^{n-1}$ to $(e \cup Y) - Y = e$ (for $e \cap Y = \varnothing$) forces u, v in S^{n-1} and $f(u) = \Phi(u) = \Phi(v) = f(v)$. Hence $u \sim v$. Lemma 8.6 applies (for $e \cup Y$ is compact Hausdorff) to show that $[u] \mapsto h(u)$ is a homeomorphism $Y_f = D^n \coprod_f Y \to e \cup Y$. □

Remark. A relative homeomorphism Φ as in the statement of this theorem is also called a **characteristic map**. This extends our earlier usage, which allowed only maps with values in $D^n \coprod_f Y$.

EXERCISES

8.15. Let K be an n-dimensional simplicial complex, and let s be an n-simplex in K. Show that $s \cup |K^{(n-1)}|$ may be viewed as a space obtained from $|K^{(n-1)}|$ by attaching an n-cell. (*Hint*: Let e be the open n-simplex $e = s - \dot{s}$.)

8.16. If Y is a singleton, show that the space obtained from Y by attaching an n-cell is S^n, hence $S^n = e^0 \cup e^n$ (disjoint union) (where e^i denotes an i-cell). (See Example 8.14.)

Theorem 8.8.

(i) *For each $n \geq 1$, $\mathbf{R}P^n$ is obtained from $\mathbf{R}P^{n-1}$ by attaching an n-cell; moreover, there is a disjoint union*

$$\mathbf{R}P^n = e^0 \cup e^1 \cup \cdots \cup e^n,$$

where e^i denotes an i-cell.

(ii) *For each $n \geq 1$, $\mathbf{C}P^n$ is obtained from $\mathbf{C}P^{n-1}$ by attaching a $2n$-cell; moreover, there is a disjoint union*

$$\mathbf{C}P^n = e^0 \cup e^2 \cup \cdots \cup e^{2n}.$$

(iii) *For each $n \geq 1$, $\mathbf{H}P^n$ is obtained from $\mathbf{H}P^{n-1}$ by attaching a $4n$-cell; moreover, there is a disjoint union*

$$\mathbf{H}P^n = e^0 \cup e^4 \cup \cdots \cup e^{4n}.$$

PROOF. (i) If $x = (x_1, \ldots, x_{n+1}) \in S^n$, denote its equivalence class in $\mathbf{R}P^n$ by $[x] = [x_1, \ldots, x_{n+1}]$. Define

$$e = \{[x_1, \ldots, x_{n+1}] \in \mathbf{R}P^n: x_{n+1} \neq 0\}.$$

The complement Y of e in $\mathbf{R}P^n$ is just (the standard imbedded copy of) $\mathbf{R}P^{n-1}$. Also, e is an n-cell, for $e \approx \mathbf{R}^n$ via $[x_1, \ldots, x_{n+1}] \mapsto (x_{n+1}^{-1}x_1, \ldots, x_{n+1}^{-1}x_n)$, and $\mathbf{R}^n \approx D^n - S^{n-1}$. By Theorem 8.7, it suffices to find a relative homeomorphism $\Phi: (D^n, S^{n-1}) \to (e \cup Y, Y) = (\mathbf{R}P^n, \mathbf{R}P^{n-1})$. Let $u = (u_1, \ldots, u_n) \in D^n$ (so $\|u\| \leq 1$), and define

$$\Phi(u) = [u_1, \ldots, u_n, \sqrt{1 - \|u\|^2}].$$

It is easy to see that Φ has the required properties.[4] Finally, the decomposition of $\mathbf{R}P^n$ as a disjoint union of cells follows by induction on n.

(ii) and (iii). Imitate the proof in (i) after identifying complex numbers as ordered pairs of real numbers in the first case and quaternions as ordered quadruples of real numbers in the second case. □

Homology and Attaching Cells

There is a close relation between $H_*(Y)$ and $H_*(Y_f)$; before displaying it, we need a technical lemma.

Definition. A Hausdorff space X is **locally compact** if, for each $x \in X$ and every open set U containing x, there exists an open set W with $\overline{W}$ compact and $x \in W \subset \overline{W} \subset U$.

Lemma 8.9. *If $v: X \to X'$ is an identification and if Z is locally compact Hausdorff, then $v \times 1: X \times Z \to X' \times Z$ is also an identification.*

Proof. It suffices to prove that if U' is a subset of $X' \times Z$ for which $U = (v \times 1)^{-1}(U')$ is open in $X \times Z$, then U' is itself open. Choose $(x', z) \in U'$; if $vx = x'$, then $(v \times 1)(x, z) = (x', z)$ and $(x, z) \in U$. Since U is open in $X \times Z$, there are open sets V in X and J in Z with $(x, z) \in V \times J \subset U$. Local compactness of Z provides an open set W in Z with $z \in W \subset \overline{W} \subset J$ with $\overline{W}$ compact. Of course, $\{x\} \times \overline{W} \subset U$. Define

$$A = \{\alpha \in X: \{\alpha\} \times \overline{W} \subset U\};$$

note that $x \in A$. We claim that A is open in X. Fix $\alpha \in A$. For each $\zeta \in \overline{W}$, there are open sets L_ζ in X and N_ζ in Z with $(\alpha, \zeta) \in L_\zeta \times N_\zeta \subset U$. The family $\{N_\zeta: \xi \in \overline{W}\}$ is an open cover of the compact set $\overline{W}$; let $\{N_1, \ldots, N_m\}$ be a finite subcover. Then, for $1 \le i \le m$, we have $L_i \times N_i \subset U$ (if $N_i = N_\zeta$, we define $L_i = L_\zeta$); moreover, $\alpha \in \bigcap L_i$, and $\overline{W} \subset \bigcup N_i$. Now $L = \bigcap L_i$ is an open set containing α, and $L \times N_i \subset U$ for all i. Therefore $L \times \overline{W} \subset \bigcup (L \times N_i) \subset U$; hence $\alpha \in L \subset A$ and A is open in X.

Now observe that, for $\beta \in X$, we have $\{\beta\} \times \overline{W} \subset U = (v \times 1)^{-1}(U')$ if and only if $\{v(\beta)\} \times \overline{W} \subset U'$. In particular, $\beta \in A$ if and only if $\{v(\beta)\} \times \overline{W} \subset \overline{U}$, from which it follows that $v^{-1}v(A) = A$. Since v is an identification, $v(A)$ is open in X', hence $v(A) \times W \subset U'$ is an open neighborhood of (x', z). Therefore U' is open. □

Lemma 8.9′ (Tube Lemma). *Let X and Y be topological spaces with Y compact. If $x_0 \in X$ and U is an open subset of $X \times Y$ containing $\{x_0\} \times Y$, then there is*

[4] Note that the attaching map $\varphi = \Phi | S^{n-1}$ is just $(u_1, \ldots, u_n) \mapsto [u_1, \ldots, u_n, 0]$.

an open neighborhood L of x_0 in X with

$$\{x_0\} \times Y \subset L \times Y \subset U.$$

PROOF. In the proof of Lemma 8.9, replace α by x_0 and $\overline{W}$ by Y. □

Corollary 8.10. *If $f, g: (X, A) \to (Y, B)$ are homotopic (as maps of pairs), then the induced maps $\bar{f}, \bar{g}: (X/A, *) \to (Y/B, *)$ are homotopic (as maps of pointed spaces).*

PROOF. If $F: (X \times \mathbf{I}, A \times \mathbf{I}) \to (Y, B)$ is a homotopy from f to g, then it induces a function $\bar{F}: (X/A) \times \mathbf{I} \to Y/B$ making the following diagram commute:

$$\begin{array}{ccc} X \times \mathbf{I} & \xrightarrow{F} & Y \\ {\scriptstyle p \times 1}\downarrow & & \downarrow{\scriptstyle q} \\ (X/A) \times \mathbf{I} & \xrightarrow[\bar{F}]{} & Y/B, \end{array}$$

where p and q are identifications. Since qF is continuous, $\bar{F}(p \times 1)$ is continuous. But $p \times 1$ is an identification (Lemma 8.9 applies because $\mathbf{I}$ is compact Hausdorff); by Theorem 1.8, $\bar{F}$ is continuous, as desired. □

If brackets denote the Hom set in the homotopy category of pairs, then this last corollary gives (the known fact) $[(\mathbf{I}, \dot{\mathbf{I}}), (X, x_0)] = [(S^1, 1), (X, x_0)]$; that is, either Hom set can be used to describe $\pi_1(X, x_0)$.

Theorem 8.11. *Let $n \geq 1$ and assume that Y_f is obtained from a Hausdorff space Y by attaching an n-cell via f. Then there is an exact sequence*

$$\cdots \longrightarrow H_p(S^{n-1}) \xrightarrow{f_*} H_p(Y) \xrightarrow{i_*} H_p(Y_f) \longrightarrow H_{p-1}(S^{n-1}) \longrightarrow \cdots$$
$$\cdots \longrightarrow H_0(S^{n-1}) \longrightarrow \mathbf{Z} \oplus H_0(Y) \longrightarrow H_0(Y_f) \longrightarrow 0,$$

where $i: Y \hookrightarrow Y_f$ is the inclusion.

PROOF. Let $v: D^n \coprod Y \to Y_f$ be the natural map, and let $\Phi = v|D^n: (D^n, S^{n-1}) \to (Y_f, Y)$ be the characteristic map; let $e = \Phi(D^n - S^{n-1})$ (an open n-cell in Y_f), and let U' be the open n-disk in D^n with center the origin and radius $\frac{1}{2}$. Since Φ is a relative homeomorphism and e is open, we have $U = \Phi(U')$ open in Y_f. Define $V = Y_f - \Phi(0)$; thus $\{U, V\}$ is an open cover of Y_f. There is an exact (Mayer–Vietoris) sequence

$$\cdots \to H_p(U \cap V) \to H_p(U) \oplus H_p(V) \to H_p(Y_f) \to H_{p-1}(U \cap V) \to \cdots.$$

For $p > 0$, we have $H_p(U) = 0$ because U is contractible. From Theorem 6.3, one sees that the homomorphisms $H_p(U \cap V) \to H_p(V)$ and $H_p(V) \to H_p(Y_f)$, for $p > 0$, are induced by inclusions. Let us examine $U \cap V$ and V. The first is an open punctured n-disk and hence has the same homotopy type as S^{n-1}.

The second space V has Y as a deformation retract: define $F: V \times \mathbf{I} \to V$ by

$$F(v, t) = \begin{cases} v & \text{if } v \in Y \\ \Phi((1-t)z + tz/\|z\|) & \text{if } v = \Phi(z) \in e. \end{cases}$$

Note that F is well defined because Y_f is the disjoint union $e \cup Y$. To see that F is continuous, consider first the following diagram:

$$\begin{array}{ccc} \{(D^n - \{0\}) \amalg Y\} \times \mathbf{I} & \xrightarrow{h} & (D^n - \{0\}) \amalg Y \\ {\scriptstyle v' \times 1}\downarrow & \searrow^{v'h} & \downarrow{\scriptstyle v'} \\ V \times \mathbf{I} & \xrightarrow[F]{} & V = Y_f - \{\Phi(0)\}, \end{array}$$

where v' is the restriction of the natural map $D^n \amalg Y \to Y_f$ and h is defined by $(x, t) \mapsto (1-t)x + tx/\|x\|$ for $x \in D^n - \{0\}$ and $t \in \mathbf{I}$, and by $(y, t) \mapsto y$ for $y \in Y$ and $t \in \mathbf{I}$. Since $h(x, t) = x$ for all $x \in S^{n-1}$ and all $t \in \mathbf{I}$, it is easy to see that $v'h$ is constant on the fibers of $v' \times 1$; it follows from Corollary 1.9 and Lemma 8.9 (for $\mathbf{I}$ is locally compact) that F is continuous.

We now see that the displayed sequence (save for recognizing that, for $p > 0$, the maps $H_p(S^{n-1}) \to H_p(Y)$ are induced by f) is exact; when $p = 0$, we are looking at the end of the Mayer–Vietoris sequence with $H_0(U)$ replaced by $\mathbf{Z}$ (for U is path connected).

Finally, consider the following commutative diagram of spaces (vertical maps are inclusions and horizontal maps are restrictions of Φ):

$$\begin{array}{ccc} U - \{0\} & \xrightarrow{\Phi'} & U \cap V \\ {\scriptstyle \alpha}\downarrow & & \downarrow{\scriptstyle k} \\ D^n - \{0\} & \xrightarrow{\Phi''} & V \\ {\scriptstyle \beta}\uparrow & & \uparrow{\scriptstyle j} \\ S^{n-1} & \xrightarrow[f]{} & Y \end{array}$$

Note that the inclusions α, β, and j are homotopy equivalences because the respective subspaces are deformation retracts; it follows from Exercise 4.11 that α_*, β_*, and j_* are isomorphisms. The map Φ'_* is also an isomorphism, for Φ' is a homeomorphism. Therefore the following diagram is commutative:

$$\begin{array}{ccc} H_p(U \cap V) & \xrightarrow{k_*} & H_p(V) \\ \uparrow & & \uparrow \\ H_p(S^{n-1}) & \xrightarrow[f_*]{} & H_p(Y), \end{array}$$

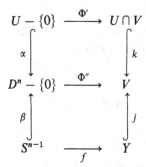

where the vertical maps are the isomorphisms $\Phi'_*\alpha_*^{-1}\beta_*$ and j_*. This completes the proof. □

Corollary 8.12. *Suppose that $n \geq 2$, and let Y_f be the space obtained from a compact Hausdorff space Y by attaching an n-cell via f.*

(i) *If $p \neq n, n-1$, then*

$$H_p(Y) \cong H_p(Y_f).$$

(ii) *There is an exact sequence*

$$0 \longrightarrow H_n(Y) \xrightarrow{i_*} H_n(Y_f) \longrightarrow H_{n-1}(S^{n-1}) \xrightarrow{f_*} H_{n-1}(Y) \longrightarrow H_{n-1}(Y_f);$$

moreover, the last map is a surjection if $n \geq 3$.

Proof. Immediate from the theorem and the computation of $H_*(S^{n-1})$. The last remark about surjectivity is true because the next term in the sequence is $H_{n-2}(S^{n-1})$, and this vanishes when $n \geq 3$. □

Theorem 8.13.

$$H_p(\mathbf{C}P^n) = \begin{cases} \mathbf{Z} & \text{if } p = 0, 2, 4, \ldots, 2n \\ 0 & \text{otherwise.} \end{cases}$$

$$H_p(\mathbf{H}P^n) = \begin{cases} \mathbf{Z} & \text{if } p = 0, 4, 8, \ldots, 4n \\ 0 & \text{otherwise.} \end{cases}$$

Proof. We prove that the formula for $H_*(\mathbf{C}P^n)$ is correct by induction on $n \geq 0$. All is well for $n = 0$ because $\mathbf{C}P^0$ is a point; since $\mathbf{C}P^1 \approx S^2$, the formula holds for $n = 1$ as well. We may now consider $\mathbf{C}P^{n+1}$ for $n \geq 1$. By Theorem 8.8(ii), $\mathbf{C}P^{n+1}$ is obtained from $\mathbf{C}P^n$ by attaching a $2(n+1)$-cell; since $n \geq 1$, $2n + 2 > 3$, and so we may use the full statement of Corollary 8.12. Thus, for $p \neq 2n+2, 2n+1$, we have

$$H_p(\mathbf{C}P^n) \cong H_p(\mathbf{C}P^{n+1}).$$

By induction, the left side is nonzero only for even $p \leq 2n$, in which case it is $\mathbf{Z}$. For $p = 2n+2, 2n+1$, there is an exact sequence

$$0 \to H_{2n+2}(\mathbf{C}P^n) \to H_{2n+2}(\mathbf{C}P^{n+1}) \to \mathbf{Z} \to H_{2n+1}(\mathbf{C}P^n) \to H_{2n+1}(\mathbf{C}P^{n+1}) \to 0.$$

Since $H_{2n+2}(\mathbf{C}P^n) = 0 = H_{2n+1}(\mathbf{C}P^n)$ (by induction), it follows that $H_{2n+2}(\mathbf{C}P^{n+1}) = \mathbf{Z}$ and $H_{2n+1}(\mathbf{C}P^{n+1}) = 0$, as desired.

The quaternionic case is similar, using the facts that $\mathbf{H}P^1 \approx S^4$ and Theorem 8.8(iii) that $\mathbf{H}P^{n+1}$ is obtained from $\mathbf{H}P^n$ by attaching a $4(n+1)$-cell. □

Theorem 8.11 and its corollary are not strong enough to allow computation of $H_*(\mathbf{R}P^n)$ (arguing as above breaks down when $n = 2$). Before introducing *cellular theory*, which will greatly assist computations (indeed it wll simplify the proof of Theorem 8.13), let us give more applications of Theorem 8.11.

EXAMPLE 8.6.

$$H_p(\mathbf{R}P^2) = \begin{cases} \mathbf{Z} & \text{if } p = 0 \\ \mathbf{Z}/2\mathbf{Z} & \text{if } p = 1 \\ 0 & \text{if } p \geq 2. \end{cases}$$

We already know (Example 7.13) that $H_p(\mathbf{R}P^2) = 0$ for $p \geq 3$, that $H_0(\mathbf{R}P^2) = \mathbf{Z}$, and that rank $H_2(\mathbf{R}P^2) =$ rank $H_1(\mathbf{R}P^2)$ (the last fact follows from our computation of the Euler–Poincaré characteristic: $\chi(\mathbf{R}P^2) = 1$). Regard $\mathbf{R}P^2$ as the space obtained from $\mathbf{R}P^1 = S^1$ by attaching a 2-cell via f, where $f(e^{i\theta}) = e^{2i\theta}$ (thereby identifying antipodal points). The attaching map f has degree 2, and so the induced map $f_*\colon H_1(S^1) \to H_1(S^1)$ is multiplication by 2. Now Theorem 8.11 gives exactness of

$$H_2(\mathbf{R}P^1) \longrightarrow H_2(\mathbf{R}P^2) \longrightarrow H_1(S^1) \xrightarrow{f_*} H_1(\mathbf{R}P^1) \xrightarrow{i_*} H_1(\mathbf{R}P^2) \longrightarrow H_0(S^1);$$

this can be rewritten (since $\mathbf{R}P^1 = S^1$) as

$$0 \longrightarrow H_2(\mathbf{R}P^2) \longrightarrow \mathbf{Z} \xrightarrow{2} \mathbf{Z} \xrightarrow{i_*} H_1(\mathbf{R}P^2) \longrightarrow \mathbf{Z}.$$

Since multiplication by 2 is monic, $H_2(\mathbf{R}P^2) = 0$; since rank $H_1(\mathbf{R}P^2) =$ rank $H_2(\mathbf{R}P^2) = 0$, it follows that $H_1(\mathbf{R}P^2)$ is torsion, and so exactness shows i_* is surjective (because $H_1(\mathbf{R}P^2)$ is torsion and $\mathbf{Z}$ is torsion-free). Therefore $H_1(\mathbf{R}P^2) = \mathbf{Z}/2\mathbf{Z}$. Note that the generator of $H_1(\mathbf{R}P^2)$ arises from the obvious 1-cycle, namely, the image of f. (Of course, this result agrees with our computation in Example 7.15.)

EXAMPLE 8.7. If T is the torus, then

$$H_p(T) = \begin{cases} \mathbf{Z} & \text{if } p = 0, 2 \\ \mathbf{Z} \oplus \mathbf{Z} & \text{if } p = 1 \\ 0 & \text{if } p \geq 3. \end{cases}$$

We already know (Example 7.12) that $H_p(T) = 0$ for $p \geq 3$, that $H_0(T) = \mathbf{Z}$, and that rank $H_2(T) + 1 =$ rank $H_1(T)$ (for $\chi(T) = 0$). The construction of T as a quotient space of $\mathbf{I} \times \mathbf{I}$ by identifying parallel edges exhibits T as being

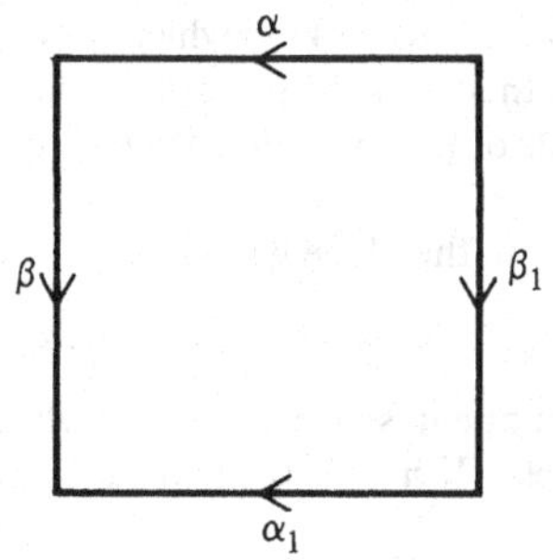

obtained from the wedge $S^1 \vee S^1$ by attaching a 2-cell. More precisely, let $\Phi: \mathbf{I} \times \mathbf{I} \to T$ be the natural map, let $\partial(\mathbf{I} \times \mathbf{I})$ denote the perimeter of $\mathbf{I} \times \mathbf{I}$ (which we identify with S^1), and let $f = \Phi|\partial(\mathbf{I} \times \mathbf{I})$. Note that $\Phi(\partial(\mathbf{I} \times \mathbf{I})) = S^1 \vee S^1$, and so we may regard f as a function $S^1 \to S^1 \vee S^1$. Now Theorem 6.22 says that (the class of) $\alpha * \beta * \alpha_1^{-1} * \beta_1^{-1}$ is a generator of $H_1(S^1)$; moreover, $f\alpha = f\alpha_1$ and $f\beta = f\beta_1$. Therefore composing f with either projection $S^1 \vee S^1 \to S^1$ yields maps $S^1 \to S^1$, namely, $f\alpha * f\alpha_1^{-1}$ and $f\beta * f\beta_1^{-1}$. Since each of these maps has degree 0 (Theorems 3.16 and 6.20), it follows that $f_*: H_1(S^1) \to H_1(S^1 \vee S^1) \cong H_1(S^1) \oplus H_1(S^1)$ is the zero map. But Theorem 8.11 gives exactness of

$$0 = H_2(S^1 \vee S^1) \to H_2(T) \to H_1(S^1) \xrightarrow{f_*} H_1(S^1 \vee S^1) \xrightarrow{i_*} H_1(T) \to H_0(S^1).$$

Since $f_* = 0$, the map $H_2(T) \to H_1(S^1)$ is an isomorphism and $H_2(T) = \mathbf{Z}$. Also, $H_1(T)$ must be torsion free (because the two flanking terms $H_1(S^1 \vee S^1)$ and $H_0(S^1)$ are) and of rank 2 (rank $H_2(T) + 1 =$ rank $H_1(T)$); hence $H_1(T) = \mathbf{Z} \oplus \mathbf{Z}$. Note that i_* must be an isomorphism, so that the two obvious circles on the torus are independent generating 1-cycles, as one expects. (Of course, this result agrees with our computation in Example 7.14.)

These examples do not yet indicate the power of "cellular homology" that will be apparent by the end of this chapter.

Exercises

8.17. If K is the Klein bottle, prove that

$$H_p(K) = \begin{cases} \mathbf{Z} & \text{if } p = 0 \\ \mathbf{Z} \oplus \mathbf{Z}/2\mathbf{Z} & \text{if } p = 1 \\ 0 & \text{if } p \geq 2. \end{cases}$$

(*Hint*: Show that K arises by attaching a 2-cell to $S^1 \vee S^1$; compute the induced map $f_*: H_1(S^1) \to H_1(S^1 \vee S^1)$ as in Example 8.7.)

*8.18. Let W be a $4h$-gon in the plane whose edges are labeled

$$\alpha_1, \beta_1, \alpha_1^{-1}, \beta_1^{-1}, \ldots, \alpha_h, \beta_h, \alpha_h^{-1}, \beta_h^{-1},$$

and let M be the quotient space of W in which the edges are identified according to the labels. (If $h = 1$, then M is the torus; if $h = 0$, one defines M by identifying all the boundary points of W to a point, hence $M \approx S^2$.) One calls h the number of **handles** of M.

Let W' be a $2n$-gon in the plane whose edges are labeled

$$\alpha_1, \alpha_1, \alpha_2, \alpha_2, \ldots, \alpha_n, \alpha_n,$$

and let M' be the quotient space of W' in which the edges are identified according to the labels. (If $n = 1$, then M' is $\mathbf{R}P^2$.) One calls n the number of **crosscaps** of M'.

(i) Prove that M (respectively, M') is obtained from a wedge of h (respectively, n) circles by attaching a 2-cell (here $h \geq 0$ and $n \geq 1$).
(ii) Prove that $H_2(M) = \mathbf{Z}$, that $H_1(M)$ is free abelian of rank $2h$, and that $\chi(M) = h$.
(iii) Prove that $H_2(M') = 0$, that rank $H_1(M') = n - 1$, and that $\chi(M') = 2 - n$.
(iv) Use the method of adequate subcomplexes for these computations.

Definition. An ***n*-manifold** is a Hausdorff space M such that each point in M has a neighborhood homeomorphic to $\mathbf{R}^n$.

It is a remarkable fact that every compact connected 2-manifold is homeomorphic to either M or M' as defined in Exercise 8.18; in the first case, M is called a **sphere with** h **handles** and is **orientable**; in the second case, M' is **nonorientable**.

If M_1 and M_2 are compact connected 2-manifolds, choose open sets U_i in M_i (for $i = 1, 2$) with $U_i \approx \mathbf{R}^2$, and choose (closed) disks D_i in U_i. Define a new space $M_1 \# M_2$ (the **connected sum**) by removing the interiors of D_1 and D_2 and then gluing M_1 and M_2 together at the boundaries of the D_i. More precisely, choose a homeomorphism $k: \dot{D}_1 \to \dot{D}_2 \subset M_2$ (of course, $\dot{D}_i \approx S^1$) and let $M_1 \# M_2 = M_1 \coprod_k M_2$. It can be shown that $M_1 \# M_2$ can be viewed as a (compact connected) 2-manifold and that, to homeomorphism, it is independent of the several choices. The $\#$ operation is commutative and associative; also, S^2 acts as a unit, that is, for every (compact connected) 2-manifold X, we have $S^2 \# X \approx X$. The "remarkable fact" mentioned above is proved in [Massey (1967), Chap. 1]. In more detail, it is first shown that such a 2-manifold X is homeomorphic to either S^2, $T = S^1 \times S^1$ (the torus), $\mathbf{R}P^2$, or a connected sum of several copies of the latter two. Then one sees that $\mathbf{R}P^2 \# T \approx \mathbf{R}P^2 \# \mathbf{R}P^2 \# \mathbf{R}P^2$ (this last relation explains why there is no mixture of tori and projective planes needed in expressing X as a connected sum; one can also prove [Massey (1967), p. 9] that the Klein bottle is homeomorphic to $\mathbf{R}P^2 \# \mathbf{R}P^2$). If X is orientable, then either $X \approx S^2$ or $X \approx M_1 \# \cdots \# M_g$, where each $M_i = T$; if X is nonorientable, then $X \approx M_1 \# \cdots \# M_g$, where each $M_i = \mathbf{R}P^2$. The number g of "summands" is called the **genus** of X. To see that g is an invariant of X, one first proves that two compact connected 2-manifolds X and X' are homeomorphic if and only if (1) both are orientable or both are nonorientable and (2) $\chi(X) = \chi(X')$, where $\chi(X)$ is the Euler–Poincaré characteristic of X. Next, one sees that $g = \frac{1}{2}(2 - \chi(X))$ when X is orientable and $g = 2 - \chi(X)$ when χ is nonorientable. The invariance of g is thus a consequence of the invariance of χ (Theorem 7.15 proves the invariance of χ for simplicial complexes, and compact 2-manifolds can be triangulated). The fundamental groups of these surfaces are the surface groups of Corollary 7.46: If X is orientable of genus g, then $\pi_1(X)$ has a presentation of the first type in that corollary; if X is nonorientable of genus g, then $\pi_1(X)$ has a presentation of the second type (see [Massey (1967), pp. 131–132] or [Seifert-Threlfall, p. 176]).

CW Complexes

This section introduces an important class of spaces that contains all (possibly infinite) simplicial complexes. The definition may appear at first to be rather technical, but we shall see that such spaces are built in stages: attach a (possibly infinite) family of 1-cells to a discrete space; attach a family of 2-cells to the result; then attach 3-cells, 4-cells, and so on. Since we allow attaching infinitely many cells, let us begin by discussing an appropriate topology.

Definition. Let X be a set covered by subsets A_j, where j lies in some (possibly infinite) index set J, that is, $X = \bigcup_{j \in J} A_j$. Assume the following:

(i) each A_j is a topological space;
(ii) for each $j, k \in J$, the topologies of A_j and of A_k agree on $A_j \cap A_k$;
(iii) for each $j, k \in J$, the intersection $A_j \cap A_k$ is closed in A_j and in A_k.

Then the **weak topology** on X **determined** by $\{A_j: j \in J\}$ is the topology whose closed sets are those subsets F for which $F \cap A_j$ is closed in A_j for every $j \in J$.

It is easy to see that each A_j is a closed subset of X when X is given the weak topology; moreover, each A_j, as a subspace of X, retains its original topology. If the index set J is finite, then there is only one topology on X compatible with conditions (i), (ii), and (iii), and so it must be the weak topology.

EXERCISES

8.19. If X has the weak topology determined by $\{A_j: j \in J\}$, then a subset U of X is open if and only if $U \cap A_j$ is open in A_j for every $j \in J$.

8.20. If X has the weak topology determined by $\{A_j: j \in J\}$ and if Y is a closed subspace of X, then Y has the weak topology determined by $\{Y \cap A_j: j \in J\}$.

EXAMPLE 8.8. If $\{X_j: j \in J\}$ is a family of topological spaces, then their **coproduct** $\coprod X_j$ is their disjoint union equipped with the weak topology determined by $\{X_j: j \in J\}$. The reader may check that each X_j is both open and closed in the coproduct.

EXAMPLE 8.9. If $\{(X_j, x_j): j \in J\}$ is a family of pointed topological spaces, then their **wedge** $\bigvee X_j$ is the quotient space of $\coprod X_j$ in which all the (closed) basepoints x_j are identified ($\bigvee X_j$ is thus a pointed space with basepoint the identified family of basepoints). Each X_j is imbedded in $\bigvee X_j$, and $\bigvee X_j$ has the weak topology determined by these subspaces.

EXAMPLE 8.10. Let $X = \bigvee_{i \geq 1} S_i^1$ (where $S_i^1 \approx S^1$) have basepoint b, and let Y be the subspace of $\mathbf{R}^2$ consisting of the circles C_n, $n \geq 1$, where C_n has center $(0, 1/n)$ and radius $1/n$. Now X and Y are not homeomorphic. For each $n \geq 1$,

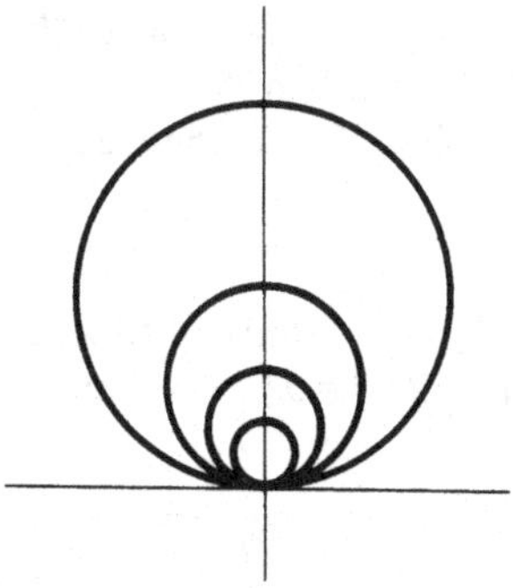

choose $x_n \in C_n - \{\text{origin}\}$, and define $F = \{x_n: n \geq 1\}$. Now $F \cap C_n = \{x_n\}$, so that $F \cap C_n$ is closed in C_n. Thus F is a closed subset of $\bigvee S_i^1 = X$. On the other hand, F is not a closed subset of Y, for $(0, 0) \notin F$ and $(0, 0)$ is a limit point of F in Y.

A similar argument, using a compact neighborhood of the origin, shows that X is not homeomorphic to the subspace Z of $\mathbf{R}^2$, which is the union of the circles B_n having center $(0, n)$ and radius n. Of course, Y and Z are not homeomorphic because Y is compact and Z is not.

EXAMPLE 8.11. Let K be an abstract simplicial complex with vertex set V (which may be infinite). Define $\mathbf{I}^{(V)}$ to be the set of all functions $\varphi: V \to \mathbf{I}$ that are zero for all but a finite number of $v \in V$ ($\mathbf{I}^{(V)}$ is a subspace of the cartesian product $\mathbf{I}^V$, the latter consisting of all functions $V \to \mathbf{I}$). In particular, for each $v \in V$, define $\tilde{v} \in \mathbf{I}^{(V)}$ by

$$\tilde{v}(u) = \begin{cases} 1 & \text{if } u = v \\ 0 & \text{if } u \neq v. \end{cases}$$

(Informally, we identify the function $\tilde{v}$ with the vertex v.) If $\sigma = \{v_0, \ldots, v_n\}$ is an n-simplex in K, define $\tilde{\sigma} \subset \mathbf{I}^{(V)}$ as the family of all convex combinations of $\{\tilde{v}_0, \ldots, \tilde{v}_n\}$ (it is easy to see that the subspace $\tilde{\sigma}$ is homeomorphic to Δ^n via $\sum \lambda_i \tilde{v}_i \mapsto (\lambda_0, \ldots, \lambda_n)$, the barycentric coordinates). Finally, define

$$|K| = \bigcup_{\sigma \in K} \tilde{\sigma}$$

and equip $|K|$ with the weak topology determined by $\{\tilde{\sigma}: \sigma \in K\}$. (Note that, when K is infinite, $|K|$ is not a subspace of $\mathbf{I}^{(V)} \subset \mathbf{I}^V$.) One calls $|K|$ the **geometric realization** of K. The reader should check that, when K is finite, this definition coincides with our earlier definition (in Chapter 7).

Lemma 8.14. *Let X have the weak topology determined by a family of subsets $\{A_j: j \in J\}$. For any topological space Y, a function $f: X \to Y$ is continuous if and only if $f|A_j$ is continuous for every $j \in J$.*

PROOF. As always, f continuous implies that each of its restrictions is continuous. Conversely, if G is closed in Y, then $f^{-1}(G) \cap A_j = (f|A_j)^{-1}(G)$ is closed in A_j for every $j \in J$. Since X has the weak topology, $f^{-1}(G)$ is closed in X and f is continuous. □

EXAMPLE 8.8′. Let $\{X_j: j \in J\}$ be a family of topological spaces, let Y also be a topological space, and let $\{f_j: j \in J\}$ be a family of continuous functions, where $f_j: X_j \to Y$. Define

$$f: \coprod X_j \to Y$$

as follows: if $x \in \coprod X_j$, then there is a unique X_j containing x; define $f(x) = f_j(x)$. Since $f|X_j = f_j$ for all j, it follows from Lemma 8.14 that f is continuous. One often uses the notation

$$f = \coprod f_j.$$

EXAMPLE 8.9′. Let $\{(X_j, x_j): j \in J\}$ be a family of pointed spaces, let (Y, y_0) also be a pointed space, and let $\{f_j: j \in J\}$ be a family of continuous pointed maps, where $f_j: (X_j, x_j) \to (Y, y_0)$. Define

$$f: (\bigvee X_j, b) \to (Y, y_0)$$

as follows: if $x \in \bigvee X_j$ and $x \neq b$, then there is a unique X_j containing x; define $f(x) = f_j(x)$; if $x = b$, define $f(x) = y_0$. Since $f|X_j = f_j$ for all j, it follows from Lemma 8.14 that f is continuous (one can also prove continuity of f using Theorem 1.8). One often uses the notation

$$f = \bigvee f_j.$$

Definition. Assume that a topological space X is a disjoint union of cells: $X = \bigcup \{e: e \in E\}$. For each $k \geq 0$, the ***k*-skeleton** $X^{(k)}$ of X is defined by

$$X^{(k)} = \bigcup \{e \in E: \dim(e) \leq k\}.$$

Of course, $X^{(0)} \subset X^{(1)} \subset X^{(2)} \subset \cdots$ and $X = \bigcup_{k \geq 0} X^{(k)}$.

Definition. A **CW complex** is an ordered triple (X, E, Φ), where X is a Hausdorff space, E is a family of cells in X, and $\Phi = \{\Phi_e: e \in E\}$ is a family of maps, such that

(1) $X = \bigcup \{e: e \in E\}$ (disjoint union);
(2) for each k-cell $e \in E$, the map $\Phi_e: (D^k, S^{k-1}) \to (e \cup X^{(k-1)}, X^{(k-1)})$ is a relative homeomorphism;
(3) if $e \in E$, then its closure $\bar{e}$ is contained in a finite union of cells in E;
(4) X has the weak topology determined by $\{\bar{e}: e \in E\}$.

If (X, E, Φ) is a CW complex, then X is called a **CW space**, (E, Φ) is called a **CW decomposition** of X, and $\Phi_e \in \Phi$ is called the **characteristic map** of e. One should regard a CW space X as a generalized polyhedron (Examples 8.11 and 8.12 below), and one should regard (E, Φ) as a generalized triangulation of X.

Remarks. (1) Axiom (1) says that the cells E partition X.

(2) Axiom (2) says that each k-cell e arises from attaching a k-cell to $X^{(k-1)}$ via the attaching map $\Phi_e|S^{k-1}$.

(3) Axiom (3) is called **closure finiteness**; the letters CW are the initials of "closure finiteness" and "weak topology".

(4) Axiom (4) says that a subset A of X is closed if and only if $A \cap \bar{e}$ is closed in $\bar{e}$ for every $e \in E$. Moreover, Lemma 8.14 implies that if Y is any topological space and $f: X \to Y$ any function, then f is continuous if and only if $f|\bar{e}$ is continuous for every $e \in E$.

(5) Just as a polyhedron may have many triangulations, a CW space may have many CW decompositions.

Definition. A CW complex (X, E, Φ) is **finite** if E is a finite set.

If (X, E, Φ) is a finite CW complex, then axioms (3) and (4) in the definition of CW complex are redundant. The reader interested in this case only can shorten many of the coming proofs.

EXAMPLE 8.12. Let X be a compact polyhedron and let (K, h) be a triangulation of X, where K is a finite simplicial complex and $h: |K| \to X$ is a homeomorphism. For each simplex $\sigma \in K$, let $\sigma^\circ = \sigma - \dot{\sigma}$ denote the corresponding open simplex, and define $E = \{h(\sigma^\circ): \sigma \in K\}$. It is clear that E is a partition of X. If $K^{(n-1)}$ denotes the $(n-1)$-skeleton of K, and if $\sigma \in K$ is an n-simplex, then define $\Phi_\sigma = (h|\sigma) \circ \alpha_\sigma$, where $\alpha_\sigma: (D^n, S^{n-1}) \to (\sigma, \dot{\sigma})$ is some homeomorphism. Because simplexes intersect in faces (or not at all),

$$\Phi_\sigma: (D^n, S^{n-1}) \to (\sigma, \dot{\sigma}) \to (h(\sigma^\circ) \cup h(|K^{(n-1)}|), h(|K^{(n-1)}|))$$

is a relative homeomorphism (indeed each Φ_σ is a homeomorphism from D^n to $\Phi_\sigma(D^n)$). Since K is finite, it follows that X is a finite CW complex.

The geometric realization $|K|$ of a (possibly infinite) simplicial complex K is a CW complex. The definition of (E, Φ) is as above; by definition (Example 8.11), $|K|$ has the weak topology determined by the closures of its cells, and it is straightforward to check that $|K|$ is closure finite.

EXAMPLE 8.13. If f is a real-valued function on a manifold M and if $a \in \mathbf{R}$, then $M^a = \{x \in M: f(x) \leq a\}$. A basic result of Morse theory (see [Milnor (1963)]) is that if f is a differentiable function on a manifold M with no "degenerate critical points" and if each M^a is compact, then M has the same homotopy type as a CW complex. Indeed it is shown in [Lundell and Weingram, p. 135] that every separable manifold has the same homotopy type as a CW complex.

EXAMPLE 8.14. Regard S^n as a subspace of $\mathbf{R}^{n+1}$. For each $n \geq 1$, define $\Phi: (D^n, S^{n-1}) \to (S^n, p)$, where $p = (0, \ldots, 0, 1) \in S^n$, by

$$x \mapsto (2\sqrt{1 - \|x\|^2}\,x, 2\|x\|^2 - 1).$$

If we denote an i-cell by e^i, then the map Φ allows one to view S^n as a CW complex with $E = \{e^0, e^n\}$. Of course, S^0 also has a CW decomposition with two cells, namely, $\{e_1^0, e_2^0\}$.

EXAMPLE 8.15. Recall that $S^n = E_+^n \cup E_-^n$ (upper and lower closed hemispheres) and $E_+^n \cap E_-^n = S^{n-1}$ (the equator). There are thus two n-cells e_1 and e_2 with $\bar{e}_1 = E_+^n$ and $\bar{e}_2 = E_-^n$; one concludes by induction that S^n has a CW decomposition with two i-cells in every dimension $0 \leq i \leq n$.

EXAMPLE 8.16. $\mathbf{R}P^n$ has a CW decomposition $\{e^0, e^1, \ldots, e^n\}$ (Theorem 8.8(i)).

EXAMPLE 8.17. $\mathbf{R}P^\infty$ has a CW decomposition with one i-cell in every dimension $i \geq 0$ (by definition, $\mathbf{R}P^\infty = \bigcup \mathbf{R}P^n$ has the weak topology determined by the family $\{\mathbf{R}P^n: n \geq 0\}$).

EXAMPLE 8.18. $\mathbf{C}P^n$ has a CW decomposition $\{e^0, e^2, \ldots, e^{2n}\}$ (Theorem 8.8(ii)).

EXAMPLE 8.19. $\mathbf{H}P^n$ has a CW decomposition $\{e^0, e^4, \ldots, e^{4n}\}$ (Theorem 8.8(iii)).

Definition. Let (X, E, Φ) be a CW complex. If $E' \subset E$, define

$$|E'| = \bigcup \{e: e \in E'\} \subset X,$$

and define $\Phi' = \{\Phi_e: e \in E'\}$. Call $(|E'|, E', \Phi')$ a **CW subcomplex** if im $\Phi_e \subset |E'|$ for every $e \in E'$.

If $E' \subset E$ and $X' = |E'|$, then it is easy to see that every CW subcomplex is itself a CW complex (once one observes that $(X')^{(k)} = X^{(k)} \cap X'$ for all $k \geq 0$). It is also easy to see that any union and any intersection of CW subcomplexes is again a CW subcomplex.

Henceforth we may not display all necessary ingredients, and we may say that (X, E), or even X, is a CW complex; similarly, we may say that (X', E'), or even X', is a CW subcomplex. The next lemma is just Exercise 8.14.

Lemma 8.15. *If (X, E) is a CW complex and if $e \in E$ is a k-cell (where $k > 0$) with characteristic map Φ_e, then $\bar{e} = \operatorname{im} \Phi_e = \Phi_e(D^k)$.*

PROOF. Since Φ_e is continuous,

$$\Phi_e(D^k) = \Phi_e(\overline{D^k - S^{k-1}}) \subset \overline{\Phi_e(D^k - S^{k-1})} = \bar{e}.$$

For the reverse inclusion, observe that compactness of D^k gives compactness of $\Phi_e(D^k)$. Since X is Hausdorff, $\Phi_e(D^k)$ is a closed subset of X containing $e = \Phi_e(D^k - S^{k-1})$, and so $\bar{e} \subset \Phi_e(D^k)$. □

EXAMPLE 8.20. If (X, E) is a CW complex and $E' \subset E$, then $|E'|$ is a CW subcomplex if and only if $\bar{e} \subset |E'|$ for every $e \in E'$. Hence, if E' is a family of k-cells in E, for some fixed $k > 0$, then $|E'| \cup X^{(k-1)}$ is a CW subcomplex (for axiom (2) gives $\bar{e} \subset e \cup X^{(k-1)}$).

EXAMPLE 8.21. Every k-skeleton $X^{(k)}$ is a CW subcomplex (this is the special case of the previous example for which $E' = \varnothing$).

The following technical lemma will be useful.

Lemma 8.16. *Let (X, E, Φ) be an ordered triple satisfying axiom (1) and axiom (2) of the definition of CW complex, and let $\varphi\colon \coprod_{e\in E} D^{n(e)} \to X$ be the map $\varphi = \coprod_{e\in E} \Phi_e$. Then X has the weak topology determined by $\{\bar{e}: e \in E\}$ if and only if φ is an identification.*

PROOF. Assume that X has the weak topology. Since φ is a continuous surjection, it suffices to show that if $C \subset X$ and $\varphi^{-1}(C)$ is closed in $\coprod_e D^{n(e)}$, then C is closed in X. Now $\varphi^{-1}(C) \cap D^{n(e)}$ is compact, being a closed subset of $D^{n(e)}$. On the other hand,

$$\begin{aligned} \varphi^{-1}(C) \cap D^{n(e)} &= \Phi_e^{-1}(C) \cap D^{n(e)} \\ &= \Phi_e^{-1}(C) \cap \Phi_e^{-1}(\bar{e}) \quad \text{(Lemma 8.15)} \\ &= \Phi_e^{-1}(C \cap \bar{e}). \end{aligned}$$

Thus $C \cap \bar{e}$ is compact, hence closed in $\bar{e}$. As X has the weak topology, C is closed in X.

Conversely, assume that φ is an identification. Let $C \subset X$ be such that $C \cap \bar{e}$ is closed in $\bar{e}$ for all $e \in E$. Then $\varphi^{-1}(C) \cap D^{n(e)}$ is closed in $D^{n(e)}$ for all $e \in E$. Since $\coprod D^{n(e)}$ has the weak topology determined by $\{D^{n(e)}: e \in E\}$, $\varphi^{-1}(C)$ is closed in $\coprod D^{n(e)}$; since φ is an identification, C is closed in X. Therefore X has the weak topology. □

Lemma 8.17. *Let (X, E) be a CW complex, and let E' be a finite subset of E. Then $|E'|$ is a CW subcomplex if and only if $|E'|$ is closed.*

PROOF. If $|E'|$ is a CW subcomplex, then $\bar{e} \subset |E'|$ for every $e \in E'$. Hence $|E'| = \bigcup\{e: e \in E'\} = \bigcup\{\bar{e}: e \in E'\}$ is closed, being a finite union of closed sets. Conversely, if $|E'|$ is closed and $e \in E'$, then $e \subset |E'|$ and $\bar{e} \subset |E'|$; hence $|E'|$ is a CW subcomplex. □

Lemma 8.18. *If (X, E) is a CW complex and $e \in E$, then the closure $\bar{e}$ is contained in a finite CW subcomplex.*

PROOF. We proceed by induction on $n = \dim(e)$; the statement is obviously true when $n = 0$. If $n > 0$, then Lemma 8.15 gives

$$\bar{e} - e = \Phi_e(D^n) - e \subset (e \cup X^{(n-1)}) - e \subset X^{(n-1)}.$$

By axiom (3), $\bar{e}$ meets only finitely many cells other than e, say, $e_1, \ldots, e_m$, and we have just seen that $\dim(e_i) \le n - 1$ for all i. By induction, there is a finite CW subcomplex X_i containing $\bar{e}_i$, for $i = 1, \ldots, m$, and each X_i is closed, by

Lemma 8.17. But $\bar{e} \subset e \cup X_1 \cup \cdots \cup X_m$, so that this union of finitely many cells is closed and hence is a finite CW subcomplex. □

Theorem 8.19. *If (X, E) is a CW complex, then every compact subset K of X lies in a finite CW subcomplex. Therefore, a CW space X is compact if and only if (X, E) is a finite CW complex for every CW decomposition E.*

PROOF. For each $e \in E$ with $K \cap e \neq \emptyset$, choose a point $a_e \in K \cap e$, and let A be the set comprised of all such a_e. For each $e \in E$, Lemma 8.18 says that there is a finite CW subcomplex X_e containing $\bar{e}$. Therefore $A \cap \bar{e} \subset A \cap X_e$ is a finite set and hence is closed in $\bar{e}$. Since X has the weak topology, A is closed in X; indeed the same argument shows that every subset of A is closed in X, hence A is discrete. But A is also compact, being a closed subset of K. Thus A is finite, so that K meets only finitely many $e \in E$, say, $e_1, \ldots, e_m$. By Lemma 8.18, there are finite CW subcomplexes X_i with $\bar{e}_i \subset X_i$ for $i = 1, \ldots, m$. It follows that K is contained in the finite CW subcomplex $\bigcup X_i$. □

Lemma 8.20. *If (X, E) is a CW complex, then a subset A of X is closed if and only if $A \cap X'$ is closed in X' for every finite CW subcomplex X' in X.*

PROOF. If A is closed in X, then $A \cap X'$ is certainly closed in X'. Conversely, for each $e \in E$, choose a finite subcomplex X_e containing $\bar{e}$. By hypothesis, $A \cap X_e$ is closed in X_e; it follows that $A \cap \bar{e} = (A \cap X_e) \cap \bar{e}$ is closed in X_e and hence is closed in the smaller set $\bar{e}$. Therefore A is closed in X because X has the weak topology determined by all $\bar{e}$. □

It follows that a CW complex has the weak topology determined by the family of its finite CW subcomplexes. The next result generalizes Lemma 8.17 by removing the finiteness hypothesis.

Theorem 8.21. *Let (X, E) be a CW complex and let E' be a (possibly infinite) subset of E. Then $|E'|$ is a CW subcomplex if and only if $|E'|$ is closed.*

PROOF. If $|E'|$ is closed and $e \in E'$, then $\bar{e} \subset |E'|$ and $|E'|$ is a CW subcomplex. Conversely, assume that $|E'| = X'$ is a CW subcomplex. It suffices, by Lemma 8.20, to show that $X' \cap Y$ is closed in Y for every finite CW subcomplex Y of X. Now $X' \cap Y$ is a finite union of cells, say, $X' \cap Y = e_1 \cup \cdots \cup e_m$. As $X' \cap Y$ is a CW subcomplex, $\bar{e}_i \subset X' \cap Y$ for all i; hence $X' \cap Y = \bar{e}_1 \cup \cdots \cup \bar{e}_m$, and so $X' \cap Y$ is closed in Y (even in X). □

Corollary 8.22. *Let (X, E) be a CW complex and, for some fixed $n > 0$, let E' be a family of n-cells in E.*

(i) $X' = |E'| \cup X^{(n-1)}$ *is closed in* X;
(ii) *every n-skeleton $X^{(n)}$ is closed in* X;

(iii) *every n-cell e is open in* $X^{(n)}$;
(iv) $X^{(n)} - X^{(n-1)}$ *is an open subset of* $X^{(n)}$.

PROOF. (i) We know that X' is a CW subcomplex (Example 8.20), hence it is closed.

(ii) If $n > 0$, this is the special case of (i) in which $E' = \varnothing$; if $n = 0$, this is Exercise 8.22.

(iii) This is the special case of (i) in which E' consists of every n-cell in E except e.

(iv) Immediate from (iii). □

Theorem 8.23. *Let X be a CW complex.*

(i) *Every path component of X is a CW subcomplex, hence is closed.*
(ii) *The path components of X are closed and open.*
(iii) *The path components of X are the components of X.*
(iv) *X is connected if and only if X is path connected.*

PROOF. (i) Since X is a disjoint union of cells, each of which is path connected, it follows that each path component A is a union of cells. If e is an n-cell with $e \subset A$, then $\bar{e} = \Phi_e(D^n)$ is also path connected, and so $\bar{e} \subset A$. Therefore A is a CW subcomplex and hence is closed.

(ii) Let A be a path component of X, and let B be the union of the other path components. Since B is a union of CW subcomplexes, it is a CW subcomplex and hence is closed. As B is the complement of A, we see that A is open.

(iii) Let A be a path component of X and let Y be the component of X containing A. Since A is closed and open, it follows that $A = Y$ (lest $Y = A \cup (Y - A)$ be a disconnection).

(iv) Immediate from (iii). □

The $\sin(1/x)$ space is connected but not path connected; there are thus Hausdorff spaces (even compact subsets of the plane) that are not CW spaces.

EXERCISES

8.21. A space is called **compactly generated** if it is Hausdorff and it has the weak topology determined by its compact subsets. Prove that every CW complex is compactly generated.

*8.22. If (X, E) is a CW complex, then $X^{(0)}$ is a discrete closed subset of X. (*Hint*: If A is any subset of $X^{(0)}$, then $A \cap \bar{e}$ is finite for every $e \in E$. Note that $\{1/n \colon n \geq 1\}$ is a discrete subspace of $\mathbf{R}$ that is not a closed subset of $\mathbf{R}$.)

*8.23. Show that a CW complex X has the weak topology determined by the family of its skeletons $\{X^{(n)} \colon n \geq 0\}$. Conclude that a set U is open in X if and only if $U \cap X^{(n)}$ is open in $X^{(n)}$ for every $n \geq 0$.

*8.24. Let (X, E) be a CW complex, and for fixed $n > 0$, let E_n be the family of all n-cells in E. Show that

$$X^{(n)} = \left(\coprod_{E_n} D^n\right) \amalg_f X^{(n-1)},$$

where $f = \coprod_{e \in E_n} (\Phi_e | S^{n-1})$. (As in Exercise 8.12(ii), we identify $X^{(n-1)}$ with its image in the space obtained from $X^{(n-1)}$ by attaching $\coprod D^n$ via f.)

*8.25. Show that both the torus and the Klein bottle have CW decompositions of the form $\{e^0, e_1^1, e_2^1, e^2\}$, that is, one 0-cell, two 1-cells, and one 2-cell.

8.26. (i) Show that neither union of tangent circles in Example 8.10 is a CW space.
(ii) Show that the subspace X of $\mathbf{R}$, namely,

$$X = \{0\} \cup \{1/n: n \geq 1\},$$

is not a CW space.

8.27. Define the **dimension** of a CW complex (X, E) to be

$$\dim X = \sup\{\dim(e): e \in E\}.$$

If E' is another CW decomposition of X, show that (X, E) and (X, E') have the same dimension. (*Hint*: See the proof of Theorem 7.1.) Conclude that $\dim X$ is independent of the CW decomposition of X.

8.28. Show that a CW complex X is connected if and only if its 1-skeleton $X^{(1)}$ is connected.

8.29. Let (X, E) be a CW complex, and Y be any space. Prove that a function $f: X \to Y$ is continuous if and only if $f\Phi_e$ is continuous for all $e \in E$.

The next theorem is the generalization of Lemma 8.6 (where we attached one cell), which characterizes CW complexes as spaces obtained by a sequence of attaching spaces. More important, this theorem provides an inductive method of constructing CW complexes.

Theorem 8.24. *Let X be a space, and let*

$$X^0 \subset X^1 \subset X^2 \subset \cdots$$

be a sequence of subsets with $X = \bigcup_{n \geq 0} X^n$. Assume the following:

(i) *X^0 is discrete*;
(ii) *for each $n > 0$, there is a (possibly empty) index set A_n and a family of continuous functions $\{f_\alpha^{n-1}: S^{n-1} \to X^{n-1} | \alpha \in A_n\}$ so that*

$$X^n = \left(\coprod_{\alpha} D^n\right) \amalg_f X^{n-1},$$

where $f = \coprod f_\alpha^{n-1}$;
(iii) *X has the weak topology determined by $\{X^n: n \geq 0\}$.*

If Φ^n_α denotes the (usual) composite

$$D^n \to \coprod_\alpha D^n \to \left(\coprod_\alpha D^n\right) \amalg X^{n-1} \to \left(\coprod_\alpha D^n\right) \amalg_f X^{n-1},$$

then (X, E, Φ) is a CW complex, where

$$E = X^0 \cup \bigcup_{n \geq 1} \{\Phi^n_\alpha(D^n - S^{n-1}) \colon \alpha \in A_n\}$$

and

$$\Phi = \{\textit{constant maps to } X^0\} \cup \bigcup_{n \geq 1} \{\Phi^n_\alpha \colon \alpha \in A_n\}.$$

Remarks. (1) The converse of this theorem is contained in Exercises 8.22, 8.23, and 8.24.

(2) Often the most difficult part of verifying that a space X is a CW complex is checking that it is Hausdorff; this theorem is a way to avoid this problem.

PROOF (after Maunder). Let us show that X is Hausdorff. Let x, y be distinct points in X, and let n be the least integer with both x and y in X^n; we may assume that $x \in e^n_\alpha = \Phi^n_\alpha(D^n - S^{n-1})$. We claim that there are disjoint sets U_n, V_n that are open in X^n and with $x \in U_n$ and $y \in V_n$. If $n = 0$, such sets exist because X^0 is discrete. If $n > 0$, and $B(\varepsilon)$ denotes the closed disk in D^n with radius ε and center $(\Phi^n_\alpha)^{-1}(x)$, then one may choose $V_n = X^n - \Phi^n_\alpha(B(\varepsilon))$ for suitable ε. Next, we show by induction on $k \geq n$ that there exist disjoint subsets U_k, V_k, open in X^k, with $U_k \subset U_{k+1}$ and $V_k \subset V_{k+1}$, and with $U_k \cap X^n = U_n$ and $V_k \cap X^n = V_n$. Given U_k and V_k, observe that, for each $(k+1)$-cell e, the sets $\Phi_e^{-1}(U_k)$ and $\Phi_e^{-1}(V_k)$ are disjoint open sets in $S^k \subset D^{k+1}$. Define

$$A_e = \{z \in D^{k+1} - S^k \colon \|z\| > \tfrac{1}{2} \text{ and } z/\|z\| \in \Phi_e^{-1}(U_k)\}$$

and

$$B_e = \{z \in D^{k+1} - S^k \colon \|z\| > \tfrac{1}{2} \text{ and } z/\|z\| \in \Phi_e^{-1}(V_k)\}.$$

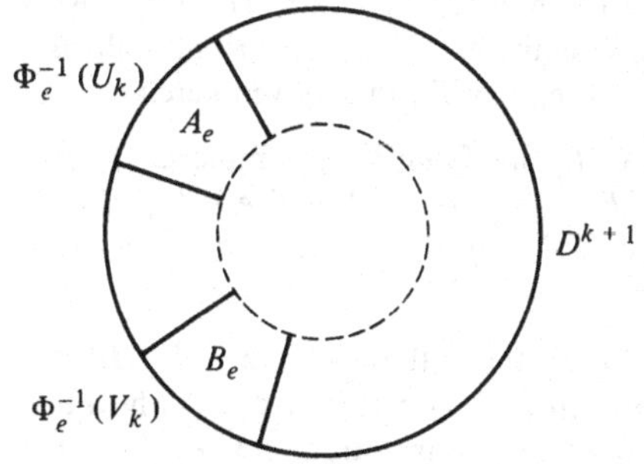

Now define $U_{k+1} = U_k \cup (\bigcup \Phi_e(A_e))$ and $V_{k+1} = V_k \cup (\bigcup \Phi_e(B_e))$, where e varies over all $(k+1)$-cells in X. Clearly, $U_{k+1} \cap V_{k+1} = \varnothing$, and $U_{k+1} \cap X^n = U_k \cap X^n = U_n$ (similarly, $V_{k+1} \cap X^n = V_n$). Also, U_{k+1} (and V_{k+1}) is open in X^{k+1}, using condition (ii) and Exercise 8.12(i). Finally, define $U = \bigcup_{k \geq n} U_k$ and $V = \bigcup_{k \geq n} V_k$. Both U and V are open in X, because X has the weak topology determined by $\{X^n: n \geq 0\}$, by condition (iii). Therefore X is Hausdorff.

Let us now verify the four axioms in the definition of CW complex. It is plain that axiom (1) and axiom (2) are satisfied. We prove, by induction on n, that each X^n is a CW complex. If $n = 0$, this follows from X^0 being discrete. Assume that $n > 0$. To see that X^n is closure finite, let e be an i-cell in X^n (hence $i \leq n$). If $i < n$, then induction shows that $\bar{e}$ meets only finitely many cells. If $i = n$, then $\bar{e} = \Phi_\alpha^n(D^n)$ for some α (the proof of Lemma 8.15 requires that X be Hausdorff). But $\Phi_\alpha^n(S^{n-1})$ is a compact subset of X^{n-1}, which is a CW complex, by induction; by Theorem 8.19, there is a finite CW subcomplex Y containing $\Phi_\alpha^n(S^{n-1})$. Therefore

$$\bar{e} = \Phi_\alpha^n(D^n) = e \cup \Phi_\alpha^n(S^{n-1}) \subset e \cup Y,$$

and so $\bar{e}$ meets only finitely many cells. We prove, by induction on $n \geq 0$, that X^n has the weak topology determined by $\{\bar{e}: \dim(e) \leq n\}$. Of course, the discrete space X^0 has the weak topology. Since X^n is a quotient space of $(\coprod D^n) \coprod X^{n-1}$, the result follows from the inductive hypothesis and Lemma 8.16.

We now know that X^n is a CW complex for every n. To see that X is a CW complex, one can quickly check axioms (1), (2), and (3). To check the weak topology, assume that $Z \subset X$ and $Z \cap \bar{e}$ is closed in $\bar{e}$ for every cell e. In particular, $Z \cap \bar{e}$ is closed in $\bar{e}$ for every i-cell e with $i \leq n$; hence $Z \cap X^n$ is closed in X^n for every n. Condition (iii) now gives Z closed in X. □

Exercises

8.30. (i) Show that there exist finite CW complexes that are not polyhedra. (*Hint*: Attach a 2-cell to S^1 with an attaching map resembling $x \sin(1/x)$.)

(ii) Prove that every finite CW complex has the same homotopy type as a polyhedron. (*Hint*: Use Theorem 8.24, induction, and the simplicial approximation theorem. See [Lundell and Weingram, p. 131].)

8.31. If $\{X_\lambda: \lambda \in \Lambda\}$ is a family of CW complexes with basepoint, then their wedge $\bigvee X_\lambda$ is also a CW complex. (*Note*: This result follows at once from Theorem 8.27 below, but a direct proof can be given here.)

*8.32. If (X, E) and (X', E) are finite CW complexes, then $(X \times X', E'')$ is a CW complex, where $E'' = \{e \times e': e \in E \text{ and } e' \in E'\}$. (*Hint*: If e is an i-cell and e' is a j-cell, then $e \times e'$ is an $(i+j)$-cell and $\overline{e \times e'} - e \times e' = [(\bar{e} - e) \times \bar{e}'] \cup [\bar{e} \times (\bar{e}' - e')]$.)

*8.33. If (X, E) is a CW complex, then so is $X \times \mathbf{I}$. (*Hint*: View $\mathbf{I}$ as a CW complex having two 0-cells, a^0, b^0, and one 1-cell c^1. Show that $E'' = \{e \times a^0, e \times b^0, e \times c^1: e \in E\}$ is a CW decomposition of $X \times \mathbf{I}$. In particular, show that $X \times \mathbf{I}$ has the weak topology determined by E''.)

Remark. There exist CW complexes (X_1, E_1) and (X_2, E_2) with $X_1 \times X_2$ not a CW space (see [Dowker]). It is known (see [Maunder, p. 282] that if X_1 and X_2 are CW spaces, then so is $X_1 \times X_2$ if either one of the X_i is locally compact or if both X_1 and X_2 have only a countable number of cells.

The inductive construction of open sets given in the proof of Theorem 8.24, that is, the "thickening" of U_k to U_{k+1}, can be modified and used again.

Theorem 8.25. *Every CW complex X is locally path connected.*

PROOF. Let $x \in X$ and let U be an open neighborhood of x; it suffices to find an open path connected set V with $x \in V \subset U$. Let n be the smallest integer with $x \in X^{(n)}$, and let e_0 be the n-cell containing x. We prove, by induction on $k \geq n$, that there exist path connected subsets V_k in $X^{(k)}$ with $V_{k+1} \cap X^{(k)} = V_k$, with $x \in V_k \subset X^{(k)} \cap U$, and with V_k open in $X^{(k)}$. The base of the induction requires two cases: if $n > 0$, then V_n exists because e_0 is homeomorphic to the locally path connected space $D^n - S^{n-1}$; if $n = 0$, define $V_0 = e_0 = \{x\}$, which is an open set in the discrete space $X^{(0)}$.

For the inductive step, assume that V_k exists as in the inductive statement. If T is any open path connected subset of S^k, and if $0 < \varepsilon < 1$, define "thickenings"

$$B(T, \varepsilon) = \{z \in D^{k+1} - S^k \colon \|z\| > \varepsilon \text{ and } z/\|z\| \in T\}.$$

It is easy to see that every $B(T, \varepsilon)$ is an open path connected subset of D^{k+1} with $T \cup B(T, \varepsilon) \subset \overline{B(T, \varepsilon)}$. Moreover, if W is any open subset of D^{k+1} and $\sigma \in W \cap S^k$, then there exists an open path connected set $T(\sigma)$ with $\sigma \in T(\sigma) \subset W \cap S^k$ and there exists an ε with $0 < \varepsilon < 1$ such that $B(T(\sigma), \varepsilon) \subset W$. For every $(k+1)$-cell e, $\Phi_e^{-1}(U \cap X^{(k+1)})$ is an open subset of D^{k+1}; moreover, $\Phi_e^{-1}(V_k)$ is an open subset (in S^k) contained in $S^k \cap \Phi_e^{-1}(U \cap X^{(k+1)})$. Define

$$V_{k+1} = V_k \cup \bigcup_e \bigcup_{\sigma \in \Phi_e^{-1}(V_k)} \Phi_e(B(T(\sigma), \varepsilon)),$$

where $\sigma \in T(\sigma) \subset \Phi_e^{-1}(V_k)$ and ε (which depends on $T(\sigma)$) are chosen so that $B(T(\sigma), \varepsilon) \subset \Phi_e^{-1}(U \cap X^{(k+1)})$. Now V_{k+1} is path connected: it is the union of the path connected subsets $\Phi_e(T(\sigma) \cup B(T(\sigma), \varepsilon))$ (for $B(T(\sigma), \varepsilon) \subset T(\sigma) \cup B(T(\sigma), \varepsilon) \subset \overline{B(T(\sigma), \varepsilon)}$ and $\bigcup_\sigma T(\sigma) = \Phi_e^{-1}(V_k)$) each of which meets the path connected set V_k contained in the union. Plainly, $V_{k+1} \cap X^{(k)} = V_k$, and $x \in V_{k+1} \subset U \cap X^{(k+1)}$. Finally, V_{k+1} is open in $X^{(k+1)}$, by Exercise 8.12(iii) (V_{k+1} is the image of

$$\coprod_e \left(\Phi_e^{-1}(V_k) \cup \bigcup_{\sigma \in \Phi_e^{-1}(V_k)} B(T(\sigma), \varepsilon)) \amalg X^{(k)} \right).$$

Define $V = \bigcup V_k$. Then $x \in V \subset U$, V is open in X, and V is path connected. □

This result coupled with Corollary 1.20 gives another proof of Theorem 8.23(iii). We continue investigating topological properties of CW complexes.

Theorem 8.26. *Every CW complex X is a normal space.*

PROOF. If A and B are disjoint closed subsets of X, it suffices to find a continuous $f: X \to \mathbf{I}$ with $f(A) = \{0\}$ and $f(B) = \{1\}$. We shall prove by induction on n that there exist continuous maps $f_n: X^{(n)} \to \mathbf{I}$ with $f_n(A \cap X^{(n)}) = \{0\}$ and $f_n(B \cap X^{(n)}) = \{1\}$. Given such maps f_n, one can define the desired map f by setting $f|X^{(n)} = f_n$.

If $n = 0$, then f_0 exists because $X^{(0)}$ is discrete. Assume that $n > 0$ and let e be an n-cell in X with characteristic map Φ_e. Now $\Phi_e^{-1}(A) = \Phi_e^{-1}(A \cap \bar{e})$ and $\Phi_e^{-1}(B) = \Phi_e^{-1}(B \cap \bar{e})$ are disjoint closed subsets of D^n. If $h_e = f_{n-1} \circ (\Phi_e|S^{n-1})$, then $h_e: S^{n-1} \to \mathbf{I}$ is a continuous map with $h_e(\Phi_e^{-1}(A) \cap S^{n-1}) = \{0\}$ and $h_e(\Phi_e^{-1}(B) \cap S^{n-1}) = \{1\}$. Extend h_e to h_e' defined on $\Phi_e^{-1}(A) \cup \Phi_e^{-1}(B) \cup S^{n-1}$ by defining $h_e'(\Phi_e^{-1}(A)) = \{0\}$ and $h_e'(\Phi_e^{-1}(B)) = \{1\}$. By the Tietze extension theorem, there is a continuous $f_e: D^n \to \mathbf{I}$ extending h_e'. Finally, define $f_n: X^{(n)} \to \mathbf{I}$ as the extension of f_{n-1} with $f_n|\bar{e} = f_e$ for all n-cells e in X. It is easy to see that f_n has the necessary properties. □

It is known that CW complexes are paracompact and perfectly normal (see [Lundell and Weingram, p. 54]).

Theorem 8.27. *If X is a CW complex and Y is a CW subcomplex, then X/Y is a CW complex.*

PROOF. Let us prove that X/Y is Hausdorff. Let $\nu: X \to X/Y$ be the natural map, let $* = \nu(Y)$, and let $\nu(x)$, $\nu(z)$ be distinct points in X/Y. If neither $\nu(x)$ nor $\nu(z)$ equals $*$, then they can be separated by open sets because X, hence $X - Y$, is Hausdorff and $\nu|X - Y$ is a homeomorphism from $X - Y$ to the subspace $(X/Y) - \{*\}$. If $\nu(z) = *$, then $x \notin Y$; since X is normal, there is a continuous $f: X \to \mathbf{I}$ with $f(x) = 0$ and $f(Y) = \{1\}$. Since f is constant on the fibers of ν, Corollary 1.9 says that f induces a continuous $f': X/Y \to \mathbf{I}$ with $f'(\nu(x)) = 0$ and $f'(*) = 1$; it follows that X/Y is Hausdorff.

Let (E, Φ) be a CW decomposition of X and let (E', Φ') be a CW decomposition of Y, where $E' \subset E$ and $\Phi' \subset \Phi$. For each $n \geq 0$, let E_n (respectively, E_n') denote the family of all n-cells in E (respectively, E'). Define the 0-cells in X/Y by

$$(X/Y)^0 = \{\nu(e): e \in E_0 - E_0'\} \cup \{*\};$$

for $n > 0$, define

$$(X/Y)^n = \{\nu(e): e \in E_n - E_n'\}.$$

Finally, define the characteristic map of $\nu(e)$ as the composite $\nu\Phi_e$.

We now verify the four axioms in the definition of CW complex.

(1) X/Y is the disjoint union of its cells; this follows easily from the corresponding property of X.

(2) Note first that, for each $e \in E_n - E_n'$, the map $\nu\Phi_e$ is a map of pairs

$(D^n, S^{n-1}) \to (\nu(e) \cup (X/Y)^{(n-1)}, (X/Y)^{(n-1)})$; moreover, since Φ_e and $\nu: (X, Y) \to (X/Y, *)$ are relative homeomorphisms (Example 8.5), it follows that $\nu\Phi_e$ is also a relative homeomorphism.

(3) If $e \in E_n - E'_n$, then $\overline{\nu(e)} = \nu\Phi_e(D^n) = \nu(\bar{e})$; since $\bar{e}$ is contained in the union of finitely many cells in X, it follows that $\overline{\nu(e)}$ is also contained in such a finite union in X/Y.

(4) Suppose that $B \subset X/Y$ is such that $B \cap \overline{\nu(e)}$ is closed in $\overline{\nu(e)}$ for every e. Then $\nu^{-1}(B \cap \overline{\nu(e)}) = \nu^{-1}(B) \cap \nu^{-1}(\overline{\nu(e)})$ is closed in $\nu^{-1}(\overline{\nu(e)}) = Y \cup \bar{e}$ for every cell e. For every cell a in X, $\nu^{-1}(B) \cap \bar{a} = \nu^{-1}(B) \cap (Y \cup \bar{a}) \cap \bar{a}$ is closed in $\bar{a}$. Since X has the weak topology determined by its cells, $\nu^{-1}(B)$ is closed in X; since ν is an identification, $B = \nu\nu^{-1}B$ is closed in X/Y. □

One can also prove this theorem using Theorem 8.24.

Definition. Let A be a subspace of X, and let $i: A \hookrightarrow X$ be the inclusion. Then A is a **strong deformation retract** of X if there is a continuous $r: X \to A$ such that $r \circ i = 1_A$ and $i \circ r \simeq 1_X$ rel A; one calls r a **strong deformation retraction**.

One can rephrase this definition as follows: There is a continuous $F: X \times \mathbf{I} \to X$ such that

(i) $F(x, 0) = x$ for all $x \in X$;
(ii) $F(x, 1) \in A$ for all $x \in X$;
(iii) $F(a, t) = a$ for all $a \in A$ and all $t \in \mathbf{I}$.

Now define $r: X \to A$ by $r(x) = F(x, 1)$ to recapture the definition.

Recall the weaker definitions already given. A subspace A is a *retract* of X if there exists a continuous $r: X \to A$ with $r \circ i = 1_A$; a subspace A is a *deformation retract* of X if $r \circ i = 1_A$ and $i \circ r \simeq 1_X$. Thus A is a strong deformation retract if A is a retract and there is a relative homotopy $i \circ r \simeq 1_X$, not merely a free homotopy.

Let X be the subset of the closed strip in $\mathbf{R}^2$ between the y-axis and the line $x = 1$, which is the union of $\mathbf{I}$ and all the line segments through the origin having slope $1/n$ for $n = 1, 2, 3, \ldots$. It can be shown that $\mathbf{I}$ is a deformation retract of X, but that $\mathbf{I}$ is not a strong deformation retract of X.

Exercise

*8.34. Let $Z \subset Y \subset X$. If Z is a strong deformation retract of Y and Y is a strong deformation retract of X, then Z is a strong deformation retract of X. More precisely, if $r_2: X \to Y$ and $r_1: Y \to Z$ are strong deformation retractions, then $r_1 r_2: X \to Z$ is a strong deformation retraction.

The next technical lemma is useful.

Lemma 8.28. *Let (X, E) be a CW complex and let (Y, E') be a CW subcomplex (where $E' \subset E$). If $M \subset X^{(k)} - (X^{(k-1)} \cup Y)$ consists of one point chosen from each k-cell in $X - Y$, then $X^{(k-1)} \cup Y$ is a strong deformation retract of $(X^{(k)} \cup Y) - M$ for every $k \geq 1$.*

Proof. There is no loss in generality in assuming that, for each k-cell e, the characteristic map $\Phi_e: D^k \to X^{(k)} \cup Y$ satisfies $\Phi_e(0) = m_e$, where $\{m_e\} = e \cap M$. Define $F: ((X^{(k)} \cup Y) - M) \times \mathbf{I} \to (X^{(k)} \cup Y) - M$ by

$$F(x, t) = \begin{cases} x & \text{if } x \in X^{(k-1)} \cup Y \\ \Phi_e[(1-t)v + tv/\|v\|] & \text{if } x = \Phi_e(v),\ v \neq 0, \text{ and } e \notin E' \end{cases}$$

(we are merely projecting $e - \{m_e\}$ onto the boundary of $\bar{e}$ by contracting along radii from m_e). It suffices to show that F is continuous.

Now $X^{(k)} \cup Y$ has the weak topology determined by its cells; it is easy to see that $(X^{(k)} \cup Y) - M$ has the weak topology determined by the cells in E' and the punctured cells $e - \{m_e\}$ for $e \in E - E'$. It follows from Exercise 8.33 that $((X^{(k)} \cup Y) - M) \times \mathbf{I}$ has the weak topology determined by all $e^\# \times a^0$, $e^\# \times b^0$, and $e^\# \times c^1$, where the cells in $\mathbf{I}$ are a^0, b^0 (the endpoints), c^1 (the open interval), and $e^\#$ is either a cell in E' or a punctured cell in $E - E'$. But, as in the proof of Theorem 8.11, the restriction of F to any of these subsets is continuous, so that the continuity of F follows from Lemma 8.14. □

Theorem 8.29. *Let X be a CW complex and let Y be a CW subcomplex. There is an open set U in X containing Y with Y a strong deformation retract of U.*

Proof (Dold). By Lemma 8.28, $X^{(k-1)} \cup Y$ is a strong deformation retract of $(X^{(k)} \cup Y) - M$ for every $k \geq 1$ (where M consists of exactly one point from each cell in X that is not in Y); let

$$r_k: (X^{(k)} \cup Y) - M \to X^{(k-1)} \cup Y$$

be a strong deformation retraction. Define $U_0 = Y$ and $U_k = r_k^{-1}(U_{k-1})$ for $k \geq 1$. Clearly, $Y \subset U_1$; moreover, U_1 is open in $(X^{(1)} \cup Y) - M$ (since $U_0 = Y$ is open in $X^{(0)} \cup Y$ because $X^{(0)}$ is discrete). It follows that U_k is open in $(X^{(k)} \cup Y) - M$ for all $k \geq 1$ and that $U_k \subset U_{k+1}$. Hence $U = \bigcup U_k$ is an open set in X containing Y (that U is open is by now a familiar argument). By Exercise 8.34, Y is a strong deformation retract of each U_k: there are continuous maps $G_k: U_k \times \mathbf{I} \to U_k$ for all $k \geq 1$ such that

$$\left.\begin{aligned} G_k(x, 0) &= x \\ G_k(x, 1) &= r_1 r_2 \cdots r_k(x) \in Y \end{aligned}\right\} \quad \text{for all } x \in U_k;$$

$$G_k(y, t) = y \quad \text{for all } y \in Y \text{ and } t \in \mathbf{I}.$$

Moreover, these G_k can be constructed, inductively, so that $G_{k+1} | U_k \times \mathbf{I} = G_k$. Finally, define $H: U \times \mathbf{I} \to U$ by $H | U_k \times \mathbf{I} = G_k$. Now H is continuous, for $U \times \mathbf{I}$ has the weak topology determined by $\{(U \times \mathbf{I}) \cap X^{(k)} \times \mathbf{I}: k \geq 1\} = \{U_k \times \mathbf{I}: k \geq 1\}$, and H exhibits Y as a strong deformation retract of U. □

Remark. More is true: given any open set W containing Y, there exists an open V with $Y \subset V \subset W$ and with Y a strong deformation retract of V ([Lundell and Weingram, p. 63]).

Lemma 8.30. *Let (X, E) be a CW complex and let $x \in X$. Then there is a CW decomposition of X having x as a* 0*-cell.*

PROOF. There exists a (unique) n-cell $e \in E$ containing x. If $n = 0$, then x is already a 0-cell. If $n > 0$, there exists $z \in D^n - S^{n-1}$ with $\Phi_e(z) = x$. By subdividing, one can regard D^n as a CW complex with cells Σ, say, and with z a 0-cell in Σ. Then $E' = (E - \{e\}) \cup \{\Phi_e(\sigma): \sigma \in \Sigma\}$ gives the desired CW decomposition of X. □

Corollary 8.31. *If X is a CW complex and $x \in X$, then there exists an open neighborhood U of x with U contractible to x.*

PROOF. By Lemma 8.30, we may assume that x is a 0-cell in X, hence $\{x\}$ is a CW subcomplex of X. Theorem 8.29 now applies. □

Remark. More is true. A space X is **locally contractible** if, for every $x \in X$, each neighborhood U of x contains an open neighborhood V of x that is contractible to x in U; that is, there exists a continuous $F: V \times \mathbf{I} \to U$ with $F(v, 0) = v$ and $F(v, 1) = x$ for all $v \in V$. Using the improved version of Theorem 8.29 cited above, one can prove that CW complexes are locally contractible.

Lemma 8.32. *Let X be a normal space, and let Y be a closed subspace. If $X \times \{0\} \cup Y \times \mathbf{I}$ is a retract of some open set U containing it, then, for every space Z, every map $H': X \times \{0\} \cup Y \times \mathbf{I} \to Z$ can be extended to $X \times \mathbf{I}$.*

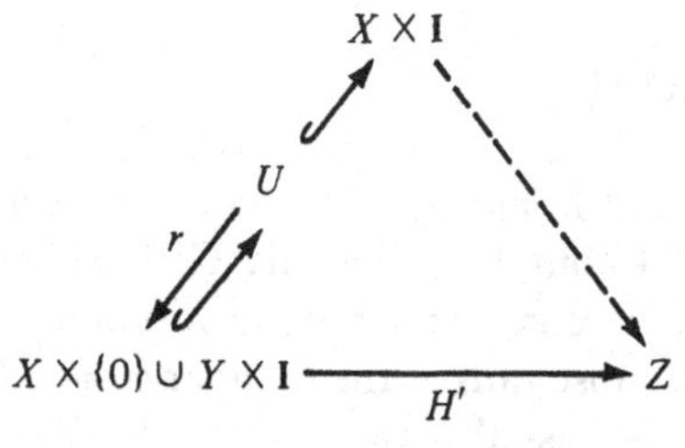

PROOF. Let $r: U \to X \times \{0\} \cup Y \times \mathbf{I}$ be a retraction. We shall construct a continuous map $u': X \times \mathbf{I} \to U$ that fixes $X \times \{0\} \cup Y \times \mathbf{I}$ pointwise; then the composite $H'ru'$ is the desired extension of H'.

For each $y \in Y$, $\{y\} \times \mathbf{I} \subset U$, so that the tube lemma (Lemma 8.9′) gives an open set W_y of X containing y such that $W_y \times \mathbf{I} \subset U$. If W is the union of these sets W_y, then W is an open set in X with $Y \subset W$ and with $W \times \mathbf{I} \subset U$. Since X is normal, the Urysohn lemma gives a continuous map $u: X \to \mathbf{I}$ with $u(Y) = 1$ and $u(X - W) = 0$.

Define $u': X \times \mathbf{I} \to X \times \mathbf{I}$ by $u'(x, t) = (x, tu(x))$. First, we show that im $u' \subset U$, so that we may assume that $u': X \times \mathbf{I} \to U$. If $x \notin W$, then $u(x) = 0$ and

$u'(x, t) = (x, 0) \in X \times \{0\} \subset U$; if $x \in W$, then $u'(x, t) = (x, tu(x)) \in W \times \mathbf{I} \subset U$. Finally, we show that u' fixes $X \times \{0\} \cup Y \times \mathbf{I}$ pointwise. Clearly, $u'(x, 0) = (x, 0)$, while $y \in Y$ implies $u'(y, t) = (y, tu(y)) = (y, t)$ because $u(y) = 1$. □

Theorem 8.33 (Homotopy Extension Theorem). *Let X be a CW complex, let Y be a CW subcomplex, and let Z be a space. For every continuous $f: X \to Z$ and every homotopy $h: Y \times \mathbf{I} \to Z$ with $h(y, 0) = f(y)$ for all $y \in Y$, there exists a homotopy $H: X \times \mathbf{I} \to Z$ with*

$$H(x, 0) = f(x) \quad \textit{for all } x \in X$$

and

$$H(y, t) = h(y, t) \quad \textit{for all } (y, t) \in Y \times \mathbf{I}.$$

PROOF. Define $H': (X \times \{0\}) \cup (Y \times \mathbf{I}) \to Z$ by $H'(x, 0) = f(x)$ for all $x \in X$ and $H'(y, t) = h(y, t)$ for all $(y, t) \in Y \times \mathbf{I}$. Since $X \times \{0\} \cup Y \times \mathbf{I}$ is a CW subcomplex of $X \times \mathbf{I}$, Theorem 8.29 provides an open neighborhood U containing it and with $X \times \{0\} \cup Y \times \mathbf{I}$ a (strong deformation) retract of U. Since X is normal, by Theorem 8.26, Lemma 8.32 applies to show that H' can be extended to $X \times \mathbf{I}$. □

For a proof of Theorem 8.33 avoiding Theorem 8.29, see [Maunder, p. 284].

There is standard terminology describing these theorems. Every CW subcomplex of a CW complex is an *absolute neighborhood retract* (ANR) (Theorem 8.29), and its inclusion is a *cofibration* (Theorem 8.33).

Cellular Homology

We now introduce cellular homology, the theory most suitable for computing homology groups of CW complexes. Given a CW decomposition E of a space X, we shall define a chain complex whose group of n-chains, for each $n \geq 0$, is a free abelian group whose rank is the number of n-cells in E. For simplicial complexes, these ranks are usually smaller than the ranks of the chain groups of simplicial theory, for the number of n-cells in a CW decomposition can be less than the number of n-simplexes in a triangulation. Thus Example 8.14 shows that the cellular chain groups for S^n (with $n > 0$) can be infinite cyclic in degree 0 and n and zero elsewhere. Knowing this, one can instantly compute $H_*(S^n)$.

One could define the cellular chain complex directly (we have not yet described the differentiations), but it is quicker for us to define it in terms of singular homology groups.

Definition. A **filtration** of a topological space X is a sequence of subspaces $\{X^n: n \in \mathbf{Z}\}$ with $X^n \subset X^{n+1}$ for all n. A filtration is **cellular** if:

(i) $H_p(X^n, X^{n-1}) = 0$ for all $p \neq n$;
(ii) for every $m \geq 0$ and every continuous $\sigma\colon \Delta^m \to X$, there is an integer n with $\operatorname{im} \sigma \subset X^n$.

Note that condition (ii), a weak version of Theorem 8.19, implies that $X = \bigcup X^n$ because every 0-simplex (namely, every point in X) lies in some X^n.

Definition. A **cellular space** is a topological space with a cellular filtration. If X and Y are cellular spaces, then a **cellular map** is a continuous function $f\colon X \to Y$ with $f(X^n) \subset Y^n$ for all $n \in \mathbf{Z}$.

It is plain that all cellular spaces and cellular maps form a category. The filtration of a CW complex by its skeletons will be seen to be cellular (Theorem 8.38). If no other cellular filtrations are mentioned, a continuous map $f\colon X \to Y$ between CW complexes is called **cellular** if

$$f(X^{(n)}) \subset Y^{(n)} \quad \text{for all } n \geq 0.$$

Definition. If X is a cellular space and $k \geq 0$, define

$$W_k(X) = H_k(X^k, X^{k-1}) \quad \text{(singular homology)};$$

define $d_k\colon W_k(X) \to W_{k-1}(X)$ as the composite $d_k = i_*\partial$,

$$\begin{array}{ccc} H_k(X^k, X^{k-1}) & & \\ \downarrow \partial & \searrow d_k & \\ H_{k-1}(X^{k-1}) & \xrightarrow[i_*]{} & H_{k-1}(X^{k-1}, X^{k-2}) \end{array}$$

where $i\colon (X^{k-1}, \varnothing) \hookrightarrow (X^k, X^{k-1})$ is the inclusion and ∂ is the connecting homomorphism arising from the long exact sequence of the pair (X^k, X^{k-1}).

Lemma 8.34. *If X is cellular space, then $(W_*(X), d)$ is a chain complex (called the* **cellular chain complex** *of the filtration of X).*

Proof. We need show only that $d_k d_{k+1} = 0$. But $d_k d_{k+1}$ is the composite

$$H_{k+1}(X^{k+1}, X^k) \to H_k(X^k) \to H_k(X^k, X^{k-1}) \to H_{k-1}(X^{k-1}) \to H_{k-1}(X^{k-1}, X^{k-2}),$$

and this is zero because the middle two arrows are adjacent arrows in the long exact sequence of the pair (X^k, X^{k-1}). □

The hypothesis that the filtration on X is cellular is not needed for Lemma 8.34; any filtration gives the same result.

Lemma 8.35. *Let X be a cellular space and let $p \geq q$.*

(i) $H_n(X^p, X^q) = 0$ *if* $q \geq n$ *or if* $n > p$.
(ii) $H_n(X, X^q) = 0$ *for all* $q \geq n$.

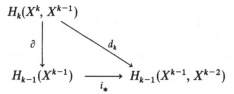

(iii) $H_n(X, X^q) \cong H_n(X^{n+1}, X^q)$ *if* $q < n$.
(iv) $H_n(X, X^{-1}) \cong H_n(X, X^{-2}) \cong H_n(X, X^{-3}) \cong \cdots$ *for all* n.

PROOF. (i) We do an induction on $p - q \geq 0$. If $p - q = 0$, then the result is true because $H_n(Y, Y) = 0$ for every space Y and every n (use the long exact sequence of the pair (Y, Y)). If $p - q > 0$, consider the following portion of the long exact sequence of the triple (X^p, X^{q+1}, X^q):

$$H_n(X^{q+1}, X^q) \to H_n(X^p, X^q) \to H_n(X^p, X^{q+1}).$$

Now $n > p \geq q + 1$ implies that $n \neq q + 1$, so that the first term is zero, by definition of cellular filtration. The inductive hypothesis does apply to the third term, for $p - (q + 1) < p - q$; moreover, we know that $n > p$, while $q \geq n$ implies that $q + 1 \geq n$; therefore $H_n(X^p, X^{q+1}) = 0$, and hence the middle term is zero, as desired.

(ii) Let $\text{cls}\,\zeta \in H_n(X, X^q)$. By condition (ii) in the definition of cellular filtration, there exists $p \geq q$ with $\zeta \in S_n(X^p)$. Hence

$$\text{cls}\,\zeta \in \text{im}(H_n(X^p, X^q) \to H_n(X, X^q)),$$

and this last (sub)group is zero, by (i).

(iii) The long exact sequence of the triple (X, X^{n+1}, X^q) contains the portion

$$H_{n+1}(X, X^{n+1}) \to H_n(X^{n+1}, X^q) \to H_n(X, X^q) \to H_n(X, X^{n+1}),$$

and the outside flanking terms are zero, by (ii).

(iv) Let $q \leq -2$. Given n, we know that $H_n(X, X^q) \cong H_n(X^{n+1}, X^q)$, by (iii). The long exact sequence of the triple (X^{n+1}, X^{-1}, X^q) contains the portion

$$H_n(X^{-1}, X^q) \to H_n(X^{n+1}, X^q) \to H_n(X^{n+1}, X^{-1}) \to H_{n-1}(X^{-1}, X^q).$$

Now $H_n(X^{-1}, X^q) = 0$ if $n \geq 0$, by (i); when $n < 0$, singular homology always vanishes. We conclude that

$$H_n(X, X^q) \cong H_n(X^{n+1}, X^q) \cong H_n(X^{n+1}, X^{-1}) \cong H_n(X, X^{-1}). \qquad \square$$

Theorem 8.36. *If X is a cellular space and $k \geq 0$, then*

$$H_k(W_*(X)) \cong H_k(X, X^{-1}).$$

PROOF (Dold). Let $k < n - 1$. By remark (4) after Theorem 5.9, the triple (X^{n+1}, X^n, X^k) gives a commutative diagram

$$\begin{array}{ccc} H_{n+1}(X^{n+1}, X^n) & \xrightarrow{\;\partial\;} & H_n(X^n) \\ \Big\downarrow{\scriptstyle \partial'} & \swarrow{\scriptstyle \lambda_*} & \\ H_n(X^n, X^k), & & \end{array}$$

where ∂, ∂' are appropriate connecting homomorphisms and λ_* is induced from the inclusion $\lambda: (X^n, \varnothing) \to (X^n, X^k)$. Since H_n is a functor, there is a commutative diagram

$$\begin{array}{ccc} & & H_n(X^n) \\ & \lambda_* \swarrow & \downarrow \mu_* \\ H_n(X^n, X^k) & \xrightarrow[j_*]{} & H_n(X^n, X^{n-1}), \end{array}$$

where all arrows are induced from inclusions. Combining these triangles gives a commutative diagram

$$\begin{array}{ccc} H_{n+1}(X^{n+1}, X^n) & \xrightarrow{\partial} & H_n(X^n) \\ \partial' \downarrow & \lambda_* \swarrow & \downarrow \mu_* \\ H_n(X^n, X^k) & \xrightarrow[j_*]{} & H_n(X^n, X^{n-1}). \end{array}$$

By definition, $d_{n+1} = \mu_* \partial$. There is thus a commutative diagram

$$\begin{array}{ccc} H_{n+1}(X^{n+1}, X^n) & & \\ \partial' \downarrow & \searrow d_{n+1} & \\ H_n(X^n, X^k) & \xrightarrow[j_*]{} & H_n(X^n, X^{n-1}). \end{array}$$

A similar argument gives commutativity of the other triangle in the diagram below:

$$\begin{array}{ccccccc} & & H_{n+1}(X^{n+1}, X^n) & & & & 0 \\ & & \partial' \downarrow & \searrow d_{n+1} & & & \downarrow \\ 0 & \longrightarrow & H_n(X^n, X^k) & \xrightarrow[j_*]{} & H_n(X^n, X^{n-1}) & \xrightarrow{\partial} & H_{n-1}(X^{n-1}, X^k) \\ & & \downarrow & & & \searrow d_n & \downarrow i_* \\ & & H_n(X^{n+1}, X^k) & & & & H_{n-1}(X^{n-1}, X^{n-2}) \\ & & \downarrow & & & & \\ & & 0 & & & & \end{array}$$

The row and two columns are each portions of appropriate exact sequences of triples, the zeros occurring by Lemma 8.35. Now

$$\begin{aligned} H_n(X, X^k) &\cong H_n(X^{n+1}, X^k) && \text{[Lemma 8.35(iii)]} \\ &\cong H_n(X^n, X^k)/\text{im}\, \partial' && \text{[exactness of first column]} \end{aligned}$$

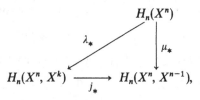

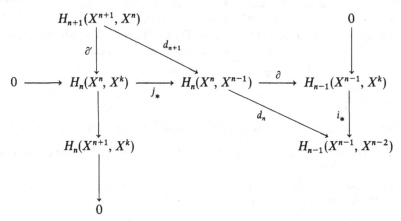

$$\begin{aligned}
&\cong \operatorname{im} j_*/\operatorname{im} j_*\partial' && [j_* \text{ is an injection}]\\
&= \ker \partial/\operatorname{im} j_*\partial' && [\text{exactness of row}]\\
&= \ker \partial/\operatorname{im} d_{n+1} && [\text{definition of } d_{n+1}]\\
&= \ker i_*\partial/\operatorname{im} d_{n+1} && [i_* \text{ is an injection}]\\
&= \ker d_n/\operatorname{im} d_{n+1} && [\text{definition of } d_n].
\end{aligned}$$

Thus $H_n(X, X^k) \cong H_n(W_*(X))$ whenever $k < n - 1$. It follows from Lemma 8.35(iv) that $H_n(X, X^{-1}) \cong H_n(W_*(X))$ for all n. □

Corollary 8.37. *If X is a cellular space with $X^{-1} = \varnothing$, then, for all k,*

$$H_k(X) \cong H_k(W_*(X)).$$ □

Let X be a CW complex with CW subcomplex Y. Define

$$X_Y^k = X^{(k)} \cup Y,$$

where (as usual) $X^{(k)}$ is the k-skeleton of X. Note that $X_Y^{-1} = Y$; in particular, $X_Y^{-1} = \varnothing$ if $Y = \varnothing$. It is plain that $X_Y^k \subset X_Y^{k+1}$ for all k, so that we have defined a filtration of X.

Notation. $W_*(X, Y)$ is the chain complex determined by the filtration of X by the X_Y^k [so that $W_k(X, Y) = H_k(X_Y^k, X_Y^{k-1})$].

Suppose that X and X' are CW complexes and that $f: X \to X'$ is a cellular map; that is, $f(X^{(k)}) \subset (X')^{(k)}$ for all $k \geq 0$. If Y and Y' are CW subcomplexes of X and X', respectively, and if f is a map of pairs [$f: (X, Y) \to (X', Y')$], then $f: X \to X'$ is cellular with respect to the filtrations X_Y^k and $(X')_{Y'}^k$. It follows that every cellular map of pairs $f: (X, Y) \to (X', Y')$ induces a chain map $f_\#: W_*(X, Y) \to W_*(X', Y')$ and hence homomorphisms

$$f_k: H_k(W_*(X, Y)) \to H_k(W_*(X', Y'))$$

for all $k \geq 0$.

Theorem 8.38. *Let X be a CW complex with CW subcomplex Y.*

(i) *The filtration of X by the subspaces X_Y^k is a cellular filtration* (*so that X is a cellular space*).
(ii) *If $W_*(X, Y)$ is the corresponding cellular chain complex, then there are isomorphisms for all $k \geq 0$,*

$$H_k(W_*(X, Y)) \cong H_k(X, Y).$$

Proof. (i) Let E be a CW decomposition of X, let $E' \subset E$ be a CW decomposition of Y, and let M consist of one point chosen from each cell in $E - E'$. If $k \geq 1$, then Lemma 8.28 says that X_Y^{k-1} is a deformation retract of $X_Y^k - M$;

for all $p \geq 0$, Exercise 5.14(iii) gives $H_p(X_Y^k - M, X_Y^{k-1}) = 0$. A portion of the long exact sequence of the triple $(X_Y^k, X_Y^k - M, X_Y^{k-1})$ is

$$H_p(X_Y^k - M, X_Y^{k-1}) \to H_p(X_Y^k, X_Y^{k-1}) \to H_p(X_Y^k, X_Y^k - M) \to H_{p-1}(X_Y^k - M, X_Y^{k-1}),$$

where the middle arrow is induced by inclusion. It follows from the two outside terms being zero that there are isomorphisms (all $p, k \geq 1$)

$$H_p(X_Y^k, X_Y^{k-1}) \cong H_p(X_Y^k, X_Y^k - M).$$

Computing closure and interior in X_Y^k, we see that $\overline{X}_Y^{k-1} \subset (X_Y^k - M)^\circ$. Hence excision applies and there are isomorphisms

$$H_p(X_Y^k, X_Y^k - M) \cong H_p(X_Y^k - X_Y^{k-1}, X_Y^k - X_Y^{k-1} - M).$$

Since $X_Y^k - X_Y^{k-1} = \coprod_\lambda \{e_\lambda^k \in E - E' : e_\lambda^k \text{ is a } k\text{-cell}\}$, Theorem 5.13 applies to give isomorphisms for all $p, k \geq 1$

$$H_p(X_Y^k - X_Y^{k-1}, X_Y^k - X_Y^{k-1} - M) \cong \sum_\lambda H_p(e_\lambda^k, e_\lambda^k - M) \tag{$*$}$$

(remember that $e_\lambda^k - M = e_\lambda^k - \{m_\lambda\}$ for some $m_\lambda \in e_\lambda^k$). Now $e_\lambda^k \approx \mathbf{R}^k$ implies that $H_p(e_\lambda^k, e_\lambda^k - M) \cong H_p(\mathbf{R}^k, \mathbf{R}^k - \{0\})$. Taking composites, we have isomorphisms for all $p, k \geq 1$

$$H_p(X_Y^k, X_Y^{k-1}) \cong \sum_\lambda H_p(\mathbf{R}^k, \mathbf{R}^k - \{0\}). \tag{$**$}$$

Since $\mathbf{R}^k - \{0\}$ has the same homotopy type as S^{k-1}, the long exact sequence of the pair $(\mathbf{R}^k, \mathbf{R}^k - \{0\})$ becomes

$$\cdots \to H_{p+1}(\mathbf{R}^k) \to H_{p+1}(\mathbf{R}^k, \mathbf{R}^k - \{0\}) \to H_p(S^{k-1}) \to H_p(\mathbf{R}^k) \to \cdots.$$

If $p \geq 1$, then both homology groups of $\mathbf{R}^k$ vanish, and there are isomorphisms

$$H_{p+1}(\mathbf{R}^k, \mathbf{R}^k - \{0\}) \cong H_p(S^{k-1}).$$

The right-hand side is zero unless $p = k - 1$, hence $H_p(\mathbf{R}^k, \mathbf{R}^k - \{0\}) = 0$ for $p \neq k$.

If $p = 1$ (and $k \neq 1$), we have exactness of

$$0 = H_1(\mathbf{R}^k) \to H_1(\mathbf{R}^k, \mathbf{R}^k - \{0\}) \to \tilde{H}_0(\mathbf{R}^k - \{0\}).$$

Since $k \neq 1$, we know that $\mathbf{R}^k - \{0\}$ is path connected; therefore $\tilde{H}_0(\mathbf{R}^k - \{0\}) = 0$ and $H_1(\mathbf{R}^k, \mathbf{R}^k - \{0\}) = 0$, as desired.

Finally, if $p = 0$, then $H_0(\mathbf{R}^k, \mathbf{R}^k - \{0\}) = 0$ when $k \neq 0$ because $\mathbf{R}^k$ is path connected and $\mathbf{R}^k - \{0\} \neq \varnothing$.

The second condition in the definition of a cellular filtration is that for every $m \geq 0$ and every continuous $\sigma: \Delta^m \to X$, there exists an integer n with $\operatorname{im} \sigma \subset X_Y^n$. This follows at once from Corollary 8.19, because $\operatorname{im} \sigma$ is compact and hence lies in some finite CW subcomplex of X.

(ii) Now that we have verified that the X_Y^k form a cellular filtration with $X_Y^{-1} = Y$, the result is immediate from Theorem 8.36. □

One can squeeze more information from the proof just given.

Theorem 8.39. *Let (X, E) be a CW complex with CW subcomplex (Y, E'). For each $k \geq 0$, the chain group $W_k(X, Y)$ is free abelian of (possibly infinite) rank r_k, where r_k is the number of k-cells in $E - E'$. In particular, $W_k(X) = W_k(X, \varnothing)$ is free abelian of rank equal to the number of k-cells in E.*

PROOF. Equation (**) in the proof of Theorem 8.38 holds for all $p, k \geq 1$; in particular, when $p = k$, it says that

$$W_k(X, Y) = H_k(X_Y^k, X_Y^{k-1}) \cong \sum_\lambda H_k(\mathbf{R}^k, \mathbf{R}^k - \{0\}),$$

where λ ranges over an index set of cardinal r_k. We later observed, when $k > 1$ (remember $p = k$ now), that $H_k(\mathbf{R}^k, \mathbf{R}^k - \{0\}) \cong H_{k-1}(S^{k-1})$. The theorem is thus proved for all $k \geq 2$, for then $H_{k-1}(S^{k-1}) \cong \mathbf{Z}$.

When $k = 1$, there is an exact sequence

$$0 = H_1(\mathbf{R}^1) \to H_1(\mathbf{R}^1, \mathbf{R}^1 - \{0\}) \to \tilde{H}_0(S^0) \to \tilde{H}_0(\mathbf{R}^1) = 0;$$

that is, there is an exact sequence

$$0 \to H_1(\mathbf{R}^1, \mathbf{R}^1 - \{0\}) \to \mathbf{Z} \to 0.$$

It follows that $H_1(\mathbf{R}^1, \mathbf{R}^1 - \{0\}) = \mathbf{Z}$, and the theorem holds in this case too.

The case $k = 0$ follows from Corollary 5.14: $H_0(X, A) \cong \sum_j H_0(X_j, A \cap X_j)$, where $\{X_j: j \in J\}$ is the set of path components of X. Here we must compute $H_0(X_Y^0, Y)$, where $X_Y^0 = X^{(0)} \cup Y$. Let $\{Y_i: i \in \mathbf{I}\}$ be the path components of Y, and let $X_1^{(0)} = \{e_\lambda^0 \in X^{(0)}: e_\lambda^0 \notin Y\}$. Since $X^{(0)}$ is discrete and Y is closed, the path components of X_Y^0 are the Y_i and the singletons $\{e_\lambda^0\}$. Thus

$$\begin{aligned} H_0(X_Y^0, Y) &\cong \sum_\lambda H_0(\{e_\lambda^0\}, Y \cap \{e_\lambda^0\}) \oplus \sum_i H_0(Y_i, Y_i \cap Y) \\ &\cong \sum_\lambda H_0(\{e_\lambda^0\}, Y \cap \{e_\lambda^0\}) \quad \text{(Theorem 5.12)} \\ &\cong \sum_\lambda H_0(\{e_\lambda^0\}) \qquad\qquad (Y \cap \{e_\lambda^0\} = \varnothing). \end{aligned}$$

We have shown that $W_0(X, Y) = H_0(X_Y^0, Y)$ is free abelian of rank r_0, where r_0 is the cardinal of $X_1^{(0)}$, the 0-cells in $E - E'$. □

EXERCISE

*8.35. If X is a CW complex, show that there is a chain map $W_*(X) \to S_*(X)$ inducing isomorphisms in homology. (*Hint*: First define an isomorphism $r_p: W_p(X) \to \sum H_p(e_\lambda, e_\lambda - M)$ as in the proof of Theorem 8.38, namely, the composite

$$\begin{aligned} W_p(X) = H_p(X^{(p)}, X^{(p-1)}) &\to H_p(X^{(p)}, X^{(p)} - M) \\ &\to H_p(X^{(p)} - X^{(p-1)}, X^{(p)} - X^{(p-1)} - M) \to \sum H_p(e_\lambda, e_\lambda - M). \end{aligned}$$

Now define $s_p: \sum H_p(e_\lambda, e_\lambda - M) \to S_p(X)$ by sending a generator of $H_p(e_\lambda, e_\lambda - M)$ into $\Phi_{e_\lambda} \circ \alpha$, where $\alpha: (\Delta^p, \dot{\Delta}^p) \to (D^p, S^{p-1})$ is a homeomorphism. Then $\{s_p r_p\}$ is the desired chain map.)

As we asserted at the outset, the chain groups of the cellular chain complex of a CW complex are free abelian groups of "small" rank. For example, the reader may now give a quick proof of Theorem 6.22, which says that a loop generates $H_1(S^1)$.

Corollary 8.40. *If X is a compact CW complex of dimension m, then*

(i) *$H_p(X)$ is f.g. for every $p \geq 0$;*
(ii) *$H_p(X) = 0$ for all $p > m$;*
(iii) *$H_m(X)$ is free abelian.*

Theorem 8.41. *Let (X, E) be a CW complex with CW subcomplex (Y, E'). Then the natural map $v: X \to X/Y$ induces isomorphisms for every $k \geq 0$*

$$v_*: H_k(X, Y) \xrightarrow{\simeq} H_k(X/Y, *) \cong \tilde{H}_k(X/Y),$$

where $*$ denotes the singleton point $v(Y)$ in X/Y.

Proof. As in Theorem 8.27, we regard X/Y as a CW complex with CW decomposition $E'' = (E - E') \cup \{*\}$. The natural map v is a cellular map of pairs $(X, Y) \to (X/Y, *)$, which maps those cells of X not in Y homeomorphically onto those cells of X/Y other than the 0-cell $*$. That v is cellular implies that v induces a chain map $v_\#: W_*(X, Y) \to W_*(X/Y, *)$. Recall the isomorphism $(*)$ in the proof of Theorem 8.38:

$$W_k(X, Y) = H_k(X_Y^k, X_Y^{k-1}) \xrightarrow{\simeq} \sum H_k(e^k, e^k - M) \quad \text{for all } k \geq 1;$$

this map is a composite of inclusions (and the injections and projections of a direct sum decomposition). Since v maps cells in $E - E'$ homeomorphically, it follows that $v_{\# k}$ is an isomorphism for all $k \geq 1$. Even $v_{\# 0}$ is an isomorphism, but now we must also observe, using Theorem 5.13, that v induces a bijection from the family of path components of $X^{(0)} \cup Y$ not containing Y to the family of path components of X/Y not containing $*$. Therefore $v_\#$ is an isomorphism of chain complexes, and hence it induces isomorphisms of the respective homology groups:

$$H_*(X, Y) \cong H_*(X/Y, *).$$

But Theorem 5.17 gives $H_*(X/Y, *) \cong \tilde{H}_*(X/Y)$. □

Corollary 8.42. *For $i = 1, 2$, let X_i be a CW complex with CW subcomplex Y_i; let $f: (X_1, Y_1) \to (X_2, Y_2)$ be a continuous (but not necessarily cellular) map of pairs that induces a homeomorphism $\bar{f}: X_1/Y_1 \xrightarrow{\simeq} X_2/Y_2$. Then f induces isomorphisms for all $k \geq 0$*

$$f_*: H_k(X_1, Y_1) \xrightarrow{\simeq} H_k(X_2, Y_2).$$

PROOF. For $i = 1, 2$, let $\nu_i: X_i \to X_i/Y_i$ be the natural map. The following diagram commutes:

$$\begin{array}{ccc} H_k(X_1, Y_1) & \xrightarrow{f_*} & H_k(X_2, Y_2) \\ \downarrow{\scriptstyle \nu_{1*}} & & \downarrow{\scriptstyle \nu_{2*}} \\ \tilde{H}_k(X_1/Y_1) & \xrightarrow[\bar{f}_*]{} & \tilde{H}_k(X_2/Y_2). \end{array}$$

Since three of the maps are isomorphisms ($\bar{f}_*$ by hypothesis and the ν_{i*} by Theorem 8.41), it follows that f_* is an isomorphism. □

Corollary 8.43 (Excision). *If X is a CW complex and Y_1, Y_2 are CW subcomplexes with $X = Y_1 \cup Y_2$, then the inclusion $i: (Y_1, Y_1 \cap Y_2) \hookrightarrow (X, Y_2)$ induces isomorphisms for all $k \geq 0$,*

$$i_*: H_k(Y_1, Y_1 \cap Y_2) \xrightarrow{\sim} H_k(X, Y_2).$$

PROOF. The inclusion induces a homeomorphism $Y_1/Y_1 \cap Y_2 \xrightarrow{\sim} X/Y_2$ (if E_i is the family of cells in Y_i, then the cells of the left side are those of $E_1 - (E_1 \cap E_2)$ and the basepoint, while the cells of the right side are those of $(E_1 \cup E_2) - E_2$ and the basepoint). The corollary now follows at once from Corollary 8.42. □

Corollary 8.44 (Mayer–Vietoris). *If X is a CW complex with CW subcomplexes Y_1 and Y_2 with $X = Y_1 \cup Y_2$, then there is an exact sequence*

$$\cdots \to H_k(Y_1 \cap Y_2) \to H_k(Y_1) \oplus H_k(Y_2) \to H_k(X) \to H_{k-1}(Y_1 \cap Y_2) \to \cdots$$

whose maps can be given explicitly.

PROOF. The usual consequence of excision (see Theorem 6.3). □

EXERCISES

8.36. (i) If X is a compact CW complex with CW subcomplex Y, then $H_k(X, Y)$ is f.g. for every $k \geq 0$.
(ii) If (X, E) is a CW complex having only finitely many cells in each dimension, then $H_k(X)$ is f.g. for every $k \geq 0$.

8.37. Show that $H_k(\mathbf{R}P^\infty)$ is cyclic (possibly zero) for every $k \geq 0$.

8.38. If $\{X_\lambda: \lambda \in \Lambda\}$ is a family of CW complexes with basepoint, show that

$$\tilde{H}_*(\bigvee X_\lambda) \cong \sum \tilde{H}_*(X_\lambda).$$

8.39. Using the CW decomposition of the torus T given in Exercise 8.25, compute

$H_*(T)$ (once again!). (*Hint*: The 1-skeleton is $S^1 \vee S^1$, so that $H_2(T, S^1 \vee S^1) = \mathbf{Z}$ and $H_1(T, S^1 \vee S^1) = 0$.)

Definition. Let (X, E) be a finite CW complex and let α_i denote the number of i-cells in E. The **Euler–Poincaré characteristic** of (X, E) is

$$\chi(X) = \sum_{i \geq 0} (-1)^i \alpha_i.$$

The remark after the proof of Theorem 7.15 shows that the Euler–Poincaré characteristic of a finite CW complex (X, E) is equal to $\sum_{i \geq 0} (-1)^i$ rank $H_i(X)$; hence this number depends only on X and not on the CW decomposition E. Moreover, since every polyhedron X may be viewed as a CW complex (determined by a triangulation of X), we see that we have generalized our previous definition from polyhedra to finite CW complexes.

Theorem 8.45. *If X and X' are finite CW complexes, then*

$$\chi(X \times X') = \chi(X)\chi(X').$$

Proof. If E and E' are CW decompositions of X and X', respectively, then we saw in Exercise 8.32 that $E'' = \{e \times e' : e \in E \text{ and } e' \in E'\}$ is a CW decomposition of $X \times X'$. If e is an i-cell and e' is a j-cell, then $e \times e'$ is, of course, an $(i + j)$-cell. The number β_k of k-cells in E'' is thus $\sum_{i+j=k} \alpha_i \alpha'_j$ (where α_i is the number of i-cells in E and α'_j is the number of j-cells in E'). But

$$\left(\sum (-1)^i \alpha_i\right)\left(\sum (-1)^j \alpha'_j\right) = \sum_{i,j} (-1)^{i+j} \alpha_i \alpha'_j = \sum_k (-1)^k \sum_{i+j=k} \alpha_i \alpha'_j = \sum (-1)^k \beta_k.$$

Therefore $\chi(X)\chi(X') = \chi(X \times X')$. □

If X and X' are polyhedra, a proof of this formula using the techniques of Chapter 7 is fussy (a triangulation of $X \times X'$ is more complicated than the CW decomposition E'' above).

Exercises

8.40. Compute $\chi(S^m \times S^n)$.

8.41. Show that $\chi(\mathbf{C}P^n) = n + 1 = \chi(\mathbf{H}P^n)$.

8.42. Show that $\chi(\mathbf{R}P^n) = \frac{1}{2}(1 + (-1)^n)$.

8.43. If X is a CW complex with CW subcomplexes Y_1 and Y_2 such that $X = Y_1 \cup Y_2$, then

$$\chi(Y_1) + \chi(Y_2) = \chi(X) + \chi(Y_1 \cap Y_2).$$

Of course, one must know both the chain groups and the differentiations to understand a chain complex; apart from its definition, we have not yet discussed the differentiations d_k of $W_*(X)$. To emphasize this point, consider

$X = \mathbf{R}P^n$. By Theorem 8.8(i), there is a CW decomposition of X of the form $\{e^0, e^1, \ldots, e^n\}$, and so each chain group $W_k(X) = W_k(X, \varnothing) = \mathbf{Z}$ for $0 \leq k \leq n$ (and $W_k(X) = 0$ for $k > n$). We conclude that $H_k(\mathbf{R}P^n)$ is cyclic for $0 \leq k \leq n$; however, without knowing the differentiations explicitly, we cannot yet say whether these cyclic groups, for any fixed k, are infinite, finite, or even zero. (On the other hand, the CW decompositions of $\mathbf{C}P^n$ and $\mathbf{H}P^n$ of Theorem 8.8 may now be used to give an instant proof of Theorem 8.13.)

As in Example 8.15, regard S^n as a CW complex having two k-cells $\{e_1^k, e_2^k\}$ for each k with $0 \leq k \leq n$. Recall that $\bar{e}_1^k$ is the closed northern hemisphere of S^k and that $\bar{e}_2^k$ is the closed southern hemisphere; thus the k-skeleton of S^n consists of the cells in S^k. By Eq. (*) in the proof of Theorem 8.38,

$$W_k(S^n) = H_k(S^k, S^{k-1}) \cong H_k(e_1^k, e_1^k - \{m_1\}) \oplus H_k(e_2^k, e_2^k - \{m_2\}),$$

where $m_i \in e_i^k$, for $i = 1, 2$, are any chosen points; choose m_2 to be the antipode of m_1. We know that each summand on the right-hand side is infinite cyclic. Let β^k be a generator of $H_k(e_1^k, e_1^k - \{m_1\})$.

As usual, let $a^k\colon S^k \to S^k$ denote the antipodal map; let us denote the antipodal map of pairs $(S^k, S^{k-1}) \to (S^k, S^{k-1})$ by A^k. Now A^k restricts to a homeomorphism $(e_1^k, e_1^k - \{m_1\}) \to (e_2^k, e_2^k - \{m_2\})$. Therefore, $A_*^k(\beta^k)$ is a generator of $H_k(e_2^k, e_2^k - \{m_2\})$, and hence $\{\beta^k, A_*^k(\beta^k)\}$ is a basis of $W_k(S^n)$.

Consider the commutative diagram

$$\begin{array}{ccc} H_k(S^k, S^{k-1}) & \xrightarrow{A_*^k} & H_k(S^k, S^{k-1}) \\ \Big\downarrow{\scriptstyle \partial_k} & & \Big\downarrow{\scriptstyle \partial_k} \\ H_{k-1}(S^{k-1}) & \xrightarrow[a_*^{k-1}]{} & H_{k-1}(S^{k-1}): \end{array}$$

$\partial_k A_*^k = a_*^{k-1} \partial_k$. Recall that Theorem 6.23 says that a_*^{k-1} is multiplication by $(-1)^k$; on the other hand, we see that A_*^k is the automorphism of $H_k(S^k, S^{k-1}) \cong \mathbf{Z} \oplus \mathbf{Z}$ which interchanges the free generators β^k and $A_*^k(\beta^k)$.

Let $v\colon S^n \to \mathbf{R}P^n$ be the usual identification, which identifies antipodal points. If we regard $\mathbf{R}P^n$ as a CW complex with the CW decomposition of Theorem 8.8(i), then v is a cellular map and it induces a chain map $v_\#\colon W_*(S^n) \to W_*(\mathbf{R}P^n)$. Since $vA^k = v$ for all k, it is plain that $v_\#(\beta^k) = v_\#(A_*^k(\beta^k))$ for all k and that their common value is a generator of the infinite cyclic group $W_k(\mathbf{R}P^n)$.

Lemma 8.46. *Using the notation above, for all $k \geq 0$ there is a basis of $W_k(S^n)$ of the form $\{\beta^k, A_*^k(\beta^k)\}$ with the following properties:*

(i) *$v_\#(\beta^k) = v_\#(A_*^k(\beta^k))$, and their common value is a generator of $W_k(\mathbf{R}P^n)$;*
(ii) *the differentiation $d_k\colon W_k(S^n) \to W_{k-1}(S^n)$, for $k > 0$, satisfies*

$$d_k(\beta^k) = \pm(A_*^{k-1}(\beta^{k-1}) + (-1)^k\beta^{k-1}).$$

PROOF. (i) This was proved above.

(ii) Identify $\beta^k \in H_k(e_1^k, e_1^k - \{m_1\})$ with its image in $H_k(S^k, S^{k-1}) = W_k(S^n)$ (Eq. $(*)$ in the proof of Theorem 8.38). Recall that $d_k\colon W_k(S^n) \to W_{k-1}(S^n)$ is the composite $d_k = i_* \partial_k$:

$$\begin{array}{ccc} H_k(S^k, S^{k-1}) & \xrightarrow{\quad d_k \quad} & H_{k-1}(S^{k-1}, S^{k-2}), \\ & {\scriptstyle \partial_k} \searrow \qquad \nearrow {\scriptstyle i_*} & \\ & H_{k-1}(S^{k-1}) & \end{array}$$

where $i\colon (S^{k-1}, \varnothing) \hookrightarrow (S^{k-1}, S^{k-2})$. Now

(1) $\quad d_k A_*^k(\beta^k) = i_* \partial_k A_*^k(\beta^k) = i_* a_*^{k-1} \partial_k(\beta^k) = (-1)^k i_* \partial_k(\beta^k) = (-1)^k d_k(\beta^k).$

It follows that γ_k defined by

$$\gamma_k = A_*^k(\beta^k) - (-1)^k \beta^k = A_*^k(\beta^k) + (-1)^{k+1}\beta^k$$

lies in $\ker d_k = Z_k(W_*(S^n))$. We claim that $Z_k(W_*(S^n)) = \langle \gamma_k \rangle$ for $1 \le k \le n$. For the other inclusion, assume that

$$\begin{aligned} 0 &= d_k(rA_*^k(\beta^k) + s\beta^k), \qquad r, s \in \mathbf{Z} \\ &= r d_k A_*^k(\beta^k) + s d_k(\beta^k) \\ &= r(-1)^k d_k(\beta^k) + s d_k(\beta^k) = [r(-1)^k + s] d_k(\beta^k). \end{aligned}$$

Since $W_{k-1}(S^n)$ is a free abelian group, either $r(-1)^k + s = 0$ or $d_k(\beta^k) = 0$. In the first case, $s = (-1)^{k+1} r$ and $rA_*^k(\beta^k) + s\beta^k = r\gamma_k \in \langle \gamma_k \rangle$, as desired. We now show that the second case cannot occur. If $d_k(\beta^k) = 0$, then Eq. (1) gives $d_k A_*^k(\beta^k) = (-1)^k d_k(\beta^k) = 0$, and so $B_{k-1}(W_*(S^n)) = \operatorname{im} d_k = 0$. If $n > k - 1 > 0$, then we contradict $H_{k-1}(S^n) = 0$ (since we have just seen that $Z_{k-1}(W_*(S^n)) = \langle \gamma_{k-1} \rangle \neq 0$). There is also a contradiction if $k = 1$, because $Z_0(W_*(S^n)) = W_0(S^n)$ is free abelian of rank 2, while $H_0(S^n) = \mathbf{Z}$ forces $B_0(W_*(S^n)) \neq 0$.

If $1 < k \le n$, then $d_k(\beta^k) \in B_{k-1}(W_*(S^n)) = Z_{k-1}(W_*(S^n)) = \langle \gamma_{k-1} \rangle$, by our computation above. Hence $d_k(\beta^k) = m\gamma_{k-1}$ for some $m \in \mathbf{Z}$; furthermore $d_k A_*^k(\beta^k) = (-1)^k m \gamma_{k-1}$. Thus $\operatorname{im} d_k \subset \langle m\gamma_{k-1} \rangle$; since $\gamma_{k-1} \neq 0$ has infinite order, $m = \pm 1$. Therefore

$$d_k(\beta^k) = \pm \gamma_{k-1} = \pm (A_*^{k-1}(\beta^{k-1}) + (-1)^k \beta^{k-1}),$$

as desired. When $k = 1$, we may compute d_1 directly. Now β^1 is the upper half circle from 1 to -1, and $\beta^0 = 1$; hence

$$d_1(\beta^1) = i_* \partial_1(\beta^1) = \operatorname{cls} 1 - \operatorname{cls}(-1),$$

as desired. □

Theorem 8.47. *If n is odd, then*

$$H_p(\mathbf{R}P^n) = \begin{cases} \mathbf{Z} & \text{if } p = 0 \text{ or } p = n \\ \mathbf{Z}/2\mathbf{Z} & \text{if } p \text{ is odd and } 0 < p < n \\ 0 & \text{otherwise.} \end{cases}$$

If n is even, then

$$H_p(\mathbf{R}P^n) = \begin{cases} \mathbf{Z} & \text{if } p = 0 \\ \mathbf{Z}/2\mathbf{Z} & \text{if } p \text{ is odd and } 0 < p < n \\ 0 & \text{otherwise.} \end{cases}$$

PROOF. Consider the commutative diagram

$$\begin{array}{ccc} W_k(S^n) & \xrightarrow{d_k} & W_{k-1}(S^n) \\ \downarrow{\scriptstyle \nu_\#} & & \downarrow{\scriptstyle \nu_\#} \\ W_k(\mathbf{R}P^n) & \xrightarrow[D_k]{} & W_{k-1}(\mathbf{R}P^n), \end{array}$$

where we are writing D_k for the differentiation in $W_*(\mathbf{R}P^n)$. Then

$$\begin{aligned} D_k \nu_\#(\beta^k) &= \nu_\# d_k(\beta^k) \\ &= \pm \nu_\#(A_*^{k-1}(\beta^{k-1}) + (-1)^k \beta^{k-1}) \\ &= \pm(1 + (-1)^k)\nu_\#(\beta^{k-1}). \end{aligned}$$

Since $\nu_\#(\beta^k)$ is a generator of $W_k(\mathbf{R}P^n)$, D_k is the zero map for odd k and is multiplication by 2 for even k. Abbreviating $W_k(\mathbf{R}P^n)$ to W_k, we see that the cellular complex is:

$$\begin{array}{ccccccccccccccc} W_* = 0 \longrightarrow & W_n & \xrightarrow{D_n} & W_{n-1} & \longrightarrow \cdots \cdots \longrightarrow & W_4 & \xrightarrow{D_4} & W_3 & \xrightarrow{D_3} & W_2 & \xrightarrow{D_2} & W_1 & \xrightarrow{D_1} & W_0 & \longrightarrow 0 \\ & \| & & \| & & \| & & \| & & \| & & \| & & \| & \\ 0 \longrightarrow & \mathbf{Z} & \xrightarrow{1+(-1)^n} & \mathbf{Z} & \longrightarrow \cdots \cdots \longrightarrow & \mathbf{Z} & \xrightarrow{2} & \mathbf{Z} & \xrightarrow{0} & \mathbf{Z} & \xrightarrow{2} & \mathbf{Z} & \xrightarrow{0} & \mathbf{Z} & \longrightarrow 0. \end{array}$$

The theorem now follows easily. □

Remark. It is possible to compute $H_*(\mathbf{R}P^n)$ by simplicial methods using an explicit triangulation of $\mathbf{R}P^n$. First, triangulate S^n using the $2(n+1)$ vertices $\varepsilon_i e_i$, where $\varepsilon_i = \pm 1$ and e_i is the $(n+1)$- tuple having $(i+1)$ st coordinate 1 and all other coordinates 0; the n-simplexes are of the form $[\varepsilon_0 e_0, \ldots, \varepsilon_n e_n]$ for every choice of signs $(\varepsilon_0, \ldots, \varepsilon_n)$. If this simplicial complex is called K, then one proves by induction that Sd K induces a triangulation of $\mathbf{R}P^n$ (under the map $S^n \to \mathbf{R}P^n$, which identifies antipodal points). This triangulation is essentially in [Hilton and Wylie, p. 133]; the reader is referred to the discussion in

[Wallace, p. 71], which contains a geometric version of Theorem 7.30; see also [Maunder, p. 140].

EXAMPLE 8.22. Let p and q be relatively prime integers. Regard S^3 as all $(z_0, z_1) \in \mathbf{C}^2$ with $|z_0|^2 + |z_1|^2 = 1$. Let $\zeta = e^{2\pi i/p}$ be a primitive pth root of unity; define $h: S^3 \to S^3$ by

$$h(z_0, z_1) = (\zeta z_0, \zeta^q z_1),$$

and define an equivalence relation on S^3 by $(z_0, z_1) \sim (z_0', z_1')$ if there exists an integer m with $h^m(z_0, z_1) = (z_0', z_1')$. The quotient space $S^3/\sim$ is called a **lens space** and is denoted by $L(p, q)$.

EXERCISES

8.44. Show that $L(p, q)$ is a compact Hausdorff space. (In Exercise 10.32, we shall see that $L(p, q)$ is a compact connected 3-manifold.)

8.45. (i) Show that $L(1, 1) = S^3$.
(ii) Show that $L(2, 1) = \mathbf{R}P^3$.
(iii) If $q \equiv q' \bmod p$, then $L(p, q) = L(p, q')$.

*8.46. (i) Show that there is a CW decomposition of S^3 having p cells in each dimension, namely, for $r = 0, 1, \ldots, p-1$,

$$e_r^0 = \{(z_0, 0) \in S^3\colon \arg(z_0) = 2\pi r/p\},$$

$$e_r^1 = \{(z_0, 0) \in S^3\colon 2\pi r/p < \arg(z_0) < 2\pi(r+1)/p\},$$

$$e_r^2 = \{(z_0, z_1) \in S^3\colon \arg(z_1) = 2\pi r/p\},$$

$$e_r^3 = \{(z_0, z_1) \in S^3\colon 2\pi r/p < \arg(z_1) < 2\pi(r+1)/p\}.$$

(Recall that if z is a nonzero complex number, then $z = \rho e^{i\theta}$ for $\rho > 0$ and $0 \le \theta < 2\pi$; one defines $\arg(z) = \theta$.)
(ii) If $v\colon S^3 \to L(p, q)$ is the natural map, show that the family of all $v(e_r^k)$ is a CW decomposition of $L(p, q)$. Conclude that $L(p, q)$ may be viewed as a CW complex having one cell in every dimension ≤ 3.

8.47. (i) Show that the CW decomposition of S^3 in the above exercise leads to a cellular chain complex $W_*(S^3)$ with differentiations:

$$d(e_r^1) = e_r^0 - e_{r+1}^0$$

$$d(e_r^2) = \textstyle\sum_{i=0}^{p-1} e_i^1$$

$$d(e_r^3) = e_r^2 - e_{r+1}^2$$

(take subscripts r mod p in the first and third formulas).
(ii) From Exercise 8.46(ii), we know that there is a CW decomposition of $L(p, q)$ with $W_k(L(p, q)) = \mathbf{Z}$ for all $k \le 3$ (and with $W_k = 0$ for $k > 3$); let γ_k denote a generator of $W_k(L(p, q))$. Use part (i) of this exercise to show that the differentiations satisfy

$$D(\gamma_2) = p\gamma_1 \quad \text{and} \quad D(\gamma_1) = 0 = D(\gamma_3).$$

(iii) Show that

$$H_k(L(p, q)) = \begin{cases} \mathbf{Z} & \text{if } k = 0, 3 \\ \mathbf{Z}/p\mathbf{Z} & \text{if } k = 1 \\ 0 & \text{otherwise.} \end{cases}$$

Lens spaces are examples that arose in investigating the Poincaré conjecture (they also enter into Milnor's counterexample to the Hauptvermutung): Is every compact simply connected manifold having the homology groups of a sphere actually homeomorphic to a sphere? (We have already mentioned this problem in Chapter 7.) A natural first question is whether two (compact connected) manifolds having the same homology groups are necessarily homeomorphic; indeed, must they have the same homotopy type? The lens spaces (which are compact 3-manifolds) settle these first questions. Note that if $p \neq p'$, then $L(p, q)$ and $L(p', q')$ do not have the same homotopy type because they have different first homology groups.

Theorem

(i) *$L(p, q)$ and $L(p, q')$ have the same homotopy type if and only if either qq' or $-qq'$ is a quadratic residue* mod p;

(ii) *$L(p, q)$ and $L(p, q')$ are homeomorphic if and only if either $q \equiv \pm q'$* mod p *or $qq' \equiv \pm 1$* mod p.

The first statement is proved in [Hilton and Wylie, p. 223] and in [Seifert and Threlfall, p. 222]; necessity of (ii) is outlined in [Munkres (1984), p. 242]; sufficiency is proved in [Brody].

The 3-manifolds $L(5, 1)$ and $L(5, 2)$ have the same homology groups, but they do not have the same homotopy type (for neither 2 nor -2 is a quadratic residue mod 5). The 3-manifolds $L(7, 1)$ and $L(7, 2)$ have the same homotopy type (for $2 \equiv 3^2$ mod 7), but they are not homeomorphic (for $2 \not\equiv \pm 1$ mod 7).

There are two general methods for computing cellular homology (aside from variations of the method used for $\mathbf{R}P^n$). One way involves selecting bases, say, $\{\alpha_i : i \in I\}$ of $W_k(X)$ and $\{\beta_j : j \in J\}$ of $W_{k-1}(X)$. Now $d_k(\alpha_i) = \sum_j [\alpha_i : \beta_j]\beta_j$, where $[\alpha_i : \beta_j]$ are certain integers called **incidence numbers**; of course, d_k is completely determined by the matrix of incidence numbers. It can be shown ([Maunder, p. 319] that $[\alpha_i : \beta_j]$ can be computed as the degree of a certain map $S^{k-1} \to S^{k-1}$ (which is a composite of maps involving the characteristic maps of α_i and of β_j).

A second approach (see [Cooke and Finney] or [Massey (1978)]) involves defining (new) cellular chain groups $C_k(X)$ as free abelian groups with bases the k-cells in a given CW decomposition of X, and then defining the differentiations d_k by specifying incidence numbers. When the CW complex is **regular**, that is, all attaching maps are homeomorphisms, then all incidence numbers are 0, 1, or -1, and there is an axiomatic description of them. In this case (which obtains, e.g., when X is a polyhedron), there is an algorithm for computing $H_*(X)$ that is essentially the same as that described for polyhedra in Chapter 7.

There are other features of CW complexes to interest us. For example, one can generalize Tietze's theorem (Corollary 7.37). If X is a finite CW complex having m 1-cells and n 2-cells, then $\pi_1(X, x_0)$ is a finitely presented group; indeed there is a presentation having m generators and n relations (see [Fuks and Rokhlin, p. 448]). One can also show that the Seifert–van Kampen theorem holds for a CW complex X and connected CW subcomplexes Y_1 and Y_2 such that $Y_1 \cup Y_2 = X$ and $Y_1 \cap Y_2$ is connected.

There is also an analogue of the simplificial approximation theorem.

Cellular Approximation Theorem. *Let X and Y be CW complexes, and let $g: X \to Y$ be continuous; suppose that $g|X'$ is a cellular map for some (possibly empty) CW subcomplex X' of X. Then there exists a cellular map $f: X \to Y$ such that $f|X' = g|X'$ and*

$$f \simeq g \text{ rel } X'.$$

There is a proof in [Maunder, p. 302] or in [Lundell-Weingram, p. 69].

CHAPTER 9

Natural Transformations

In preceding chapters, the adjective "natural" was used, always in the context of some commutative diagram. This important term will now be defined, for it will allow us to compare different functors; in particular, it will make precise the question whether two functors are isomorphic. The notion of an adjoint pair of functors, though intimately involved with naturality, will not be discussed until Chapter 11, where it will be used.

Definitions and Examples

Definition. Let $\mathscr{C}$ and $\mathscr{A}$ be categories, and let $F, G: \mathscr{C} \to \mathscr{A}$ be (covariant) functors. A **natural transformation** $\tau: F \to G$ is a one-parameter family of morphisms $\tau = \{\tau_C: F(C) \to G(C) | C \in \operatorname{obj} \mathscr{C}\}$ such that the following diagram commutes for every morphism $f: C \to C'$:

$$\begin{array}{ccc} F(C) & \xrightarrow{Ff} & F(C') \\ {\scriptstyle \tau_C}\downarrow & & \downarrow{\scriptstyle \tau_{C'}} \\ G(C) & \xrightarrow[Gf]{} & G(C'). \end{array}$$

A similar definition can be given, *mutatis mutandis*, when both functors F and G are contravariant; just reverse both horizontal arrows.

Definition. A natural transformation $\tau: F \to G$ is a **natural equivalence** if every τ_C is an equivalence. Two functors are called **isomorphic** (or **naturally equivalent**) if there is some natural equivalence between them.

EXAMPLE 9.1. If $*$ is a one-point space, say, $* = \{a\}$, then a function $h: * \to X$ is completely determined by its only value $x = h(a) \in X$; denote h by h_x. One usually identifies x and h_x even though they are distinct (e.g., we have identified the singular 0-simplexes in a space X with the points of X). More precisely, let us see that the identity functor on **Sets** is isomorphic to Hom($*$,). For each set X, define $\tau_X: X \to \mathrm{Hom}(*, X)$ by $\tau_X(x) = h_x$. If $f: X \to Y$ is a function, then the diagram below commutes:

$$\begin{array}{ccc} X & \xrightarrow{f} & Y \\ {\scriptstyle \tau_X}\downarrow & & \downarrow{\scriptstyle \tau_Y} \\ \mathrm{Hom}(*, X) & \xrightarrow[f_*]{} & \mathrm{Hom}(*, Y); \end{array}$$

if $x \in X$, then $\tau_Y f: x \mapsto h_{f(x)}$ and $f_* \tau_X: x \mapsto f \circ h_x = h_{f(x)}$. Therefore τ is a natural transformation (we let the reader check that τ is in fact a natural equivalence).

EXAMPLE 9.2. The identity functor on **Ab** is isomorphic to Hom (**Z**,). The argument is essentially that of the preceding example, for every homomorphism $f: \mathbf{Z} \to G$ is completely determined by $f(1) \in G$.

EXAMPLE 9.3. Let k be a field and let $\mathscr{C} = \mathscr{A}$ be the category of all vector spaces over k and all linear transformations. Recall that V^* denotes the dual space of a vector space V, namely, the vector space of all linear functionals on V (hence $V^* = \mathrm{Hom}(V, k)$) and that $V^{**} = (V^*)^*$ is the second dual. For $x \in V$ and $f \in V^*$, let (x, f) denote $f(x)$. For each vector space V, define $e_V: V \to V^{**}$ by $e_V(x) = (x, \quad)$, evaluation at x. The reader may check (if this has not already been seen in one's linear algebra course) that e_V is an injective linear transformation. It is also easy to check that all such e_V define a natural transformation from the identity functor to the second dual functor (which is a natural equivalence when one restricts to the subcategory of all finite-dimensional vector spaces over k).

EXERCISES

9.1. If $(S_*(X), \partial)$ is the singular complex of a space X, then we have seen, in Exercise 4.6, that S_n: **Top** $\to$ **Ab** is a functor for each fixed $n \geq 0$. For each space X, the boundary operator's complete notation is $\partial_n^X: S_n(X) \to S_{n-1}(X)$. Show that $\partial_n: S_n \to S_{n-1}$ is a natural transformation. (*Hint*: Lemma 4.8.)

9.2. For each fixed $n \geq 0$, define a functor E: **Top** $\to$ **Ab** by $EX = S_{n+1}(X \times \mathbf{I})$ and $Ef = (f \times 1)_\#$. Use Exercise 4.10 to show that the prism operator $P_n: S_n \to E$ is a natural transformation.

9.3. Recall that $\tilde{H}_1(X) \cong H_1(X, x_0)$ for any $x_0 \in X$, and regard $\tilde{H}_1$: **Top**$_*$ $\to$ **Groups** (of course, $\tilde{H}_1(X)$ is abelian, but we choose to forget this in this exercise). Show that the Hurewicz map defines a natural transformation $\pi_1 \to \tilde{H}_1$. (*Hint*: Exercise 4.13.)

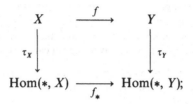

*9.4. Consider the functor $R: \mathbf{Top}^2 \to \mathbf{Top}^2$ defined on objects by $(X, A) \mapsto (A, \varnothing)$ and defined on morphisms by $f \mapsto f|A$, where $f: (X, A) \to (X', A')$. Use Theorem 5.9 to show that the connecting homomorphism defines a natural transformation $\partial: H_n \to H_{n-1} \circ R$.

9.5. For each fixed $n \geq 0$, show that subdivision, $\mathrm{Sd}_n: S_n \to S_n$, is a natural transformation, where $S_n: \mathbf{Top} \to \mathbf{Ab}$ is the nth term of the singular chain complex. (*Hint*: Exercise 6.8.)

9.6. (i) Prove that the composite of two natural transformations, when defined, is a natural transformation.
(ii) Prove that natural equivalence is an equivalence relation on the class of all (covariant) functors between a given pair of categories.

*9.7. (**Yoneda Lemma**) Let $\mathscr{C}$ be a category, let $A \in \operatorname{obj} \mathscr{C}$, and let $F: \mathscr{C} \to \mathbf{Sets}$ be a contravariant functor; let $\mathrm{Nat}(\mathrm{Hom}(\ \ , A), F)$ denote the class of all natural transformations $\mathrm{Hom}(\ \ , A) \to F$.
(i) There is a function $y: \mathrm{Nat}(\mathrm{Hom}(\ \ , A), F) \to F(A)$ given by

$$\varphi \mapsto \varphi_A(1_A).$$

(ii) There is a function $y': F(A) \to \mathrm{Nat}(\mathrm{Hom}(\ \ , A), F)$ given by $\mu \mapsto \tau$, where, for each $X \in \operatorname{obj} \mathscr{C}$, $\tau_X: \mathrm{Hom}(X, A) \to F(X)$ is defined by

$$\tau_X(f) = (Ff)(\mu).$$

(iii) y is a bijection with inverse y'.
(iv) If $B \in \operatorname{obj} \mathscr{C}$, then every natural transformation $\varphi: \mathrm{Hom}(\ \ , A) \to \mathrm{Hom}(\ \ , B)$ has the form $\varphi = (\varphi_X)$ with $\varphi_X(f) = \mu f$, where $\mu = \varphi_A(1_A)$ and $f \in \mathrm{Hom}(X, A)$.
(v) State and prove the dual version of the Yoneda lemma involving $\mathrm{Nat}(\mathrm{Hom}(A,\ \ ), G)$, where $G: \mathscr{C} \to \mathbf{Sets}$ is a covariant functor.

9.8. Call a category $\mathscr{C}$ **small** if $\operatorname{obj} \mathscr{C}$ is a set (it follows that the class of all morphisms in $\mathscr{C}$ is also a set). If $\mathscr{C}$ and $\mathscr{A}$ are categories with $\mathscr{C}$ small, show that there is a category (denoted by $\mathscr{A}^{\mathscr{C}}$) whose objects are all (covariant) functors $\mathscr{C} \to \mathscr{A}$ and whose morphisms are all natural transformations. (*Remark*: One assumes that $\mathscr{C}$ is small to guarantee that $\mathrm{Hom}(F, G)$ is a set.) A subcategory of $\mathscr{A}$ is called a **functor category**.

9.9. (i) Regard the ordered set $\mathbf{Z}$ as a category (Exercise 0.9) and show that a complex may be construed as a contravariant functor $C: \mathbf{Z} \to \mathbf{Ab}$ (with the extra condition that composites of nonidentity morphisms are zero).
(ii) If C and C' are complexes, then a chain map $f: C \to C'$ is a natural transformation.

Eilenberg–Steenrod Axioms

We are now able to state the theorem of Eilenberg and Steenrod.

Definition. A pair (X, A) of spaces, where A is a subspace of X, is called a **compact polyhedral pair** if there is a (finite) simplicial complex K, a subcomplex L, and a homeomorphism $f: |K| \to X$ with $f(|L|) = A$.

Definition. Let $\mathscr{C}$ be the category of compact polyhedral pairs. A **homology theory** (H, ∂) on $\mathscr{C}$ is a sequence of functors $H_n: \mathscr{C} \to \mathbf{Ab}$ for $n \geq 0$ and a sequence of natural transformations $\partial_n: H_n \to H_{n-1} \circ R$ (where $R: \mathscr{C} \to \mathscr{C}$ is the functor $(X, A) \mapsto (A, \varnothing)$ of Exercise 9.4) such that the following axioms hold.

Homotopy Axiom. If $f_0, f_1: (X, A) \to (Y, B)$ are homotopic (i.e., there is a homotopy $F: (X \times \mathbf{I}, A \times \mathbf{I}) \to (Y, B)$ with $F_0 = f_0$ and $F_1 = f_1$), then $H_n(f_0) = H_n(f_1): H_n(X, A) \to H_n(Y, B)$ for all $n \geq 0$.

Exactness Axiom. For every pair (X, A) with inclusions $i: (A, \varnothing) \hookrightarrow (X, \varnothing)$ and $j: (X, \varnothing) \hookrightarrow (X, A)$, there is an exact sequence

$$\cdots \to H_n(A, \varnothing) \xrightarrow{H_n(i)} H_n(X, \varnothing) \xrightarrow{H_n(j)} H_n(X, A) \xrightarrow{\partial_n} H_{n-1}(A, \varnothing) \to \cdots.$$

Excision Axiom. For every pair (X, A) and every open subset U of X with $\bar{U} \subset A^\circ$, the inclusion $(X - U, A - U) \hookrightarrow (X, A)$ induces isomorphisms

$$H_n(X - U, A - U) \xrightarrow{\simeq} H_n(X, A)$$

for all $n \geq 0$.

Dimension Axiom. If X is a one-point space, then $H_n(X, \varnothing) = 0$ for all $n > 0$. (One calls $H_0(X, \varnothing)$ the **coefficient group**.)

Since we have proved that each part of the definition holds for the singular theory, we know that homology theories with coefficient group $\mathbf{Z}$ on $\mathscr{C}$ do exist. Indeed we have even proved that such theories exist for the larger category $\mathbf{Top}^2$. Of course, one usually writes $H_n(X)$ instead of $H_n(X, \varnothing)$.

Definition. Let (H, ∂) and (H', ∂') be homology theories on $\mathscr{C}$. An **isomorphism** $\tau: (H, \partial) \to (H', \partial')$ is a sequence of natural equivalences

$$\tau_n: H_n \to H'_n, \quad \text{all } n \geq 0$$

making the following diagram commute

$$\begin{array}{ccc} H_{n+1}(X, A) & \xrightarrow{\partial} & H_n(A) \\ \downarrow{\scriptstyle\tau} & & \downarrow{\scriptstyle\tau} \\ H'_{n+1}(X, A) & \xrightarrow[\partial']{} & H'_n(A), \end{array}$$

for all pairs (X, A) and all $n \geq 0$.

Theorem (Eilenberg–Steenrod). *Any two homology theories with isomorphic coefficient groups on the category $\mathscr{C}$ of all compact polyhedral pairs are isomorphic.*[1]

Remarks. (1) A proof of this theorem (indeed of more general versions of it) can be found in [Eilenberg and Steenrod]. See also [Spanier, pp. 199–205].

[1] [Hu(1966), pp. 51–60] extends this theorem to the category of all (X, A), where X is a finite CW complex and A is a CW subcomplex.

(2) We have already seen that singular, simplicial, and cellular homology theories assign isomorphic homology groups to each compact polyhedral pair (X, A), but we have not shown the stronger result that these three theories on $\mathscr{C}$ are isomorphic; that is, the induced homomorphisms are essentially the same as well.

(3) An **extraordinary cohomology theory** on $\mathscr{C}$ is almost a homology theory: it satisfies all the conditions of the definition except the dimension axiom (on page 257, we shall introduce homology with arbitrary abelian coefficient groups). An example of an extraordinary cohomology theory is topological K-theory.

(4) There is an extension of the Eilenberg–Steenrod theorem characterizing homology theories on larger categories that contain certain noncompact pairs. This extension requires an extra axiom, **compact supports**, which is essentially Theorem 4.16. More precisely, the axiom states that if cls $z \in H_n(X, A)$, then there is a compact pair $(X', A') \subset (X, A)$ (i.e., (X', A') is a pair, X' is compact, and A' is closed in X) with cls z in the image of $j_*: H_n(X' A') \to H_n(X, A)$, where $j: (X', A') \hookrightarrow (X, A)$ is the inclusion. With this extra axiom, there is an isomorphism of any two homology theories having isomorphic coefficient groups defined on the category of all not necessarily compact polyhedral pairs.

(5) Here is the reason that the dimension axiom is so called. Given a homology theory (H, ∂), one can define an extraordinary homology theory (H', ∂') by defining $H'_n(X, A) = H_{n-1}(X, A)$ for all n and for all pairs (X, A). Since one wants a point to be zero-dimensional (and eventually that spaces X of dimension d should have $H_n(X) = 0$ for all $n > d$), the dimension axiom "tends to insure that the dimensional index should have a geometric meaning" (quotation from [Eilenberg and Steenrod, p. 12]).

(6) The only axiom guaranteeing nontriviality, that is, which forbids $H_n(X) = 0$ for all n and all X, is the dimension axiom (when we further assume that the coefficient group is nontrivial). In principle, one ought to be able to construct $H_*(X)$ from the homology of a point! The first step in this construction is the computation of the homology of spheres. In Chapter 11, we shall discuss the *suspension* ΣX of a space X.

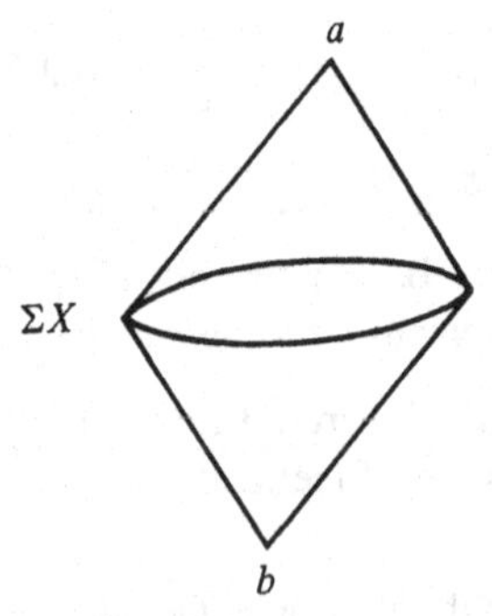

One defines ΣX as the quotient space of $X \times \mathbf{I}$ in which $X \times \{0\}$ is identified to a point, say, a, and $X \times \{1\}$ is also identified to a point, say, b. Of course, X is imbedded in ΣX as $X \times \{\frac{1}{2}\}$. The picture of ΣX reminds one of a sphere in which X is the equator, the cone over X with vertex a is the northern hemisphere, and the cone over X with vertex b is the southern hemisphere. Indeed one can prove that this picture is accurate: $\Sigma S^n \approx S^{n+1}$. Now the proof of Theorem 6.5, the computation of $H_*(S^n)$, can be adapted to prove that $\tilde{H}_p(X) \cong \tilde{H}_{p+1}(\Sigma X)$ for every space X (this is called the *suspension isomorphism*). In particular, $\mathbf{Z} \cong \tilde{H}_0(S^0) \cong \tilde{H}_1(S^1) \cong \cdots \cong \tilde{H}_n(S^n)$; the axioms produce the homology of spheres from the homology of a point.

(7) It can be shown that the excision axiom can be replaced by an exact Mayer–Vietoris sequence [Spanier, p. 208].

The Eilenberg–Steenrod theorem was very important in the development of algebraic topology. For two decades before it, there was a host of homology theories (we have discussed only three; some others are named Čech, Vietoris, cubical) designed to treat appropriate classes of problems. One was obliged to learn them all, and the subject grew quite complicated. Today one can invoke the Eilenberg–Steenrod theorem to see that the various homology theories are but different constructions of the unique theory (on compact polyhedral pairs). Besides giving a simplifying organizing principle, the Eilenberg–Steenrod theorem also introduced the possibility of axiomatic proofs in algebraic topology, which are conceptually easy to grasp.

Chain Equivalences

Definition. A chain complex C_* is called **free** if each of its terms is a free abelian group.

The main theorem in this section is a necessary and sufficient condition that a chain map between free chain complexes be a chain equivalence.

Theorem 9.1. *Let F be a free abelian group. In the diagram below with exact row, that is, g is a surjective homomorphism, there exists a homomorphism $f: F \to B$ with $gf = h$.*

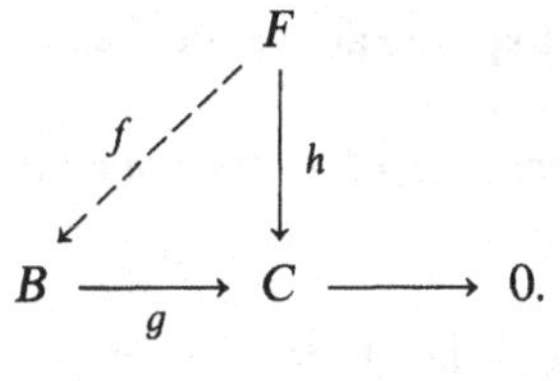

Proof. Let X be a basis of F. For each $x \in X$, choose $b_x \in B$ with $g(b_x) = h(x)$ (which is possible because g is surjective). The function $x \mapsto b_x$ defines a homomorphism $f: F \to B$ by extending by linearity. For each $x \in X$, we have

$gf(x) = g(b_x) = h(x)$; it follows that $gf = h$, because both homomorphisms agree on a set of generators of F. □

Corollary 9.2. *If F is a free abelian group and $g: B \to F$ is a surjective homomorphism from some abelian group B, then*

$$B = \ker g \oplus F',$$

where $F' \cong F$.

PROOF. Consider the diagram

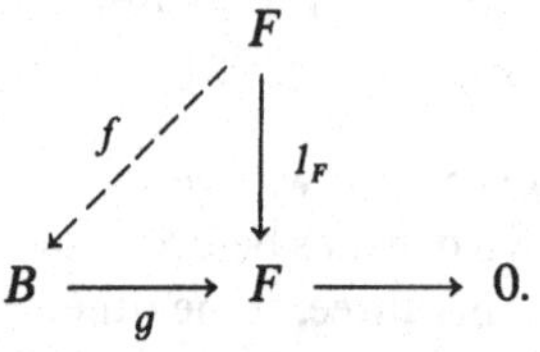

By the theorem, there is a homomorphism $f: F \to B$ with $gf = 1_F$; it follows that f is injective. But $B = \ker g \oplus \operatorname{im} f$: if $b \in B$, then $b = (b - fg(b)) + fg(b) \in \ker g + \operatorname{im} f$ (because $g(b - fg(b)) = 0$), and (it is easy to see that) $\ker g \cap \operatorname{im} f = 0$. The result follows by defining $F' = \operatorname{im} f$. □

One can rephrase the conclusion of the corollary in terms of exact sequences.

Definition. A **split exact sequence** is an exact sequence $0 \to A \xrightarrow{i} B \xrightarrow{p} C \to 0$ for which there exists a homomorphism $s: C \to B$ with $ps = 1_C$.

EXERCISE

*9.10. The following statements are equivalent.
  (i) The exact sequence $0 \to A \xrightarrow{i} B \xrightarrow{p} C \to 0$ is split.
  (ii) A is a direct summand of B; that is, there exists a subgroup C' of B with $C' \cong C$ via $p|C'$ and $B = \operatorname{im} i \oplus C'$.
  (iii) There exists a homomorphism $q: B \to A$ with $qi = 1_A$.

Corollary 9.2 thus says that an exact sequence $0 \to A \to B \to F \to 0$ with F free abelian is necessarily a split exact sequence.

Theorem 9.3. *Every subgroup H of a free abelian group F is free abelian; moreover,* rank $H \leq$ rank F.

Remark. It follows that if F is f.g., then H is f.g.

PROOF. We give two proofs: the first proof works only when F has finite rank, but it allows us to focus on essentials.

Suppose that F has finite rank n; we prove the theorem by induction on n. If $n = 1$, then $F \cong \mathbf{Z}$ and the division algorithm shows that any subgroup H of F is cyclic, hence is 0 or isomorphic to $\mathbf{Z}$. Thus H is free abelian and rank $H \leq 1 = \operatorname{rank} F$. For the inductive step, let $\{x_1, \ldots, x_n\}$ be a basis of F, let $F_n = \langle x_1, \ldots, x_{n-1} \rangle$, and let $H_n = H \cap F_n$. By induction, H_n is free abelian of rank $\leq n - 1$. Now

$$H/H_n = H/(H \cap F_n) \cong (H + F_n)/F_n \subset F/F_n \cong \mathbf{Z}.$$

By Corollary 9.2, $H = H_n$ or $H = H_n \oplus \langle h \rangle$, where $\langle h \rangle \cong \mathbf{Z}$. Therefore H is free abelian of rank $\leq n$.

We now give a second proof that does not assume the rank of F is finite. Let $\{x_k : k \in K\}$ be a basis of F, which we assume is well ordered. (That every nonempty set can be somehow well ordered is equivalent to the axiom of choice.)

For each $k \in K$, define $F_k = \sum_{j<k} \langle x_j \rangle$ and $\bar{F}_k = \sum_{j \leq k} \langle x_j \rangle$; define $H_k = H \cap F_k$ and $\bar{H}_k = H \cap \bar{F}_k$. Now $F = \bigcup \bar{F}_k$ and $H = \bigcup \bar{H}_k$; also, $H_k = H \cap F_k = \bar{H}_k \cap F_k$. Hence

$$\bar{H}_k/H_k = \bar{H}_k/(\bar{H}_k \cap F_k) \cong (\bar{H}_k + F_k)/F_k \subset \bar{F}_k/F_k \cong \mathbf{Z}.$$

By Corollary 9.2, either $\bar{H}_k = H_k$ or

$$\bar{H}_k = H_k \oplus \langle h_k \rangle, \quad \text{where } \langle h_k \rangle \cong \mathbf{Z}.$$

We claim that H is free abelian on the set of h_k's; note that it will then follow that rank $H \leq$ rank F, for the set of h_k clearly has cardinality $\leq |K| =$ rank F.

Let H° be the subgroup of H generated by the h_k. Since $F = \bigcup \bar{F}_k$, each $h \in H$ (as any element of F) lies in some $\bar{F}_k$. Let $\mu(h)$ be the least index k with $h \in \bar{F}_k$. Suppose that $H \neq H^\circ$ and consider $\{\mu(h) : h \in H \text{ and } h \notin H^\circ\}$. There is a least such index j, because K is well ordered. Choose $h' \in H$ with $\mu(h') = j$ and $h' \notin H^\circ$. Now $\mu(h') = j$ says that $h' \in H \cap \bar{F}_j$, so

$$h' = a + mh_j, \qquad a \in H_j, \qquad m \in \mathbf{Z}.$$

Therefore $a = h' - mh_j \in H$, $a \notin H^\circ$ (lest $h' \in H^\circ$), and $\mu(a) < j$, a contradiction. Hence $H = H^\circ$.

Next, we show that linear combinations of the h_k are unique. It suffices to show that if

$$m_1 h_{k_1} + \cdots + m_n h_{k_n} = 0, \qquad k_1 < \cdots < k_n,$$

then each $m_i = 0$. We may assume that $m_n \neq 0$. But then $m_n h_{k_n} \in \langle h_{k_n} \rangle \cap H_{k_n} = 0$, a contradiction. This shows that H is free abelian on the h_k. □

Theorem 9.4. *A free chain complex (A_*, ∂) is acyclic if and only if it has a contracting homotopy.*

Proof. Sufficiency is Corollary 5.4. For the converse, assume that $H_n(A_*) = 0$ for all $n \geq 0$. Now $Z_n(A_*) \subset A_n$ is free abelian, by Theorem 9.3. The differentiation $\partial_n: A_n \to A_{n-1}$ has image $B_{n-1}(A_*) = Z_{n-1}(A_*)$ (since $H_{n-1}(A_*) = 0$), so that Theorem 9.1 gives a homomorphism $s_{n-1}: Z_{n-1}(A_*) \to A_n$ with $\partial_n s_{n-1} = 1$. It follows that the map $1 - s_{n-1}\partial_n: A_n \to A_n$ has its image in $Z_n(A_*)$. Define $t_n: A_n \to A_{n+1}$ as the composite

$$t_n = s_n(1 - s_{n-1}\partial_n).$$

Then

$$\begin{aligned}\partial_{n+1}t_n + t_{n-1}\partial_n &= \partial_{n+1}s_n(1 - s_{n-1}\partial_n) + s_{n-1}(1 - s_{n-2}\partial_{n-1})\partial_n \\ &= 1 - s_{n-1}\partial_n + s_{n-1}\partial_n = 1.\end{aligned}$$

Therefore $\{t_n\}$ is a contracting homotopy of A_*. □

Definition. Let $f: (A_*, \partial) \to (B_*, \partial')$ be a chain map. The **mapping cone** of f is the chain complex $C(f)$ whose nth term is

$$C(f)_n = A_{n-1} \oplus B_n$$

and whose differentiation $D_n: C(f)_n \to C(f)_{n-1}$ is given by

$$D_n(a_{n-1}, b_n) = (-\partial_{n-1}a_{n-1}, f_{n-1}a_{n-1} + \partial'_n b_n).$$

It is convenient to write D_n in matrix form:

$$D = \begin{pmatrix} -\partial & 0 \\ f & \partial' \end{pmatrix}.$$

Lemma 9.5. *If $f: A_* \to B_*$ is a chain map between free chain complexes, then $C(f)$ is a free chain complex.*

Proof. Matrix multiplication shows that $DD = 0$, using the fact that $-f\partial + \partial' f = 0$ (because f is a chain map). The freeness of $C(f)$ follows at once from the freeness of A_* and of B_*. □

Theorem 9.6. *Let $f: A_* \to B_*$ be a chain map between free chain complexes. If $C(f)$ is acyclic, then f is a chain equivalence.*

Proof. Assume that $C(f)$ is acyclic; since the chain complexes are free, Theorem 9.4 says that $C(f)$ has a contracting homotopy. In matrix notation, there is a 2×2 matrix T with $DT + TD = I$:

$$\begin{pmatrix} -\partial & 0 \\ f & \partial' \end{pmatrix}\begin{pmatrix} \lambda & \mu \\ \sigma & \tau \end{pmatrix} + \begin{pmatrix} \lambda & \mu \\ \sigma & \tau \end{pmatrix}\begin{pmatrix} -\partial & 0 \\ f & \partial' \end{pmatrix} = \begin{pmatrix} 1 & 0 \\ 0 & 1 \end{pmatrix}.$$

Define $f' = \mu$, $s' = -\tau$, and $s = \lambda$. Then the matrix equation shows that f' is a chain map ($-\partial\mu + \mu\partial' = 0$), $s\partial + \partial s = f'f - 1$, and $s'\partial' + \partial' s' = ff' - 1$. Therefore f is a chain equivalence. □

Lemma 9.7. *If $f: A_* \to B_*$ is a chain map, then there is an exact sequence*

$$\cdots \longrightarrow H_{n+1}(C(f)) \longrightarrow H_n(A_*) \xrightarrow{f_*} H_n(B_*) \longrightarrow H_n(C(f)) \longrightarrow \cdots.$$

PROOF. Define a shifted version A_*^+ of A_* as follows:

$$(A_*^+)_n = A_{n-1} \quad \text{and} \quad \partial_n^+ = \partial_{n-1}.$$

There is a short exact sequence of chain complexes

$$0 \to B_* \xrightarrow{i} C(f) \xrightarrow{p} A_*^+ \to 0,$$

where $i: b \mapsto (0, b)$ and $p: (a, b) \mapsto a$. There results a long exact sequence of homology groups

$$\cdots \to H_{n+1}(C(f)) \to H_{n+1}(A_*^+) \xrightarrow{\Delta} H_n(B_*) \to H_n(C(f)) \to \cdots,$$

where Δ is the connecting homomorphism. Now it is easy to see that $H_{n+1}(A_*^+) = H_n(A_*)$; let us compute Δ. Consider the usual diagram

$$\begin{array}{ccc} & & A_n \oplus B_{n+1} \xrightarrow{p} A_n \to 0 \\ & & \downarrow D \\ 0 \to B_n \xrightarrow{i} & & A_{n-1} \oplus B_n \end{array}$$

If a is a cycle, then $Dp^{-1}(a) = D(a, 0) = (-\partial a, fa) = (0, fa) = i(fa)$; hence Δ: cls $a \mapsto$ cls fa, and so $\Delta = f_*$. □

Theorem 9.8. *Let A_* and B_* be free chain complexes, and let $f: A_* \to B_*$ be a chain map. Then f is a chain equivalence if and only if $f_{*n}: H_n(A_*) \to H_n(B_*)$ is an isomorphism for every n.*

PROOF. Necessity is Theorem 5.3. For sufficiency, consider the exact sequence of Lemma 9.7. Since each f_* is an isomorphism, exactness forces $H_n(C(f)) = 0$ for all n; that is, $C(f)$ is acyclic. Theorem 9.6 now applies to show that f is a chain equivalence. □

Remark. If E'_* is an adequate subcomplex of a free chain complex E_*, then Lemma 7.28 shows that the inclusion $i: E'_* \hookrightarrow E_*$ induces isomorphisms in homology. It follows from Theorem 9.8 that i is a chain equivalence.

The chain map $j: C_*(K) \to S_*(|K|)$ of Theorem 7.22 (where K is a simplicial complex) is a chain equivalence, because j_* is an isomorphism; also, the chain map $W_*(X) \to S_*(X)$ of Exercise 8.35 is a chain equivalence for every CW complex X.

Acyclic Models

The next topic, the *method of acyclic models*, is a technique of constructing chain maps and chain homotopies. The following elementary result is the heart of the so-called comparison theorem of homological algebra; its analogue in functor categories is the heart of acyclic models.

Theorem 9.9.

(i) *Consider the commutative diagram of abelian groups*

$$\begin{array}{ccccc} F & \xrightarrow{t} & G & \xrightarrow{s} & G'' \\ {\scriptstyle c}\downarrow & & \downarrow{\scriptstyle b} & & \downarrow{\scriptstyle a} \\ E' & \xrightarrow[r]{} & E & \xrightarrow[p]{} & E'' \end{array}$$

in which the bottom row is exact, $st = 0$, and F is free abelian. Then there exists a homomorphism c making the first square commute.

(ii) *Consider the diagram of abelian groups*[2]

$$\begin{array}{ccccccccccc} \cdots & \longrightarrow & F_2 & \xrightarrow{d_2} & F_1 & \xrightarrow{d_1} & F_0 & \xrightarrow{e} & \operatorname{coker} d_1 & \longrightarrow & 0 \\ & & {\scriptstyle t_2}\downarrow & & \downarrow{\scriptstyle t_1} & & \downarrow{\scriptstyle t_0} & & \downarrow{\scriptstyle f} & & \\ \cdots & \longrightarrow & E_2 & \xrightarrow[\partial_2]{} & E_1 & \xrightarrow[\partial_1]{} & E_0 & \xrightarrow[\varepsilon]{} & \operatorname{coker} \partial_1 & \longrightarrow & 0 \end{array}$$

in which the rows are chain complexes, each F_i is free abelian, and the bottom row is exact (i.e., it is an acyclic complex). Then there exists a chain map $t: F \to E$ with $fe = \varepsilon t_0$.

PROOF. (i) If we can show that im $bt \subset$ im r, then we have a diagram

$$\begin{array}{ccccc} & & F & & \\ & & \downarrow{\scriptstyle bt} & & \\ E' & \xrightarrow{r} & \operatorname{im} r & \longrightarrow & 0 \end{array}$$

to which Theorem 9.1 applies, yielding the result. Now exactness of the bottom row gives im r = ker p, so that it suffices to prove that $pbt = 0$. But $pbt = ast$, by commutativity, and $st = 0$, by hypothesis.

(ii) Construct t_i by induction on $i \geq 0$. When $i = 0$, use Theorem 9.1 with the diagram

$$\begin{array}{ccccc} & \swarrow & F_0 & & \\ & & \downarrow{\scriptstyle fe} & & \\ E_0 & \xrightarrow[\varepsilon]{} & \operatorname{coker} \partial_1 & \longrightarrow & 0. \end{array}$$

For the inductive step, use part (i). □

Definition. One says that a chain map t is **over** f in the circumstance of Theorem 9.9(ii); that is, $fe = \varepsilon t_0$.

---

[2] If $f: A \to A'$ is a homomorphism, then its cokernel is defined as

$$\operatorname{coker} f = A'/\operatorname{im} f.$$

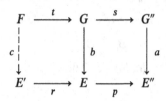

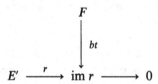

EXERCISE

9.11. In Theorem 9.9(ii), prove that t is unique in the sense that any other such chain map t' (over f) is chain homotopic to t. (*Hint*: Define $s_{-1} = 0$ and construct $s_n: F_n \to E_{n+1}$ with $\partial_{n+1} s_n + s_{n-1} d_n = t_n - t'_n$ by induction on $n \geq 0$ by using the commutative diagram

$$\begin{array}{ccccc} & & F_n & \longrightarrow & 0 \\ & & {\scriptstyle t_n - t'_n - s_{n-1}d_n} \downarrow & & \downarrow \\ E_{n+1} & \xrightarrow[\partial_{n+1}]{} & E_n & \xrightarrow[\partial_n]{} & E_{n-1}.) \end{array}$$

Definition. A functor F: **Ab** $\to$ **Ab** is **additive** if whenever $f, g: A \to B$ are homomorphisms then

$$F(f + g) = F(f) + F(g).$$

(In Theorem 5.2, we proved that H_n: **Comp** $\to$ **Ab** is additive if one makes the obvious generalization from **Ab** to **Comp**.)

EXERCISES

*9.12. Let F: **Ab** $\to$ **Ab** be an additive functor; if f is a zero homomorphism, then so is $F(f)$; if A is the zero group, then so is $F(A)$.

*9.13. Let F: **Ab** $\to$ **Ab** be an additive functor of either variance.

(i) If $0 \to A \to B \to C \to 0$ is a split exact sequence, then $0 \to FA \to FB \to FC \to 0$ is also split exact when F is covariant (and $0 \to FC \to FB \to FA \to 0$ is split exact when F is contravariant). In particular, the functored sequence is exact.

(ii) If I is a finite index set, then

$$F\left(\sum_{i \in I} A_i\right) \cong \sum_{i \in I} FA_i.$$

*9.14. Let F: **Ab** $\to$ **Ab** be an additive functor of either variance.

(i) If (A_*, ∂) is a chain complex, then $(FA_*, F\partial)$ is a chain complex.

(ii) If $f: A_* \to B_*$ is a chain map, then $Ff: FA_* \to FB_*$ is a chain map when F is covariant ($Ff: FB_* \to FA_*$ when F is contravariant).

(iii) If $f: A_* \to B_*$ is a chain equivalence, then so is Ff. Conclude that FA_* and FB_* have the same homology groups.

Definition. A category $\mathscr{C}$ with **models** $\mathscr{M}$ is an ordered pair $(\mathscr{C}, \mathscr{M})$, where $\mathscr{M}$ is a subset of obj $\mathscr{C}$. If $F: \mathscr{C} \to$ **Ab** is a functor, then an ***F*-model set** is an indexed set $\mathscr{X} = \{x_j \in FM_j: j \in J\}$, where $\{M_j: j \in J\}$ is an indexed family of models.

For every object C in $\mathscr{C}$ and every $\sigma: M_j \to C$ in $\mathscr{C}$, one has $F\sigma: FM_j \to FC$ in **Ab**, and hence $(F\sigma)(x_j) \in FC$ for every $j \in J$ and every $x_j \in \mathscr{X}$.

Definition. Let $\mathscr{C}$ be a category with models $\mathscr{M}$, and let $F: \mathscr{C} \to$ **Ab** be a functor. Then F is **free** with **base in** $\mathscr{M}$ if:

(i) FC is a free abelian group for every object C;
(ii) there is an F-model set $\mathscr{X} = \{x_j \in FM_j: j \in J\}$ such that, for every object C, the set

$$\{(F\sigma)(x_j): x_j \in \mathscr{X} \text{ and } \sigma: M_j \to C\}$$

is a basis of FC.

EXAMPLE 9.4. Fix $k \geq 0$, and consider the category **Top** with models $\mathscr{M} = \{\Delta^k\}$ (there is only one model). If $S_k: \textbf{Top} \to \textbf{Ab}$ is the functor that is the kth term of the singular complex, then $\mathscr{X} = \{\delta\}$ is an S_k-model set, where $\delta \in S_k(\Delta^k)$ is the identity map. For every space X, we know that $S_k(X)$ is a free abelian group with basis all k-simplexes $\sigma: \Delta^k \to X$. But $S_k(\sigma)(\delta) = \sigma_\#(\delta) = \sigma \circ \delta = \sigma$, so that

$$\{S_k(\sigma)(\delta): \delta \in \mathscr{X} \text{ and } \sigma: \Delta^k \to X\}$$

is a basis of $S_k(X)$. Hence S_k is free with base in $\mathscr{M} = \{\Delta^k\}$.

EXAMPLE 9.5. Let $\mathscr{K}$ be the category of simplicial complexes, let $p \geq 0$ be fixed, and let $C_p: \mathscr{K} \to \textbf{Ab}$ be the pth term of the simplicial chain complex. If $\mathscr{M} = \{\Delta^p\}$, where Δ^p is the standard p-simplex $[e_0, \ldots, e_p]$, then proceeding as in Example 9.4 does *not* show that C_p is a free functor. Let $\mathscr{X} = \{\delta\}$, where $\delta \in C_p(\Delta^p)$ is the element $\langle e_0, \ldots, e_p \rangle$. If K is a simplicial complex, then the set

$$\{C_p(\sigma)(\delta) | \sigma: \Delta^p \to K \text{ is a simplicial map}\}$$

$$= \{\langle \sigma e_0, \ldots, \sigma e_p \rangle | \sigma: \Delta^p \to K \text{ simplicial}\}$$

is too big; it does contain a basis of $C_p(K)$, but it also contains symbols $\langle v_0, \ldots, v_p \rangle$ with repeated vertices as well as symbols $\langle v_{\pi 0}, \ldots, v_{\pi p} \rangle$ for every permutation π of $\{0, 1, \ldots, p\}$.

Define $F_p: \mathscr{K} \to \textbf{Ab}$ so that $F_p(K)$ is the free abelian group having the large basis above; that is, $F_p(K)$ is the free abelian group with basis all symbols $\langle v_0, \ldots, v_p \rangle$ for which $v_0, \ldots, v_p$ are (not necessarily distinct) vertices that span a simplex in K. Then F_p is free with base $\{\Delta^p\}$.

Lemma 9.10. *Let $\mathscr{C}$ be a category with models $\mathscr{M}$, and let $F: \mathscr{C} \to \textbf{Ab}$ be a free functor with base $\mathscr{X} = \{x_j \in FM_j: j \in J\}$. If $G: \mathscr{C} \to \textbf{Ab}$ is a functor and $\mathscr{Y} = \{y_j \in GM_j: j \in J\}$ is a G-model set (same models M_j), then there exists a unique natural transformation $\tau: F \to G$ with $\tau_{M_j}(x_j) = y_j$ for all $j \in J$.*

Remark. The following diagram is a mnemonic.

$$
\begin{array}{ccc}
F & \overset{\tau}{\dashrightarrow} & G \\
\uparrow & & \uparrow \\
\mathscr{X} & \underset{x_j \mapsto y_j}{\longrightarrow} & \mathscr{Y}.
\end{array}
$$

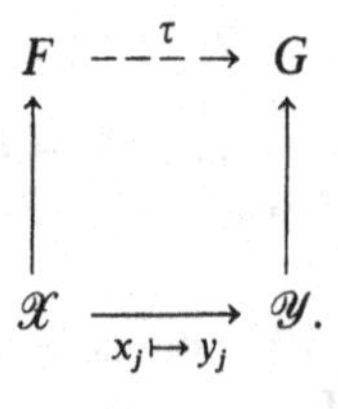

Proof. Let us prove uniqueness of τ (assuming that it exists). For fixed index j and object C, naturality of τ gives a commutative diagram for every $\sigma: M_j \to C$:

$$\begin{array}{ccc} FM_j & \xrightarrow{F\sigma} & FC \\ {\scriptstyle\tau_j}\downarrow & & \downarrow{\scriptstyle\tau_C} \\ GM_j & \xrightarrow[G\sigma]{} & GC, \end{array}$$

where τ_j abbreviates τ_{M_j}; that is, $\tau_C \circ (F\sigma) = (G\sigma) \circ \tau_j$. Hence, if $x_j \in \mathscr{X}$, the hypothesis gives $\tau_C((F\sigma)(x_j)) = (G\sigma)(\tau_j(x_j)) = (G\sigma)(y_j)$. Since the family of all $(F\sigma)(x_j)$ forms a basis of FC and hence generates FC, it follows that each homomorphism τ_C is uniquely determined. Therefore $\tau = \{\tau_C\}$ is unique.

To construct τ, define $\tau_C: FC \to GC$ by first setting $\tau_C((F\sigma)(x_j)) = (G\sigma)(y_j)$ and then extending by linearity (FC is free abelian, and we have assigned a value to each basis element). It remains to prove that all such τ_C constitute a natural transformation: if $f: C \to D$, then the following diagram commutes:

$$\begin{array}{ccc} FC & \xrightarrow{Ff} & FD \\ {\scriptstyle\tau_C}\downarrow & & \downarrow{\scriptstyle\tau_D} \\ GC & \xrightarrow[Gf]{} & GD. \end{array}$$

Since FC is free abelian, it suffices to evaluate both composites on a typical basis element, say, $(F\sigma)(x_j)$. Now

$$(Gf) \circ \tau_C: (F\sigma)(x_j) \mapsto (Gf)((G\sigma)(y_j)) = (Gf \circ G\sigma)(y_j) = (G(f\sigma))(y_j)$$

(because G is a functor); on the other hand,

$$\tau_D \circ Ff: (F\sigma)(x_j) \mapsto \tau_D((Ff \circ F\sigma)(x_j)) = \tau_D((F(f\sigma))(x_j)) = (G(f\sigma))(y_j). \quad \square$$

Lemma 9.11. *Let $\mathscr{C}$ be a category with models $\mathscr{M}$. Consider the commutative diagram of functors $\mathscr{C} \to$ **Ab** and natural transformations*

$$\begin{array}{ccccc} F & \xrightarrow{\tau} & G & \xrightarrow{\sigma} & G'' \\ {\scriptstyle\gamma}\,\vdots & & \downarrow{\scriptstyle\beta} & & \downarrow{\scriptstyle\alpha} \\ E' & \xrightarrow[\rho]{} & E & \xrightarrow[\pi]{} & E'' \end{array}$$

in which $\sigma\tau = 0$ (i.e., $\sigma_C\tau_C = 0$ for every object C), $\operatorname{im}\rho = \ker\pi$ on $\mathscr{M}$ (i.e., $\operatorname{im}\rho_M = \ker\pi_M$ for every model M), and F is free with base in $\mathscr{M}$. Then there exists a natural transformation $\gamma: F \to E'$ making the first square commute.

PROOF. By hypothesis, there exists an F-model set $\{x_j \in FM_j: j \in J\}$ that is a base for F. Now for each j, there is a commutative diagram in **Ab** satisfying the hypotheses of Theorem 9.9(i):

$$\begin{array}{ccccc} FM_j & \longrightarrow & GM_j & \longrightarrow & G''M_j \\ \downarrow & & \downarrow & & \downarrow \\ E'M_j & \longrightarrow & EM_j & \longrightarrow & E''M_j; \end{array}$$

hence each $x_j \in FM_j$ determines some $y'_j \in E'M_j$ (its image under the dashed arrow); of course, these y'_j form an E'-model set. By Lemma 9.10, there exists a natural transformation $\gamma: F \to E'$ with $\gamma_j(x_j) = y'_j$ (here γ_j abbreviates γ_{M_j}). It remains to check commutativity. Define an E-model set by setting $y_j = \rho_j y'_j$. Since both $\beta\tau$ and $\rho\gamma$ are natural transformations $F \to E$ whose M_jth component takes $x_j \mapsto y_j$, the uniqueness assertion in Lemma 9.10 gives $\beta\tau = \rho\gamma$. □

It is simplest to regard Lemma 9.11 as merely a functor version of the elementary Theorem 9.9(i). But Lemma 9.11 is stronger than this; not only is there no assumption that $\mathscr{C}$ is small (to force "functor categories" to be categories), but the most important feature is that the bottom row of the diagram is assumed exact only for models M in $\mathscr{M}$.

The theorem we seek is a version of these results with **Ab** replaced by **Comp**. Of course, if $E: \mathscr{C} \to$ **Comp** is a functor, then $H_n(EC)$ is defined for every object C; moreover, if E is **nonnegative**, that is, $E_i = 0$ for $i < 0$, then we may lengthen the complex EC as follows:

$$\cdots \to E_2C \to E_1C \to E_0C \to H_0(EC) \to 0$$

(for $H_0(EC)$ is just $\operatorname{coker}(E_1C \to E_0C)$). Finally, for every $k \geq 0$, E determines a functor $E_k: \mathscr{C} \to$ **Ab**, namely, $C \mapsto E_kC$, the kth term of the chain complex EC.

Definition. Let $E: \mathscr{C} \to$ **Comp** be a functor. An object C in $\mathscr{C}$ is called ***E*-acyclic** if $H_n(EC) = 0$ for all $n > 0$.

Theorem 9.12 (Acyclic Models). *Let $\mathscr{C}$ be a category with models $\mathscr{M}$, and let F, $E: \mathscr{C} \to$ **Comp** be nonnegative functors. For each $k \geq 0$, assume that F_k is free with base in $\mathscr{M}_k \subset \mathscr{M}$ and that each model M in $\mathscr{M}$ is E-acyclic. Then*

(i) *For every natural transformation $\varphi: H_0F \to H_0E$, there is some natural chain map $\tau: F \to E$ over φ; that is, there is a commutative diagram*

$$\begin{array}{ccccccccc} \cdots & \xrightarrow{d_2} & F_1 & \xrightarrow{d_1} & F_0 & \longrightarrow & H_0F & \longrightarrow & 0 \\ & & \downarrow{\scriptstyle \tau_1} & & \downarrow{\scriptstyle \tau_0} & & \downarrow{\scriptstyle \varphi} & & \\ \cdots & \xrightarrow[\partial_2]{} & E_1 & \xrightarrow[\partial_1]{} & E_0 & \longrightarrow & H_0E & \longrightarrow & 0. \end{array}$$

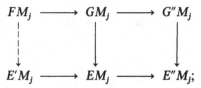

(ii) *If $\tau, \tau': F \to E$ are natural chain maps over φ, then τ and τ' are naturally chain homotopic.*

(iii) *Assume that E_k is free with base $\mathcal{M}_k \subset \mathcal{M}$ and each model M in $\mathcal{M}$ is F-acyclic. If φ is a natural equivalence, then every natural chain map $\tau: F \to E$ over φ is a natural chain equivalence.*

Remarks. (1) We shall elaborate on the term "naturally chain homotopic" in the proof.

(2) Realize what this theorem does; it constructs (natural) chain maps and, perhaps more useful, it constructs (natural) chain homotopies and chain equivalences.

(3) Recall that chain maps induce homomorphisms in homology, chain homotopic chain maps induce the same homomorphisms in homology, and chain equivalences induce isomorphisms in homology.

PROOF. (i) The statement means that for every natural transformation $\varphi: H_0F \to H_0E$ there exists a sequence of natural transformations $\tau_k: F_k \to E_k$, all $k \geq 0$, making the diagram in the statement commute (note that, for every model M, indeed for every object C in $\mathscr{C}$, the bottom row is exact at E_0C, because $H_0(EC)$ is just the cokernel of $E_1C \to E_0C$). The proof is by induction on $k \geq 0$. When $k = 0$, use Lemma 9.11 with the diagram

$$\begin{array}{ccccc} F_0 & \longrightarrow & H_0F & \longrightarrow & 0 \\ & & \downarrow{\scriptstyle \varphi} & & \downarrow{\scriptstyle 0} \\ E_0 & \longrightarrow & H_0E & \longrightarrow & 0; \end{array}$$

the inductive step also follows easily from Lemma 9.11.

(ii) Assume that both $\tau, \tau': F \to E$ are over φ; our task is to find natural transformations $s_k: F_k \to E_{k+1}$ for all $k \geq -1$ such that

$$\partial_{k+1}s_k + s_{k-1}d_k = \tau_k - \tau'_k.$$

Define $s_{-1} = 0$, and proceed by induction on $k \geq 0$ to define s_k. Let $\theta_k = \tau_k - \tau'_k$. As both τ and τ' are over φ, it is easy to see that the following diagram commutes:

$$\begin{array}{ccccc} F_0 & \xrightarrow{1} & F_0 & \longrightarrow & 0 \\ & & \downarrow{\scriptstyle \theta_0} & & \downarrow{\scriptstyle 0} \\ E_1 & \xrightarrow[\partial]{} & E_0 & \longrightarrow & H_0E. \end{array}$$

Lemma 9.11 applies at once (the bottom row is exact for every model M, indeed for every C, as noted above) to provide $s_0: F_0 \to E_1$ with $\partial s_0 = \theta_0$, as desired.

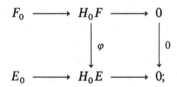

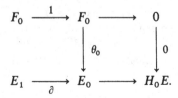

For the inductive step, consider the diagram

$$\begin{array}{ccccc} F_k & \xrightarrow{1} & F_k & \longrightarrow & 0 \\ & & \Big\downarrow \theta_k - s_{k-1}d_k & & \Big\downarrow 0 \\ E_{k+1} & \xrightarrow[\partial]{} & E_k & \xrightarrow[\partial]{} & E_{k-1} \end{array}$$

whose bottom row is exact for every model M. Now Lemma 9.11 provides $s_k: F_k \to E_{k+1}$ with $\partial s_k = \theta_k - s_{k-1}d_k$ (which is what we seek) if we can show that this diagram commutes. But, by induction,

$$\begin{aligned} \partial_k(\theta_k - s_{k-1}d_k) &= \partial_k\theta_k - (\partial_k s_{k-1})d_k \\ &= \partial\theta_k - (\theta_{k-1} - s_{k-2}d)d \\ &= \partial\theta_k - \theta_{k-1}d, \end{aligned}$$

and this last is zero because θ is a chain map.

(iii) If φ is a natural equivalence, then its inverse $\varphi^{-1}: H_0E \to H_0F$ exists. By (i), there exists a natural chain map $\sigma: E \to F$ over φ^{-1}. Therefore $\sigma\tau: F \to F$ is a natural chain map over $\varphi^{-1}\varphi = 1$, the identity natural transformation on H_0F. Obviously, the identity $1_F: F \to F$ is also a natural chain map over 1, so that (ii) gives a natural chain homotopy $\sigma\tau \simeq 1_F$. A similar argument gives $\tau\sigma \simeq 1_E$, hence $\tau: F \to E$ is a natural chain equivalence. □

Let us now review the proof of the homotopy axiom (Theorem 4.23) in the light of acyclic models. **Top** is a category with models $\mathcal{M} = \{\Delta^k: k \geq 0\}$. In Example 9.4, we saw that the singular complex S_*: **Top** → **Comp** has each term S_k: **Top** → **Ab** free with base in $\mathcal{M}$. Recall that the proof of Theorem 4.23 involved constructing a chain homotopy $P^X: \lambda_{1\#}^X \simeq \lambda_{0\#}^X$ for every space X, where $\lambda_i^X: X \to X \times \mathbf{I}$ is defined by $x \mapsto (x, i)$ for $i = 0, 1$. Define a functor E: **Top** → **Comp** by $E(X) = S_*(X \times \mathbf{I})$. In Theorem 4.19, it was shown that every convex set is acyclic; since $\Delta^k \times \mathbf{I}$ is convex, it follows that every model Δ^k is E-acyclic. Now both $\lambda_{1\#}$ and $\lambda_{0\#}$ are natural chain maps $S_* \to E$. Therefore acyclic models says that if $H_0(\lambda_1) = H_0(\lambda_0)$, then $\lambda_{1\#}$ and $\lambda_{0\#}$ are naturally chain homotopic. The equality $H_0(\lambda_1) = H_0(\lambda_0)$ is the content of Eqs. (1) and (2) in the base step ($n = 0$) of the proof of Theorem 4.23. Thus all other calculations in the proof of the homotopy axiom are necessarily routine, because they can be made once and for all in great generality.

Before giving further applications of acyclic models, we modify it to make it easier to use.

Definition. An **augmentation** of a nonnegative complex (S_*, ∂) is a surjective homomorphism $\varepsilon: S_0 \to \mathbf{Z}$ with the composite $\varepsilon\partial_1 = 0$. A chain map $f: S_* \to S'_*$ is **augmentation preserving** if there is a commutative diagram

$$
\begin{array}{ccc}
S_0 & \xrightarrow{\varepsilon} & \mathbf{Z} \\
{\scriptstyle f_0}\downarrow & & \downarrow{\scriptstyle 1} \\
S_0' & \xrightarrow[\varepsilon']{} & \mathbf{Z}.
\end{array}
$$

We have seen augmentations before, when reduced homology was introduced (then we wrote $\tilde{\partial}_0$ instead of ε, where $\tilde{\partial}_0$ adds coefficients).

Corollary 9.13. *Let $\mathscr{C}$ be a category with models $\mathscr{M}$, and let F, E be functors from $\mathscr{C}$ to the category of augmented chain complexes.*

(i) *If each F_k is free with base in $\mathscr{M}$ and each model M is* **totally E-acyclic** *(i.e., $\tilde{H}_n(EM) = 0$ for all $n \geq 0$), then there exist natural chain maps $F \to E$ that are augmentation preserving, and any two such are naturally chain homotopic.*

(ii) *If both F_k and E_k are free with bases in $\mathscr{M}$, all $k \geq 0$, and if every model is both totally E-acyclic and totally F-acyclic, then every augmentation preserving natural chain map is a natural chain equivalence.*

Proof. (i) In the proof of Theorem 9.12, replace $F_0 \to H_0F \to 0$ and $E_0 \to H_0E \to 0$ by their respective augmentations, so that $\mathbf{Z}$ now plays the role of H_0 and the identity $\mathbf{Z} \to \mathbf{Z}$ plays the role of φ: $H_0F \to H_0E$ (the only properties of H_0F and H_0E used in the proof are shared by augmentations, namely, commutativity of the square

$$
\begin{array}{ccc}
F_0 & \longrightarrow & \mathbf{Z} \\
\downarrow & & \downarrow{\scriptstyle 1} \\
E_0 & \longrightarrow & \mathbf{Z}
\end{array}
$$

and exactness of $E_1M \to E_0M \to \mathbf{Z} \to 0$ for all models M: since each model M is totally E-acyclic, $0 = \tilde{H}_0(EM) = \ker \varepsilon/\operatorname{im}(E_1M \to E_0M)$). There is thus a natural chain homotopy between any two augmentation preserving natural chain maps.

(ii) If both F and E are free and acyclic, then there are augmentation preserving natural chain maps τ: $F \to E$ and σ: $E \to F$. But the identity chain map 1: $F \to F$ is also augmentation preserving, so that uniqueness says that there is a natural chain homotopy $\sigma\tau \simeq 1_F$; similarly, $\tau\sigma \simeq 1_E$. Therefore τ and σ are natural chain equivalences. □

Exercise

*9.15. (i) The "large" simplicial chain functors F_p in Example 9.5 can be assembled (with the usual alternating sum differentiations) to form a functor

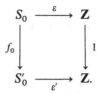

$F: \mathscr{K} \to$ **Comp**. Prove that F is naturally chain equivalent to the singular chain complex functor (restricted to the category of finite simplicial complexes).

(ii) Using Theorem 7.22, prove, for K a finite simplicial complex, that $H_*(K)$ can be computed either via singular theory, or via the large simplicial chain complex, or via oriented simplicial chain complexes. (As we observed in Example 9.5, the "oriented" functor C_* is not free, so that acyclic models does not apply to prove Theorem 7.22. In this case, Theorem 9.8 applies.)

There is a cheap variant of acyclic models called **acyclic carriers**; rather than deriving it from acyclic models, we prove it directly.

Definition. Let (S_*, ∂) be a free chain complex in which each term S_n has a given basis B_n. If $\alpha \in S_n$ and $\beta \in S_{n-1}$, then β is a **face** of α, denoted by $\beta < \alpha$, if β occurs with nonzero coefficient in $\partial\alpha$.

Let (T_*, Δ) be a chain complex, and let $\varphi: S_* \to T_*$ be a chain map. A **carrier function** for φ is a function E that assigns to each $\gamma \in \bigcup B_n$ a subcomplex $E(\gamma)$ of T_* such that, for all γ,

(i) $E(\gamma)$ is acyclic;
(ii) if $\gamma \in B_n$, then $\varphi_n(\gamma) \in E_n(\gamma) \subset T_n$;
(iii) if $\beta < \gamma$, then $E(\beta) \subset E(\gamma)$.

Carrier functions arise as follows. Let K and L be simplicial complexes, and let ξ be a function that assigns to each simplex $s \in K$ an acyclic subcomplex $\xi(s)$ of L such that $s' \subset s$ implies $\xi(s') \subset \xi(s)$. It is straightforward to check that if $\varphi: C_*(K) \to C_*(L)$ is a chain map for which $\varphi(s) \in C_n(\xi(s))$ whenever s is an n-simplex in K, then $E(s) = C_*(\xi(s))$ is a carrier function for φ.

Theorem 9.14 (Acyclic Carriers). *Let S_* be a free chain complex in which each term has a given basis, and let $\varphi: S_* \to T_*$ be a chain map into some chain complex T_*. If φ has a carrier function and if $\varphi_0: S_0 \to T_0$ is the zero map, then φ is chain homotopic to the zero chain map.*

PROOF. We prove by induction on $p \geq 0$ that there exist homomorphisms $s_p: S_p \to T_{p+1}$ such that:

(1) $\varphi_p = \partial'_{p+1} s_p + s_{p-1}\partial_p$;
(2) $s_i(\gamma) \in E(\gamma)$ for all $\gamma \in S_i$ with $i \leq p$.

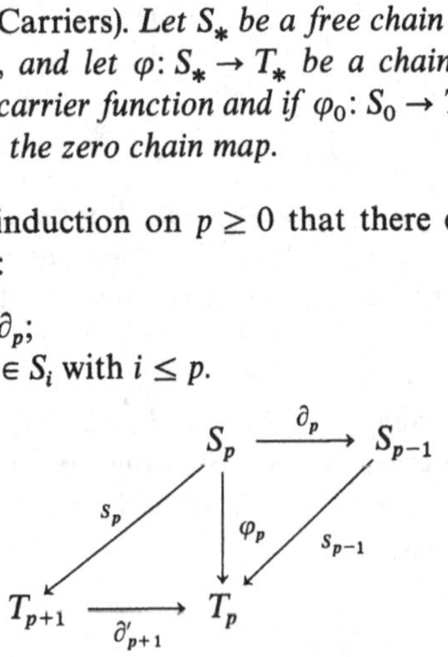

The induction begins by setting $s_{-1} = 0 = s_0$ (here one uses the hypothesis that $\varphi_0 = 0$).

Suppose, by induction, that $s_0, s_1, \ldots, s_p$ have been defined satisfying (1) and (2). To define $s_{p+1}: S_{p+1} \to T_{p+2}$, it suffices to evaluate it on any γ in the basis of S_{p+1}. The boundary $\partial\gamma$ of such a basis element γ lies in S_p, so that $s_p(\partial\gamma)$ is defined. Moreover,

$$s_p(\partial\gamma) \in \langle E(\gamma'): \gamma' < \gamma \rangle \subset E(\gamma)$$

(where $\langle\ \rangle$ means "subcomplex generated by"); the first relation holds by (2) of the inductive hypothesis; the inclusion holds by (iii) of the definition of carrier function. It follows from (ii) that

$$\varphi_{p+1}(\gamma) - s_p(\partial\gamma) \in E(\gamma).$$

Now $\varphi_{p+1}(\gamma) - s_p(\partial\gamma)$ is a cycle in the complex $E(\gamma)$:

$$\begin{aligned}\partial'(\varphi_{p+1}(\gamma) - s_p(\partial\gamma)) &= (\partial'\varphi_{p+1} - \partial' s_p \partial)(\gamma) \\ &= (\partial'\varphi_{p+1} - (\varphi_p - s_{p-1}\partial)\partial)(\gamma) \quad \text{(by (1))} \\ &= (\partial'\varphi_{p+1} - \varphi_p\partial - s\partial\partial)(\gamma) = 0,\end{aligned}$$

the last equality because φ is a chain map. Since $E(\gamma)$ is acyclic, $\varphi_{p+1}(\gamma) - s_p(\partial\gamma)$ must be a boundary: there exists a $(p+2)$-chain $\beta \in E(\gamma)_{p+2}$ with

$$\partial'(\beta) = \varphi_{p+1}(\gamma) - s_p(\partial\gamma);$$

define $s_{p+1}(\gamma) = \beta$. This last formula now reads

$$\partial' s\gamma = \varphi\gamma - s\partial\gamma,$$

and this is (1). Also, (2) holds, because β does lie in $E(\gamma)$. □

Corollary 9.15. *Let S_* and T_* be chain complexes, let S_* be free with each term having a given basis, and let φ and ψ be chain maps $S_* \to T_*$. If $\varphi_0 = \psi_0: S_0 \to T_0$ and if $\varphi - \psi$ has a carrier function, then φ and ψ are chain homotopic.*

Proof. By the theorem, $\varphi - \psi \simeq 0$. □

Lefschetz Fixed Point Theorem

Recall that when we constructed barycentric subdivision Sd in singular theory, we saw (Lemmas 6.12 and 6.13) that $\mathrm{Sd}_\#: S_*(X) \to S_*(X)$ is a chain map that induces the identity map in homology. In simplicial theory, however, $\mathrm{Sd}: K \to \mathrm{Sd}\, K$, hence the chain map $\mathrm{Sd}_\#: C_*(K) \to C_*(\mathrm{Sd}\, K)$ cannot induce the identity map in homology.

Lemma 9.16. *For every simplicial complex K, $\mathrm{Sd}_\#: C_*(K) \to C_*(\mathrm{Sd}\, K)$ induces an isomorphism in reduced homology.*

Proof. By Lemma 7.21, there is a natural chain map $j\colon \tilde{C}_*(K) \to \tilde{S}_*(|K|)$; moreover, by Theorem 7.22, j induces isomorphisms in homology. The result thus follows from commutativity of the diagram

$$\begin{array}{ccc} \tilde{C}_*(K) & \xrightarrow{\mathrm{Sd}_\#} & \tilde{C}_*(\mathrm{Sd}\,K) \\ {\scriptstyle j}\downarrow & & \downarrow{\scriptstyle j} \\ \tilde{S}_*(|K|) & \xrightarrow[\mathrm{Sd}_\#]{} & \tilde{S}_*(|\mathrm{Sd}\,K|), \end{array}$$

because $|\mathrm{Sd}\,K| = |K|$ and the other three maps in the diagram induce isomorphisms. □

It will be convenient to have an explicit description of the inverse of this isomorphism. Recall (Exercise 7.12(ii)) that there exists a simplicial map $\varphi\colon \mathrm{Sd}\,K \to K$ that is a simplicial approximation to the identity $|\mathrm{Sd}\,K| \to |K|$.

Lemma 9.17. *If K is a finite simplicial complex and $\varphi\colon \mathrm{Sd}\,K \to K$ is a simplicial approximation to the identity $|\mathrm{Sd}\,K| \to |K|$, then*

$$\varphi_* = (\mathrm{Sd}_*)^{-1}\colon \tilde{H}_*(\mathrm{Sd}\,K) \xrightarrow{\sim} \tilde{H}_*(K).$$

Proof. Let $\mathscr{A}$ be the category of all subcomplexes of K (with inclusions as the only nonidentity morphisms), and define models $\mathscr{M}$ in $\mathscr{A}$ to be all the simplexes of K. If F is the augmented (large) simplicial chain complex functor of Exercise 9.15 (restricted to $\mathscr{A}$), then each F_p is free with base in $\mathscr{M}$ and each model M is totally F-acyclic.

Define $E\colon \mathscr{A} \to$ **Comp** by setting

$$E(L) = \tilde{C}_*(\mathrm{Sd}\,L)$$

for every subcomplex L of K. Now $\mathrm{Sd}_\#\colon F \to E$ and $\varphi_\#\colon E \to F$ are augmentation preserving natural chain maps (Exercises 6.8 and 7.12(ii)), so that the composite $\varphi_\# \mathrm{Sd}_\#\colon F \to F$ is an augmentation preserving natural chain map. Since F is free with totally acyclic models, Corollary 9.13(ii) shows that $\varphi_\# \mathrm{Sd}_\#$ is naturally chain equivalent to the identity, hence $\varphi_* \mathrm{Sd}_*$ is the identity on $\tilde{H}_*(FK) = \tilde{H}_*(K)$. But Sd_* is an isomorphism, by Lemma 9.16, hence φ_* and Sd_* are inverse. □

Remark. This lemma cannot be proved using Theorem 9.9 in place of acyclic models, because the bottom row of

$$\begin{array}{ccccccccc} \cdots & \longrightarrow & C_1(K) & \xrightarrow{\partial} & C_0(K) & \xrightarrow{\varepsilon} & \mathbf{Z} & \longrightarrow & 0 \\ & & & & & & \downarrow{\scriptstyle 1} & & \\ \cdots & \longrightarrow & C_1(K) & \xrightarrow[\partial]{} & C_0(K) & \xrightarrow[\varepsilon]{} & \mathbf{Z} & \longrightarrow & 0 \end{array}$$

is not exact (it is only a chain complex), and so it is not obvious how to construct a chain homotopy between the identity and $\varphi_{\#}\,\mathrm{Sd}_{\#}$.

The purpose of these lemmas is to prove the Lefschetz fixed point theorem, which gives a sufficient condition that a continuous map f on a compact polyhedron have a fixed point.

Recall that if V is a finite-dimensional vector space over $\mathbf{Q}$ and $T\colon V \to V$ is a linear transformation, then a choice of basis of V associates a square matrix A to T; one defines the **trace** of T, denoted by tr T, to be the trace of A (namely, $\sum a_{ii}$). It is a standard argument that tr T is independent of the choice of basis (and the resulting matrix A). If $0 \to V' \hookrightarrow V \to V'' \to 0$ is a short exact sequence of vector spaces and if $T\colon V \to V$ is a linear transformation with $T(V') \subset V'$, then T induces a linear transformation T'' on V''. In fact, if $\{x_1, \ldots, x_k\}$ is a basis of V', and if one extends it to a basis $\{x_1, \ldots, x_n\}$ of V, then $T''(x_i + V') = Tx_i + V'$ for $i = k + 1, \ldots, n$ (we have identified V'' with V/V'). Moreover,

$$\operatorname{tr} T = \operatorname{tr}(T|V') + \operatorname{tr} T'',$$

for the matrix A of T with respect to the (extended) basis is

$$A = \begin{array}{|c|c|} \hline A' & * \\ \hline 0 & A'' \\ \hline \end{array},$$

where A' is the matrix of $T|V'$ with respect to $\{x_1, \ldots, x_k\}$ and A'' is the matrix of T'' with respect to $\{x_{k+1} + V', \ldots, x_n + V'\}$. The result is now clear, because tr T is just the sum of the diagonal entries of A. It is also easy to see that

$$\operatorname{tr} 1_V = \dim V.$$

The notion of trace can also be defined for endomorphisms of f.g. free abelian groups, and even for endomorphisms of arbitrary f.g. abelian groups. If G is free abelian with basis $\{x_1, \ldots, x_n\}$, then a homomorphism $f\colon G \to G$ is completely determined by the $n \times n$ matrix A over $\mathbf{Z}$, where $f(x_i) = \sum a_{ji}x_j$ and $A = [a_{ij}]$. A different choice of basis $\{y_1, \ldots, y_n\}$ of G replaces A by $P^{-1}AP$, where the ith column of P expresses y_i as a $\mathbf{Z}$-linear combination of the x_j's; it follows that tr f defined as tr A is independent of the choice of basis (and of the matrix A). Finally, if G is any f.g. abelian group and $f\colon G \to G$ is a homomorphism, then $f(tG) \subset tG$ (where tG is the torsion subgroup of G), and so f induces a homomorphism $\bar{f}\colon G/tG \to G/tG$, namely, $x + tG \mapsto f(x) + tG$. Observe that G/tG is free abelian, because it is f.g. with no (nonzero) elements of finite order.

Definition. If G is a f.g. abelian group and $f\colon G \to G$ is a homomorphism, then the **trace** of f, denoted by tr f, is defined to be tr $\bar{f}$, where $\bar{f}$ is the induced homomorphism on the f.g. free abelian group G/tG.

EXERCISES

9.16. Prove that $\operatorname{tr}(1_G) = \operatorname{rank} G$.

*9.17. If $0 \to G' \xrightarrow{i} G \xrightarrow{p} G'' \to 0$ is an exact sequence of f.g. abelian groups and $f: G \to G$ is a homomorphism with $f(G') \subset G'$, then

$$\operatorname{tr}(f|G') + \operatorname{tr} f'' = \operatorname{tr} f,$$

where $f'': G'' \to G''$ is induced by f (if $x'' \in G''$ and $x \in G$ satisfies $px = x''$, then define $f''x'' = pfx$).

Definition. Let $G_0, G_1, \ldots, G_m$ be a sequence of f.g. abelian groups and let $h = (h_0, h_1, \ldots, h_m)$, where $h_i: G_i \to G_i$ is a homomorphism for every i. The **Lefschetz number** of h, denoted by $\lambda(h)$, is

$$\lambda(h) = \sum_{i=0}^{m} (-1)^i \operatorname{tr} h_i.$$

EXAMPLE 9.6. Let K be an m-dimensional (finite) simplicial complex, and let $f: K \to K$ be a simplicial map. Let $f_\# = (f_{0\#}, \ldots, f_{m\#})$, where $f_{i\#}: C_i(K) \to C_i(K)$ is the ith term of the chain map $f_\#$. Then

$$\lambda(f_\#) = \sum_{i=0}^{m} (-1)^i \operatorname{tr} f_{i\#}.$$

EXAMPLE 9.7. In the above example, let $f = 1_K$. Then $\operatorname{tr} f_{i\#} = \operatorname{rank} C_i(K)$, so that $\lambda(1_K)$ is the Euler–Poincaré characteristic of K.

EXAMPLE 9.8. If $f: K \to K$ (as in Example 9.6), let $f_* = (f_{0*}, \ldots, f_{m*})$, where $f_{i*}: H_i(K) \to H_i(K)$. Then

$$\lambda(f_*) = \sum_{i=0}^{m} (-1)^i \operatorname{tr} f_{i*}.$$

Lemma 9.18. *Let C be a chain complex of the form $0 \to C_m \to \cdots \to C_0 \to 0$ in which each C_i is f.g., and let $f: C \to C$ be a chain map. Then*

$$\lambda(f) = \lambda(f_*),$$

where $f_{i*}: H_i(C) \to H_i(C)$.

PROOF. Imitate the proof of the corresponding result for the Euler–Poincaré characteristic (Theorem 7.15), using Exercise 9.17 at appropriate moments. □

It follows that if $f: X \to X$ is a continuous map on a compact polyhedron, then $\lambda(f)$ defined by $\lambda(f_*)$ is a well defined number, independent of any triangulation of X.

Theorem 9.19 (Lefschetz). *Let X be a compact polyhedron and let $f: X \to X$ be continuous. If $\lambda(f) \neq 0$, then f has a fixed point.*

PROOF. Assume that f has no fixed points, so that compactness of X provides $\delta > 0$ with $\|x - f(x)\| \geq \delta$ for all $x \in X$. Let $X = |K|$ for some finite simplicial complex K, and choose n so that mesh $\mathrm{Sd}^n K < \frac{1}{2}\delta$. Choose t so that there is a simplicial approximation $g\colon \mathrm{Sd}^{n+t} K \to \mathrm{Sd}^n K$ to f. If $h\colon \mathrm{Sd}^{n+t} K \to \mathrm{Sd}^n K$ is a simplicial approximation to the identity $|\mathrm{Sd}^{n+t} K| \to |\mathrm{Sd}^n K|$, then $|g| \simeq f \simeq f|h|$ and $|g|_* = f_*|h|_*$. Iterated application of Lemma 9.17 gives $h_* = (\mathrm{Sd}^t_*)^{-1}$, hence

$$g_\# \mathrm{Sd}^t_\#\colon C_*(\mathrm{Sd}^n K) \to C_*(\mathrm{Sd}^n K)$$

is a chain map inducing f_* (actually, since Lemma 9.17 applies only to reduced homology, this is so for all subscripts $n > 0$; however, it holds trivially when $n = 0$ because Sd is the identity on 0-simplexes).

If σ is a p-simplex in $\mathrm{Sd}^n K$, then $g_p \mathrm{Sd}^t_p(\sigma) = \sum m_i \tau_i$, where $m_i \in \mathbf{Z}$ and each τ_i is a p-simplex. If, for every simplex σ, none of the τ_i is σ, then the definition of trace gives $\lambda(g_\# \mathrm{Sd}^t_\#) = 0$, hence $\lambda(f_*) = \lambda(g_* \mathrm{Sd}^t_*) = 0$, by Lemma 9.18. Suppose, on the contrary, that some $\tau_{i_0} = \sigma$ for some p and some p-simplex σ. Since $\mathrm{Sd}^t_p(\sigma) = \sum m_i \rho_i$, where each ρ_i is a p-simplex with $|\rho_i| \subset |\sigma|$, it follows that $g_\# \rho_{i_0} = |g|(\rho_{i_0}) \subset |\sigma|$. Hence there is $x \in |\sigma|$ with $|g|(x) \in |\sigma|$ (namely, any $x \in |\rho_{i_0}|$); that is, $\|x - |g|(x)\| \leq \text{mesh } \mathrm{Sd}^n K < \frac{1}{2}\delta$. But Exercise 7.10 (essentially the definition of simplicial approximation) gives $\||g|(x) - f(x)\| < \frac{1}{2}\delta$. Thus

$$\|x - f(x)\| \leq \|x - |g|(x)\| + \||g|(x) - f(x)\| < \delta,$$

and this contradicts the definition of δ. □

Corollary 9.20. *Let X be a path connected compact polyhedron for which $H_n(X)$ is finite for every $n > 0$. Then every $f\colon X \to X$ has a fixed point.*

PROOF. Since rank $H_n(X) = 0$ for $n > 0$, it follows that $\bar{f}_{n*} = 0$ for all $n > 0$ where $\bar{f}_{n*}$ is the homomorphism induced by f_{n*} on $H_n(X)/tH_n(X)$; therefore $\mathrm{tr}\, f_{n*} = 0$ for all $n > 0$ and

$$\lambda(f) = \mathrm{tr}\, f_{0*}.$$

Since X is path connected, $H_0(X) \cong \mathbf{Z}$ and $\mathrm{tr}\, f_{0*} \neq 0$ (indeed f_{0*} is the identity, by Theorem 4.14(iii), and so $\mathrm{tr}\, f_{0*} = 1$). The result is now immediate from the Lefschetz theorem. □

Corollary 9.21. *If n is even, every $f\colon \mathbf{R}P^n \to \mathbf{R}P^n$ has a fixed point.*

PROOF. We saw in Theorem 8.47 that $H_q(\mathbf{R}P^n) = 0$ or $\mathbf{Z}/2\mathbf{Z}$ for all $q > 0$. □

Corollary 9.22. *If X is a compact contractible polyhedron, then every $f\colon X \to X$ has a fixed point.*

PROOF. Immediate from Corollary 9.20, because $H_n(X) = 0$ for all $n > 0$. □

Of course, Corollary 9.22 implies the Brouwer fixed point theorem.

Corollary 9.23. *If X is a compact contractible polyhedron with more than one point, then there is no multiplication $\mu: X \times X \to X$ making X a topological group.*

PROOF. Suppose X were a topological group. Choose $y \in X$ with $y \neq 1$; then $\varphi: X \to X$ defined by $\varphi(x) = xy$ $(= \mu(x, y))$ is a continuous map (even a homeomorphism) having no fixed points. □

Corollary 9.24 (= Theorem 6.23). *If $n \geq 1$, then the antipodal map $a: S^n \to S^n$ has degree $(-1)^{n+1}$.*

PROOF. By definition, degree $a = d$, where $a_{n*}: H_n(S^n) \to H_n(S^n)$ is multiplication by d. Thus tr $a_{n*} = d =$ degree a, so that

$$\lambda(a) = 1 + (-1)^n d.$$

But $\lambda(a) = 0$ because the antipodal map has no fixed points; therefore $d = (-1)^{n+1}$. □

There is a survey article [Bing] in which the following simple example is given. Let X denote a circle in the plane with a spiral converging to it.

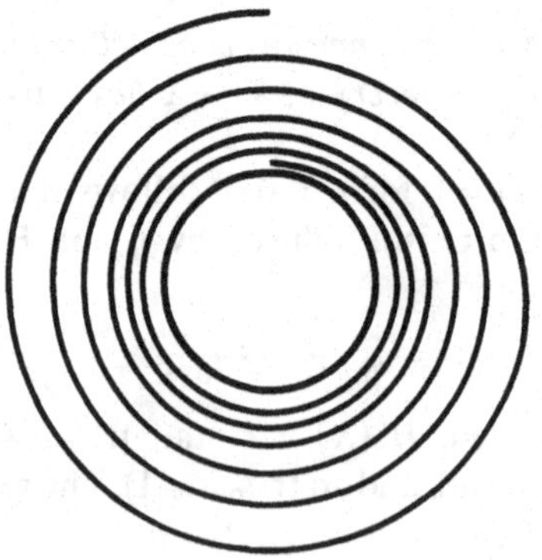

Then CX, the cone on X, is a contractible space that does not have the fixed point property.

EXERCISE

9.18. If $f: S^n \to S^n$ is a continuous map that is not a homotopy equivalence, then f has a fixed point.

Tensor Products

The last topic in this chapter answers the question: What is $H_*(X \times Y)$ in terms of $H_*(X)$ and $H_*(Y)$? The ultimate answer is quite satisfactory, and it involves a nice mixture of algebra and topology. First, we need the notion of tensor product (which we shall define) and some results from homological algebra (which we shall quote). The link with algebraic topology is the Eilenberg–Zilber theorem, whose heart is an application of acyclic models.

Definition. Let A and B be abelian groups. Their **tensor product**, denoted by $A \otimes B$, is the abelian group having the following presentation:

Generators: $A \times B$, that is, all ordered pairs (a, b).
Relations: $(a + a', b) = (a, b) + (a', b)$ and $(a, b + b') = (a, b) + (a, b')$ for all $a, a' \in A$ and all $b, b' \in B$.

If F is the free abelian group with basis $A \times B$ and if N is the subgroup of F generated by all relations, then $A \otimes B = F/N$. We denote the coset $(a, b) + N$ by $a \otimes b$. Observe that a typical element of $A \otimes B$ thus has an expression of the form $\sum m_i(a_i \otimes b_i)$ for $m_i \in \mathbf{Z}$. Indeed one can dispense with the m_i because of Exercise 9.20 below.

EXERCISES

9.19. $a \otimes 0 = 0 = 0 \otimes b$ for all $a \in A$ and $b \in B$.

*9.20. If $m \in \mathbf{Z}$, then $m(a \otimes b) = (ma) \otimes b = a \otimes (mb)$.

*9.21. If A is torsion, then $A \otimes \mathbf{Q} = 0$. (*Hint*: If $a \in A$, then $ma = 0$ for some $m > 0$; if $q \in \mathbf{Q}$, then $a \otimes q = a \otimes m(q/m) = ma \otimes (q/m) = 0$.)

*9.22. If A and B are finite abelian groups whose orders are relatively prime, then $A \otimes B = 0$.

A definition by generators and relations, though displaying elements, is difficult to work with; for example, it is usually unclear whether or not a given element is zero. A worse defect is that one does not understand what purpose the construction is to serve.

Definition. Let A, B, and G be abelian groups. A **bilinear function** $\varphi: A \times B \to G$ is a function such that

$$\varphi(a + a', b) = \varphi(a, b) + \varphi(a', b)$$

and

$$\varphi(a, b + b') = \varphi(a, b) + \varphi(a, b')$$

for all $a, a' \in A$ and all $b, b' \in B$.

The natural map $v: A \times B \to A \otimes B$ taking (a, b) into $a \otimes b$ is bilinear.

The next result states that $A \otimes B$ is a group (indeed is the only group) that converts bilinear functions into ordinary (linear) homomorphisms.

Theorem 9.25.

(i) *Given any abelian group G and any bilinear map $\varphi: A \times B \to G$, there exists a unique homomorphism $f: A \otimes B \to G$ making the following diagram commute:*

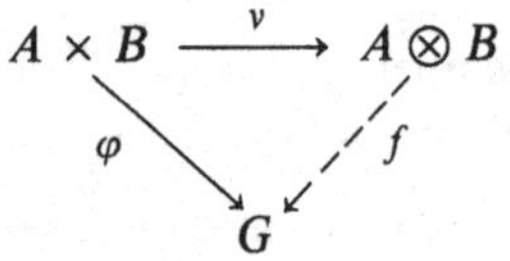

(v is the natural map $(a, b) \mapsto a \otimes b$).

(ii) *$A \otimes B$ is the only group with this property; that is, if T is an abelian group and $\eta: A \times B \to T$ is a bilinear map such that the diagram*

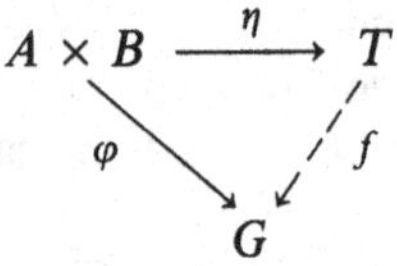

always has a unique "completion" f, then $T \cong A \otimes B$.

PROOF. (i) Recall that $A \otimes B = F/N$, where F is free abelian with basis $A \times B$ (and N is generated by certain relations). Consider the diagram

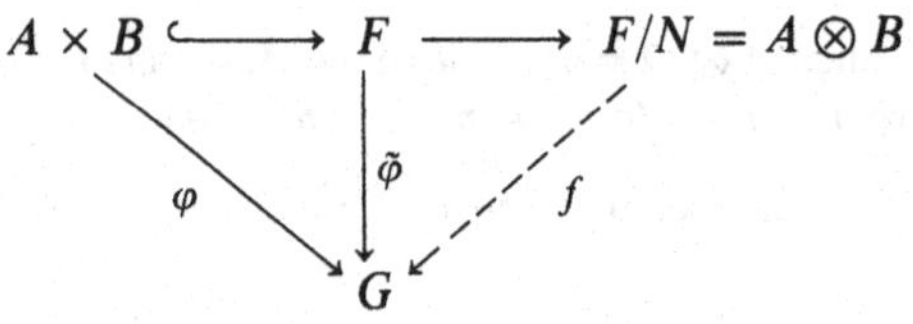

Define $\tilde{\varphi}: F \to G$ by extending by linearity. The relations N are such that $N \subset \ker \tilde{\varphi}$, and so $\tilde{\varphi}$ induces a homomorphism $f: F/N \to G$, namely, $f: (a, b) + N \mapsto \tilde{\varphi}(a, b) = \varphi(a, b)$. In other words, $f(a \otimes b) = \varphi(a, b)$ for every $(a, b) \in A \times B$. Such a homomorphism f is unique, for the set of all $a \otimes b$ generates $A \otimes B$.

(ii) Consider the following diagram:

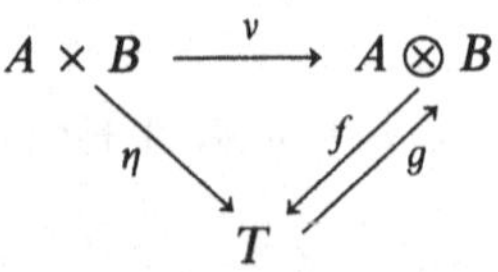

By hypothesis, there exist homomorphisms $f: A \otimes B \to T$ and $g: T \to A \otimes B$ with $fv = \eta$ and $g\eta = v$. Now consider the diagram

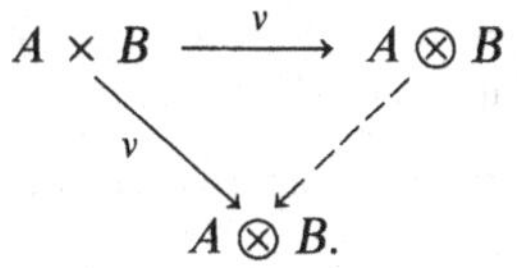

Both gf and the identity complete the diagram, so that uniqueness of the completion gives gf = identity. A similar diagram shows that $fg = 1_T$, hence f and g are isomorphisms. □

Theorem 9.26. *Let $f: A \to A'$ and $g: B \to B'$ be homomorphisms.*

(i) *There is a unique homomorphism $A \otimes B \to A' \otimes B'$, denoted by $f \otimes g$, with $a \otimes b \mapsto fa \otimes gb$ for every $a \in A$ and $b \in B$.*

(ii) *If $f': A' \to A''$ and $g': B' \to B''$ are homomorphisms, then $(f' \otimes g') \circ (f \otimes g) = (f' \circ f) \otimes (g' \circ g)$.*

Proof. (i) The function $\varphi: A \times B \to A' \otimes B'$ defined by $\varphi(a, b) = fa \otimes gb$ is easily seen to be bilinear. By Theorem 9.25(i), there is a unique homomorphism $A \otimes B \to A' \otimes B'$ with $a \otimes b \mapsto \varphi(a, b) = fa \otimes gb$.

(ii) Both maps complete the diagram

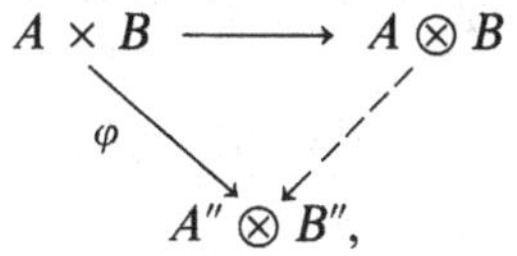

where $\varphi(a, b) = f'(f(a)) \otimes g'(g(b))$. □

Corollary 9.27. *Let A be a fixed abelian group. There is a functor $T = T_A$:* **Ab** $\to$ **Ab** *such that $T(B) = A \otimes B$ and $T(f) = 1_A \otimes f$.*

Proof. That T preserves composites follows from Theorem 9.26(ii):

$$(1_A \otimes f') \circ (1_A \otimes f) = 1_A \otimes f'f;$$

that $1_A \otimes 1_B = 1_{A \otimes B}$ follows from Theorem 9.25(i). □

One usually denotes the functor T_A by $A \otimes$ —.

Exercises

9.23. For a fixed abelian group B, show that there is a functor $F = F_B$: **Ab** $\to$ **Ab** such that $F(A) = A \otimes B$ and $F(g) = g \otimes 1_B$. (One usually denotes this functor by — $\otimes B$.)

9.24. (i) Prove that there is an isomorphism $A \otimes B \to B \otimes A$ taking $a \otimes b \mapsto b \otimes a$.
(ii) For any abelian group A, the functors $A \otimes$ — and — $\otimes A$ are isomorphic.

*9.25. Prove that the tensor product functor T_A (and F_B) is additive. Conclude that if $f: B \to B'$ is the zero map ($f = 0$), then $T(f) = 0$, and that if $B = \{0\}$, then $T(B) = \{0\}$.

*9.26. If $f: B \to B$ is multiplication by an integer m, that is, $f(b) = mb$ for all $b \in B$, then $1_A \otimes f$ is also multiplication by m.

*9.27. (i) For every abelian group A, there is an isomorphism $\tau_A: \mathbf{Z} \otimes A \to A$ with $n \otimes a \mapsto na$.

(ii) Show that the family of all τ_A comprise a natural equivalence between $\mathbf{Z} \otimes \_\_$ and the identity functor on **Ab**.

We remind the reader of a property of direct sums of abelian groups. Suppose that $\{B_j: j \in J\}$ is a (possibly infinite) family of abelian groups and that $\{f_j: B_j \to G \mid j \in J\}$ is a family of homomorphisms into some abelian group G. There exists a unique homomorphism $f: \sum B_j \to G$ with $f|B_j = f_j$ for all $j \in J$.

Theorem 9.28. *There is an isomorphism* $A \otimes \sum B_j \to \sum (A \otimes B_j)$ *with* $a \otimes (b_j) \mapsto (a \otimes b_j)$.

Proof. First, the function $\eta: A \times \sum B_j \to \sum (A \otimes B_j)$ defined by $(a, (b_j)) \mapsto (a \otimes b_j)$ is bilinear. Consider the diagram with φ bilinear

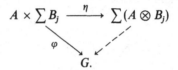

$$\begin{array}{ccc} A \times \sum B_j & \xrightarrow{\eta} & \sum (A \otimes B_j) \\ & {}_{\varphi}\searrow \quad \swarrow & \\ & G. & \end{array}$$

For each j, define $\varphi_j: A \times B_j \to G$ by $(a, b_j) \mapsto \varphi(a, \bar{b}_j)$, where $\bar{b}_j \in \sum B_j$ has b_j in the jth coordinate and 0 elsewhere. It is easy to see that each φ_j is bilinear, so there exists a homomorphism $f_j: A \otimes B_j \to G$ with $a \otimes b_j \mapsto \varphi(a, \bar{b}_j)$. Our remarks about direct sums show that there is a homomorphism $f: \sum (A \otimes B_j) \to G$ with $\sum a \otimes b_j \to \sum f_j(a \otimes b_j) = \sum \varphi(a, \bar{b}_j) = \varphi(a, \sum \bar{b}_j) = \varphi(a, (b_j))$. It follows that $f\eta = \varphi$, so Theorem 9.25(ii) gives $A \otimes \sum B_j \cong \sum (A \otimes B_j)$. But this last theorem not only asserts that an isomorphism exists; it also constructs one (with a commutative diagram). The reader can now show that this isomorphism does send $a \otimes (b_j)$ into $(a \otimes b_j)$. □

Note that Exercise 9.13 shows only that there is some isomorphism $A \otimes \sum B_j \cong \sum (A \otimes B_j)$ whenever there are only finitely many summands, and so it is a weaker result than Theorem 9.28.

Universal Coefficients

If (C_*, ∂) is a complex, then so is $(C_* \otimes G, \partial \otimes 1_G)$ (for any fixed abelian group G), because the composite of any two consecutive maps in

$$\cdots \longrightarrow C_{n+1} \otimes G \xrightarrow{\partial_{n+1} \otimes 1} C_n \otimes G \xrightarrow{\partial_n \otimes 1} C_{n-1} \otimes G \longrightarrow \cdots$$

is zero, thanks to the additivity of $\_\_ \otimes G$ (Exercise 9.25).

Definition. Let (X, A) be a pair of spaces and let G be an abelian group. If $(S_*(X, A), \partial)$ is the singular chain complex of (X, A), then the singular complex with **coefficients** G is the complex

$$\to S_{n+1}(X, A) \otimes G \xrightarrow{\partial \otimes 1} S_n(X, A) \otimes G \xrightarrow{\partial \otimes 1} S_{n-1}(X, A) \otimes G \to \cdots.$$

The nth **homology group** of (X, A) with **coefficients** G is

$$H_n(X, A; G) = \ker(\partial_n \otimes 1)/\operatorname{im}(\partial_{n+1} \otimes 1).$$

The word "coefficient" is suggested by the definition of tensor product, for a typical n-chain in $S_n(X, A) \otimes G$ has the form $\sum \beta_i \otimes g_i$, where $\beta_i \in S_n(X, A)$ and $g_i \in G$; the elements g_i do resemble coefficients.

Here is one way such a construction arises in topology. Given a field F and a space X, construct a vector space analogue of $S_*(X)$, call it $S_*(X, F)$, as follows: $S_n(X, F)$ is the F-vector space with basis all n-simplexes in X; the differentiations are F-linear transformations defined on basis vectors as the usual alternating sum. Note that the subgroups of cycles and of boundaries are F-vector spaces, since they are, respectively, kernels and images of F-linear transformations. It follows that the homology groups are now F-vector spaces. This can be convenient, because it allows one to use linear algebra. We have already alluded to this in Chapter 7 when we mentioned how to find the Betti numbers of finite polyhedra in terms of ranks and nullities of certain matrices.

Coefficients make a more serious appearance in *obstruction theory* (see [Spanier, Chap. 8]) where the coefficients are certain homotopy groups. But the major reason one needs coefficients is for spectral sequences, the most powerful method of computing homology groups (see [McCleary]). In the very statements of its theorems, one sees terms of the form $H_p(X, H_q(Y))$, homology groups of X with coefficients $H_q(Y)$.

For every abelian group G, $H_*(\ ; G)$ is a homology theory with coefficient group G. The proof is straightforward, using Theorem 9.29 (the interested reader may look at the corresponding result for cohomology, Theorems 12.3, 12.4, 12.9, and 12.10).

The first question is the relation of homology with coefficients to ordinary homology. The optimistic guess is that $H_n(X, A; G) \cong H_n(X, A) \otimes G$; unfortunately, this is always true only for certain G, namely, G torsion-free, and so it is usually false. This question eventually leads to an algebraic question: How does $\_\_ \otimes G$ affect exact sequences?

Theorem 9.29. *If $B' \xrightarrow{i} B \xrightarrow{p} B'' \to 0$ is an exact sequence of abelian groups, then for every abelian group A, there are exact sequences*

$$A \otimes B' \xrightarrow{1 \otimes i} A \otimes B \xrightarrow{1 \otimes p} A \otimes B'' \longrightarrow 0$$

and

$$B' \otimes A \xrightarrow{i \otimes 1} B \otimes A \xrightarrow{p \otimes 1} B'' \otimes A \longrightarrow 0.$$

PROOF. (i) $\operatorname{im}(1 \otimes i) \subset \ker(1 \otimes p)$.

It suffices to prove that $(1 \otimes p)(1 \otimes i) = 0$; but this composite is equal to $1 \otimes pi = 1 \otimes 0 = 0$.

(ii) $\ker(1 \otimes p) \subset \operatorname{im}(1 \otimes i)$.

If we denote $\operatorname{im}(1 \otimes i)$ by E, then $1 \otimes p$ induces a map $\bar{p}\colon (A \otimes B)/E \to A \otimes B''$ given by $a \otimes b + E \mapsto a \otimes pb$ (since $E \subset \ker(1 \otimes p)$, by (i)). It is easily seen that $1 \otimes p = \bar{p}\pi$, where $\pi\colon A \otimes B \to (A \otimes B)/E$ is the natural map.

If $\bar{p}$ is an isomorphism, then

$$\ker(1 \otimes p) = \ker \bar{p}\pi = \ker \pi = E = \operatorname{im}(1 \otimes i),$$

as desired. Let us construct a map $A \otimes B'' \to (A \otimes B)/E$ inverse to $\bar{p}$. The function $f\colon A \times B'' \to (A \otimes B)/E$ defined by

$$f(a, b'') = a \otimes b + E,$$

where $pb = b''$, is well defined: such an element $b \in B$ exists because p is a surjection; if $pb_1 = b'' = pb$, then $b_1 - b \in \ker p = \operatorname{im} i$, hence $b_1 - b = ib'$ for some $b' \in B'$, and

$$a \otimes b_1 - a \otimes b = a \otimes (b_1 - b) = (1 \otimes i)(a \otimes b') \in \operatorname{im}(1 \otimes i) = E.$$

Now f is easily seen to be bilinear, so that Theorem 9.25(i) gives a homomorphism $\bar{f}\colon A \otimes B'' \to (A \otimes B)/E$ with $\bar{f}(a \otimes b'') = a \otimes b + E$ (where $pb = b''$). It is plain that $\bar{f}$ and $\bar{p}$ are inverse functions.

(iii) $1 \otimes p$ is a surjection.

If $\sum a_i \otimes b_i'' \in A \otimes B''$, then surjectivity of p provides elements $b_i \in B$, for all i, with $pb_i = b_i''$, and

$$(1 \otimes p)(\textstyle\sum a_i \otimes b_i) = \sum a_i \otimes b_i''.$$

Proof of exactness of the second sequence is similar to that just given. □

Note that there is no zero at the left, nor need there be one even under the extra hypothesis that i is injective.

EXAMPLE 9.9. Consider the short exact sequence $0 \to \mathbf{Z} \xrightarrow{i} \mathbf{Q} \xrightarrow{p} \mathbf{Q}/\mathbf{Z} \to 0$, and let G be a torsion group. Now Exercise 9.27(i) shows that $\mathbf{Z} \otimes G \cong G$, while Exercise 9.21 shows that $\mathbf{Q} \otimes G = 0$. Thus there can be no injection $\mathbf{Z} \otimes G \to \mathbf{Q} \otimes G$, and so, in particular, $i \otimes 1$ is not injective.

Corollary 9.30.

(i) *Let $m > 0$. For any abelian group G,*

$$(\mathbf{Z}/m\mathbf{Z}) \otimes G \cong G/mG.$$

(ii) *If m, n are integers with $(m, n) = d$ (i.e., gcd $= d$), then*

$$(\mathbf{Z}/m\mathbf{Z}) \otimes (\mathbf{Z}/n\mathbf{Z}) \cong \mathbf{Z}/d\mathbf{Z}.$$

PROOF. (i) Apply $\_\_ \otimes G$ to the short exact sequence

$$0 \to \mathbf{Z} \xrightarrow{m} \mathbf{Z} \xrightarrow{p} \mathbf{Z}/m\mathbf{Z} \to 0$$

(where the first map is multiplication by m) to obtain exactness of

$$G \xrightarrow{m} G \xrightarrow{p \otimes 1} (\mathbf{Z}/m\mathbf{Z}) \otimes G \longrightarrow 0$$

(we have used Exercises 9.26 and 9.27). The first isomorphism theorem now applies: $G/\ker(p \otimes 1) \cong \operatorname{im} p \otimes 1$. But $\ker(p \otimes 1) = \operatorname{im} m = mG$, and $\operatorname{im} p \otimes 1 = (\mathbf{Z}/m\mathbf{Z}) \otimes G$.

(ii) The proof that G cyclic of order n implies G/mG cyclic of order $d = (m, n)$ is left to the reader. □

Observe that Exercise 9.22 is a consequence of this corollary.

Corollary 9.31. *If A and B are known f.g. abelian groups, then $A \otimes B$ is also known.*

PROOF. A and B, being f.g. abelian groups, have decompositions as direct sums of cyclic groups. When we say these groups are "known", we mean that we know such decompositions of each. The result now follows from Corollary 9.30 and Theorem 9.28. □

EXERCISES

*9.28. (i) Let F and F' be free abelian groups with bases $\{x_j : j \in J\}$ and $\{x'_k : k \in K\}$, respectively. Then $F \otimes F'$ is free abelian with basis $\{x_j \otimes x'_k : j \in J, k \in K\}$. (*Hint*: Theorem 9.28.)

(ii) If F and F' are f.g. free abelian groups, then $F \otimes F'$ is a f.g. free abelian group and rank $F \otimes F' = (\text{rank } F)(\text{rank } F')$.

9.29. Compute $A \otimes B$, where $A = \mathbf{Z} \oplus \mathbf{Z} \oplus \mathbf{Z}/6\mathbf{Z} \oplus \mathbf{Z}/5\mathbf{Z}$ and $B = \mathbf{Z}/3\mathbf{Z} \oplus \mathbf{Z}/5\mathbf{Z}$.

Evaluating $\ker(A' \otimes G \to A \otimes G)$, where $0 \to A' \to A \to A'' \to 0$ is a short exact sequence of abelian groups, is one of the basic problems of homological algebra.

Definition. For each abelian group A, choose an exact sequence $0 \to R \xrightarrow{i} F \to A \to 0$ with F free abelian. For any abelian group B, define

$$\operatorname{Tor}(A, B) = \ker(i \otimes 1_B).$$

Note that R is free abelian, by Theorem 9.3. Choosing bases of F and of R thus gives a presentation of A by generators and relations.

(We can view this construction in a sophisticated way. If we delete A, then $C_* = 0 \to R \to F \to 0$ is a chain complex, as is $C_* \otimes B = 0 \to R \otimes B \to F \otimes B \to 0$ (just attach a sequence of zeros). Hence $\operatorname{Tor}(A, B) = H_1(C_* \otimes B)$. For fixed B, $\operatorname{Tor}(\quad, B)$ is even a (covariant) functor. If $f\colon A \to A'$ is a homomorphism, then Theorem 9.9 asserts the existence of the dashed arrows making the diagram below commutative.

$$\begin{array}{ccccccccc} 0 & \longrightarrow & R & \longrightarrow & F & \longrightarrow & A & \longrightarrow & 0 \\ & & \downarrow & & \downarrow & & \downarrow{\scriptstyle f} & & \\ 0 & \longrightarrow & R' & \longrightarrow & F' & \longrightarrow & A' & \longrightarrow & 0. \end{array}$$

The dashed arrows constitute a chain map over f; moreover, after being tensored by 1_B, they constitute a chain map $C_* \otimes B \to C'_* \otimes B$ (where C'_* is the complex $0 \to R' \to F' \to 0$). One defines $f_*\colon \text{Tor}(A, B) \to \text{Tor}(A', B)$ as the homomorphism $H_1(C_* \otimes B) \to H_1(C'_* \otimes B)$ induced by this chain map.)

Of course, there is an obvious question. Is the definition of Tor independent of the choice of exact sequence $0 \to R \to F \to A \to 0$? The answer is "yes". In fact one can even work on the second variable: if $0 \to R' \xrightarrow{j} F' \to B \to 0$ is exact, then $\ker(i \otimes 1_B) \cong \ker(1_A \otimes j)$. Proofs of these facts can be found in any book on homological algebra.

The reader may yearn for a less sophisticated description of Tor(A, B). Here is a presentation of it. As generators, take all symbols $\langle a, m, b\rangle$, where $a \in A$, $b \in B$, $m \in \mathbf{Z}$, and $ma = 0 = mb$. These generators are subject to the following relations:

$$\langle a, m, b + b'\rangle = \langle a, m, b\rangle + \langle a, m, b'\rangle \quad \text{if } ma = 0 = mb = mb';$$

$$\langle a + a', m, b\rangle = \langle a, m, b\rangle + \langle a', m, b\rangle \quad \text{if } ma = ma' = 0 = mb;$$

$$\langle a, mn, b\rangle = \langle ma, n, b\rangle \quad \text{if } mna = 0 = nb;$$

$$\langle a, mn, b\rangle = \langle a, m, nb\rangle \quad \text{if } ma = 0 = mnb.$$

With this description of Tor(A, B), it is easy to define the map f_* induced by $f\colon A \to A'$; send the coset of $\langle a, m, b\rangle$ into the coset of $\langle fa, m, b\rangle$. It also follows that Tor(A, B) is a torsion group for all A and B (this is the etymology of Tor).

Here are the basic properties of Tor.

For each fixed abelian group B, Tor$(\;\,, B)$: **Ab** $\to$ **Ab** is an additive (covariant) functor satisfying the following:

[Tor 1]. If $0 \to A' \to A \to A'' \to 0$ is a short exact sequence, then there is an exact sequence

$$0 \to \text{Tor}(A', B) \to \text{Tor}(A, B) \to \text{Tor}(A'', B) \to A' \otimes B \to A \otimes B \to A'' \otimes B \to 0.$$

[Tor 2]. If A is torsion-free, then Tor$(A, B) = 0$ for any B.
[Tor 3]. $\text{Tor}(\sum A_j, B) \cong \sum \text{Tor}(A_j, B)$ and $\text{Tor}(A, \sum B_i) \cong \sum \text{Tor}(A, B_i)$.
[Tor 4]. $\text{Tor}(\mathbf{Z}/m\mathbf{Z}, B) \cong B[m] = \{b \in B\colon mb = 0\}$.
[Tor 5]. $\text{Tor}(A, B) \cong \text{Tor}(B, A)$ for all A and B.

Using these properties, one can compute Tor(A, B) whenever A and B are f.g. abelian groups. Indeed [Tor 4] shows that Tor(A, B) is finite in this case.

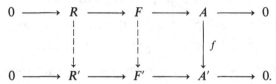

Remark. As with tensor product, fixing the first variable of Tor gives an additive functor $\text{Tor}(A, \quad): \mathbf{Ab} \to \mathbf{Ab}$, and the value of $\text{Tor}(A, \quad)$ on B is (isomorphic to) the value of $\text{Tor}(\quad, B)$ on A.

Proofs of the properties [Tor i], $1 \leq i \leq 5$ (and of the theorem mentioned in the remark), can be found in books on homological algebra.

Let A be an abelian group, and let $0 \to R \to F \to A \to 0$ be an exact sequence with F free abelian. For any abelian group B, $\text{Tor}(F, B) = 0$, by [Tor 2], and so the exact sequence given by [Tor 1] shows that $\text{Tor}(A, B) \cong \ker(R \otimes B \to F \otimes B)$; we have recaptured the definition of Tor.

Exercises

9.30. (i) For any abelian group G, prove that $\mathbf{Q} \otimes G$ is a vector space over $\mathbf{Q}$. (*Hint*: $\mathbf{Q} \otimes G$ is an abelian group admitting scalar multiplication by rational numbers.) Conclude that $\dim \mathbf{Q} \otimes G$ is defined.

(ii) If $0 \to A' \to A \to A'' \to 0$ is an exact sequence of abelian groups, then $0 \to \mathbf{Q} \otimes A' \to \mathbf{Q} \otimes A \to \mathbf{Q} \otimes A'' \to 0$ is an exact sequence of vector spaces. Conclude that

$$\dim \mathbf{Q} \otimes A = \dim \mathbf{Q} \otimes A' + \dim \mathbf{Q} \otimes A''.$$

(*Hint*: Use [Tor 1] and [Tor 2].)

9.31. Compute $\text{Tor}(A, B)$, where $A = \mathbf{Z} \oplus \mathbf{Z} \oplus \mathbf{Z}/6\mathbf{Z} \oplus \mathbf{Z}/5\mathbf{Z}$ and $B = \mathbf{Z}/3\mathbf{Z} \oplus \mathbf{Z}/5\mathbf{Z}$.

*9.32. For any abelian group G, prove that $\text{rank } G = \dim \mathbf{Q} \otimes G$. (*Hint*: Exercise 9.21.)

*9.33. If $0 \to B' \to B \to B'' \to 0$ is an exact sequence of abelian groups and if A is torsion-free, then

$$0 \to A \otimes B' \to A \otimes B \to A \otimes B'' \to 0$$

is also exact.

*9.34. If F, H are abelian groups with F free abelian, and if $a \in F$ and $h \in H$ are nonzero, then $a \otimes h \neq 0$ in $F \otimes H$.

We are now able to compute homology with coefficients.

Theorem 9.32 (Universal Coefficients Theorem for Homology).

(i) *For every space X and every abelian group G, there are exact sequences for all $n \geq 0$:*

$$0 \to H_n(X) \otimes G \xrightarrow{\alpha} H_n(X; G) \to \text{Tor}(H_{n-1}(X), G) \to 0,$$

where $\alpha: (\text{cls } z) \otimes g \mapsto \text{cls}(z \otimes g)$.

(ii) *This sequence splits; that is,*

$$H_n(X; G) \cong H_n(X) \otimes G \oplus \operatorname{Tor}(H_{n-1}(X), G).$$

Remark. The value of the first statement is that one has an explicit formula for an isomorphism $H_n(X) \otimes G \xrightarrow{\sim} H_n(X; G)$ in the special case when $\operatorname{Tor}(H_{n-1}(X), G)$ vanishes.

PROOF. (i) We prove a more general result. If (C_*, ∂) is a free chain complex, then there are exact sequences for all $n \geq 0$:

$$0 \to H_n(C_*) \otimes G \xrightarrow{\alpha} H_n(C_* \otimes G) \to \operatorname{Tor}(H_{n-1}(C_*), G) \to 0,$$

where α: $(\operatorname{cls} z) \otimes g \mapsto \operatorname{cls}(z \otimes g)$ (note that if z is a cycle in C_n, then $z \otimes g$ is a cycle in $C_n \otimes G$ for every $g \in G$). The theorem follows by specializing C_* to $S_*(X)$.

The definition of cycles and boundaries of C_* gives exact sequences for every n:

$$0 \longrightarrow Z_n \xhookrightarrow{i_n} C_n \xrightarrow{d_n} B_{n-1} \to 0, \tag{1}$$

where i_n is the inclusion and d_n differs from ∂_n only in its target; there is a commutative diagram

$$\begin{array}{ccc} C_n & \xrightarrow{\partial_n} & C_{n-1}. \\ & {\scriptstyle d_n}\searrow \quad \nearrow{\scriptstyle j_{n-1}} & \\ & B_{n-1} & \end{array}$$

Since B_{n-1} is a subgroup of the free abelian group C_{n-1}, Theorem 9.3 shows that B_{n-1} is free abelian; by (the rephrasing of) Corollary 9.2, Eq. (1) is a split exact sequence. Exercise 9.13(i) now applies to show that

$$0 \longrightarrow Z_n \otimes G \xrightarrow{i_n \otimes 1} C_n \otimes G \xrightarrow{d_n \otimes 1} B_{n-1} \otimes G \longrightarrow 0 \tag{2}$$

is a (split) exact sequence.

If Z_* is the subcomplex of C_* whose nth term is Z_n, then the differentiations in Z_* are restrictions of ∂_n and hence are zero; it follows that the differentiations in $Z_* \otimes G$ are zero. Define B_*^+ to be the chain complex whose nth term is B_{n-1} (sic) and with all differentiations zero; it follows that the differentiations in $B_*^+ \otimes G$ are zero. Assembling the exact sequences (2) gives an exact sequence of complexes

$$0 \longrightarrow Z_* \otimes G \xrightarrow{i \otimes 1} C_* \otimes G \xrightarrow{d \otimes 1} B_*^+ \otimes G \longrightarrow 0,$$

and this sequence begets a long exact sequence of abelian groups (exact triangle)

$$\cdots \to H_{n+1}(B_*^+ \otimes G) \xrightarrow{\Delta_{n+1}} H_n(Z_* \otimes G) \xrightarrow{(i \otimes 1)_*} H_n(C_* \otimes G) \xrightarrow{(d \otimes 1)_*} H_n(B_*^+ \otimes G) \xrightarrow{\Delta_n} H_{n-1}(Z_* \otimes G) \to \cdots,$$

where Δ_n is the connecting homomorphism. Since $B_*^+ \otimes G$ and $Z_* \otimes G$ have zero differentiations, Exercise 5.6 gives

$$H_n(Z_* \otimes G) = (Z_* \otimes G)_n = Z_n \otimes G$$

and

$$H_n(B_*^+ \otimes G) = (B_*^+ \otimes G)_n = B_{n-1} \otimes G.$$

The long exact sequence can thus be rewritten as

$$\cdots \to B_n \otimes G \xrightarrow{\Delta_{n+1}} Z_n \otimes G \xrightarrow{(i \otimes 1)_*} H_n(C_* \otimes G) \xrightarrow{(d \otimes 1)_*} B_{n-1} \otimes G \xrightarrow{\Delta_n} Z_{n-1} \otimes G \to \cdots.$$

For each n, there is thus an exact sequence

$$0 \to (Z_n \otimes G)/\text{im } \Delta_{n+1} \xrightarrow{\alpha} H_n(C_* \otimes G) \xrightarrow{(d \otimes 1)_*} \ker \Delta_n \to 0, \quad (3)$$

where α is induced by $(i \otimes 1)_*$; that is,

$$\alpha\colon z \otimes g + \text{im } \Delta_{n+1} \mapsto (i \otimes 1)_*(z \otimes g) = \text{cls}(z \otimes g).$$

Let us evaluate Δ_n (and Δ_{n+1}) and the two outside terms in Eq. (3). Consider the usual diagram for the connecting homomorphism:

$$\begin{array}{ccccccc} & & & & C_n \otimes G & \xrightarrow{d \otimes 1} & B_{n-1} \otimes G \longrightarrow 0 \\ & & & & \Big\downarrow \partial \otimes 1 & & \\ 0 & \longrightarrow & Z_{n-1} \otimes G & \xrightarrow{i \otimes 1} & C_{n-1} \otimes G & & \end{array}$$

On any generator $b_{n-1} \otimes g$ of $B_{n-1} \otimes G$, we have

$$\Delta_n(b_{n-1} \otimes g) = (i \otimes 1)^{-1}(\partial \otimes 1)(d \otimes 1)^{-1}(b_{n-1} \otimes g)$$

$= b_{n-1} \otimes g$ regarded as an element of $Z_{n-1} \otimes G$; thus $\Delta_n = j_{n-1} \otimes 1$, where $j_{n-1}\colon B_{n-1} \hookrightarrow Z_{n-1}$ is the inclusion. We may thus rewrite the exact sequence (3) as

$$0 \to (Z_n \otimes G)/\text{im}(j_n \otimes 1) \xrightarrow{\alpha} H_n(C_* \otimes G) \to \ker j_{n-1} \otimes 1 \to 0. \quad (3')$$

The definition of homology gives exact sequences for every n:

$$0 \longrightarrow B_{n-1} \xrightarrow{j_{n-1}} Z_{n-1} \longrightarrow H_{n-1}(C_*) \longrightarrow 0.$$

By [Tor 1], there is an exact sequence

$$\text{Tor}(Z_{n-1}, G) \to \text{Tor}(H_{n-1}(C_*), G) \to B_{n-1} \otimes G \xrightarrow{j_{n-1} \otimes 1} Z_{n-1} \otimes G \to H_{n-1}(C_*) \otimes G \to 0.$$

Since Z_{n-1} is torsion-free (it is a subgroup of the free abelian group C_{n-1}), [Tor 2] says that $\text{Tor}(Z_{n-1}, G) = 0$. Therefore

$$\ker(j_{n-1} \otimes 1) \cong \text{Tor}(H_{n-1}(C_*), G)$$

and, by replacing $n - 1$ by n,

$$(Z_n \otimes G)/\text{im}(j_n \otimes 1) = \text{coker}(j_n \otimes 1) = H_n(C_*) \otimes G. \tag{4}$$

The exact sequence (3′) can thus be rewritten as

$$0 \to H_n(C_*) \otimes G \xrightarrow{\alpha} H_n(C_* \otimes G) \to \text{Tor}(H_{n-1}(C_*), G) \to 0. \tag{5}$$

(ii) It remains to show that the exact sequence (3′), hence (5), is split. Examining elements gives the string of inclusions

$$\text{im}(\partial_{n+1} \otimes 1) \subset Z_n \otimes G \subset \ker(\partial_n \otimes 1) \subset C_n \otimes G.$$

Now $Z_n \otimes G$ is a direct summand of $C_n \otimes G$ (because the exact sequence (2) splits); *a fortiori*, it is a direct summand of $\ker(\partial_n \otimes 1)$. It follows that $(Z_n \otimes G)/\text{im}(\partial_{n+1} \otimes 1)$ is a direct summand of $\ker(\partial_n \otimes 1)/\text{im}(\partial_{n+1} \otimes 1) = H_n(C_* \otimes G)$. But $\text{im}(\partial_{n+1} \otimes 1) = \text{im}(j_n \otimes 1)$, so that $(Z_n \otimes G)/\text{im}(\partial_{n+1} \otimes 1) \cong H_n(C_*) \otimes G$ (by (4)), and the result now follows from Exercise 9.10. □

Remarks. (1) There are stronger forms of the universal coefficients theorem (and also a contravariant version to be discussed in Chapter 12); this weaker version is satisfactory almost always.

(2) The name of the theorem is well chosen, because it reminds one that homology with any coefficient group G can be computed from ordinary homology.

(3) Note that Tor delayed his entrance until the last act of the proof.

Exercises

9.35. If G is a torsion-free abelian group, then $(\text{cls } z) \otimes g \mapsto \text{cls}(z \otimes g)$ is an isomorphism $H_n(X) \otimes G \to H_n(X; G)$.

9.36. For every positive integer m and every space X,

$$H_n(X; \mathbf{Z}/m\mathbf{Z}) \cong H_n(X) \otimes \mathbf{Z}/m\mathbf{Z} \oplus H_{n-1}(X)[m],$$

where, for an abelian group H, one defines

$$H[m] = \{h \in H : mh = 0\}.$$

Conclude that if $H_{n-1}(X)$ is torsion-free, then

$$H_n(X; \mathbf{Z}/m\mathbf{Z}) \cong H_n(X) \otimes \mathbf{Z}/m\mathbf{Z}.$$

(When p is a prime, one calls $H_*(X; \mathbf{Z}/p\mathbf{Z})$ **homology mod *p*.**)

Example 9.10. Although homology with coefficients was defined only for singular homology, the proof of the universal coefficients theorem also applies to the simplicial and cellular homology theories as well, since they have been defined using free chain complexes (see the remark after Theorem 9.8 and Exercise 9.14(iii)).

Example 9.11. Ordinary homology can be regarded as homology with coefficients in $\mathbf{Z}$, for $H_n(X) \otimes \mathbf{Z} \cong H_n(X)$, by Example 9.2, and $\text{Tor}(H_{n-1}(X), \mathbf{Z}) = 0$, by [Tor 2]. An easier way to see this, however, is to return to the definition of homology: applying the functor $\_\_ \otimes \mathbf{Z}$ to a chain complex does not change anything (Example 9.2).

EXAMPLE 9.12. If G is the additive group of either **Q**, **R**, or **C**, then

$$H_n(X; G) \cong H_n(X) \otimes G,$$

because G is torsion-free in each case, and so [Tor 2] gives $\mathrm{Tor}(H_{n-1}(X), G) = 0$. ($H_*(X; \mathbf{Q})$ is called **rational homology**, $H_*(X; \mathbf{R})$ is called **real homology**, and $H_*(X; \mathbf{C})$ is called **complex homology**.)

One can simplify the discussion of the Lefschetz number if one uses rational homology. By Example 9.12, $H_n(X; \mathbf{Q})$ is a vector space over **Q**; moreover, if $f: X \to X$ is continuous, then $f_*: H_n(X; \mathbf{Q}) \to H_n(X; \mathbf{Q})$ can be seen to be a linear transformation, and so the trace of f_* is now the usual trace of linear algebra. A similar simplification occurs in our discussion, in Chapter 7, of computing homology; if one wants only Betti numbers, then all is linear algebra.

Eilenberg–Zilber Theorem and the Künneth Formula

The long algebraic interlude began with the problem of computing $H_*(X \times Y)$. The main result shows that $S_*(X \times Y)$ is determined by $S_*(X)$ and $S_*(Y)$.

Definition. Let (C_*, d) and (G_*, ∂) be nonnegative chain complexes. Their **tensor product** $C_* \otimes G_*$ is the (nonnegative) chain complex whose term of degree $n \geq 0$ is

$$(C_* \otimes G_*)_n = \sum_{i+j=n} C_i \otimes G_j$$

and whose differentiation $D_n: (C_* \otimes G_*)_n \to (C_* \otimes G_*)_{n-1}$ is defined on generators by

$$D_n(c_i \otimes g_j) = dc_i \otimes g_j + (-1)^i c_i \otimes \partial g_j, \qquad i + j = n.$$

Since $i + j = n$ and both d and ∂ lower degrees by 1, we have $D_n(c_i \otimes g_j) \in (C_* \otimes G_*)_{n-1}$. The sign in the definition of D_n is present to force $D_{n-1}D_n = 0$, as the reader can easily check.

EXERCISES

9.37. If $\lambda: C_* \to C'_*$ and $\mu: E_* \to E'_*$ are chain maps, then $\lambda \otimes \mu: C_* \otimes E_* \to C'_* \otimes E'_*$ is a chain map, where

$$(\lambda \otimes \mu)_n = \sum_{i+j=n} \lambda_i \otimes \mu_j.$$

9.38. If $\lambda, \lambda': (C_*, d) \to (C'_*, d')$ are chain homotopic and if $\mu, \mu': E_* \to E'_*$ are chain homotopic, then $\lambda \otimes \mu$ and $\lambda' \otimes \mu'$ are chain homotopic. (*Hint*: First show that $\lambda \otimes \mu$ and $\lambda' \otimes \mu$ are chain homotopic via $s \otimes \mu$, where $d's + sd = \lambda - \lambda'$.)

*9.39. If C_* is chain equivalent to C'_* and if E_* is chain equivalent to E'_*, then $C_* \otimes E_*$ is chain equivalent to $C'_* \otimes E'_*$.

*9.40. Let $0 \to S'_* \to S_* \to S''_* \to 0$ be a short exact sequence of nonnegative complexes. If E_* is a nonnegative free chain complex, then $0 \to S'_* \otimes E_* \to S_* \otimes E_* \to S''_* \otimes E_* \to 0$ is exact. (*Hint*: Exercise 9.33.)

Theorem 9.33 (Eilenberg–Zilber). *For topological spaces X and Y, there is a (natural) chain equivalence* $\zeta: S_*(X \times Y) \to S_*(X) \otimes S_*(Y)$, *unique to chain homotopy, hence*

$$H_n(X \times Y) \cong H_n(S_*(X) \otimes S_*(Y))$$

for all $n \geq 0$.

Proof. Let **Top** × **Top** denote the category with objects all ordered pairs of topological spaces (A, B) (we do not demand that B be a subspace of A), with morphisms all ordered pairs of continuous maps, and with coordinatewise composition. Let $\mathcal{M}$ be the set of all (Δ^p, Δ^q), $p, q \geq 0$. Define functors F, E: **Top** × **Top** → **Comp** by $F(X, Y) = S_*(X \times Y)$ and $E(X, Y) = S_*(X) \otimes S_*(Y)$. We show that both F and E are free and acyclic.

For fixed $p \geq 0$, define an F_p-model set $\mathscr{X}_p$ to be the singleton $\{d^p\}$, where $d^p: \Delta^p \to \Delta^p \times \Delta^p$ is the diagonal $x \mapsto (x, x)$ (note that $d^p \in F_p(\Delta^p, \Delta^p) = S_p(\Delta^p \times \Delta^p)$). If A_1 and A_2 are spaces and $\sigma: \Delta^p \to A_1 \times A_2$ is a p-simplex, then there are continuous maps $\sigma_i: \Delta^p \to A_i$, for $i = 1, 2$, with $\sigma = (\sigma_1 \times \sigma_2) \circ d^p$ (define $\sigma_i = p_i \circ \sigma$, where $p_i: A_1 \times A_2 \to A_i$ is the projection, for $i = 1, 2$). Conversely, given any pair of continuous maps $\sigma_i: \Delta^p \to A_i$, then $(\sigma_1 \times \sigma_2) \circ d^p$ is a p-simplex in $A_1 \times A_2$. It follows that F_p is free with base $\mathscr{X}_p$. Since $\Delta^p \times \Delta^p$ is convex, we see that the model (Δ^p, Δ^p) is F-acyclic.

Let us now consider the functor E. Exercise 9.28(i) shows that $S_p(X) \otimes S_q(Y)$ is free abelian with basis all symbols $\sigma \otimes \tau$, where $\sigma: \Delta^p \to X$ and $\tau: \Delta^q \to Y$ are continuous. By Example 9.4, the functor S_p is free with (singleton) base $\{\delta^p\}$, where $\delta^p: \Delta^p \to \Delta^p$ is the identity (of course, $\delta^p \in S_p(\Delta^p)$). It follows easily that E_n is free with base in $\mathcal{M}$: indeed the E-model set $\mathscr{Y}_n = \{\delta^p \otimes \delta^q \in S_p(\Delta^p) \otimes S_q(\Delta^q): p + q = n\}$ serves. To check acyclicity, recall that each Δ^p is S_p-acyclic (i.e., $H_m(\Delta^p) = H_m(S_*(\Delta^p)) = 0$ for all $m \geq 1$), so that the free chain complex $S_*(\Delta^p)$ is chain equivalent to $\mathbf{Z}_*$, the chain complex with $\mathbf{Z}$ concentrated in degree 0 (Theorem 9.8). Therefore the model (Δ^p, Δ^q) is E-acyclic, because $E(\Delta^p, \Delta^q) = S_*(\Delta^p) \otimes S_*(\Delta^q)$ is chain equivalent to $\mathbf{Z}_* \otimes \mathbf{Z}_* \cong \mathbf{Z}_*$, by Exercise 9.39.

Define $\varphi: H_0F \to H_0E$ as follows. For a pair of spaces A_1 and A_2, $F_0(A_1, A_2) = S_0(A_1 \times A_2)$ is the free abelian group on all ordered pairs (a_1, a_2) (where $a_i \in A_i$), while $E_0(A_1, A_2) = (S_*(A_1) \otimes S_*(A_2))_0$ is the free abelian group on all symbols $a_1 \otimes a_2$. It is easy to see that the maps $\varphi_{A_1, A_2}: F_0(A_1, A_2) \to E_0(A_1, A_2)$, defined by $(a_1, a_2) \mapsto a_1 \otimes a_2$, induce isomorphisms $H_0F(A_1, A_2) \to H_0E(A_1, A_2)$ which constitute a natural equivalence φ. By acyclic models (Theorem 9.12(iii)), there is a natural chain map $\tau: F \to E$, unique to homotopy, that is a natural chain equivalence. □

To use the Eilenberg–Zilber theorem, it is necessary to solve the algebraic problem of computing the homology groups of the tensor product of two complexes. The proof below (which I learned from [Vick]) reduces the problem to the universal coefficients theorem.

Lemma 9.34. *Let A_* and G_* be nonnegative chain complexes. If every differentiation in A_* is zero, then*

$$H_n(A_* \otimes G_*) \cong \sum_{i \geq 0} H_n(A_i \otimes G_*^i),$$

where G_*^i is G_* "shifted by i", that is,

$$(G_*^i)_n = G_{n-i}.$$

PROOF. Recall that $D_n: (A_* \otimes G_*)_n \to (A_* \otimes G_*)_{n-1}$ is defined by $a_i \otimes g_j \mapsto da_i \otimes g_j + (-1)^i a_i \otimes \partial g_j$, where $i + j = n$, where d is the differentiation in A_*, and where ∂ is the differentiation in G_*. As $d = 0$, by hypothesis, we have

$$H_n(A_* \otimes G_*) = \frac{\ker D_n}{\operatorname{im} D_{n+1}} = \sum_i \left(\frac{\ker 1 \otimes \partial_{n-i}}{\operatorname{im} 1 \otimes \partial_{n+1-i}} \right).$$

For each fixed i, there is thus the shifting described in the statement. □

Lemma 9.35. *If C_* and G_* are nonnegative free chain complexes, then*

$$H_n(C_* \otimes G_*) \cong H_n(H_*(C_*) \otimes G_*),$$

where $H_*(C_*)$ is regarded as a chain complex in which every differentiation is zero.

PROOF. Let B_*, Z_* be the subcomplexes of C_* whose terms are boundaries and cycles, respectively; each of these complexes has all differentiations zero, hence the quotient complex $H_* = H_*(C_*)$ may also be viewed as a complex with zero differentiations. As in the proof of the universal coefficients theorem, let B_*^+ denote the complex (with zero differentiations) with nth term B_{n-1}. There are two short exact sequences of complexes:

$$0 \to Z_* \xrightarrow{i} C_* \xrightarrow{d} B_*^+ \to 0; \tag{1}$$

$$0 \to B_* \xrightarrow{j} Z_* \xrightarrow{p} H_* \to 0 \tag{2}$$

(here p is the natural map and i, j are inclusions). Since each term of G_* is free abelian, Exercise 9.40 gives exactness of

$$0 \longrightarrow Z_* \otimes G_* \xrightarrow{i \otimes 1} C_* \otimes G_* \xrightarrow{d \otimes 1} B_*^+ \otimes G_* \longrightarrow 0 \tag{3}$$

and

$$0 \longrightarrow B_* \otimes G_* \xrightarrow{j \otimes 1} Z_* \otimes G_* \xrightarrow{p \otimes 1} H_* \otimes G_* \longrightarrow 0. \tag{4}$$

Next, the exact sequences

$$0 \longrightarrow Z_n \xrightarrow{i_n} C_n \xrightarrow{d_n} B_{n-1} \longrightarrow 0$$

must split, since Corollary 9.2 applies because B_{n-1} is free abelian (Theorem 9.3); there are thus homomorphisms $q_n: C_n \to Z_n$ with $q_n i_n = 1_{Z_n}$. Define $\varphi_n: C_n \to H_n$ as the composite

$$\varphi_n = p_n q_n.$$

The map $\varphi: C_* \to H_*$ is a chain map: if $c \in C_{n+1}$, then

$$\begin{aligned} \varphi dc &= p_n q_n dc \\ &= p_n(dc) \quad \text{(because } dc \in Z_n \text{ and hence is fixed by } q_n) \\ &= 0 \qquad\quad \text{(because } dc \text{ is a boundary).} \end{aligned}$$

On the other hand, $d'\varphi c = 0$, where d' is the (zero) differentiation of H_*, and so $\varphi d = d'\varphi$.

Consider the following diagram:

$$\begin{array}{ccccccccc} H_{n+1}(B_*^+ \otimes G_*) & \xrightarrow{\Delta} & H_n(Z_* \otimes G_*) & \xrightarrow{(i\otimes 1)_*} & H_n(C_* \otimes G_*) & \xrightarrow{(d\otimes 1)_*} & H_n(B_*^+ \otimes G_*) & \xrightarrow{\Delta} & H_{n-1}(Z_* \otimes G_*) \\ \downarrow \alpha & & \downarrow 1 & & \downarrow \beta & & \downarrow \alpha' & & \downarrow 1 \\ H_n(B_* \otimes G_*) & \xrightarrow[(j\otimes 1)_*]{} & H_n(Z_* \otimes G_*) & \xrightarrow[(p\otimes 1)_*]{} & H_n(H_* \otimes G_*) & \xrightarrow[D]{} & H_{n-1}(B_* \otimes G_*) & \xrightarrow[(j\otimes 1)_*]{} & H_{n-1}(Z_* \otimes G_*). \end{array}$$

The rows are exact, because they arise by applying the exact triangle to sequences (3) and (4); thus D and Δ are connecting homomorphisms. Since $H_{n+1}(B_*^+ \otimes G_*) = H_n(B_* \otimes G_*)$, we may define α and α' to be identities. Finally, define β to be $(\varphi \otimes 1)_*$.

We claim that each of these squares commutes up to sign. If this is so, then a trivial modification of the five lemma shows that β is an isomorphism, and the proof is complete. Each verification is routine; for example, let us prove that the first (and fourth) squares actually commute. Let $b_i^+ \otimes g_{n-i}$ be a cycle in $(B_*^+)_i \otimes G_{n-i} = B_{i-1} \otimes G_{n-i}$; hence $0 = db_i^+ \otimes g_{n-i} + (-1)^i b_i^+ \otimes \partial g_{n-i} = (-1)^i b_i^+ \otimes \partial g_{n-i}$, because $db_i^+ = 0$. By Exercise 9.34, it follows that $\partial g_{n-i} = 0$ (for B_{i-1} is free abelian). Hence $\Delta \operatorname{cls}(b_i^+ \otimes g_{n-i}) = \operatorname{cls}(D(c_i \otimes g_{n-i}))$, where $c_i \in C_i$ and $dc_i = b_i^+$. Now

$$D(c_i \otimes g_{n-i}) = dc_i \otimes g_{n-i} + (-1)^i c_i \otimes \partial g_{n-i} = b_i^+ \otimes g_{n-i}$$

(since $dc_i = b_i^+$ and $\partial g_{n-i} = 0$). We have checked commutativity for a set of generators of $H_{n+1}(B_*^+ \otimes G_*)$. The commutativity to sign of the other two squares is left to the reader. □

Theorem 9.36 (Künneth Theorem).

(i) *If C_* and G_* are nonnegative free chain complexes, then there are exact sequences for all n:*

$$0 \to \sum_{i+j=n} H_i(C_*) \otimes H_j(G_*) \xrightarrow{\alpha} H_n(C_* \otimes G_*) \to \sum_{p+q=n-1} \operatorname{Tor}(H_p(C_*), H_q(G_*)) \to 0,$$

where α: $\operatorname{cls}(z_i) \otimes (\operatorname{cls} z_j') \mapsto \operatorname{cls}(z_i \otimes z_j')$.

(ii) *This exact sequence splits; that is,*

$$H_n(C_* \otimes G_*) \cong \sum_{i+j=n} H_i(C_*) \otimes H_j(G_*) \oplus \sum_{p+q=n-1} \operatorname{Tor}(H_p(C_*), H_q(G_*)).$$

PROOF. By Lemmas 9.35 and 9.34,

$$H_n(C_* \otimes G_*) \cong H_n(H_*(C_*) \otimes G_*) \cong \sum_i H_n(H_i(C_*) \otimes G_*^i).$$

By the universal coefficients theorem, Theorem 9.32 (actually, by the more general isomorphism given in its proof), there are split exact sequences for all n, i,

$$0 \to H_i(C_*) \otimes H_n(G_*^i) \xrightarrow{\alpha} H_n(H_i(C_*) \otimes G_*^i) \to \operatorname{Tor}(H_i(C_*), H_{n-1}(G_*^i)) \to 0;$$

that is, there are split exact sequences

$$0 \to H_i(C_*) \otimes H_{n-i}(G_*) \xrightarrow{\alpha} H_n(H_i(C_*) \otimes G_*^i) \to \operatorname{Tor}(H_i(C_*), H_{n-i-1}(G_*)) \to 0.$$

If $i > n$, we have $H_n(G_*^i) = 0$. Taking the direct sum over all $i \geq 0$ now gives the result. □

There are more sophisticated proofs allowing one to prove the Künneth theorem with no (freeness) condition on the nonnegative chain complex G_*. There is an immediate proof of the universal coefficients theorem from this more general Künneth theorem. Given an abelian group G, define G_* as the chain complex with G concentrated in degree 0: $G_0 = G$ and $G_n = 0$ if $n \neq 0$ (all differentiations are necessarily zero). By Exercise 5.6, $H_0(G_*) = G$ and $H_n(G_*) = 0$ for $n \neq 0$. The tensor product $C_* \otimes G_*$ in this case is just $C_* \otimes G$, and the Künneth theorem simplifies to

$$H_n(C_* \otimes G) \cong H_n(C_*) \otimes G \oplus \operatorname{Tor}(H_{n-1}(C_*), G),$$

as claimed.

Combining the Eilenberg–Zilber theorem with the Künneth theorem yields the result we have been seeking.

Theorem 9.37 (Künneth Formula).[3] *For every pair of topological spaces X and Y and for every integer $n \geq 0$, there is a split exact sequence*

---

[3] Here is the original form of the Künneth formula. If X and Y are compact polyhedra, then

$$b_n(X \times Y) = \sum_{i+j=n} b_i(X) b_j(Y),$$

where $b_i(X)$ is the ith Betti number of X. This follows from Theorem 9.37 once one observes that, for any f.g. abelian groups A and B, the group $\operatorname{Tor}(A, B)$ is finite, and hence it contributes nothing to the calculation of Betti numbers.

$$0 \to \sum_{i+j=n} H_i(X) \otimes H_j(Y) \xrightarrow{\alpha''} H_n(X \times Y) \to \sum_{p+q=n-1} \mathrm{Tor}(H_p(X), H_q(Y)) \to 0,$$

where α'': $(\mathrm{cls}\ z_i) \otimes (\mathrm{cls}\ z'_j) \mapsto \mathrm{cls}(\zeta'(z_i \otimes z'_j))$ *and* ζ': $S_*(X) \otimes S_*(Y) \to S_*(X \times Y)$ *is the inverse of an Eilenberg–Zilber chain equivalence. Hence*

$$H_n(X \times Y) \cong \sum_{i+j=n} H_i(X) \otimes H_j(Y) \oplus \sum_{p+q=n-1} \mathrm{Tor}(H_p(X), H_q(Y)).$$

PROOF. The Künneth theorem gives a split exact sequence with middle term $H_n(S_*(X) \otimes S_*(Y))$, and the Eilenberg–Zilber theorem identifies this term with $H_n(X \times Y)$. □

This theorem is especially useful when X and Y are compact polyhedra, better, finite CW complexes, for then each of their homology groups is a f.g. abelian group. Thus, if the homology groups of X and Y are known, then Corollary 9.31 and the cited properties of Tor are adequate for computing $H_n(X \times Y)$.

EXAMPLE 9.13. Let m, n be positive integers. If $m \neq n$, then

$$H_p(S^m \times S^n) = \begin{cases} \mathbf{Z} & \text{if } p = 0, m, n, m+n \\ 0 & \text{otherwise.} \end{cases}$$

If $m = n$, then

$$H_p(S^m \times S^m) = \begin{cases} \mathbf{Z} & \text{if } p = 0, 2m \\ \mathbf{Z} \oplus \mathbf{Z} & \text{if } p = m \\ 0 & \text{otherwise.} \end{cases}$$

(Note that this example agrees with our earlier computation of the homology groups of the torus $S^1 \times S^1$.)

EXAMPLE 9.14. If $X = S^1 \vee S^2 \vee S^3$ (wedge), then we saw in Exercise 7.26 that

$$H_p(X) = \begin{cases} \mathbf{Z} & \text{if } p = 0, 1, 2, 3 \\ 0 & \text{otherwise.} \end{cases}$$

It follows from Example 9.13 that X and $S^1 \times S^2$ have the same homology groups; however, they do not have the same homotopy type (see Exercise 9.46 below).

EXAMPLE 9.15. If $X = \mathbf{R}P^3 \times \mathbf{R}P^2$, then Theorem 8.47 with the Künneth formula gives

$$H_p(X) = \begin{cases} \mathbf{Z} & \text{if } p = 0 \\ \mathbf{Z}/2\mathbf{Z} \oplus \mathbf{Z}/2\mathbf{Z} & \text{if } p = 1 \\ \mathbf{Z}/2\mathbf{Z} & \text{if } p = 2 \\ \mathbf{Z} \oplus \mathbf{Z}/2\mathbf{Z} & \text{if } p = 3 \\ 0 & \text{if } p \geq 4. \end{cases}$$

EXERCISES

9.41. If X and Y are acyclic, then $X \times Y$ is acyclic.

9.42. If X and Y are path connected, then

$$H_1(X \times Y) = H_1(X) \oplus H_1(Y)$$

and

$$H_2(X \times Y) = H_2(X) \oplus [H_1(X) \otimes H_1(Y)] \oplus H_2(Y).$$

9.43. Compute $H_*(K \times \mathbf{R}P^n)$, where K is the Klein bottle.

9.44. Compute $H_*(\mathbf{R}P^n \times S^m)$.

9.45. Compute $H_*(\mathbf{R}P^n \times \mathbf{R}P^m)$.

*9.46. Prove that $S^1 \vee S^2 \vee S^3$ does not have the same homotopy type as $S^1 \times S^2$.

*9.47. Show that $S^1 \times S^1$ and $S^2 \vee S^1 \vee S^1$ have the same homology groups. (In Example 12.8, we shall see that these two spaces do not have the same homotopy type.)

9.48. (i) Show that $\mathbf{R}P^3$ and $\mathbf{R}P^2 \vee S^3$ have the same homology groups. (*Hint*: Exercise 7.26(ii).)
(ii) Show that $\mathbf{R}P^3$ and $\mathbf{R}P^2 \vee S^3$ do not have the same homotopy type.
(iii) Show that $\mathbf{R}P^3 \times \mathbf{R}P^2$ and $(\mathbf{R}P^2 \vee S^3) \times \mathbf{R}P^2$ have the same homology groups and the same fundamental group. (These spaces do not have the same homotopy type.)

9.49. Compute $H_*(T^r)$, where T^r is the r-torus, that is, T^r is the cartesian product of r copies of S^1.

CHAPTER 10

Covering Spaces

When first computing $\pi_1(S^1)$, we looked to winding numbers for inspiration. Every closed path f in S^1 at 1 ($1 = e^{2\pi i 0} \in S^1$) suggested the picture

$$\begin{array}{ccc} & & \mathbf{R} \\ & & \big\downarrow \exp \\ \mathbf{I} & \xrightarrow[f]{} & S^1. \end{array}$$

We proved two preliminary results: the lifting lemma (Lemma 3.14) says that every (not necessarily closed) path f possesses a "lifting" $\tilde{f}\colon \mathbf{I} \to \mathbf{R}$ that is unique once $\tilde{f}(0)$ is specified

$$\begin{array}{ccc} & & \mathbf{R} \\ & \overset{\tilde{f}}{\nearrow} & \big\downarrow \exp \\ \mathbf{I} & \xrightarrow[f]{} & S^1; \end{array}$$

the covering homotopy lemma (Corollary 3.15) says that if $g\colon \mathbf{I} \to S^1$ is a path and if $f \simeq g$ rel $\dot{\mathbf{I}}$, then $\tilde{f} \simeq \tilde{g}$ rel $\dot{\mathbf{I}}$. More precisely, if $F\colon \mathbf{I} \times \mathbf{I} \to S^1$ is continuous (i.e., F is a homotopy), then one can lift the homotopy: there exists a continuous $\tilde{F}$ making the following diagram commute:

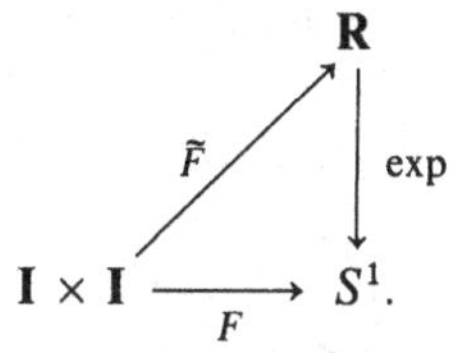

These two lemmas were used (in Theorem 3.16) to show that the degree function $d: \pi_1(S^1, 1) \to \mathbf{Z}$, defined by $[f] \mapsto \tilde{f}(1)$, is an isomorphism. In the first section, we shall extend this proof by replacing the exponential map $\exp: \mathbf{R} \to S^1$ by certain maps between more general spaces. What emerges is a tool for computing fundamental groups as well as an analogue of Galois theory! Moreover, many important constructions (e.g., fibrations, "killing" homotopy groups) can be viewed as generalizations of covering spaces.

Basic Properties

Definition. Let $\tilde{X}$ and X be topological spaces and let $p: \tilde{X} \to X$ be continuous. An open set U in X is **evenly covered** by p if $p^{-1}(U)$ is a disjoint union of open sets S_i in $\tilde{X}$, called **sheets**, with $p|S_i: S_i \to U$ a homeomorphism for every i.

The exponential map $\exp: \mathbf{R} \to S^1$ provides an example: the open set $U = S^1 - \{-1\}$ is evenly covered by exp, where $\exp(t) = e^{2\pi i t}$: indeed

$$(\exp)^{-1}(U) = \bigcup_{n \in \mathbf{Z}} (n - \tfrac{1}{2}, n + \tfrac{1}{2}),$$

so that the sheets here are open intervals.

Definition. If X is a topological space, then an ordered pair $(\tilde{X}, p)$ is a **covering space** of X if:

(i) $\tilde{X}$ is a path connected topological space;
(ii) $p: \tilde{X} \to X$ is continuous;
(iii) each $x \in X$ has an open neighborhood $U = U_x$ that is evenly covered by p.

The map p is called the **covering projection**,[1] and an open set that is evenly covered by p is called ***p*-admissible** or, more simply, **admissible**.

[1] A covering projection is an example of a local homeomorphism, defined as follows. A continuous map $f: Y \to X$ is a **local homeomorphism** if each $y \in Y$ has an open neighborhood V with $f(V)$ open in X and with $f|V: V \to f(V)$ a homeomorphism.

It is clear that the admissible open sets comprise an open cover of X. The picture to keep in mind is

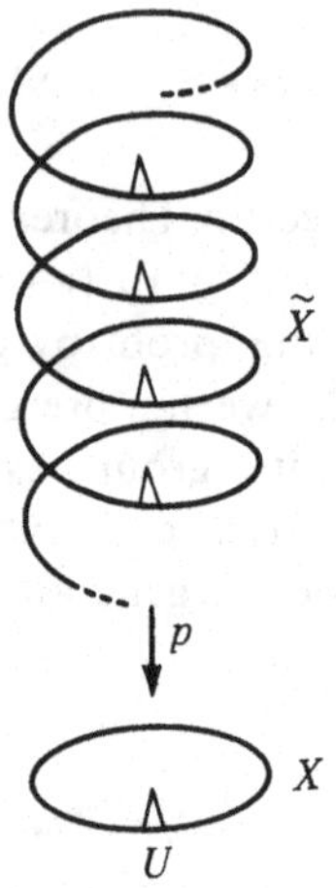

Lemma 10.1. *Let $(\tilde{X}, p)$ be a covering space of X. Then p is an open continuous surjection and hence is an identification; moreover, X is path connected.*

Proof. If $x \in X$ and $U = U_x$ is admissible, then $p(p^{-1}(U)) = U$ and $x \in \operatorname{im} p$; hence p is a surjection. To see that p is open, let V be an open set in $\tilde{X}$ and let $x \in p(V)$; let U be an admissible open set containing x, let $\tilde{x} \in p^{-1}(x) \cap V$, and let $\tilde{U}$ be the sheet over U containing $\tilde{x}$. Then $\tilde{U} \cap V$ is an open set in $\tilde{U}$ (containing $\tilde{x}$), and so $p(\tilde{U} \cap V)$ is an open subset of U containing x; therefore $p(V)$ is open. Finally, an open continuous map is an identification; moreover, a continuous image of a path connected space is path connected. □

Remark. The covering projection $p: \tilde{X} \to X$ need not be closed. For example, the discrete set $\{n + 1/n : n \geq 3\}$ is a closed subset of $\mathbf{R}$ whose image under exp is not closed in S^1.

Exercises

10.1. Show that $(\mathbf{R}, \exp)$ is a covering space of S^1.

10.2. If S^1 is regarded as a multiplicative topological group and if $k \in \mathbf{Z} - \{0\}$, then the map $p_k: S^1 \to S^1$ given by $z \mapsto z^k$ is continuous. Prove that (S^1, p_k) is a covering space of S^1.

*10.3. Prove that (S^n, p) is a covering space of $\mathbf{R}P^n$, where p is the map identifying antipodal points.

*10.4. Let X be a wedge of two circles, say, $X = A \vee B$, and let x_0 denote their point of tangency. Let $q: S^1 \to X$ be the identification map that identifies -1 and $+1$ (so we may assume that $q(-1) = x_0 = q(1)$).

(i) Show that (S^1, q) is not a covering space of X.
(ii) Let $\tilde{X}$ be a doubly infinite sequence of tangent circles:

$\tilde{X} = \cdots$

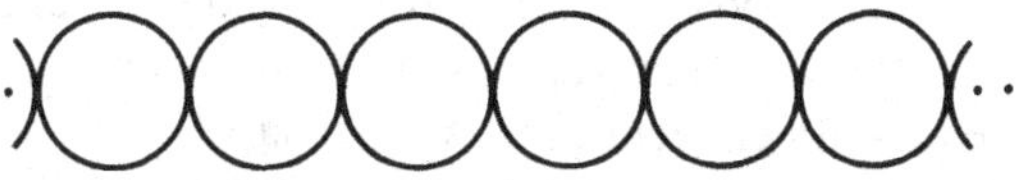

$\cdots,$

with points of tangency $\tilde{x}_k$ for $k \in \mathbf{Z}$. Define $p: \tilde{X} \to X$ to be the map with $p(\tilde{x}_k) = x_0$ for every k and with the restriction of p to every circle being q (we have been imprecise). Prove that $(\tilde{X}, p)$ is a covering space of X.

*10.5. If $(\tilde{X}, p)$ is a covering space of X and if $x_0 \in X$, then the fiber $p^{-1}(x_0)$ is a discrete subset of $\tilde{X}$.

*10.6. Let $(\tilde{X}, p)$ be a covering space of X. Prove that if X is either Hausdorff or locally compact or locally path connected or is an n-manifold, then so is $\tilde{X}$. Indeed any "local" property of X is inherited by $\tilde{X}$.

*10.7. Let $p: \tilde{X} \to X$ be continuous and let U be an open set in X that is evenly covered by p. If V is an open subset of U, then V is also evenly covered by p.

10.8. (i) If $(\tilde{X}_i, p_i)$ is a covering space of X_i, for $i = 1, 2$, then $(\tilde{X}_1 \times \tilde{X}_2, p_1 \times p_2)$ is a covering space of $X_1 \times X_2$. (*Hint*: If U_1 is p_1-admissible and U_2 is p_2-admissible, then $U_1 \times U_2$ is $(p_1 \times p_2)$-admissible.) Conclude that the plane is a covering space of the torus.
(ii) Prove that an infinite cylinder $\mathbf{R} \times S^1$ is a covering space of the torus. (*Hint*: For any path connected space X, $(X, 1_X)$ is a covering space of X.)

*10.9. Consider the commutative diagram

$$\begin{array}{ccc} \tilde{Y} & \xrightarrow{\beta} & \tilde{X} \\ {\scriptstyle q}\downarrow & & \downarrow{\scriptstyle p} \\ Y & \xrightarrow[\alpha]{} & X \end{array}$$

in which β and α are homeomorphisms and $(\tilde{X}, p)$ is a covering space of X. Show that $(\tilde{Y}, q)$ is a covering space of Y.

Theorem 10.2. *Let G be a path connected topological group, and let H be a discrete normal subgroup of G. If $p: G \to G/H$ is the natural homomorphism, then (G, p) is a covering space of G/H.*

Remark. We know, by Exercise 3.23, that G/H is a topological group.

PROOF. Let us first show that p is an open map. If V is open in G, then $p(V) = \{Hx: x \in V\}$. Hence

$$p^{-1}p(V) = \bigcup_{x \in V} Hx = \bigcup_{h \in H} hV.$$

Now each hV is open in G (because $g \mapsto hg$ is a homeomorphism $G \to G$), and so $p^{-1}p(V)$ is open in G; since p is an identification, it follows that $p(V)$ is open in G/H.

Since H is discrete, every subset of H is closed in H, and so every subset of H is open in H. In particular, there is an open set W in G with $W \cap H = \{1\}$, where 1 is the identity element of G. As the map $G \times G \to G$ given by $(x, y) \mapsto xy^{-1}$ is continuous, there is an open neighborhood V of 1 with $VV^{-1} \subset W$ (recall that $VV^{-1} = \{ab^{-1}: a, b \in V\}$). Define $U = p(V)$; since p is open, U is an open neighborhood of 1 in G/H. We claim that U is evenly covered by p. As we saw above,

$$p^{-1}(U) = p^{-1}p(V) = \bigcup_{h \in H} hV,$$

where each hV is open in G. The sets of the form hV, where $h \in H$, are pairwise disjoint: if h, k are distinct elements of H and $hV \cap kV \neq \varnothing$, then there are elements $v, w \in V$ with $hv = kw$; hence $vw^{-1} = k^{-1}h \in VV^{-1} \cap H \subset W \cap H = \{1\}$, a contradiction. Finally, $p|hV$ is a homeomorphism from hV to U. We already know that $p|hV$ is an open continuous map; $p|hV$ is a surjection, since $p(hV) = p(h)p(V) = p(V) = U$ (because $h \in H = \ker p$); $p|hV$ is an injection, because if $p(hv) = p(hw)$ (where $v, w \in V$), then $p(v) = p(w)$ and $vw^{-1} \in VV^{-1} \cap H = \{1\}$.

It is now easy to see that if $\bar{x} \in G/H$, then $\bar{x}U$ is an open neighborhood of $\bar{x}$ in G/H that is evenly covered by p. Therefore (G, p) is a covering space of G/H. □

Note that if (G, p) is a covering space of G/H, then $H = \ker p$ is just the fiber over 1; by Exercise 10.5, H must be discrete.

After giving a uniqueness result, we shall show that the lifting lemma and the covering homotopy lemma (which we have proved for $(\mathbf{R}, \exp)$) hold for arbitrary covering spaces.

Lemma 10.3. *Let $(\tilde{X}, p)$ be a covering space of X, let Y be a connected space, and let $f: (Y, y_0) \to (X, x_0)$ be continuous. Given $\tilde{x}_0$ in the fiber over x_0, there is at most one continuous $\tilde{f}: (Y, y_0) \to (\tilde{X}, \tilde{x}_0)$ with $p\tilde{f} = f$.*

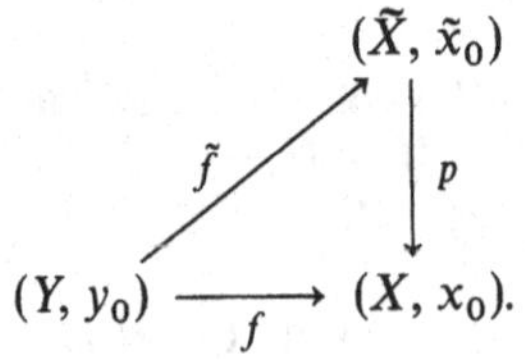

Proof. Suppose that $f': (Y, y_0) \to (\tilde{X}, \tilde{x}_0)$ satisfies $pf' = f$. Let

$$A = \{y \in Y: \tilde{f}(y) = f'(y)\}$$

and

$$B = \{y \in Y: \tilde{f}(y) \neq f'(y)\}.$$

Clearly, $Y = A \cup B$, $A \cap B = \varnothing$, and $A \neq \varnothing$ (because $y_0 \in A$). If we show that A and B are open, then the connectivity of Y will force $B = \varnothing$, hence $\tilde{f} = f'$.

Let $a \in A$. Let U be an admissible neighborhood of $f(a)$, and let S be the sheet over U containing $\tilde{f}(a) = f'(a)$. Of course, $W = \tilde{f}^{-1}(S) \cap f'^{-1}(S)$ is an open neighborhood of a in Y. Indeed $W \subset A$: if $w \in W$, then $\tilde{f}(w)$ and $f'(w)$ lie in S, and so $p\tilde{f}(w) = f(w) = pf'(w)$; hence $\tilde{f}(w) = f'(w)$ because $p|S$ is a homeomorphism. Therefore $w \in A$, $W \subset A$, and A is open.

Were $\tilde{X}$ Hausdorff, one could use a standard result that A is closed, and this would complete the proof. Without this assumption, we argue as follows. If $b \in B$, then let V be an admissible neighborhood of $f(b)$. If both $\tilde{f}(b)$ and $f'(b)$ lie in the same sheet over V, then the argument above gives $\tilde{f}(b) = f'(b)$, contradicting the fact that $b \in B$. Hence $\tilde{f}(b) \in S$ and $f'(b) \in S'$, where S, S' are distinct sheets. But now $W' = \tilde{f}^{-1}(S) \cap f'^{-1}(S')$ is an open neighborhood of b, and one can check quickly that $W' \subset B$. Therefore B is open. □

Theorem 10.4 (Lifting Lemma). *Let $(\tilde{X}, p)$ be a covering space of X and let $f\colon (\mathbf{I}, 0) \to (X, x_0)$ be a path. If $\tilde{x}_0$ is in the fiber over x_0, then there exists a unique $\tilde{f}\colon (\mathbf{I}, 0) \to (\tilde{X}, \tilde{x}_0)$* with $p\tilde{f} = f$.

PROOF. In light of Lemma 10.3, $\mathbf{I}$ connected implies the uniqueness of any such $\tilde{f}$. We now prove that $\tilde{f}$ exists. Suppose that $[a, b] \subset \mathbf{I}$ is such that $f([a, b]) \subset U$, where U is an admissible neighborhood of $x = f(a)$. If $\tilde{x}$ lies in the fiber over x, then $\tilde{x}$ lies in a unique sheet, say, S. It is easy to see that $\tilde{g}\colon ([a, b], a) \to (\tilde{X}, \tilde{x})$ defined by $\tilde{g} = (p|S)^{-1} \circ (f|[a, b])$ satisfies $p\tilde{g} = f|[a, b]$.

For each $t \in \mathbf{I}$, let U_t be an admissible neighborhood of $f(t)$. Now $\{f^{-1}(U_t)\colon t \in \mathbf{I}\}$, being an open cover of the compact metric space $\mathbf{I}$, has a Lebesgue number λ. This means that if $0 < \delta < \lambda$ and Y is a subset of $\mathbf{I}$ of diameter less than δ, then $Y \subset f^{-1}(U_t)$ for some $t \in \mathbf{I}$; that is, $f(Y) \subset U_t$. Partition $\mathbf{I}$ with points $t_1 = 0, t_2, \ldots, t_m = 1$, where $t_{i+1} - t_i < \delta$ for $1 \le i \le m - 1$. By our initial remarks, there is a continuous $\tilde{g}_1\colon [0, t_2] \to \tilde{X}$ with $p\tilde{g}_1 = f|[0, t_2]$ and $\tilde{g}_1(0) = \tilde{x}_0$. Similarly, there is a continuous $\tilde{g}_2\colon [t_2, t_3] \to \tilde{X}$ with $p\tilde{g}_2 = f|[t_2, t_3]$ and $\tilde{g}_2(t_2) = \tilde{g}_1(t_2)$; indeed, for $1 \le i \le m - 2$, there is a continuous $\tilde{g}_{i+1}\colon [t_{i+1}, t_{i+2}] \to \tilde{X}$ with $p\tilde{g}_{i+1} = f|[t_{i+1}, t_{i+2}]$ and $\tilde{g}_{i+1}(t_{i+1}) = \tilde{g}_i(t_{i+1})$. By the gluing lemma (Lemma 1.1), we may assemble the functions g_i into a continuous function $\tilde{f}\colon \mathbf{I} \to \tilde{X}$, where $\tilde{f}(t) = \tilde{g}_i(t)$ if $t \in [t_i, t_{i+1}]$. □

A stronger version of the covering homotopy lemma holds, and its proof is essentially that of the special case.

Theorem 10.5 (Covering Homotopy Theorem).[2] *Let $(\tilde{X}, p)$ be a covering space of X, and let Y be any space. Consider the diagram of continuous maps*

---

[2] Anticipating terminology not yet introduced, this theorem says that a covering projection $p\colon \tilde{X} \to X$ is a **fibration**.

Suppose one defines $g\colon Y \to X$ by $g(y) = F(y, 1)$ and $\tilde{g}\colon Y \to \tilde{X}$ by $\tilde{g}(y) = \tilde{F}(y, 1)$. Then $p\tilde{g} = g$ and $\tilde{F}\colon \tilde{f} \simeq \tilde{g}$. Therefore, if $f \simeq g$, then their respective liftings $\tilde{f}$ and $\tilde{g}$ are also homotopic.

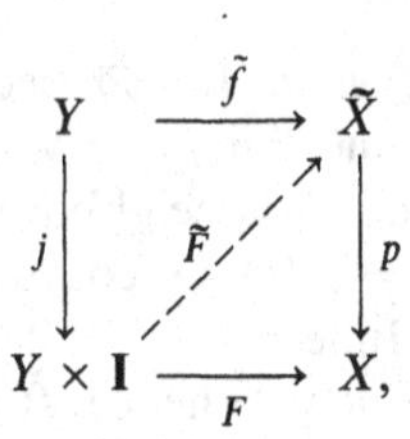

where $j(y) = (y, 0)$ *for all* $y \in Y$. *Then there exists a continuous map* $\tilde{F}: Y \times \mathbf{I} \to \tilde{X}$ *making the diagram commute; moreover, if* Y *is connected, then* $\tilde{F}$ *is unique.*

PROOF. Note that if Y is connected, then $Y \times \mathbf{I}$ is also connected, and so Lemma 10.3 gives the uniqueness of $\tilde{F}$ (because, for any $y \in Y$, we have $\tilde{F}(y, 0) = \tilde{f}(y)$).

We show first that it suffices to work locally: we shall show that $\tilde{F}$ exists if each $y \in Y$ has an open neighborhood N_y such that there is a continuous $\tilde{F}_y$ making the following diagram commute

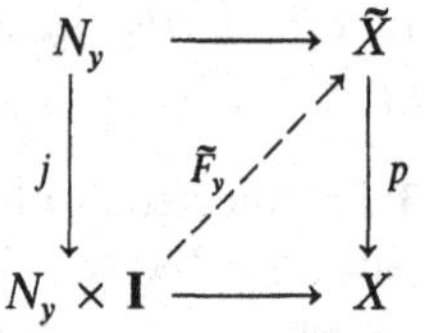

(the horizontal maps are restrictions of $\tilde{f}$ and F, respectively). Since $\{N_y \times \mathbf{I}: y \in Y\}$ is an open cover of $Y \times \mathbf{I}$, it suffices to show that the $\tilde{F}_y$ agree on overlaps (Lemma 1.1′). Suppose that $y' \in N_y \cap N_z$. Then

$$\tilde{F}_y(y', 0) = \tilde{f}(y') = \tilde{F}_z(y', 0);$$

moreover, if $t \in \mathbf{I}$, then

$$p\tilde{F}_y(y', t) = F(y', t) = p\tilde{F}_z(y', t).$$

Thus both $\tilde{F}_y$ and $\tilde{F}_z$ are liftings of $F|\{y'\} \times \mathbf{I}$ which agree on $(y', 0)$. Since $\{y'\} \times \mathbf{I}$ is connected, Lemma 10.3 gives $\tilde{F}_y|\{y'\} \times \mathbf{I} = \tilde{F}_z|\{y'\} \times \mathbf{I}$; as y' is an arbitrary element of $N_y \cap N_z$, it follows that $\tilde{F}_y$ and $\tilde{F}_z$ agree on $(N_y \cap N_z) \times \mathbf{I} = (N_y \times \mathbf{I}) \cap (N_z \times \mathbf{I})$, as desired.

We now construct the neighborhoods N_y and the maps $\tilde{F}_y$. For each $y \in Y$ and each $t \in \mathbf{I}$, let U_t be an admissible neighborhood of $F(y, t)$ in X; since F is continuous, there are open neighborhoods M_y and $\mathbf{I}_t$ of y and t, respectively, with $F(M_y \times \mathbf{I}_t) \subset U_t$. Compactness of $\mathbf{I}$ implies that finitely many $\mathbf{I}_t$'s cover $\mathbf{I}$; denote them by $\mathbf{I}_1, \ldots, \mathbf{I}_n$. If we define $N_y = \bigcap_{v=1}^{n} M_v$, then N_y is an open neighborhood of y. Also, there are numbers $0 = t_0 < t_1 < \cdots < t_m = 1$ in $\mathbf{I}$ with $[t_{i-1}, t_i]$ contained in some $\mathbf{I}_v$ (depending on i), where $v = 1, \ldots, n$. Hence $N_y \times [t_{i-1}, t_i] \subset M_v \times \mathbf{I}_v$, and $F(N_y \times [t_{i-1}, t_i])$ is contained in some admissible open set in X (which depends on i).

It suffices to construct continuous maps $G_i: N_y \times [t_{i-1}, t_i] \to \tilde{X}$ for $i = 1, \ldots, m$, such that

(i) $pG_i = F|N_y \times [t_{i-1}, t_i]$,
(ii) $G_1(y', 0) = \tilde{f}(y')$ for all $y' \in N_y$,
and
(iii) $G_{i-1}(y', t_{i-1}) = G_i(y', t_{i-1})$ for all $y' \in N_y$ and all i,

because such maps G_i can be glued together giving $\tilde{F}_y\colon N_y \times \mathbf{I} \to \tilde{X}$, as desired. To define G_1, let U be an admissible open set with $F(N_y \times [0, t_1]) \subset U$, let $\{S_\lambda\colon \lambda \in \Lambda\}$ be the sheets in $\tilde{X}$ over U, and let $V_\lambda = \tilde{f}^{-1}(S_\lambda)$ in Y. Note that $\{V_\lambda\colon \lambda \in \Lambda\}$ is a disjoint open cover of N_y. Define G_1 as the composite

$$V_\lambda \times [0, t_1] \xrightarrow{F} U \xrightarrow{(p|S_\lambda)^{-1}} S_\lambda.$$

It is easy to see that G_1 satisfies (i) and (ii). Assuming that G_{i-1} exists, a similar construction gives G_i, and so the proof is completed by induction. □

Corollary 10.6 (Covering Homotopy Lemma). *Let $(\tilde{X}, p)$ be a covering space of X. Let x_0, x_1 be points in X, let $f, g\colon \mathbf{I} \to X$ be paths in X from x_0 to x_1, and let $\tilde{x}_0$ be in the fiber over x_0.*

(i) *If $F\colon \mathbf{I} \times \mathbf{I} \to X$ is a relative homotopy $F\colon f \simeq g$ rel $\dot{\mathbf{I}}$, then there exists a unique continuous $\tilde{F}\colon \mathbf{I} \times \mathbf{I} \to \tilde{X}$ with $p\tilde{F} = F$ and $\tilde{F}(0, 0) = \tilde{x}_0$.*
(ii) *If $\tilde{f}$, $\tilde{g}$ are the liftings of f, g, respectively, with $\tilde{f}(0) = \tilde{x}_0 = \tilde{g}(0)$, then $\tilde{f}(1) = \tilde{g}(1)$ and $\tilde{F}\colon \tilde{f} \simeq \tilde{g}$ rel $\dot{\mathbf{I}}$.*

Remark. Statement (ii) is often called the **monodromy theorem**.

PROOF. (i) This follows from Theorem 10.5 if we set $Y = \mathbf{I}$.

(ii) If we define $\tilde{F}_0\colon \mathbf{I} \to \tilde{X}$ by $\tilde{F}_0(t) = \tilde{F}(t, 0)$, then $p\tilde{F}_0 = f$ and $\tilde{F}_0(0) = \tilde{F}(0, 0) = \tilde{x}_0$; the lifting lemma (Theorem 10.4) gives $\tilde{F}_0 = \tilde{f}$. Next, $\tilde{F}|\{0\} \times \mathbf{I}$ is a path in $\tilde{X}$ lying over the constant path at x_0 and starting at $\tilde{x}_0$; Theorem 10.4 gives $\tilde{F}(0, t) = \tilde{x}_0$ for all $t \in \mathbf{I}$. Similarly, $\tilde{F}|\{1\} \times \mathbf{I}$ is the constant path at $\tilde{f}(1)$. Finally, if $\tilde{F}_1\colon \mathbf{I} \to \tilde{X}$ is defined by $\tilde{F}_1(t) = \tilde{F}(t, 1)$, then $p\tilde{F}_1 = g$ and $\tilde{F}_1(0) = \tilde{F}(0, 1) = \tilde{x}_0$ (since $\tilde{F}|\{0\} \times \mathbf{I}$ is constant). Hence $\tilde{F}_1 = \tilde{g}$. Therefore $\tilde{g}(1) = \tilde{F}_1(1) = \tilde{F}(1, 1) = \tilde{f}(1)$, and $\tilde{F}\colon \tilde{f} \simeq \tilde{g}$ rel $\dot{\mathbf{I}}$, as desired. □

Recall that the fundamental group π_1 is a functor from pointed spaces to groups; in particular, if $\varphi\colon (X', x_0') \to (X, x_0)$ is continuous, then there is a homomorphism $\varphi_*\colon \pi_1(X', x_0') \to \pi_1(X, x_0)$ defined by $[f'] \mapsto [\varphi f']$.

Theorem 10.7. *If $(\tilde{X}, p)$ is a covering space of X, then*

$$p_*\colon \pi_1(\tilde{X}, \tilde{x}_0) \to \pi_1(X, x_0)$$

is an injection (where $\tilde{x}_0$ lies in the fiber over x_0).

PROOF. Let $\tilde{f}\colon (\mathbf{I}, \dot{\mathbf{I}}) \to (\tilde{X}, \tilde{x}_0)$ be a closed path in $\tilde{X}$ at $\tilde{x}_0$. If $p_*[\tilde{f}] = [p\tilde{f}] = 1$ in $\pi_1(X, x_0)$, then there is a relative homotopy $F\colon p\tilde{f} \simeq c$ rel $\dot{\mathbf{I}}$, where c is the constant path at x_0. By Corollary 10.6, there is a lifting $\tilde{F}\colon \tilde{f} \simeq \tilde{c}$ rel $\dot{\mathbf{I}}$, where

$\tilde{c}$ is the constant path at $\tilde{x}_0$, hence $[\tilde{f}] = 1$ in $\pi_1(\tilde{X}, \tilde{x}_0)$. Therefore p_* is an injection. □

In the special case of the covering space $(\mathbf{R}, \exp)$ of S^1, the covering homotopy lemma enables one to define the degree map $d: \pi_1(S^1, 1) \to \mathbf{Z}$ (the fiber over 1) by $[f] \mapsto \tilde{f}(1)$, where $\tilde{f}$ is the lifting of f with $\tilde{f}(0) = 0 \in \mathbf{R}$. For an arbitrary covering space $(\tilde{X}, p)$ of X, there is a function

$$\pi_1(X, x_0) \times Y \to Y,$$

where Y is the fiber over x_0, given by

$$([f], y) \mapsto \tilde{f}(1),$$

where $\tilde{f}$ is the lifting of f with $\tilde{f}(0) = y \in Y \subset \tilde{X}$. Corollary 10.6 shows that this function is well defined, for it is independent of the choice of path in the path class $[f]$. Fixing $y \in Y$ thus gives a function $\pi_1(X, x_0) \to Y$ that generalizes the degree function d. Since a fiber Y may not be equipped with a group structure (as is the case for $(\mathbf{R}, \exp)$ or, more generally, for topological groups), these generalized degree functions are not homomorphisms.

Definition. Let G be a group and let Y be a set (topological space). Then G **acts** on Y if there is a (continuous) function $G \times Y \to Y$, denoted by $(g, y) \mapsto gy$, such that

$$(gg')y = g(g'y)$$

and

$$1y = y$$

for all $y \in Y$ and $g, g' \in G$ (here 1 is the identity element of G). Call Y a ***G*-set** (***G*-space**) if G acts on Y. One says that G acts **transitively** on Y if, for each y, $y' \in Y$, there exists $g \in G$ with $gy = y'$; call Y a **transitive *G*-set** (***G*-space**) in this case.

Let a group G act on a set Y. For each $g \in G$, the function on Y defined by $y \mapsto gy$ is a permutation of Y (its inverse is $y \mapsto g^{-1}y$); moreover, if G acts on a topological space Y, then $y \mapsto gy$ is a homeomorphism.

Definition. Let a group G act on a set Y, and let $y \in Y$. Then the **orbit** of y is

$$o(y) = \{gy: g \in G\} \subset Y,$$

and the **stabilizer** of y (also called the **isotropy subgroup** of y) is

$$G_y = \{g \in G: gy = y\} \subset G.$$

It is easy to see that G_y is a subgroup of G. Note that G acts transitively on Y if and only if $o(y) = Y$ for every $y \in Y$.

Lemma 10.8. *If a group G acts on a set Y and if $y \in Y$, then*

$$|o(y)| = [G : G_y].$$

In particular, if G acts transitively, then $|Y| = [G : G_y]$.

Proof. The following statements are equivalent: $gy = hy$; $g^{-1}hy = y$; $g^{-1}h \in G_y$; $gG_y = hG_y$. If $G//G_y$ denotes the family of left cosets of G_y in G, then it follows that $\varphi: o(y) \to G//G_y$ given by $\varphi(gy) = gG_y$ is a well defined function that is injective. Since φ is obviously a surjection, it is a bijection. □

Remark. There is another way that Y can be a G-set: if there is a function $G \times Y \to Y$, denoted by $(g, y) \mapsto yg$, such that

$$y(gg') = (yg)g'$$

and

$$y1 = y$$

for all $y \in Y$ and $g, g' \in G$. Call Y a **right G-set** if such a function exists; call Y a **left G-set** when the original definition holds.

We are forced to consider both types of G-sets because of our choice of notation. When f and g are paths, then $f * g$ means first traverse f and then g; when f and g are functions, then their composite $f \circ g$ means first apply g and then f. There is no real problem here, because one can convert a right G-set into a left G-set by defining

$$gy = yg^{-1}.$$

Note that this does work, because

$$g(g'y) = g(yg'^{-1}) = (yg'^{-1})g^{-1} = y(g'^{-1}g^{-1}) = y(gg')^{-1} = (gg')y.$$

Theorem 10.9. *Let $(\tilde{X}, p)$ be a covering space of X, let $x_0 \in X$, and let Y be the fiber over x_0.*

(i) *$\pi_1(X, x_0)$ acts transitively on Y.*
(ii) *If $\tilde{x}_0 \in Y$, then the stabilizer of $\tilde{x}_0$ is $p_*\pi_1(\tilde{X}, \tilde{x}_0)$.*
(iii) *$|Y| = [\pi_1(X, x_0): p_*\pi_1(\tilde{X}, \tilde{x}_0)]$.*

Proof. (i) Let us first show that Y is a (right) $\pi_1(X, x_0)$-set. If $[f] \in \pi_1(X, x_0)$ and $\tilde{x} \in Y$, then $\tilde{x}[f]$ is defined as $\tilde{f}(1)$, where $\tilde{f}$ is the (unique) lifting of f with $\tilde{f}(0) = \tilde{x}$. By Corollary 10.6, this definition does not depend on the choice of representative in the path class $[f]$.

It is easy to see that $\tilde{x}[f] = \tilde{x}$ when f is the constant path at x_0, for then $\tilde{f}$ is the constant path at $\tilde{x}$. Suppose that $[g] \in \pi_1(X, x_0)$. Let $\tilde{f}$ be the lifting of f with $\tilde{f}(0) = \tilde{x}$; let $\tilde{g}$ be the lifting of g with $\tilde{g}(0) = \tilde{f}(1)$. Then $\tilde{f} * \tilde{g}$ is the lifting of $f * g$ that begins at $\tilde{x}$, and it ends at $\tilde{g}(1)$. It follows easily that $\tilde{x}[f * g] = (\tilde{x}[f])[g]$.

Choose $\tilde{x}_0 \in Y$, and let $\tilde{x}$ be any point in Y. Since $\tilde{X}$ is path connected (this is the first time we have recognized this property of $\tilde{X}$), there is a path $\tilde{\lambda}$ in $\tilde{X}$ from $\tilde{x}_0$ to $\tilde{x}$. Now $p\tilde{\lambda}$ is a closed path in X at x_0 whose lifting with initial point $\tilde{x}_0$ is visibly $\tilde{\lambda}$. Thus $[p\tilde{\lambda}] \in \pi_1(X, x_0)$, and $\tilde{x}_0[p\tilde{\lambda}] = \tilde{\lambda}(1) = \tilde{x}$. It follows that $\pi_1(X, x_0)$ acts transitively on Y.

(ii) If f is a closed path in X at x_0, let $\tilde{f}$ be the lifting of f with $\tilde{f}(0) = \tilde{x}_0$. If $[f] \in \pi_1(X, x_0)_{\tilde{x}_0}$, the stabilizer of $\tilde{x}_0$, then $\tilde{x}_0 = \tilde{x}_0[f] = \tilde{f}(1)$; hence $[\tilde{f}] \in \pi_1(\tilde{X}, \tilde{x}_0)$ and $[f] = [p\tilde{f}] \in p_*\pi_1(\tilde{X}, \tilde{x}_0)$. For the reverse inclusion, assume that $[f] = [p\tilde{g}]$ for some $[\tilde{g}] \in \pi_1(\tilde{X}, \tilde{x}_0)$. Then $\tilde{f} = \tilde{g}$ (for both lift f and both have initial point $\tilde{x}_0$), and so $\tilde{f}(1) = \tilde{g}(1) = \tilde{x}_0$. Therefore $\tilde{x}_0[f] = \tilde{f}(1) = \tilde{x}_0$, and $[f]$ lies in the stabilizer of $\tilde{x}_0$.

(iii) This now follows from Lemma 10.8. □

Theorem 10.10. *Let $(\tilde{X}, p)$ be a covering space of X, let $x_0, x_1 \in X$, and let Y_0, Y_1 be the fibers over x_0, x_1, respectively. Then*

$$|Y_0| = |Y_1|.$$

Proof. Choose $\tilde{x}_0 \in Y_0$ and $\tilde{x}_1 \in Y_1$, let $\tilde{\lambda}$ be a path in $\tilde{X}$ from $\tilde{x}_0$ to $\tilde{x}_1$, and let $\lambda = p\tilde{\lambda}$ denote the corresponding path in X from x_0 to x_1. It is easy to see that the following diagram commutes:

$$\begin{array}{ccc} \pi_1(\tilde{X}, \tilde{x}_0) & \xrightarrow{\Sigma} & \pi_1(\tilde{X}, \tilde{x}_1) \\ \downarrow{\scriptstyle p_*} & & \downarrow{\scriptstyle p_*} \\ \pi_1(X, x_0) & \xrightarrow{\sigma} & \pi_1(X, x_1). \end{array}$$

Here the top map Σ sends $[\tilde{f}] \mapsto [\tilde{\lambda}^{-1} * \tilde{f} * \tilde{\lambda}]$, and the bottom map σ sends $[f] \mapsto [\lambda^{-1} * f * \lambda]$. Since these maps are isomorphisms and p_* is an injection, it follows that Σ induces a bijection between cosets: $[\pi_1(X, x_0): p_*\pi_1(\tilde{X}, \tilde{x}_0)] = [\pi_1(X, x_1): p_*\pi_1(\tilde{X}, \tilde{x}_1)]$. Theorem 10.9(iii) now gives the result. □

We have just proved that all the fibers in a covering space have the same cardinal. Since each fiber is discrete, it follows that any two fibers are homeomorphic.

Definition. The **multiplicity** of a covering space $(\tilde{X}, p)$ of X is the cardinal of a fiber. If the multiplicity is m, one also says that $(\tilde{X}, p)$ is an ***m*-sheeted** covering space of X, or that $(\tilde{X}, p)$ is an ***m*-fold cover** of X.

Corollary 10.11. *If $n \geq 2$, then $\pi_1(\mathbf{R}P^n) \cong \mathbf{Z}/2\mathbf{Z}$.*

Proof. We know that (S^n, p) is a covering space of $\mathbf{R}P^n$ (Exercise 10.3) of multiplicity 2; therefore $[\pi_1(\mathbf{R}P^n, x_0): p_*\pi_1(S^n, \tilde{x}_0)] = 2$. By Corollary 7.6, S^n is simply connected for $n \geq 2$. Therefore $|\pi_1(\mathbf{R}P^n, x_0)| = 2$ and $\pi_1(\mathbf{R}P^n, x_0) \cong \mathbf{Z}/2\mathbf{Z}$. □

Corollary 10.12. *Let $(\tilde{X}, p)$ be a covering space of X, let $x_0 \in X$, and let Y be the fiber over x_0.*

(i) *If $\tilde{x}_0, \tilde{x}_1 \in Y$, then $p_*\pi_1(\tilde{X}, \tilde{x}_0)$ and $p_*\pi_1(\tilde{X}, \tilde{x}_1)$ are conjugate subgroups of $\pi_1(X, x_0)$.*

(ii) *If S is a subgroup of $\pi_1(X, x_0)$ that is conjugate to $p_*\pi_1(\tilde{X}, \tilde{x}_0)$ for some $\tilde{x}_0 \in Y$, then there exists $\tilde{x}_1 \in Y$ with $S = p_*\pi_1(\tilde{X}, \tilde{x}_1)$.*

Remark. Since $\tilde{X}$ is path connected, we know that $\pi_1(\tilde{X}, \tilde{x}_0) \cong \pi_1(\tilde{X}, \tilde{x}_1)$, so that their images under the injection p_* are isomorphic. This corollary asserts that these images are even conjugate.

PROOF. (i) Recall the commutative diagram from Theorem 10.10 (with $Y_0 = Y_1$):

$$\begin{array}{ccc} \pi_1(\tilde{X}, \tilde{x}_0) & \xrightarrow{\Sigma} & \pi_1(\tilde{X}, \tilde{x}_1) \\ {\scriptstyle p_*}\downarrow & & \downarrow{\scriptstyle p_*} \\ \pi_1(X, x_0) & \xrightarrow{\sigma} & \pi_1(X, x_0); \end{array}$$

here $\Sigma: [\tilde{f}] \mapsto [\tilde{\lambda}^{-1} * \tilde{f} * \tilde{\lambda}]$ and $\sigma: [f] \mapsto [\lambda^{-1} * f * \lambda]$, where $\tilde{\lambda}$ is a path in $\tilde{X}$ from $\tilde{x}_0$ to $\tilde{x}_1$, and $\lambda = p\tilde{\lambda}$. Now

$$p_*\Sigma\pi_1(\tilde{X}, \tilde{x}_0) = p_*\pi_1(\tilde{X}, \tilde{x}_1) = \sigma p_*\pi_1(\tilde{X}, \tilde{x}_0),$$

so that the two subgroups are conjugate by $[\lambda] \in \pi_1(X, x_0)$ (note that λ is a closed path in X at x_0 since both $\tilde{x}_0, \tilde{x}_1 \in Y$).

(ii) Suppose that $S = [\lambda^{-1}]p_*\pi_1(\tilde{X}, \tilde{x}_0)[\lambda]$ for some closed path λ in X at x_0. Let $\tilde{\lambda}$ be the path in $\tilde{X}$ lying over λ for which $\tilde{\lambda}(0) = \tilde{x}_0$. Note that $\tilde{\lambda}(1) \in Y$ (because $p\tilde{\lambda} = \lambda$), say, $\tilde{\lambda}(1) = \tilde{x}_1$. Using the commutative diagram in Theorem 10.10,

$$S = \sigma p_*\pi_1(\tilde{X}, \tilde{x}_0) = p_*\Sigma\pi_1(\tilde{X}, \tilde{x}_0) = p_*\pi_1(\tilde{X}, \tilde{x}_1),$$

as desired. □

Definition. A covering space $(\tilde{X}, p)$ of X is **regular** if $p_*\pi_1(\tilde{X}, \tilde{x}_0)$ is a normal subgroup of $\pi_1(X, x_0)$ for every $x_0 \in X$.

If $(\tilde{X}, p)$ is a regular covering space of X, then $p_*\pi_1(\tilde{X}, \tilde{x}_0) = p_*\pi_1(\tilde{X}, \tilde{x}_1)$ for every $\tilde{x}_0, \tilde{x}_1$ in the same fiber. If $\tilde{X}$ is simply connected, then $(\tilde{X}, p)$ is regular.

EXERCISES

10.10. Let $(\tilde{X}, p)$ be an m-sheeted covering space of X, where m is prime. If $\tilde{X}$ is simply connected, prove that $\pi_1(X, x_0) \cong \mathbf{Z}/m\mathbf{Z}$.

10.11. It is known that $(m, \varphi(m)) = 1$, where φ is the Euler φ-function, if and only if every group of order m is cyclic. Prove that if $(\tilde{X}, p)$ is an m-sheeted covering space of X, where $(m, \varphi(m)) = 1$, and if $\tilde{X}$ is simply connected, then $\pi_1(X, x_0) \cong \mathbf{Z}/m\mathbf{Z}$.

10.12. Let $(\tilde{X}, p)$ be an m-sheeted covering space of X (we allow m to be an infinite cardinal). If U is an admissible open set in X, so that $p^{-1}(U) = \bigcup_{i \in I} S_i$, then $|I| = m$.

*10.13. Let (X, x_0) be a pointed space, let $(\tilde{X}, p)$ be a covering space of X, and let $Y = p^{-1}(x_0)$. Let $\theta: \pi_1(X, x_0) \to S_Y$ (where S_Y is the symmetric group on Y) be the homomorphism corresponding to the action of $\pi_1(X, x_0)$ on the fiber, namely, $\theta([f]): \tilde{x} \mapsto \tilde{x}[f]$.
(i) Show that $\ker \theta = \bigcap_{\tilde{x} \in Y} p_* \pi_1(\tilde{X}, \tilde{x})$. (The quotient $\pi_1(X, x_0)/\ker \theta$ is called the **monodromy group** of $(\tilde{X}, p)$.)
(ii) If $\tilde{X}$ is simply connected, then θ is an injection.

10.14. Let G be a simply connected topological group, and let H be a discrete normal subgroup. Prove that $\pi_1(G/H, 1) \cong H$. (*Remark*: This is Exercise 3.24, whose solution should now be clearer.)

*10.15. If $\pi_1(X, x_0)$ is abelian, then every covering space of X is regular.

*10.16. Let $(\tilde{Y}, q)$ and $(\tilde{X}, p)$ be covering spaces of X. If there exists a continuous $h: \tilde{Y} \to \tilde{X}$ with $ph = q$, then h is a surjection. (*Hint*: Use unique path lifting.)

*10.17. Let $(\tilde{X}, p)$ be a covering space of X, let $x_0 \in X$, and let $\tilde{x}_0 \in p^{-1}(x_0)$. If f is a closed path in X at x_0 and if $\tilde{f}$ is the lifting of f with $\tilde{f}(0) = \tilde{x}_0$, then $[f] \in p_* \pi_1(\tilde{X}, \tilde{x}_0)$ if and only if $\tilde{f}$ is a closed path in $\tilde{X}$ at $\tilde{x}_0$.

Covering Transformations

In this section, we investigate maps between covering spaces of a space X. Let us begin by recalling the covering homotopy theorem (Theorem 10.5). If $(\tilde{X}, p)$ is a covering space of X and if $f: Y \to X$ is a continuous map that has a lifting $\tilde{f}: Y \to \tilde{X}$, then any homotopy starting with f lifts to a homotopy starting at $\tilde{f}$. Thus, if $f \simeq g$ and f has a lifting $\tilde{f}$, then g has a lifting $\tilde{g}$ and $\tilde{f} \simeq \tilde{g}$. If $Y = \mathbf{I}$, then Theorem 10.4 says that every $f: \mathbf{I} \to X$ does have a lifting; the next result gives a necessary and sufficient condition for $f: Y \to X$ to have a lifting.

Theorem 10.13 (Lifting Criterion). *Let Y be connected and locally path connected, and let $f: (Y, y_0) \to (X, x_0)$ be continuous. If $(\tilde{X}, p)$ is a covering space of X, then there exists a unique $\tilde{f}: (Y, y_0) \to (\tilde{X}, \tilde{x}_0)$ (where $\tilde{x}_0 \in p^{-1}(x_0)$) lifting f if and only if $f_* \pi_1(Y, y_0) \subset p_* \pi_1(\tilde{X}, \tilde{x}_0)$.*

Proof. Lemma 10.3 allows us to consider only existence. Assume that a lifting $\tilde{f}$ does exist: $p\tilde{f} = f$ and $\tilde{f}(y_0) = \tilde{x}_0$. Then

$$f_* \pi_1(Y, y_0) = p_* \tilde{f}_* \pi_1(Y, y_0) \subset p_* \pi_1(\tilde{X}, \tilde{x}_0).$$

The converse is less obvious. By Corollary 1.21, Y is path connected. Let $y \in Y$ and let $h: \mathbf{I} \to Y$ be a path from y_0 to y; thus fh is a path from $f(y_0) = x_0$ to $f(y)$.

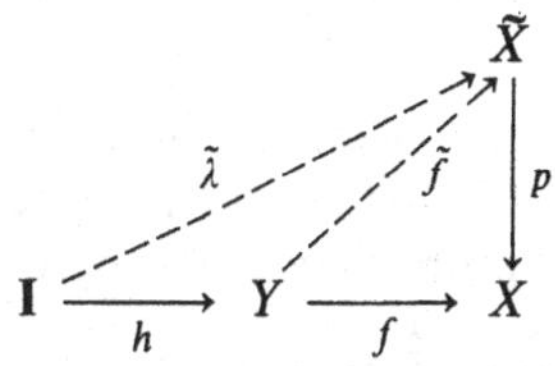

By Theorem 10.4, there is a unique path $\tilde{\lambda}$ in $\tilde{X}$ that lifts fh and with $\tilde{\lambda}(0) = \tilde{x}_0$. We propose to define $\tilde{f}: Y \to \tilde{X}$ by $\tilde{f}(y) = \tilde{\lambda}(1)$. If $\tilde{f}$ is well defined, then $p\tilde{f}(y) = p\tilde{\lambda}(1) = fh(1) = f(y)$.

We claim that $\tilde{\lambda}(1)$ is independent of the choice of path h. Choose another path h_1 from y_0 to y, and let $\tilde{\lambda}_1$ be the path in $\tilde{X}$ lifting fh_1 for which $\tilde{\lambda}_1(0) = \tilde{x}_0$. Now $h * h_1^{-1}$ is a closed path in Y at y_0, hence $f \circ (h * h_1^{-1}) = (f \circ h) * (f \circ h_1^{-1})$ is a closed path in X at x_0. Since

$$[(f \circ h) * (f \circ h_1^{-1})] = f_*[h * h_1^{-1}] \in f_*\pi_1(Y, y_0) \subset p_*\pi_1(\tilde{X}, \tilde{x}_0)$$

(the inclusion is the hypothesis), there exists a closed path $\tilde{g}$ in $\tilde{X}$ at $\tilde{x}_0$ with

$$(f \circ h) * (f \circ h_1^{-1}) \simeq p\tilde{g} \text{ rel } \dot{\mathbf{I}}.$$

Hence

$$(f \circ h) * (f \circ h_1^{-1}) * p\tilde{\lambda}_1 \simeq p\tilde{g} * p\tilde{\lambda}_1 \text{ rel } \dot{\mathbf{I}},$$

and

$$f \circ h \simeq p \circ (\tilde{g} * \tilde{\lambda}_1) \text{ rel } \dot{\mathbf{I}},$$

because $p\tilde{\lambda}_1 = f \circ h_1$.

By Theorem 10.5, the covering homotopy theorem,

$$\tilde{\lambda} \simeq \tilde{g} * \tilde{\lambda}_1 \text{ rel } \dot{\mathbf{I}}$$

and $\tilde{\lambda}(1) = (\tilde{g} * \tilde{\lambda}_1)(1) = \tilde{\lambda}_1(1)$, as desired.

It remains to prove that $\tilde{f}: Y \to \tilde{X}$ is continuous. Let $y \in Y$, let $\tilde{x} = \tilde{f}(y)$, and let $\tilde{U}_1$ be an open neighborhood of $\tilde{x}$; we must find an open neighborhood V of y with $\tilde{f}(V) \subset \tilde{U}_1$. Let $x = p\tilde{x} \in X$, let U be an admissible open neighborhood of x, and let S be the sheet over U containing $\tilde{x}$. Replacing $\tilde{U}_1$ by $\tilde{U}_1 \cap S$ if necessary, we may assume that $\tilde{U}_1 \subset S$ (remember that S is an open set in $\tilde{X}$).

Since p is an open map, the set U_1 defined by $U_1 = p(\tilde{U}_1)$ is an open neighborhood of x with $U_1 \subset U$; since f is continuous, $f^{-1}(U_1)$ is an open neighborhood of y in Y. By Corollary 1.19, Y locally path connected implies that there is an open path connected V with $y \in V \subset f^{-1}(U_1)$. We claim that $\tilde{f}(V) \subset \tilde{U}_1$, which will complete the proof. Let $h: \mathbf{I} \to Y$ be a path from y_0 to y, and let $\tilde{\lambda}$ be the lifting of fh with $\tilde{\lambda}(0) = \tilde{x}_0$. If $v \in V$, then there is a path $h_2: \mathbf{I} \to V$ from y to v; thus $h_2(\mathbf{I}) \subset V \subset f^{-1}(U_1)$ and $fh_2(\mathbf{I}) \subset U_1$. Let $\tilde{\mu}: \mathbf{I} \to \tilde{X}$ be the lifting of fh_2 with $\tilde{\mu}(0) = \tilde{x}$. Since $U_1 \subset U$ and U is admissible, it follows that $\tilde{\mu} = (p|S)^{-1}(fh_2)$, hence $\tilde{\mu}(1) \in S \cap \tilde{U}_1 = \tilde{U}_1$. Now $\tilde{\lambda}(1) = \tilde{x} = \tilde{\mu}(0)$, so that $\tilde{\lambda} * \tilde{\mu}$ is defined. But $p(\tilde{\lambda} * \tilde{\mu}) = p\tilde{\lambda} * p\tilde{\mu} = fh * fh_2 = f(h * h_2)$, where $h * h_2$ is a path from y_0 to v; moreover, $(\tilde{\lambda} * \tilde{\mu})(0) = \tilde{\lambda}(0) = \tilde{x}_0$. Therefore $\tilde{f}(v) = (\tilde{\lambda} * \tilde{\mu})(1) = \tilde{\mu}(1) \in \tilde{U}_1$, as desired. □

The lifting criterion is a fine example of a good theorem of algebraic topology: a topological result (here the existence of a certain continuous map) is equivalent to an algebraic problem (is one subgroup contained in another).

Corollary 10.14. *Let Y be simply connected and locally path connected, and let $f\colon (Y, y_0) \to (X, x_0)$ be continuous. If $(\tilde{X}, p)$ is a covering space of X and if $\tilde{x}_0 \in p^{-1}(x_0)$, then there exists a unique $\tilde{f}\colon (Y, y_0) \to (\tilde{X}, \tilde{x}_0)$ lifting f.*

Proof. Since Y is simply connected, $\pi_1(Y, y_0) = \{1\}$, and so $f_*\pi_1(Y, y_0) = \{1\} \subset p_*\pi_1(\tilde{X}, \tilde{x}_0)$. □

This corollary applies, in particular, to $Y = S^n$, $n \geq 2$.

Remark. Recall (Exercise 3.8) that a simply connected space need not be locally path connected.

Corollary 10.15. *Let X be connected and locally path connected, and let $(\tilde{X}, p)$ and $(\tilde{Y}, q)$ be covering spaces of X. Choose basepoints $x_0 \in X$, $\tilde{x}_0 \in \tilde{X}$, and $\tilde{y}_0 \in \tilde{Y}$ with $p\tilde{x}_0 = x_0 = q\tilde{y}_0$.*

If $q_*\pi_1(\tilde{Y}, \tilde{y}_0) = p_*\pi_1(\tilde{X}, \tilde{x}_0)$, then there exists a unique continuous $h\colon (\tilde{Y}, \tilde{y}_0) \to (\tilde{X}, \tilde{x}_0)$ with $ph = q$, and h is a homeomorphism.

Proof. The existence and uniqueness of h are guaranteed by the theorem; we need check only that h is a homeomorphism. Consider the commutative diagram

$$\begin{array}{ccc} \tilde{Y} & \xrightarrow{h} & \tilde{X} \\ & {\scriptstyle q}\searrow \quad \swarrow{\scriptstyle p} & \\ & X. & \end{array}$$

The theorem also guarantees a continuous $k\colon (\tilde{X}, \tilde{x}_0) \to (\tilde{Y}, \tilde{y}_0)$ with $qk = p$. The composite hk and the identity $1_{\tilde{X}}$ both complete the diagram

$$\begin{array}{ccc} \tilde{X} & \longrightarrow & \tilde{X} \\ & {\scriptstyle p}\searrow \quad \swarrow{\scriptstyle p} & \\ & X; & \end{array}$$

uniqueness of the completion gives $hk = 1_{\tilde{X}}$. Similarly, $kh = 1_{\tilde{Y}}$, and h is a homeomorphism (with inverse k). □

The next theorem augments the lifting criterion (Theorem 10.13).

Theorem 10.16. *Let X be connected and locally path connected, and let $(\tilde{X}, p)$ and $(\tilde{Y}, q)$ be covering spaces of X. Choose basepoints $x_0 \in X$, $\tilde{x}_0 \in \tilde{X}$, and $\tilde{y}_0 \in \tilde{Y}$ with $p\tilde{x}_0 = x_0 = q\tilde{y}_0$.*

If $q_*\pi_1(\tilde{Y}, \tilde{y}_0) \subset p_*\pi_1(\tilde{X}, \tilde{x}_0)$, then there exists a unique continuous $h: (\tilde{Y}, \tilde{y}_0) \to (\tilde{X}, \tilde{x}_0)$ with $ph = q$. Moreover, $(\tilde{Y}, h)$ is a covering space of $\tilde{X}$, and so $\tilde{X}$ is a quotient space of $\tilde{Y}$.

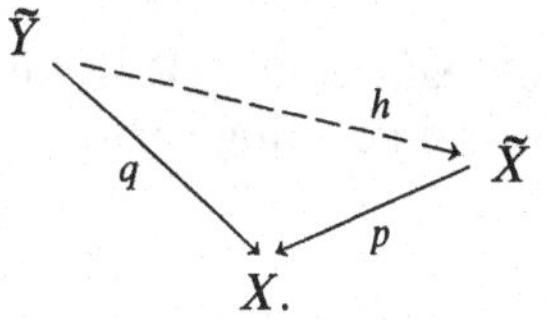

PROOF. By the definition of covering space, both $\tilde{X}$ and $\tilde{Y}$ are path connected; by Exercise 10.6, X locally path connected implies that both $\tilde{X}$ and $\tilde{Y}$ are locally path connected. Since $q_*\pi_1(\tilde{Y}, \tilde{y}_0) \subset p_*\pi_1(\tilde{X}, \tilde{x}_0)$, the lifting criterion provides a unique continuous $h: (\tilde{Y}, \tilde{y}_0) \to (\tilde{X}, \tilde{x}_0)$ such that $ph = q$. It remains to prove that $(\tilde{Y}, h)$ is a covering space of $\tilde{X}$; Lemma 10.1 will then apply to show that h is an identification, hence $\tilde{X}$ is a quotient space of $\tilde{Y}$.

Let $\tilde{x} \in \tilde{X}$, and let $x = p\tilde{x} \in X$. Let U_1 be a p-admissible open neighborhood of x and let U_2 be a q-admissible open neighborhood of x. Then $U_1 \cap U_2$ is an open neighborhood of x, and, since X is locally path connected, there is an open path connected U with $x \in U \subset U_1 \cap U_2$. By Exercise 10.7, U is evenly covered by p and by q. Hence $p^{-1}(U) = \bigcup S_j$, where the S_j are sheets in $\tilde{X}$; let $S = S_{j_0}$ be the sheet containing $\tilde{x}$. It suffices to prove that S is evenly covered by h (for h is a surjection, by Exercise 10.16).

Now $q^{-1}(U) = \bigcup T_k$, where the T_k are sheets in $\tilde{Y}$; thus the T_k are open, pairwise disjoint, and $q|T_k: T_k \to U$ are homeomorphisms, hence each T_k is path connected. For each k,

$$ph(T_k) = q(T_k) = U,$$

so that

$$h(T_k) \subset p^{-1}(U) = \bigcup S_j.$$

Since $h(T_k)$ is path connected and the S_j are open, pairwise disjoint, it follows that either $h(T_k) \subset S$ or $h(T_k) \cap S = \varnothing$. Therefore $h^{-1}(S)$ is the disjoint union of those T_k such that $h(T_k) \subset S$. Finally, if $h(T_k) \subset S$, then there is a commutative diagram

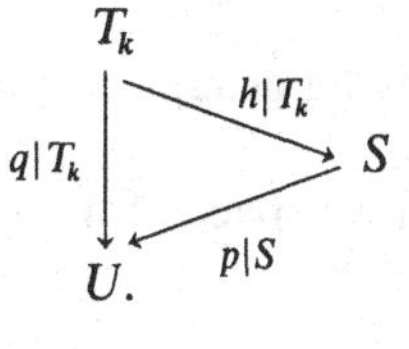

Since $q|T_k$ and $p|S$ are homeomorphisms, it follows that $h|T_k$ is a homeomorphism. We have shown that S is evenly covered by h. □

Definition. A **universal covering space** of X is a covering space $(\tilde{X}, p)$ with $\tilde{X}$ simply connected.

Often one abuses notation and says that $\tilde{X}$ is a universal covering space of X when $\tilde{X}$ is simply connected.

EXAMPLE 10.1. The space of real numbers **R** is a universal covering space of S^1, and the plane is a universal covering space of the torus. If $n \geq 2$, S^n is a universal covering space of $\mathbf{R}P^n$.

The reason for the adjective "universal" is provided by the next theorem. We defer the question of the existence of universal covering spaces (see Theorem 10.34).

Theorem 10.17. *Let X be connected and locally path connected, and let $(\tilde{Y}, q)$ be a covering space of X. If $(\tilde{X}, p)$ is a universal covering space of X, then there exists a unique continuous $h: \tilde{X} \to \tilde{Y}$ making the following diagram commute:*

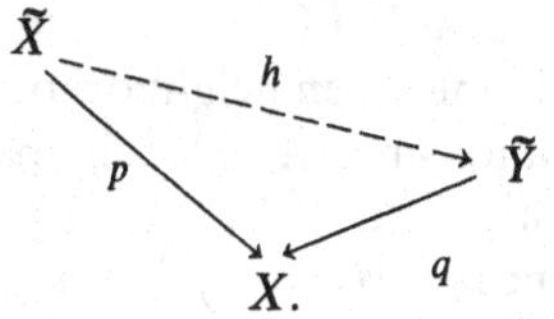

PROOF. Since X is locally path connected, Exercise 10.6 says that $\tilde{X}$ is locally path connected. Corollary 10.14 now gives the result. □

A standard argument shows that a universal covering space, if it exists, is unique to homeomorphism. The converse of Theorem 10.17 is not true unless some mild restrictions are imposed on X (see Corollary 10.35).

The fundamental group has already been seen to be intimately related to covering spaces. When comparing covering spaces $(\tilde{Y}, q)$ and $(\tilde{X}, p)$ of a space X, one considers diagrams of the form

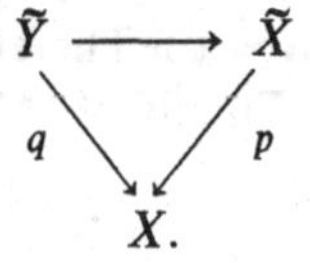

This leads one to the following group.

Definition. If $(\tilde{X}, p)$ is a covering space of X, then a **covering transformation** (or **deck transformation**) is a homeomorphism $h: \tilde{X} \to \tilde{X}$ with $ph = p$; that is, the following diagram commutes:

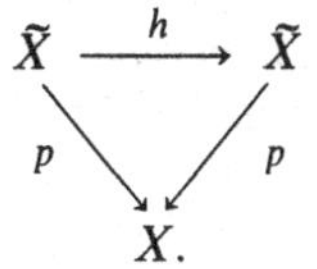

Define $\text{Cov}(\tilde{X}/X)$ as the set of all covering transformations of $\tilde{X}$.

It is easy to see that $\text{Cov}(\tilde{X}/X)$ is a group under composition of functions. Before continuing, we mention an analogy between groups of covering transformations and Galois groups. Suppose that F is a subfield of a field E. Recall that

$$\text{Gal}(E/F) = \{\text{automorphisms } \sigma\colon E \to E \mid \sigma \text{ fixes } F \text{ pointwise}\}.$$

If $i\colon F \hookrightarrow E$ is the inclusion, then an automorphism σ of E lies in $\text{Gal}(E/F)$ if and only if the following diagram commutes:

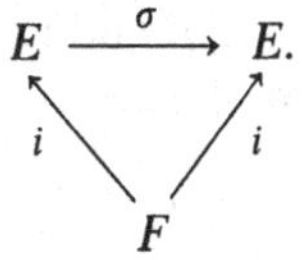

Since all arrows are reversed, one might expect that covering transformations give a "co-Galois theory"; that is, there may be "duals" for covering spaces of the usual results for Galois groups. In this analogy, universal covering spaces will play the role of algebraic closures (see Theorem 10.52).

In light of Theorem 10.9, the next result suggests that $\text{Cov}(\tilde{X}/X)$ resembles $\pi_1(X, x_0)$.

Theorem 10.18. *Let X be connected and locally path connected, and let $x_0 \in X$. Then a covering space $(\tilde{X}, p)$ of X is regular if and only if* $\text{Cov}(\tilde{X}/X)$ *acts transitively on the fiber over x_0.*

PROOF. Let $\tilde{x}_0, \tilde{x}_1 \in p^{-1}(x_0)$. If $(\tilde{X}, p)$ is regular, then Corollary 10.12 gives $p_*\pi_1(\tilde{X}, \tilde{x}_0) = p_*\pi_1(\tilde{X}, \tilde{x}_1)$. By Corollary 10.15, there is a homeomorphism $h\colon (\tilde{X}, \tilde{x}_0) \to (\tilde{X}, \tilde{x}_1)$ with $ph = p$; thus $h \in \text{Cov}(\tilde{X}/X)$ and $h(\tilde{x}_0) = \tilde{x}_1$, as desired.

Conversely, assume that $\text{Cov}(\tilde{X}/X)$ acts transitively on $p^{-1}(x_0)$: if $\tilde{x}_0, \tilde{x}_1 \in p^{-1}(x_0)$, then there exists $h \in \text{Cov}(\tilde{X}/X)$ with $h(\tilde{x}_0) = \tilde{x}_1$. Now $h_*\pi_1(\tilde{X}, \tilde{x}_0) = \pi_1(\tilde{X}, \tilde{x}_1)$. Since $p = ph$, it follows that $p_* = p_*h_*$, hence

$$p_*\pi_1(\tilde{X}, \tilde{x}_0) = p_*h_*\pi_1(\tilde{X}, \tilde{x}_0) = p_*\pi_1(\tilde{X}, \tilde{x}_1).$$

By Corollary 10.12, $p_*\pi_1(\tilde{X}, \tilde{x}_0)$ is a normal subgroup of $\pi_1(X, x_0)$, and so $(\tilde{X}, p)$ is regular. $\square$

Theorem 10.19. *Let $(\tilde{X}, p)$ be a covering space of X.*

(i) *If $h \in \operatorname{Cov}(\tilde{X}/X)$ and $h \neq 1_{\tilde{X}}$, then h has no fixed points.*
(ii) *If $h_1, h_2 \in \operatorname{Cov}(\tilde{X}/X)$ and there exists $\tilde{x} \in \tilde{X}$ with $h_1(\tilde{x}) = h_2(\tilde{x})$, then $h_1 = h_2$.*

PROOF. (i) Suppose that there exists $\tilde{x} \in \tilde{X}$ with $h(\tilde{x}) = \tilde{x}$; let $x = p\tilde{x}$. Consider the diagram

$$\begin{array}{ccc} (\tilde{X}, \tilde{x}) & \dashrightarrow & (\tilde{X}, \tilde{x}) \\ & {\scriptstyle p}\searrow \quad \swarrow{\scriptstyle p} & \\ & (X, x). & \end{array}$$

By Lemma 10.3, there is at most one way to complete this diagram so that it commutes. Since both h and $1_{\tilde{X}}$ complete it, $h = 1_{\tilde{X}}$, a contradiction.

(ii) The map $h_1^{-1}h_2 \in \operatorname{Cov}(\tilde{X}/X)$ has a fixed point, namely, $\tilde{x}$, and so $h_1^{-1}h_2 = 1_{\tilde{X}}$, by (i); therefore $h_1 = h_2$. □

Definition. Two covering spaces $(\tilde{Y}, q)$ and $(\tilde{X}, p)$ of a space X are **equivalent** if there exists a homeomorphism $\varphi\colon \tilde{Y} \to \tilde{X}$ making the following diagram commute:

$$\begin{array}{ccc} \tilde{Y} & \xrightarrow{\varphi} & \tilde{X} \\ & {\scriptstyle q}\searrow \quad \swarrow{\scriptstyle p} & \\ & X. & \end{array}$$

Theorem 10.20. *Let X be locally path connected, and let $x_0 \in X$. Let $(\tilde{Y}, q)$ and $(\tilde{X}, p)$ be covering spaces of X, and let $\tilde{x}_0 \in p^{-1}(x_0)$ and $\tilde{y}_0 \in q^{-1}(x_0)$. Then $(\tilde{Y}, q)$ and $(\tilde{X}, p)$ are equivalent if and only if $q_*\pi_1(\tilde{Y}, \tilde{y}_0)$ and $p_*\pi_1(\tilde{X}, \tilde{x}_0)$ are conjugate subgroups of $\pi_1(X, x_0)$.*

PROOF. Assume that $(\tilde{Y}, q)$ and $(\tilde{X}, p)$ are equivalent, and let $\varphi\colon \tilde{Y} \to \tilde{X}$ be a homeomorphism with $p\varphi = q$. Then $\varphi(\tilde{y}_0) \in p^{-1}(x_0)$ and $q_*\pi_1(\tilde{Y}, \tilde{y}_0) = p_*\pi_1(\tilde{X}, \varphi\tilde{y}_0)$. By Corollary 10.12(i), $p_*\pi_1(\tilde{X}, \varphi\tilde{y}_0)$ and $p_*\pi_1(\tilde{X}, \tilde{x}_0)$ are conjugate subgroups of $\pi_1(X, x_0)$.

Conversely, assume that $q_*\pi_1(\tilde{Y}, \tilde{y}_0)$ and $p_*\pi_1(\tilde{X}, \tilde{x}_0)$ are conjugate subgroups of $\pi_1(X, x_0)$. By Corollary 10.12(ii), there exists $\tilde{x}_1 \in p^{-1}(x_0)$ such that $q_*\pi_1(\tilde{Y}, \tilde{y}_0) = p_*\pi_1(\tilde{X}, \tilde{x}_1)$. The lifting criterion provides a continuous map $\varphi\colon \tilde{Y} \to \tilde{X}$ with $p\varphi = q$, and Corollary 10.15 says that φ is a homeomorphism. Therefore $(\tilde{Y}, q)$ and $(\tilde{X}, p)$ are equivalent. □

Recall that if $(\tilde{X}, p)$ is a covering space of X and if $x_0 \in X$, then $\pi_1(X, x_0)$ acts transitively on the fiber $p^{-1}(x_0)$: if $[f] \in \pi_1(X, x_0)$ and $\tilde{x} \in p^{-1}(x_0)$, then $\tilde{x}[f] = \tilde{f}(1)$, where $\tilde{f}$ is the lifting of f with $\tilde{f}(0) = \tilde{x}$.

Definition. Let G be a group, and let Y and Z be G-sets. A function $\varphi\colon Y \to Z$ is a ***G*-map** (one also says that φ is ***G*-equivariant**) if

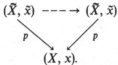

$$\varphi(gy) = g\varphi(y)$$

for all $g \in G$ and $y \in Y$. A **G-isomorphism** is a G-map that is also a bijection. Let Aut(Y) denote the group (under composition) of all G-isomorphisms from Y to itself.

We are going to prove that a covering space $(\tilde{X}, p)$ of a space X (with basepoint x_0) is completely determined by the fiber $p^{-1}(x_0)$ viewed as a $\pi_1(X, x_0)$-set. Another group-theoretic lemma is needed.

Let G be a group, let H be a (not necessarily normal) subgroup of G, and let $G//H$ denote the family of all left cosets of H in G. Now G acts on $G//H$ by left translation: if $a \in G$ and $gH \in G//H$, then $a: gH \mapsto agH$. It is easy to see that $G//H$ is a transitive G-set and that H is the stabilizer of the coset H.

Lemma 10.21.

(i) *If X is a transitive G-set and H is the stabilizer of a point, then X is G-isomorphic to $G//H$, the family of all left cosets of H in G on which G acts by left translation.*
(ii) *If H and K are subgroups of a group G, then $G//H$ and $G//K$ are G-isomorphic if and only if H and K are conjugate in G.*

PROOF. (i) Let $x_0 \in X$, and let $H = G_{x_0}$. For each $x \in X$, transitivity provides an element $g_x \in G$ with $g_x x_0 = x$. The routine argument that $\theta: H \to G//H$, defined by $\theta(x) = g_x H$, is a well defined bijection is left to the reader. To see that θ is a G-isomorphism, let $a \in G$ and $x \in X$. Now $x = g_x x_0$ and $ax = g_{ax} x_0$; hence $ax = ag_x x_0$, and so $g_{ax}^{-1} a g_x \in G_{x_0} = H$. Thus $g_{ax}H = ag_x H$. But $\theta(ax) = g_{ax}H$ and $a\theta(x) = ag_x H$, so that $\theta(ax) = a\theta(x)$, as desired.

(ii) Assume that $\theta: G//H \to G//K$ is a G-isomorphism. There exists $g \in G$ with $\theta(H) = gK$. If $h \in H$, then

$$gK = \theta(H) = \theta(hH) = h\theta(H) = hgK.$$

Therefore $g^{-1}hg \in K$ and $g^{-1}Hg \subset K$. Since $\theta(g^{-1}H) = g^{-1}\theta(H) = g^{-1}gK = K$, we see that $\theta^{-1}(K) = g^{-1}H$. The argument above now gives $gKg^{-1} \subset H$, hence $g^{-1}Hg = K$.

For the converse, choose $g \in G$ with $g^{-1}Hg = K$. Observe that the following are equivalent for $a, b \in G$: $aH = bH$; $a^{-1}b \in H$; $g^{-1}a^{-1}bg \in g^{-1}Hg = K$; $agK = bgK$. We conclude that the function $\theta: G//H \to G//K$ given by $\theta(aH) = agK$ is a well defined injection. Clearly, θ is onto, because $b \in G$ implies $bK = \theta(bg^{-1}H)$. Finally, θ is a G-map, because $\theta(abH) = (ab)gK$ and $a\theta(bH) = a(bgK)$. □

Corollary 10.22. *Let X be locally path connected, and let $x_0 \in X$. Two covering spaces $(\tilde{X}, p)$ and $(\tilde{Y}, q)$ of X are equivalent if and only if the fibers $p^{-1}(x_0)$ and $q^{-1}(x_0)$ are isomorphic $\pi_1(X, x_0)$-sets.*

Proof. Choose $\tilde{x}_0 \in p^{-1}(x_0)$ and $\tilde{y}_0 \in q^{-1}(x_0)$. By Theorem 10.20, $(\tilde{X}, p)$ and $(\tilde{Y}, q)$ are equivalent if and only if $p_*\pi_1(\tilde{X}, \tilde{x}_0)$ and $q_*\pi_1(\tilde{Y}, \tilde{y}_0)$ are conjugate subgroups of $\pi_1(X, x_0)$. By Theorem 10.9, the fiber $p^{-1}(x_0)$ is a transitive $\pi_1(X, x_0)$-set, and $p_*\pi_1(\tilde{X}, \tilde{x}_0)$ is the stabilizer of $\tilde{x}_0$; similarly, $q^{-1}(x_0)$ is a transitive $\pi_1(X, x_0)$-set and $q_*\pi_1(\tilde{Y}, \tilde{y}_0)$ is the stabilizer of $\tilde{y}_0$. It follows from the lemma that $p_*\pi_1(\tilde{X}, \tilde{x}_0)$ and $q_*\pi_1(\tilde{Y}, \tilde{y}_0)$ are conjugate subgroups of $\pi_1(X, x_0)$ if and only if the fibers are $\pi_1(X, x_0)$-isomorphic. □

This last corollary explains why so much of the theory of permutation groups appears in this context.

Lemma 10.23. *Let a group G act transitively on a set Y, and let $x, y \in Y$. Then the stabilizers G_x and G_y are equal if and only if there exists $\varphi \in \text{Aut}(Y)$ with $\varphi(x) = y$.*

Proof. Assume that there exists $\varphi \in \text{Aut}(Y)$ with $\varphi(x) = y$. If $h \in G_x$, then $hx = x$ and $\varphi(hx) = \varphi(x) = y$; on the other hand, $\varphi(hx) = h\varphi(x) = hy$, and so h fixes y. Therefore $G_x \subset G_y$; the reverse inclusion is proved similarly.

Conversely, assume that $G_x = G_y$. If $z \in Y$, then there exists $g \in G$ with $z = gx$; define φ: $Y \to Y$ by $\varphi(z) = \varphi(gx) = gy$. Now φ is well defined, because if $gx = g_1x$, then $g^{-1}g_1x = x$, hence $g^{-1}g_1 \in G_x = G_y$, and so $gy = g_1y$. Also, φ is a G-map, because $\varphi(hz) = \varphi(hgx) = hgy = h\varphi(z)$. Finally, φ is a bijection: its inverse is θ: $Y \to Y$, where $\theta(g'y) = g'x$. □

Lemma 10.24. *Let $(\tilde{X}, p)$ be a covering space of X, where X is locally path connected; let $x_0 \in X$, and recall that $p^{-1}(x_0)$ is a transitive $\pi_1(X, x_0)$-set. Given $\tilde{x}_0, \tilde{x}_1 \in p^{-1}(x_0)$, there exists $h \in \text{Cov}(\tilde{X}/X)$ with $h(\tilde{x}_0) = \tilde{x}_1$ if and only if there exists $\varphi \in \text{Aut}(p^{-1}(x_0))$ with $\varphi(\tilde{x}_0) = \tilde{x}_1$.*

Proof. If there exists $h \in \text{Cov}(\tilde{X}/X)$ with $h(\tilde{x}_0) = \tilde{x}_1$, then the lifting criterion (with h and with h^{-1}) gives $p_*\pi_1(\tilde{X}, \tilde{x}_0) = p_*\pi_1(\tilde{X}, \tilde{x}_1)$; the converse follows from Corollary 10.15. Since $p_*\pi_1(\tilde{X}, \tilde{x}_0)$ is the stabilizer of $\tilde{x}_0$, by Theorem 10.9(ii), h exists if and only if the stabilizers of $\tilde{x}_0$ and $\tilde{x}_1$ coincide. But, by Lemma 10.23, these stabilizers coincide if and only if there exists $\varphi \in \text{Aut}(p^{-1}(x_0))$ with $\varphi(\tilde{x}_0) = \tilde{x}_1$. □

Lemma 10.25. *Let $(\tilde{X}, p)$ be a covering space of X, where X is locally path connected. Let $x_0 \in X$, and let the fiber $p^{-1}(x_0)$ be viewed as a $\pi_1(X, x_0)$-set. Then $h \mapsto h|p^{-1}(x_0)$ is an isomorphism*

$$\text{Cov}(\tilde{X}/X) \cong \text{Aut}(p^{-1}(x_0)).$$

Proof. Denote $p^{-1}(x_0)$ by Y. If $h \in \text{Cov}(\tilde{X}/X)$, then it is plain that $h(Y) = Y$ and that $h|Y$: $Y \to Y$ is a bijection. To see that $h|Y$ is a $\pi_1(X, x_0)$-isomorphism, consider $[f] \in \pi_1(X, x_0)$ and $\tilde{x} \in Y$. Now

$$h([f]\tilde{x}) = h\tilde{f}(1),$$

where $\tilde{f}$ is the lifting of f with $\tilde{f}(0) = \tilde{x}$. On the other hand,

$$[f]h(\tilde{x}) = \tilde{f}_1(1),$$

where $\tilde{f}_1$ is the lifting of f with $\tilde{f}_1(0) = h(\tilde{x})$. But $ph\tilde{f} = p\tilde{f} = f$ and $h\tilde{f}(0) = h(\tilde{x})$, so that uniqueness gives $\tilde{f}_1 = h\tilde{f}$. Hence $h([f]\tilde{x}) = [f]h(\tilde{x})$, as desired.

Clearly, $h \mapsto h|Y$ is a homomorphism. By Theorem 10.19(i), this map is an injection. To see that this map is a surjection, let $\varphi \in \text{Aut}(Y)$. If $\tilde{x} \in Y$, then Lemma 10.24 provides $h \in \text{Cov}(\tilde{X}/X)$ with $h(\tilde{x}) = \varphi(\tilde{x})$. Since $\pi_1(X, x_0)$ acts transitively on Y, for each $\tilde{x}_1 \in Y$ there exists $[f] \in \pi_1(X, x_0)$ with $\tilde{x}_1 = [f]\tilde{x}$. Therefore

$$h(\tilde{x}_1) = h([f]\tilde{x}) = [f]h(\tilde{x}) = [f]\varphi(\tilde{x}) = \varphi([f]\tilde{x}) = \varphi(\tilde{x}_1),$$

and so $h|Y = \varphi$, as desired. □

Recall that if H is a subgroup of a group G, then its **normalizer** is the subgroup

$$N_G(H) = \{g \in G: gHg^{-1} = H\}.$$

Note that H is a normal subgroup of $N_G(H)$; moreover, if H is a normal subgroup of G, then $N_G(H) = G$.

Lemma 10.26. *Let G be a group acting transitively on a set Y, and let $y_0 \in Y$. Then*

$$\text{Aut}(Y) \cong N_G(G_0)/G_0,$$

where G_0 is the stabilizer of y_0.

Proof. Let $\varphi \in \text{Aut}(Y)$. Since G acts transitively on Y, there is $g \in G$ with $\varphi(y_0) = gy_0$. First, we show that $g \in N_G(G_0)$. If $h \in G_0$, then $hy_0 = y_0$ and

$$gy_0 = \varphi(y_0) = \varphi(hy_0) = h\varphi(y_0) = hgy_0;$$

hence $y_0 = g^{-1}hgy_0$ and $g^{-1}hg \in G_0$, as desired. Second, if $\varphi(y_0) = gy_0 = g_1 y_0$, then $g^{-1}g_1$ fixes y_0 and $g_1 G_0 = gG_0$. Therefore the function

$$\Gamma: \text{Aut}(Y) \to N_G(G_0)/G_0$$

defined by

$$\Gamma(\varphi) = g^{-1}G_0,$$

where $\varphi(y_0) = gy_0$, is a well defined function.

To see that Γ is a homomorphism, let $\theta \in \text{Aut}(Y)$ and let $\theta(y_0) = g'y_0$. Now $\theta\varphi(y_0) = \theta(gy_0) = g\theta(y_0) = gg'y_0$, so that $\Gamma(\theta\varphi) = (gg')^{-1}G_0$. On the other hand, $\Gamma(\theta)\Gamma(\varphi) = g'^{-1}G_0 g^{-1}G_0 = g'^{-1}g^{-1}G_0$. Since $(gg')^{-1} = g'^{-1}g^{-1}$, it follows that Γ is a homomorphism (the reason for the inverse in the definition of Γ is now apparent).

Assume that $\Gamma(\varphi) = G_0$. Then $\varphi(y_0) = y_0$, hence $\varphi(hy_0) = h\varphi(y_0) = hy_0$ for every $h \in G$; that is, φ fixes every element in Y of the form hy_0. As G acts transitively, $\varphi = 1_Y$, and so Γ is an injection.

Finally, assume that $g \in N_G(G_0)$. The function $\varphi: Y \to Y$ given by $\varphi(y) = hgy_0$, where $y = hy_0$, is easily seen to be a well defined G-automorphism of Y. As $\Gamma(\varphi) = g^{-1}G_0$, it follows that Γ is a surjection and hence that Γ is an isomorphism. □

Theorem 10.27. *Let $(\tilde{X}, p)$ be a covering space of X, where X is locally path connected. Then, for $x_0 \in X$ and $\tilde{x}_0 \in p^{-1}(x_0)$,*

$$\mathrm{Cov}(\tilde{X}/X) \cong N_\pi(p_*\pi_1(\tilde{X}, \tilde{x}_0))/p_*\pi_1(\tilde{X}, \tilde{x}_0),$$

where π denotes $\pi_1(X, x_0)$.

PROOF. By Lemma 10.25, $\mathrm{Cov}(\tilde{X}/X) \cong \mathrm{Aut}(p^{-1}(x_0))$, where the fiber $p^{-1}(x_0)$ is viewed as a transitive $\pi_1(X, x_0)$-set (Theorem 10.9(i)). The stabilizer of $\tilde{x}_0$ is $p_*\pi_1(\tilde{X}, \tilde{x}_0)$, by Theorem 10.9(ii). The theorem now follows from Lemma 10.26. □

Corollary 10.28. *Let $(\tilde{X}, p)$ be a regular covering space of X, where X is locally path connected. Then, for $x_0 \in X$ and $\tilde{x}_0 \in p^{-1}(x_0)$,*

$$\mathrm{Cov}(\tilde{X}/X) \cong \pi_1(X, x_0)/p_*\pi_1(\tilde{X}, \tilde{x}_0),$$

the monodromy group of the regular covering space.

PROOF. Since $(\tilde{X}, p)$ is a regular covering space of Y, $p_*\pi_1(\tilde{X}, \tilde{x}_0)$ is a normal subgroup of $\pi_1(X, x_0)$, and so $p_*\pi_1(\tilde{X}, \tilde{x}_0) = p_*\pi_1(\tilde{X}, y)$ for all $y \in p^{-1}(x_0)$. □

Corollary 10.29. *Let $(\tilde{X}, p)$ be a universal covering space of X, where X is locally path connected. Then, for $x_0 \in X$,*

$$\mathrm{Cov}(\tilde{X}/X) \cong \pi_1(X, x_0).$$

PROOF. Since $\tilde{X}$ is simply connected, $\pi_1(\tilde{X}, \tilde{x}_0) = \{1\}$ for every $\tilde{x}_0 \in p^{-1}(x_0)$, and so $p_*\pi_1(\tilde{X}, \tilde{x}_0) = \{1\}$. □

Observe that the last result gives a description of the fundamental group of X, which requires no choice of basepoint.

EXAMPLE 10.2. We use Corollary 10.29 to give another proof that $\pi_1(S^1, 1) \cong \mathbf{Z}$. Since $\mathbf{R}$ is simply connected, $(\mathbf{R}, \exp)$ is a universal covering space of S^1 (of course, S^1 is locally path connected); hence $\mathrm{Cov}(\mathbf{R}/S^1) \cong \pi_1(S^1, 1)$. Let $h: \mathbf{R} \to \mathbf{R}$ be a homeomorphism with $\exp(h(x)) = \exp(x)$; then $h(x) = x + n(x)$, where $n(x) \in \mathbf{Z}$ (by definition, $\exp(x) = e^{2\pi i x}$). Hence $n(x) = h(x) - x$ is a continuous map $\mathbf{R} \to \mathbf{Z}$; as $\mathbf{R}$ is connected and $\mathbf{Z}$ is discrete, $n(x)$ is constant, say, $n(x) \equiv n$.

Therefore $h(x)$ is the translation $x \mapsto x + n$. It is clear that all such translations under composition form an infinite cyclic group.

EXERCISES

10.18. Let $(\tilde{X}, p)$ be a universal covering space of X, where X is locally path connected. If $x_0 \in X$, give an explicit isomorphism $\text{Cov}(\tilde{X}/X) \to \pi_1(X, x_0)$.

10.19. In Exercises 8.6 and 8.7, it is shown that complex and quaternionic projective spaces are quotient spaces of spheres. Are these spheres universal covering spaces?

10.20. If G is a simply connected and locally path connected topological group, if H is a discrete normal subgroup, and if $p: G \to G/H$ is the natural map, then every continuous $\varphi: G \to G$ with $p\varphi = p$ has the form $\varphi(x) = xh_0$ for some $h_0 \in H$. (*Hint*: Adapt the argument in Example 10.2.)

10.21. Let $(\tilde{X}, p)$ be a covering space of X, where X is locally path connected. Prove that $(\tilde{X}, p)$ is regular if and only if, for each closed path $f: \mathbf{I} \to X$, either every lifting $\tilde{f}$ of f is a closed path or no lifting $\tilde{f}$ of f is a closed path. (*Hint*: Exercise 10.17.)

10.22. Let $(\tilde{X}, p)$ be a covering space of X, where X is locally path connected. If $(\tilde{X}, p)$ is regular, then the monodromy group of $(\tilde{X}, p)$ (see Exercise 10.13) is isomorphic to $\text{Cov}(\tilde{X}/X)$.

10.23. If X is an H-space (*a fortiori*, if X is a topological group), then every covering space of X is regular. (*Hint*: Exercise 10.15.)

Existence

When does a space X possess a universal covering space $(\tilde{X}, p)$? More generally, given $x_0 \in X$ and a subgroup G of $\pi_1(X, x_0)$, when does there exist a covering space $(\tilde{X}, p)$ of X (and a point $\tilde{x}_0 \in p^{-1}(x_0)$) with $G = p_*\pi_1(\tilde{X}, \tilde{x}_0)$?

Definition. Let G be a subgroup of $\pi_1(X, x_0)$ and let $P(X, x_0)$ be the family of all paths f in X with $f(0) = x_0$. Define $f_1 \sim f_2$ (more precisely, $f_1 \sim f_2 \bmod G$) by

(i) $f_1(1) = f_2(1)$;
(ii) $[f_1 * f_2^{-1}] \in G$.

Lemma 10.30. *If G is a subgroup of $\pi_1(X, x_0)$, then the relation $f_1 \sim f_2$ is an equivalence relation on $P(X, x_0)$.*

PROOF. Reflexivity holds because $1 \in G$; symmetry holds because $g \in G$ implies that $g^{-1} \in G$; transitivity holds because $g, h \in G$ implies that $gh \in G$. □

Definition. Let (X, x_0) be a pointed space and let G be a subgroup of $\pi_1(X, x_0)$. Denote the equivalence class of $f \in P(X, x_0)$ by $\langle f \rangle_G$, and define $\tilde{X}_G$ as the set of all such equivalence classes. If e_0 is the constant path at x_0, define $\tilde{x}_0 = \langle e_0 \rangle_G \in \tilde{X}_G$. Finally, define a function $p: \tilde{X}_G \to X$ by $\langle f \rangle_G \mapsto f(1)$.

It is obvious that $p(\tilde{x}_0) = x_0$. We shall prove that, with some mild conditions on X, the set $\tilde{X}_G$ can be topologized so that $(\tilde{X}_G, p)$ is a covering space of X with $p_* \pi_1(\tilde{X}_G, \tilde{x}_0) = G$.

Definition. If $f \in P(X, x_0)$ and U is an open neighborhood of $f(1)$, then a **continuation of** f **in** U is a path $F \in P(X, x_0)$ of the form $F = f * \lambda$, where $\lambda(0) = f(1)$ and $\lambda(\mathbf{I}) \subset U$.

Definition. Let $\tilde{x} = \langle f \rangle_G$, and let U be an open neighborhood of x in X. Then

$$(U, \tilde{x}) = (U, \langle f \rangle_G) = \{\langle F \rangle_G \in \tilde{X}_G: F \text{ is a continuation of } f \text{ in } U\}.$$

Note that if $f \sim f'$ and $\lambda(0) = f(1)$, then $f * \lambda \sim f' * \lambda$.

Recall that a family $\mathscr{B}$ of subsets of a set Y is a **basis** for a topology if:

(B1) for each $y \in Y$, there is $B \in \mathscr{B}$ with $y \in B$;
(B2) if $B_1, B_2 \in \mathscr{B}$ and if $y \in B_1 \cap B_2$, then there is $B_3 \in \mathscr{B}$ with $y \in B_3 \subset B_1 \cap B_2$.

The corresponding topology on Y is the family of all unions of sets in $\mathscr{B}$.

One may rephrase Corollary 1.19 by saying that a space is locally path connected if and only if it has a basis of path connected subsets.

Lemma 10.31. *Let (X, x_0) be a pointed topological space, and let G be a subgroup of $\pi_1(X, x_0)$. Then the subsets $(U, \tilde{x})$ form a basis for a topology on $\tilde{X}_G$ for which the function $p: \tilde{X}_G \to X$ is continuous. Moreover, if X is path connected, then p is a surjection.*

PROOF. Let $\tilde{x} = \langle f \rangle_G \in \tilde{X}_G$, and let e be the constant path in X at $f(1)$. For every open neighborhood U of $f(1)$, the function $F = f * e$ is a continuation of f in U. Therefore (B1) holds, for $\tilde{x} = \langle f \rangle_G = \langle F \rangle_G \in (U, \tilde{x})$.

We show that if $\tilde{y} \in (U, \tilde{x})$, then $(U, \tilde{x}) = (U, \tilde{y})$. Now $\tilde{y} = \langle F \rangle_G = \langle f * \lambda \rangle_G$, where $\lambda(0) = f(1)$ and $\lambda(\mathbf{I}) \subset U$. If $\tilde{z} \in (U, \tilde{x})$, then $\tilde{z} = \langle F' \rangle_G = \langle f * \mu \rangle_G$, where $\mu(0) = f(1)$ and $\mu(\mathbf{I}) \subset U$. Hence $F' \sim f * \mu \sim (f * \lambda) * (\lambda^{-1} * \mu) \sim F * (\lambda^{-1} * \mu)$; since $(\lambda^{-1} * \mu)(0) = F(1)$ and $(\lambda^{-1} * \mu)(\mathbf{I}) \subset U$, we have $\tilde{z} = \langle F' \rangle_G = \langle F * (\lambda^{-1} * \mu) \rangle_G \in (U, \tilde{y})$ and $(U, \tilde{x}) \subset (U, \tilde{y})$. The reverse inclusion is proved similarly. To prove (B2), assume that $\tilde{z} \in (U, \tilde{x}) \cap (V, \tilde{y})$; then $(U, \tilde{x}) = (U, \tilde{z})$ and $(V, \tilde{y}) = (V, \tilde{z})$, and it is easy to see that $\tilde{z} \in (U \cap V, \tilde{z}) \subset (U, \tilde{z}) \cap (V, \tilde{z})$.

To prove that $p: \tilde{X}_G \to X$ is continuous, let $\tilde{x} \in \tilde{X}$ and let U be an open neighborhood of $p\tilde{x}$ in X. Then it is easy to see that $p((U, \tilde{x})) \subset U$. Finally, if X is path connected, then for each $x \in X$, there is a path f in X from x_0 to x, and $p(\tilde{x}) = x$, where $\tilde{x} = \langle f \rangle_G$. □

Lemma 10.32. *Let (X, x_0) be a pointed space, and let G be a subgroup of $\pi_1(X, x_0)$. Every path f in X beginning at x_0 can be lifted to a path $\tilde{f}$ in $\tilde{X}_G$ beginning at $\tilde{x}_0$ and ending at $\langle f \rangle_G$.*

PROOF. For $t \in \mathbf{I}$, define $f_t: \mathbf{I} \to X$ by $f_t(s) = f(ts)$. Each f_t is a path in X beginning at x_0; that is, $f_t \in P(X, x_0)$, $f_0 = e_0$ (the constant path at x_0), and $f_1 = f$. Define $\tilde{f}: \mathbf{I} \to \tilde{X}$ by

$$\tilde{f}(t) = \langle f_t \rangle_G.$$

Observe that $\tilde{f}(0) = \langle f_0 \rangle_G = \langle e_0 \rangle_G = \tilde{x}_0$ and that $\tilde{f}(1) = \langle f_1 \rangle_G = \langle f \rangle_G$. Moreover, for each $t \in \mathbf{I}$, we have $p\tilde{f}(t) = p\langle f_t \rangle_G = f_t(1) = f(t)$, that is, $p\tilde{f} = f$. It remains to prove that $\tilde{f}$ is continuous.

Let $t_0 \in \mathbf{I}$ and let $(U, \tilde{f}(t_0))$ be a basic open set containing $\tilde{f}(t_0)$. Since f is continuous, there is an open interval V of t_0 in $\mathbf{I}$ with $f(V) \subset U$; we claim that $\tilde{f}(V) \subset (U, \tilde{f}(t_0))$, that is, if $t \in V$, then f_t is a continuation of f_{t_0} in U. It is straightforward to show that $f_t = f_{t_0} * \lambda$ for some path λ with $\lambda(0) = f_{t_0}(1) = f(t_0)$ and with $\lambda(\mathbf{I}) \subset U$: if $t > t_0$, then let $\lambda = f|[t_0, t]$ suitably reparametrized so that its domain is $\mathbf{I}$; if $t < t_0$, reparametrize $f^{-1}|[t, t_0]$. □

Corollary 10.33. *If (X, x_0) is a pointed space and G is a subgroup of $\pi_1(X, x_0)$, then $\tilde{X}_G$ is path connected.*

PROOF. For each $\tilde{x} = \langle f \rangle_G \in \tilde{X}_G$, there is a path in $\tilde{X}_G$ from $\tilde{x}_0$ to $\tilde{x}$. □

There is a necessary condition that a locally path connected space X have a universal covering space $(\tilde{X}, p)$. If $x \in X$, then Exercise 10.7 allows us to assume that x has a path connected admissible open neighborhood U. Let $\tilde{x} \in p^{-1}(x)$, and let S be the sheet lying over U that contains $\tilde{x}$. There is a commutative diagram

$$\begin{array}{ccc} \pi_1(S, \tilde{x}) & \longrightarrow & \pi_1(\tilde{X}, \tilde{x}) \\ {\scriptstyle (p|S)_*}\downarrow & & \downarrow{\scriptstyle p_*} \\ \pi_1(U, x) & \longrightarrow & \pi_1(X, x), \end{array}$$

where the horizontal maps are induced by inclusions. Since $\pi_1(\tilde{X}, \tilde{x}) = \{1\}$ (because $\tilde{X}$ is simply connected) and $(p|S)_*$ is an isomorphism (because $p|S$ is a homeomorphism), it follows that $\pi_1(U, x) \to \pi_1(X, x)$ is the trivial map.

Definition. A space X is **semilocally 1-connected**[3] if each $x \in X$ has an open neighborhood U so that $i_*: \pi_1(U, x) \to \pi_1(X, x)$ is the trivial map (where $i: U \hookrightarrow X$ is the inclusion).

EXAMPLE 10.3. Every simply connected space is semilocally 1-connected.

[3] A space X is called **locally 1-connected** if, for each $x \in X$, every neighborhood N of x contains a neighborhood U of x with $i_*: \pi_1(U, x) \to \pi_1(N, x)$ trivial. Compare this definition with that of locally path connected (which could be called **locally 0-connected**).

Example 10.4. If each point $x \in X$ has a contractible open neighborhood, then X is semilocally 1-connected. By Corollary 8.31, every CW complex (and hence every simplicial complex) is semilocally 1-connected.

Example 10.5. A cartesian product of infinitely many circles is connected and locally path connected, but it is not semilocally 1-connected (see [Spanier, p. 84]).

One can rephrase the definition. A space X is semilocally 1-connected if each $x \in X$ has an open neighborhood U with the following property: every closed path in U at x is nullhomotopic in X.

Theorem 10.34. *Let (X, x_0) be a pointed space and let G be a subgroup of $\pi_1(X, x_0)$. If X is connected, locally path connected, and semilocally 1-connected, then $(\tilde{X}_G, p)$ is a covering space of X and $p_*\pi_1(\tilde{X}_G, \tilde{x}_0) = G$.*

Proof. Let $x \in X$. Since X is semilocally 1-connected, there is an open neighborhood W of x with every closed path in W at x nullhomotopic in X. Since X is locally path connected, there is an open path connected neighborhood U of x with $x \in U \subset W$; of course, every closed path in U at x is nullhomotopic in X. We shall show that U is evenly covered by p, and this will show that $(\tilde{X}_G, p)$ is a covering space of X (for we already know that $\tilde{X}_G$ is path connected and p is a continuous surjection).

Let $\tilde{x} \in p^{-1}(x)$, so that $\tilde{x} = \langle f \rangle_G$, where f is a path in X from x_0 to x. To prove that U is evenly covered by p, we shall show that $(U, \tilde{x})$ is the sheet over U containing $\tilde{x}$. First, $p|(U, \tilde{x})\colon (U, \tilde{x}) \to U$ is a surjection. If $y \in U$, there exists a path λ in U from x to y (because U is path connected). Then $f * \lambda$ is a continuation of f in U with $(f * \lambda)(1) = y$; hence $\langle f * \lambda \rangle_G \in (U, \tilde{x})$ and $p(\langle f * \lambda \rangle_G) = (f * \lambda)(1) = y$. Second, $p|(U, \tilde{x})$ is an injection. Suppose that $\tilde{y}$, $\tilde{z} \in (U, \tilde{x})$ and $p(\tilde{y}) = p(\tilde{z})$. Now $\tilde{z} = \langle f * \mu \rangle_G$, where $\mu(0) = f(1) = x$ and $\mu(\mathbf{I}) \subset U$; similarly, $\tilde{y} = \langle f * \lambda \rangle_G$, where $\lambda(0) = x$ and $\lambda(\mathbf{I}) \subset U$. Since $p(\tilde{y}) = p(\tilde{z})$, we have $\lambda(1) = \mu(1)$, so that $\lambda * \mu^{-1}$ is a closed path in U at x. By the choice of U, $\lambda * \mu^{-1}$ is nullhomotopic in X. Hence $f * \lambda * \mu^{-1} * f^{-1}$ is nullhomotopic in X; that is, $[f * \lambda * \mu^{-1} * f^{-1}] = 1$ in $\pi_1(X, x_0)$. Therefore $[f * \lambda * \mu^{-1} * f^{-1}] \in G$, and so $\langle f * \lambda \rangle_G = \langle f * \mu \rangle_G$, that is, $\tilde{y} = \tilde{z}$. Third, $p|(U, \tilde{x})$ is an open map. Every neighborhood $\tilde{W}$ of $\tilde{x}$ in $\tilde{X}_G$ contains an open set of the form $(U, \tilde{x})$, where U is as chosen in the first paragraph. But, for such U, we know that $p((U, \tilde{x})) = U$ (because $p|(U, \tilde{x})$ is a surjection). It follows that $p|(U, \tilde{x})\colon (U, \tilde{x}) \to U$ is a homeomorphism.

Next, we show that $p^{-1}(U) = \bigcup_{\tilde{x}} (U, \tilde{x})$. Clearly, $p^{-1}(U)$ contains the union. For the reverse inclusion, let $\tilde{y} \in \tilde{X}_G$ be such that $p(\tilde{y}) \in U$, that is, $\tilde{y} = \langle f \rangle_G$ and $f(1) \in U$. Since U is path connected, there is a path λ in U from $f(1)$ to x. Then $f * \lambda$ is a continuation of f in U, so that $\tilde{x}$ defined by $\tilde{x} = \langle f * \lambda \rangle_G$ lies in the fiber over x. Now $(f * \lambda) * \lambda^{-1}$ is a continuation of $f * \lambda$ in U, so that $\langle (f * \lambda) * \lambda^{-1} \rangle_G \in (U, \tilde{x})$. But $\tilde{y} = \langle f \rangle_G = \langle (f * \lambda) * \lambda^{-1} \rangle_G$.

As each $(U, \tilde{x})$ is open in $\tilde{X}_G$, it remains to prove that the sheets are pairwise disjoint. In the proof of Lemma 10.31, we showed that if $\tilde{y} \in (U, \tilde{x})$, then

$(U, \tilde{x}) = (U, \tilde{y})$. If $\tilde{x}_1, \tilde{x}_2 \in p^{-1}(x)$, and if there exists $\tilde{y} \in (U, \tilde{x}_1) \cap (U, \tilde{x}_2)$, then $(U, \tilde{x}_i) = (U, \tilde{y})$ for $i = 1, 2$, and $(U, \tilde{x}_1) = (U, \tilde{x}_2)$. We have proved that $(\tilde{X}_G, p)$ is a covering space of X.

Finally, let us show that $p_* \pi_1(\tilde{X}_G, \tilde{x}_0) = G$. Let $[f] \in \pi_1(X, x_0)$. Since $(\tilde{X}_G, p)$ is a covering space of X, there exists a unique lifting $\tilde{f}$ of f with $\tilde{f}(0) = \tilde{x}_0$. In Lemma 10.32, however, we constructed such a lifting, namely, $\tilde{f}(t) = \langle f_t \rangle_G$, where f_t is a path from x_0 to $f(t)$. By Exercise 10.17, $[f] \in p_* \pi_1(\tilde{X}_G, \tilde{x}_0)$ if and only if $\tilde{f}$ is a closed path at $\tilde{x}_0$, that is, $\tilde{f}(0) = \tilde{f}(1) = \tilde{x}_0$. But $\tilde{f}(0) = \tilde{x}_0 = \langle e_0 \rangle_G$, where e_0 is the constant path in X at x_0, while $\tilde{f}(1) = \langle f_1 \rangle_G = \langle f \rangle_G$. Hence $\tilde{f}(0) = \tilde{f}(1)$ if and only if $f \sim e_0$. But $f \sim e_0$ if and only if $[f] = [f * e_0^{-1}] \in G$. Therefore $p_* \pi_1(\tilde{X}_G, \tilde{x}_0) = G$, as desired. □

Corollary 10.35. *Let X be a connected, locally path connected, semilocally 1-connected space.*[4] *Every covering space $(\tilde{Y}, q)$ of X is equivalent to a covering space of the form $(\tilde{X}_G, p)$.*

PROOF. Choose a basepoint $x_0 \in X$ and let $\tilde{y}_0 \in \tilde{Y}$ lie in the fiber over x_0. If $G = q_* \pi_1(\tilde{Y}, \tilde{y}_0)$, then $p_* \pi_1(\tilde{X}_G, \tilde{x}_0) = G$, and so Theorem 10.20 applies to show that $(\tilde{Y}, q)$ is equivalent to $(\tilde{X}_G, p)$. □

Corollary 10.36. *Let X be a connected, locally path connected, semilocally 1-connected space. If $(\tilde{X}, p)$ is a covering space of X, then every open contractible set U in X is evenly covered by p.*

PROOF. In the proof of the theorem, we saw that if U is an open path connected set in X for which every closed path in U is nullhomotopic in X, then U is evenly covered by p (indeed, if $x \in U$, then

$$p^{-1}(U) = \bigcup_{\tilde{x} \in p^{-1}(x)} (U, \tilde{x})).$$

In particular, contractible open sets are evenly covered in every covering space of the form $(\tilde{X}_G, p)$. The result follows from Corollary 10.35. □

Corollary 10.37. *Let X be connected and locally path connected. Then X has a universal covering space if and only if X is semilocally 1-connected.*

PROOF. Sufficiency follows immediately from the theorem; necessity was proved in our discussion given just before the definition of semilocally 1-connected. □

We repeat the description of the elements of $\tilde{X}_G$ when it is simply connected, that is, when $G = \{1\}$: they are the equivalence classes of $P(X, x_0)$ defined by the relation $f \sim g$ if $f(1) = g(1)$ and $f * g^{-1}$ is nullhomotopic in X.

---

[4] Perhaps such spaces should be called *triply connected*!

Theorem 10.38. *Every connected* CW *complex has a universal covering space.*

PROOF. CW complexes are locally path connected (Theorem 8.25) and semi-locally 1-connected (Example 10.4). □

It follows immediately that connected polyhedra have universal covering spaces. Let us give a direct proof of this, avoiding the fussy proofs of Theorem 8.25 and Corollary 8.31.

Lemma 10.39. *A locally contractible space X is locally path connected and semilocally* 1*-connected.*

PROOF. Let $x \in X$, let U be an open neighborhood of x, and let $V \subset U$ be an open neighborhood of x, which is contractible to x in U; that is, let $F: V \times \mathbf{I} \to U$ be a continuous map with $F(v, 0) = v$ and $F(v, 1) = x$ for all $v \in V$. If $v_0 \in V$, then $f(t) = F(v_0, t)$ is a path in U from v_0 to x. It follows that X is locally path connected (use the definition of locally path connected rather than its characterization, Corollary 1.19).

With the same notation as in the first paragraph, it is easy to see that if $i: V \hookrightarrow U$ is the inclusion, then $i_*: \pi_1(V, x) \to \pi_1(U, x)$ is trivial. It follows that if $j: V \hookrightarrow X$ is the inclusion, then $j_*: \pi_1(V, x) \to \pi_1(X, x)$ is trivial; hence X is semilocally 1-connected. □

Theorem 10.40. *Every polyhedron is locally contractible, and every connected polyhedron has a universal covering space.*

PROOF. The second half of the statement follows from the first, in light of Lemma 10.39 and Corollary 10.37.

Let $x \in X$ and let U be an open neighborhood of x. By Exercise 7.12(iii), we may assume that there is a simplicial complex K with $|K| = X$ and with $x \in \mathrm{Vert}(K)$. Define $F: \mathrm{st}(x) \times \mathbf{I} \to |K|$ by $F(w, t) = tx + (1 - t)w$, where $w \in \mathrm{st}(x)$ (Exercise 7.7(ii) guarantees that such convex combinations make sense). Note that F is a deformation in $|K|$ of $\mathrm{st}(x)$ to x and that $F(\{x\} \times \mathbf{I}) = \{x\} \subset U$. By the tube lemma (Lemma 8.9′), there is an open neighborhood V of x such that $F(V \times \mathbf{I}) \subset U$. Replacing V by $V \cap U$ if necessary, we may assume that $V \subset U$. It follows that X is locally contractible. □

Though we have not proved it, we remind the reader that CW complexes are locally contractible.

Corollary 10.41. *Every connected n-manifold has a universal covering space (which is also an n-manifold).*

PROOF. We have already remarked that n-manifolds are locally contractible, hence locally path connected and semilocally 1-connected, by Lemma 10.39.

In Exercise 10.6, we observed that any covering space of an n-manifold is itself an n-manifold. □

In Exercise 10.6, we observed that every covering space $\tilde{X}$ inherits local properties of X. Let us prove that other properties of the base space X may lift to properties of covering spaces $(\tilde{X}, p)$ of X.

Theorem 10.42. *Every covering space $(\tilde{X}, p)$ of a connected, locally path connected, semilocally 1-connected topological group X can be equipped with a multiplication making $\tilde{X}$ a topological group and p a homomorphism.*

PROOF. Let e be the identity element of X. By Corollary 10.35, we may assume that $\tilde{X} = \tilde{X}_G$ for some subgroup G of $\pi_1(X, e)$. Let $m: X \times X \to X$ be the given multiplication in the topological group X, and write $m(x, y) = x \circ y$ for $x, y \in X$. If $f, g \in P(X, e)$, define a product $f \circ g$ by pointwise multiplication:

$$(f \circ g)(t) = f(t) \circ g(t) \quad \text{for all } t \in \mathbf{I}.$$

Note that $f \circ g$ is continuous, being the composite of the continuous functions $f \times g$ and m. Since $f(0) = e = g(0)$, it follows that $(f \circ g)(0) = e$, and so $f \circ g \in P(X, e)$. We propose to define multiplication in $\tilde{X}_G$ by

$$\langle f \rangle_G \langle g \rangle_G = \langle f \circ g \rangle_G, \tag{1}$$

but we need some preliminary results to prove that this is well defined.

Let $\alpha_1, \alpha_2, \beta_1, \beta_2$ be paths in X that agree when necessary: $\alpha_1(1) = \beta_1(0)$, $\alpha_2(1) = \beta_2(0)$, and $\alpha_1(1) \circ \alpha_2(1) = \beta_1(0) \circ \beta_2(0)$. Evaluating at $t \in \mathbf{I}$ gives

$$(\alpha_1 * \beta_1) \circ (\alpha_2 * \beta_2) = (\alpha_1 \circ \alpha_2) * (\beta_1 \circ \beta_2). \tag{2}$$

If, as usual, $\alpha^{-1}(t) = \alpha(1 - t)$, then evaluating at $t \in \mathbf{I}$ gives

$$(\alpha_1 \circ \alpha_2)^{-1} = \alpha_1^{-1} \circ \alpha_2^{-1}. \tag{3}$$

Let us denote the pointwise inverse of α by $\bar{\alpha}$:

$$\bar{\alpha}(t) = (\alpha(t))^{-1}.$$

Now suppose that α and β are closed paths at e; we claim that

$$\alpha \circ \beta \simeq \alpha * \beta \text{ rel } \dot{\mathbf{I}}. \tag{4}$$

To see this, consider the continuous map $F: \mathbf{I} \times \mathbf{I} \to X$ defined by $F(s, t) = \alpha(st) \circ \beta(s)$. The following picture displays F on the boundary of $\mathbf{I} \times \mathbf{I}$:

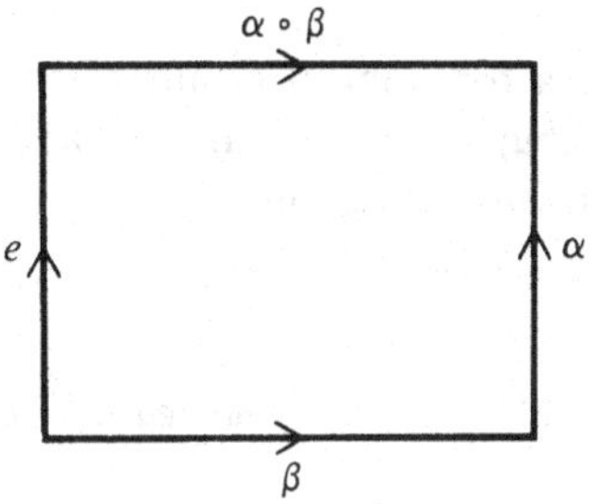

By Exercise 3.4(iii), $\beta \simeq (\alpha \circ \beta) * \alpha^{-1}$ rel $\dot{\mathbf{I}}$, hence $\beta * \alpha \simeq \alpha \circ \beta$ rel $\dot{\mathbf{I}}$. Since $\pi_1(X, e)$ is abelian, by Corollary 3.21, it follows that $\alpha * \beta \simeq \alpha \circ \beta$ rel $\dot{\mathbf{I}}$.

Let us now show that formula (1) does not depend on the choice of paths in $\langle f \rangle_G$ and $\langle g \rangle_G$. Suppose that $f_1 \in \langle f \rangle_G$ and $g_1 \in \langle g \rangle_G$: thus $f_1(1) = f(1)$ and $[f * f_1^{-1}] \in G$; $g_1(1) = g(1)$ and $[g * g_1^{-1}] \in G$. Now $(f \circ g)(1) = f(1) \circ g(1) = (f_1 \circ g_1)(1)$. Moreover,

$$\begin{aligned}
[(f \circ g) * (f_1 \circ g_1)^{-1}] &= [(f \circ g) * (f_1^{-1} \circ g_1^{-1})], \quad \text{by (3)} \\
&= [(f * f_1^{-1}) \circ (g * g_1^{-1})], \quad \text{by (2)} \\
&= [(f * f_1^{-1}) * (g * g_1^{-1})], \quad \text{by (4)} \\
&= [f * f_1^{-1}][g * g_1^{-1}] \in G.
\end{aligned}$$

Therefore $\langle f \rangle_G \langle g \rangle_G = \langle f_1 \rangle_G \langle g_1 \rangle_G$, as desired.

Define $\tilde{e} \in P(X, e)$ to be the constant path at e. It is easy to see that $\tilde{X}_G$ is a group with identity $\tilde{e}$ and with $[\bar{f}]$ the inverse of $[f]$. Since $p(\langle f \rangle_G) = f(1)$, it follows at once that p is a homomorphism.

It remains to prove that $\tilde{X}_G$ is a topological group. To see that inversion $\tilde{X}_G \to \tilde{X}_G$ is continuous, let $(U, \langle \bar{f} \rangle_G)$ be a basic open neighborhood of $\langle \bar{f} \rangle_G$ (thus U is an admissible open neighborhood of $f(1)^{-1}$). Since X is a topological group, $U^{-1} = \{x^{-1} : x \in U\}$ is an open neighborhood of $f(1)$; moreover, we may assume that U^{-1} is admissible (for any open subset of an admissible open set is admissible). But inversion carries $(U^{-1}, \langle f \rangle_G)$ inside $(U, \langle \bar{f} \rangle_G)$, and hence it is continuous. To see that multiplication is continuous, let W be an admissible open neighborhood of $f(1)g(1)$, and let U, V be admissible open neighborhoods of $f(1)$, $g(1)$, respectively, such that $U \circ V = \{u \circ v : u \in U, v \in V\} \subset W$. Then multiplication carries $(U, \langle f \rangle_G) \times (V, \langle g \rangle_G)$ inside $(W, \langle f \rangle_G \langle g \rangle_G)$, and so $\tilde{X}_G$ is a topological group. □

EXERCISES

10.24. Let X be a topological group that is connected, locally path connected, and semilocally 1-connected, and let G be a subgroup of $\pi_1(X, e)$, where e is the identity of X. If f is a closed path in X at e, show that $[\bar{f}] = [f^{-1}]$ and $\langle \bar{f} \rangle_G = \langle f^{-1} \rangle_G$.

10.25. Let X be an H-space that is connected, locally path connected, and semilocally 1-connected, and let G be a subgroup of $\pi_1(X, e)$, where e is a homotopy identity in X. Prove that $\tilde{X}_G$ is an H-space and that p "preserves" multiplication.

Remark. A **Lie group** is a topological group whose underlying space is an n-manifold and whose group operations are real analytic. Covering spaces of connected Lie groups are also Lie groups.

Theorem 10.43.

(i) *Every covering space $(\tilde{X}, p)$ of a connected CW complex (X, E, Φ) can be* *equipped with a CW decomposition so that $\tilde{X}$ is a CW complex with* $\dim \tilde{X} = \dim X$ *and p is a cellular map.*[5]

(ii) *Every covering space $(\tilde{X}, p)$ of a connected polyhedron X can be equipped with a triangulation so that $\tilde{X}$ is a polyhedron with* $\dim \tilde{X} = \dim X$ *and p is a simplicial map.*

Remark. Since covering spaces of compact spaces need not be compact, one is thus obliged to consider infinite CW complexes and infinite simplicial complexes.

Proof. (i) Let I be a set indexing the points in a fiber: if $x \in X$, then $p^{-1}(x) = \{\tilde{x}_i : i \in I, p(\tilde{x}_i) = x\}$. For each $e \in E$, let $x_e = \Phi_e(0) \in e$. Since D^n is simply connected, the lifting criterion (Theorem 10.13) provides continuous maps $\tilde{\Phi}_{ei}: (D^n, 0) \to (\tilde{X}, \tilde{x}_i)$, all $e \in E$ and $i \in I$, with $p\tilde{\Phi}_{ei} = \Phi_e$ and $p(\tilde{x}_i) = x_e$.

$$\begin{array}{ccc} & & (\tilde{X}, \tilde{x}_i) \\ & \overset{\tilde{\Phi}_{ei}}{\nearrow} & \downarrow p \\ (D^n, 0) & \xrightarrow[\Phi_e]{} & (X, x_e). \end{array}$$

Denote $\tilde{\Phi}_{ei}(D^n - S^{n-1})$ by $\tilde{e}_i$. Define

$$\tilde{\Phi} = \{\tilde{\Phi}_{ei}: D^n \to \tilde{X} \mid e \in E, i \in I, n = n(e)\},$$

$$\tilde{E} = \{\tilde{e}_i : e \in E, i \in I\}$$

and

$$\tilde{X}^{(n)} = \bigcup \{\tilde{e}_i \in \tilde{E} : \dim(\tilde{e}_i) \leq n\}.$$

If $(\tilde{X}, \tilde{E}, \tilde{\Phi})$ is a CW complex, then $\dim \tilde{X} = \dim X$ and p is a cellular map (indeed, since $p\tilde{\Phi}_{ei} = \Phi_e$, a relative homeomorphism, it is easy to see that $p|\tilde{e}_i : \tilde{e}_i \to e$ is a homeomorphism).

We show by induction on $n \geq 0$ that $\tilde{X}^{(n)}$ is a CW complex; the argument in the last paragraph of the proof of Theorem 8.24 shows that this implies that $\tilde{X}$ is a CW complex. The induction begins because $\tilde{X}^{(0)}$ is discrete ($(\tilde{X}, p)$ is a covering space).

Assume that $n > 0$; let us check the axioms in the definition of CW complex.

(1) If $\tilde{y} \in \tilde{X}$, let $y = p(\tilde{y})$, let e be the cell in X containing y, and let f be a path in e from y to x_e. For each $\tilde{y}_i \in p^{-1}(y)$, there is a lifting $\tilde{f}_i$ of f that is a path in $\tilde{e}_i$ from $\tilde{y}_i$ to $\tilde{x}_i$. But $\tilde{y} = \tilde{y}_i$ for some i, so that $\tilde{y} \in \tilde{e}_i \subset \tilde{X}$. Therefore $\tilde{X} = \bigcup_{n \geq 0} \tilde{X}^{(n)}$. (Of course, $\tilde{X}^{(n)}$ is the union of cells, by its very definition.)

To see that this is a disjoint union, consider cells e, a in $\tilde{X}^{(n)}$, and suppose

[5] More is true: the appropriate restriction of p is a homeomorphism from each cell in $\tilde{X}$ to a cell in X.

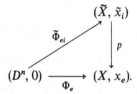

that $\tilde{e}_i \cap \tilde{a}_j \neq \varnothing$ for some i, j. By induction, we may assume that $\tilde{e}_i$ is an n-cell and hence is open in $\tilde{X}^{(n)}$ (Corollary 8.22(iii)). If $e \neq a$, then $p(\tilde{e}_i \cap \tilde{a}_j) \subset p\tilde{e}_i \cap p\tilde{a}_j = e \cap a = \varnothing$, hence $\tilde{e}_i \cap \tilde{a}_j = \varnothing$; if $e = a$, then $\tilde{e}_i \cap \tilde{e}_j = \varnothing$ for $i \neq j$ because Lemma 10.36 says that e is evenly covered by p.

(2) If $\dim(e) = n$, then $\Phi_e(S^{n-1}) \subset X^{(n-1)}$. Since $p\tilde{\Phi}_{ei} = \Phi_e$, it follows that $\tilde{\Phi}_{ei}(S^{n-1}) \subset \tilde{X}^{(n-1)}$, so that $\tilde{\Phi}_{ei}$ is a map of pairs $(D^n, S^{n-1}) \to (\tilde{e}_i \cup \tilde{X}^{(n-1)}, \tilde{X}^{(n-1)})$. Furthermore, each $\tilde{\Phi}_{ei}$ is a relative homeomorphism because each Φ_e is.

(3) We use the following commutative diagram to check that $\tilde{X}^{(n)}$ has the weak topology determined by the closures of its cells.

$$\begin{array}{ccc} \coprod D^{n(\tilde{e}_i)} & \xrightarrow{\tilde{\varphi}} & \tilde{X} \\ \Big\downarrow q & & \Big\downarrow p \\ \coprod D^{n(e)} & \xrightarrow[\varphi]{} & X. \end{array}$$

Here $\tilde{\varphi} = \coprod_{e,i} \tilde{\Phi}_{ei}$, $\varphi = \coprod_e \Phi_e$, and $q = \coprod q_{ei}$, where $q_{ei}\colon D^{n(\tilde{e}_i)} \to D^{n(e)}$ acts as the identity. By Lemma 8.16, $\tilde{X}$ has the weak topology if and only if $\tilde{\varphi}$ is an identification. Suppose that $\tilde{B}$ is a subset of $\tilde{X}$ with $\tilde{\varphi}^{-1}(\tilde{B})$ open; since $\tilde{\varphi}$ is a continuous surjection, $\tilde{\varphi}$ is an identification if each such $\tilde{B}$ is open in $\tilde{X}$. Let $\tilde{x} \in \tilde{B}$, let $x = p(\tilde{x})$, let U be an admissible open neighborhood of x, and let $\tilde{U}$ be the sheet over U containing $\tilde{x}$. Then $\tilde{B}$ is open if and only if each such $\tilde{B} \cap \tilde{U}$ is open in $\tilde{X}$. Changing notation if necessary, we may thus assume that $\tilde{B} \subset \tilde{U}$, where $p|\tilde{U}$ is a homeomorphism. Now $\tilde{\varphi}^{-1}(\tilde{B})$ is open; since q is an open map, $q\tilde{\varphi}^{-1}(\tilde{B})$ is open. We claim that $q\tilde{\varphi}^{-1}(\tilde{B}) = \varphi^{-1}p(\tilde{B})$. If this claim is correct, then $\varphi q\tilde{\varphi}^{-1}(\tilde{B}) = \varphi\varphi^{-1}p(\tilde{B}) = p(\tilde{B})$ is open in X, because φ is an identification. It follows that $(p|\tilde{U})^{-1}(\tilde{B}) = \tilde{B}$ is open in $\tilde{X}$, for $p|\tilde{U}$ is a homeomorphism, and this will complete the proof.

Assume that $\tilde{\varphi}(z) \in \tilde{B}$. Commutativity of the diagram gives $\varphi q(z) = p\tilde{\varphi}(z) \in p(\tilde{B})$; hence $q\tilde{\varphi}^{-1}(\tilde{B}) \subset \varphi^{-1}p(\tilde{B})$. For the reverse inclusion, let $z \in \varphi^{-1}p(\tilde{B})$, so that $\varphi(z) \in p(\tilde{B})$. Now $z \in D^{n(e)}$, say; choose a path f in $D^{n(e)}$ from z to 0. Hence φf is a path in e from $\varphi(z)$ to x_e. Let $\tilde{g}$ be a lifting of φf with $\tilde{g}(0) \in \tilde{B}$; of course, $\tilde{g}(1) = \tilde{x}_i$ for some i. But $\tilde{\varphi}q_{ei}^{-1}f$ is also a lifting of φf (for $p\tilde{\varphi}q_{ei}^{-1}f = \varphi q q_{ei}^{-1}f = \varphi f$), which ends at $\tilde{x}_i$. By uniqueness of path lifting (here we lift the reverse of φf), it follows that $\tilde{\varphi}q_{ei}^{-1}f = \tilde{g}$, and so $\tilde{\varphi}q_{ei}^{-1}f(0) = \tilde{g}(0) \in \tilde{B}$. But $\tilde{\varphi}q_{ei}^{-1}f(0) = \tilde{\varphi}q_{ei}^{-1}(z)$, and so $z \in q\tilde{\varphi}^{-1}(\tilde{B})$, as desired.

(4) $\tilde{X}^{(n)}$ is closure finite, for if $\dim(\tilde{e}_i) = n$, then the closure of $\tilde{e}_i$ is contained in $\tilde{e}_i \cup \tilde{\Phi}_{ei}(S^{n-1})$. Since $\tilde{\Phi}_{ei}(S^{n-1})$ is compact, it is contained in a finite CW subcomplex of $\tilde{X}^{(n-1)}$; it follows that the closure of $\tilde{e}_i$ meets only finitely many cells.

(ii) If X is a polyhedron, then one can adapt the proof above replacing the word "cell" everywhere by "open simplex". The straightforward and simpler details are left to the reader. □

Corollary 10.44. *If X is a connected graph, then its universal covering space is a tree.*

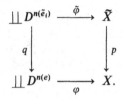

PROOF. A connected graph is, by definition, a connected one-dimensional simplicial complex. The universal covering space $\tilde{X}$ of X is thus a simply connected graph. It is easy to see that $\tilde{X}$ can have no circuits, hence $\tilde{X}$ is a tree. □

Corollary 10.45. *Let X be a compact connected CW complex, and let $(\tilde{X}, p)$ be a j-sheeted covering space of X for some integer j. Then $\tilde{X}$ is compact and*

$$\chi(\tilde{X}) = j\chi(X).$$

PROOF. The proof of the theorem shows that there is a CW decomposition of $\tilde{X}$ having precisely j i-cells for each i-cell in X. Thus, if $\tilde{\alpha}_i$, respectively α_i, denotes the number of i-cells in $\tilde{X}$, respectively X, then $\tilde{\alpha}_i = j\alpha_i$ for all i. Since X has only finitely many cells, it follows that $\tilde{X}$ has only finitely many cells and hence is compact. Moreover, the definition of the Euler–Poincaré characteristic is

$$\chi(\tilde{X}) = \sum(-1)^i\tilde{\alpha}_i = j\sum(-1)^i\alpha_i = j\chi(X). \qquad \square$$

Here are some applications to group theory; for deeper applications, see [Massey (1967)].

Theorem 10.46. *Every subgroup G of a free group F is itself free.*

PROOF. Let $\{x_i: i \in I\}$ be a basis of F, and let X be a wedge of $|I|$ circles. By Corollary 7.35, $\pi_1(X, x_0) \cong F$ (where x_0 is a basepoint of X). Now the covering space $(\tilde{X}_G, p)$ of X has fundamental group isomorphic (via p_*) to G. Theorem 10.43(ii) says that $\tilde{X}_G$ is a (connected) one-dimensional simplicial complex, and Corollary 7.35 says that its fundamental group is free. □

Theorem 10.47. *A free group F of rank 2 contains a subgroup that is not finitely generated.*

PROOF. In Exercise 10.4(ii), we exhibited a covering space $(\tilde{X}, p)$ of $S^1 \vee S^1$ that is a doubly infinite sequence of tangent circles. If one regards $\tilde{X}$ as a simplicial complex, then there is a maximal tree whose complement is the union of the open upper semicircles in $\tilde{X}$. By Corollary 7.35, $\pi_1(\tilde{X}, \tilde{x}_0)$ is free of infinite rank. But $\pi_1(\tilde{X}, \tilde{x}_0)$ is isomorphic to a subgroup of $\pi_1(S^1 \vee S^1, x_0)$, which is free of rank 2. □

One can show that the commutator subgroup of a free group of rank ≥ 2 is free of infinite rank.

Theorem 10.48. *Let F be a free group of finite rank n, and let G be a subgroup of finite index j. Then G is a free group of finite rank; indeed*

$$\text{rank } G = jn - j + 1.$$

PROOF. If Γ is a finite graph (i.e., a finite one-dimensional simplicial complex), let $e(\Gamma)$ denote the number of edges in Γ and let $v(\Gamma)$ denote the number of vertices in Γ. If T is a finite tree, then $e(T) = v(T) - 1$: since T is contractible, $\chi(T) = 1$; on the other hand, $\chi(T) = v(T) - e(T)$ (there are elementary proofs of this equality). It follows that if T is a maximal tree in a finite graph Γ, then the number of edges in $\Gamma - T$ is $e(\Gamma) - e(T) = e(\Gamma) - v(T) + 1$. Since T is a maximal tree, $v(T) = v(\Gamma)$ (Lemma 7.33). Therefore, if Γ is a finite graph, then $\pi_1(\Gamma, x_0)$ is free of rank $e(\Gamma) - v(\Gamma) + 1$ (Corollary 7.35).

If X is a wedge of n circles, then it is easy to see that $\chi(X) = 1 - n$. Let $(\tilde{X}_G, p)$ be the covering space of X corresponding to G (we identify F with $\pi_1(X, *)$). Since $[F : G] = j$, Theorem 10.9(ii) says that $\tilde{X}_G$ is a j-sheeted covering space. Therefore

$$\begin{aligned} e(\tilde{X}_G) - v(\tilde{X}_G) + 1 &= -\chi(\tilde{X}_G) + 1 \\ &= -j\chi(X) + 1 \quad \text{by Corollary 10.45} \\ &= -j(1 - n) + 1 \\ &= jn - j + 1. \end{aligned}$$

as claimed. □

EXERCISES

In each of the following exercises, the space X is connected, locally path connected, and semilocally 1-connected.

10.26. If X is compact and $(\tilde{X}, p)$ is a finite-sheeted covering space of X, then $\tilde{X}$ is compact.

10.27. If j is a positive integer and x_0 is a basepoint in X, then the number of j-sheeted covering spaces of X is the number of subgroups of $\pi_1(X, x_0)$ having index j. (*Remark*: There is a group-theoretic result that could be used in conjunction with this exercise. If G is a finite abelian group, then the number of subgroups of G having index j is equal to the number of subgroups of G having order j.)

10.28. If j is a positive integer and X is a finite CW complex, then there are only finitely many j-sheeted covering spaces of X. (*Hint*: Use the group-theoretic result that a finitely generated group has only finitely many subgroups of index j.)

Orbit Spaces

If $(\tilde{X}, p)$ is a covering space of X, then both $\pi_1(X, x_0)$ and $\mathrm{Cov}(\tilde{X}/X)$ act on the fiber $p^{-1}(x_0)$; moreover, if $(\tilde{X}, p)$ is a regular covering space, then $\mathrm{Cov}(\tilde{X}/X) \cong \pi_1(X, x_0)/p_*\pi_1(\tilde{X}, \tilde{x}_0)$. Let us now concentrate on groups acting on $\tilde{X}$ instead of on fibers. Plainly, $\mathrm{Cov}(\tilde{X}/X)$ acts on $\tilde{X}$; moreover, if $(\tilde{X}, p)$ is regular, then there is a surjection $\pi_1(X, x_0) \to \mathrm{Cov}(\tilde{X}/X)$ (the isomorphism above displays $\mathrm{Cov}(\tilde{X}/X)$ as a quotient group of $\pi_1(X, x_0)$), which shows that $\pi_1(X, x_0)$ acts on $\tilde{X}$ as well.

Definition. If a group G acts on a space Y, then the **orbit space** Y/G is the set of all orbits of G,

$$Y/G = \{o(y) \colon y \in Y\},$$

regarded as a quotient space of Y via the identification $\nu \colon y \mapsto o(y)$.

The next pair of lemmas will be used in proving an analogue of the fundamental theorem of Galois theory.

Lemma 10.49. *Let $(\tilde{X}, p)$ be a regular covering space of X, where X is connected and locally path connected, and let $G = \operatorname{Cov}(\tilde{X}/X)$. There exists a homeomorphism $\varphi \colon X \to \tilde{X}/G$ making the following diagram commute:*

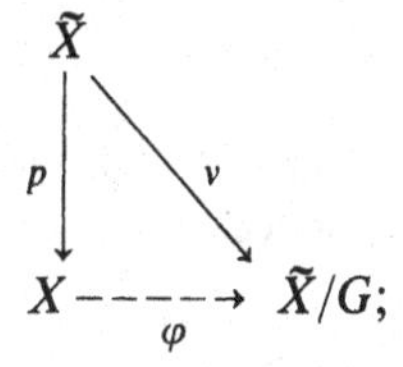

moreover, $(\tilde{X}, \nu)$ is a covering space of $\tilde{X}/G$.

PROOF. If $x \in X$, choose $\tilde{x} \in p^{-1}(x)$, and define φ by

$$\varphi(x) = \nu(\tilde{x}) = o(\tilde{x}).$$

(1) φ is well defined.

Assume that $\tilde{x}_1 \in p^{-1}(x)$. The hypotheses allow us to use Theorem 10.18, so that $\operatorname{Cov}(\tilde{X}/X)$ acts transitively on $p^{-1}(x)$. There exists $g \in \operatorname{Cov}(\tilde{X}/X)$ with $g(\tilde{x}) = \tilde{x}_1$, and so $o(\tilde{x}) = o(\tilde{x}_1)$, as required.

(2) φ is a bijection.

Commutativity of the diagram and the surjectivity of ν imply that φ is surjective. To see that φ is injective, assume that $\varphi(x) = \varphi(y)$. Then there exists $\tilde{x} \in p^{-1}(x)$ and $\tilde{y} \in p^{-1}(y)$ with $o(\tilde{x}) = o(\tilde{y})$; that is, there exists $g \in \operatorname{Cov}(\tilde{X}/X)$ with $\tilde{x} = g(\tilde{y})$. Hence

$$x = p(\tilde{x}) = pg(\tilde{y}) = p\tilde{y} = y$$

(recall that $pg = p$ for every covering transformation g).

(3) φ is continuous.

If U is open in $\tilde{X}/G$, then the continuity of ν shows that $p^{-1}\varphi^{-1}(U) = \nu^{-1}(U)$ is open in $\tilde{X}$. But p is an open map, so that $p(p^{-1}\varphi^{-1}(U)) = \varphi^{-1}(U)$ is open in X.

(4) φ is open.

If V is open in X, then $\nu^{-1}\varphi(V) = p^{-1}(V)$ is open in $\tilde{X}$; since ν is an identification, $\varphi(V)$ is open in $\tilde{X}/G$.

We have shown that φ is a homeomorphism. That $(\tilde{X}, \nu)$ is a covering space of $\tilde{X}/G$ now follows from Exercise 10.9. □

Let us generalize the notion of equivalence of covering spaces.

Definition. Let $(\widetilde{X}, p)$ and $(\widetilde{Y}, q)$ be covering spaces of X and Y, respectively. These covering spaces are **equivalent** if there exist homeomorphisms φ and ψ making the following diagram commute:

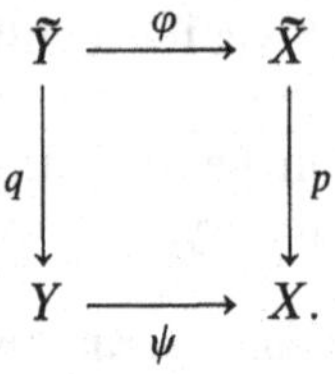

If $X = Y$ and $\psi = 1_X$, then we have the old definition of equivalence. The conclusion of Lemma 10.49 can now be restated: the covering spaces $(\widetilde{X}, p)$ and $(\widetilde{X}, \nu)$ are equivalent.

Lemma 10.50. *Let X be connected and locally path connected, and consider the commutative diagram of covering spaces*

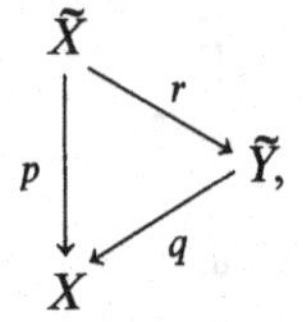

where $(\widetilde{X}, p)$ and $(\widetilde{X}, r)$ are regular; let $G = \mathrm{Cov}(\widetilde{X}/\widetilde{Y})$ and let $H = \mathrm{Cov}(\widetilde{X}/X)$. Then there is a commutative diagram

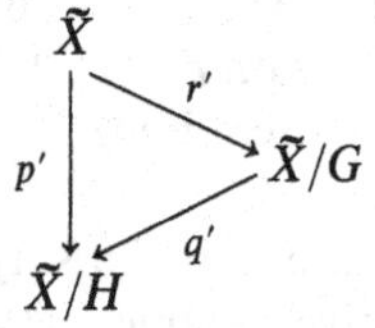

of covering spaces, each of which is equivalent to the corresponding covering space in the original diagram.

Proof. By Lemma 10.49, $(\widetilde{X}, r)$ is equivalent to $(\widetilde{X}, r')$ and $(\widetilde{X}, p)$ is equivalent to $(\widetilde{X}, p')$, where r' and p' are natural maps that send a point into its orbit. Lemma 10.49 does not apply to the third covering space because $(\widetilde{Y}, q)$ need not be regular.

Now $G = \mathrm{Cov}(\widetilde{X}/\widetilde{Y}) \subset H = \mathrm{Cov}(\widetilde{X}/X)$: if $\varphi\colon \widetilde{X} \to \widetilde{X}$ is a homeomorphism with $r\varphi = r$, then $p\varphi = qr\varphi = qr = p$. It follows that, for each $\tilde{x} \in \widetilde{X}$, the G-orbit of $\tilde{x}$ is contained in the H-orbit of $\tilde{x}$. Define $q'\colon \widetilde{X}/G \to \widetilde{X}/H$ to be the function that sends a G-orbit into the H-orbit containing it; it is clear that $q'r' = p'$. Note that q' is continuous: if U is open in X, then $r'^{-1}q'^{-1}(U) = p^{-1}(U)$ is open in $\widetilde{X}$; since r' is an identification (because $(\widetilde{X}, r')$ is a covering space), $r'(r'^{-1}q'^{-1}(U)) = q'^{-1}(U)$ is open. Finally, $(\widetilde{X}/G, q')$ is a covering space of $\widetilde{X}/H$ equivalent to $(\widetilde{Y}, q)$, by Exercise 10.9. □

Corollary 10.51. *Let X be a connected, locally path connected, semilocally 1-connected space, and let $(\tilde{X}, p)$ be its universal covering space. Every covering space $(\tilde{Y}, q)$ of X is equivalent to $(\tilde{X}/G, v)$ for some subgroup G of* $\mathrm{Cov}(\tilde{X}/X)$.

PROOF. By Theorem 10.17, there exists a continuous map $r: \tilde{X} \to \tilde{Y}$ making the following diagram commute:

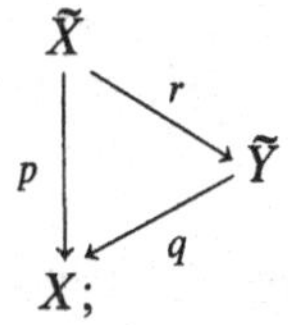

moreover, $(\tilde{X}, r)$ is a covering space of $\tilde{Y}$, by Theorem 10.16. Since $\tilde{X}$ is simply connected, both $(\tilde{X}, p)$ and $(\tilde{X}, r)$ are regular covering spaces. Therefore Lemma 10.50 applies to show that $(\tilde{Y}, q)$ is equivalent to $(\tilde{X}/G, v)$, where $G = \mathrm{Cov}(\tilde{X}/\tilde{Y})$. □

There are set-theoretic problems arising from an attempt to consider all the covering spaces of a space X: the totality of all covering spaces equivalent to a fixed covering space $(\tilde{X}, p)$ is a proper class and not a set. The same problem arises in Galois theory; there are too many field extensions of a given field F unless one restricts attention to only those inside a given algebraic closure of F. In light of the last corollary, let us regard "all" covering spaces of a space X to be of the form $(\tilde{X}/G, v)$, where $(\tilde{X}, p)$ is a universal covering space of X and G is a subgroup of $\mathrm{Cov}(\tilde{X}/X)$.

Theorem 10.52. *Let X be a connected, locally path connected, semilocally 1-connected space, and let $(\tilde{X}, p)$ be its universal covering space. Denote the family of all covering spaces of X of the form $(\tilde{X}/G, v)$, where G is a subgroup of* $\mathrm{Cov}(\tilde{X}/X)$, *by $\mathcal{Q}$, and denote the family of all subgroups of* $\mathrm{Cov}(\tilde{X}/X)$ by $\mathcal{S}$.

Then $\Phi: \mathcal{Q} \to \mathcal{S}$ defined by $(\tilde{Y}, q) \mapsto \mathrm{Cov}(\tilde{X}/\tilde{Y})$ and $\Psi: \mathcal{S} \to \mathcal{Q}$ defined by $G \mapsto (\tilde{X}/G, v)$ are bijections inverse to one another.

Remark. Recall that Corollary 10.29 gives an isomorphism $\mathrm{Cov}(\tilde{X}/X) \cong \pi_1(X, x_0)$. Therefore this theorem shows that the covering spaces of X are classified by the subgroups of the fundamental group of X. (Also see Theorem 10.20.)

PROOF. Let us evaluate both composites $\Phi\Psi$ and $\Psi\Phi$ to see that they are identities. If $G \subset \mathrm{Cov}(\tilde{X}/X)$, then $\Phi\Psi(G) = \mathrm{Cov}(\tilde{X}/(\tilde{X}/G))$; call this last group G^*. Note that G^* consists of all homeomorphisms $h: \tilde{X} \to \tilde{X}$ making the following diagram commute:

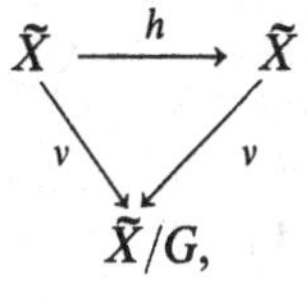

where $v: \tilde{X} \to \tilde{X}/G$ is the natural map. Is $G = G^*$? If $g \in G$ and $\tilde{x} \in \tilde{X}$, then $o(\tilde{x}) = o(g\tilde{x})$ (definition of orbit), so that $vg = v$; hence $g \in G^*$ and $G \subset G^*$. For the reverse inclusion, let $h \in G^*$, so that $vh = v$. If $\tilde{x} \in \tilde{X}$, then $o(\tilde{x}) = o(h(\tilde{x}))$, so that, by definition of G-orbit, there exists $g \in G$ with $g(h(\tilde{x})) = \tilde{x}$. Since $g \in G^*$ (by the first inclusion), it follows that $gh \in G^*$. By Theorem 10.19(i), $gh = 1_{\tilde{X}}$, and $h = g^{-1} \in G$.

Finally, $\Psi\Phi$ is the composite $(\tilde{X}/G, v) \mapsto \operatorname{Cov}(\tilde{X}/(\tilde{X}/G)) = G^* \mapsto (\tilde{X}/G^*, v^*)$. But we have just seen that $G = G^*$, so that $v = v^*$ and $\Psi\Phi$ is also an identity. □

Corollary 10.53. *Let X be a connected, locally path connected, semilocally 1-connected space, and let $(\tilde{X}, p)$ be its universal covering space. If G is a subgroup of* $\operatorname{Cov}(\tilde{X}/X)$ $(\cong \pi_1(X, x_0))$, *then*

$$\pi_1(\tilde{X}/G, *) \cong G.$$

Proof. By Corollary 10.29, $\pi_1(\tilde{X}/G, *) \cong \operatorname{Cov}(\tilde{X}/(\tilde{X}/G))$. In the proof of the theorem, however, we saw that the latter group is just G. □

The theorem reverses the viewpoint adopted earlier: instead of beginning with X and constructing $\tilde{X}$, one can also start with $\tilde{X}$ and construct X (as an orbit space). Let us pursue this further. Let $(\tilde{X}, p)$ be a covering space of X, let U be an admissible open set in X, and let S be a sheet in $\tilde{X}$ lying over U. Suppose that $h \in \operatorname{Cov}(\tilde{X}/X)$ and that $h(S) \cap S \neq \varnothing$. If $\tilde{x} \in h(S) \cap S$, then there is $\tilde{y} \in S$ with $\tilde{x} = h(\tilde{y})$. Hence $p\tilde{x} = ph(\tilde{y}) = p\tilde{y}$, so that both $\tilde{x}$ and $\tilde{y}$ lie in the fiber over $p\tilde{x}$. Since $p|S$ is a homeomorphism, it follows that $\tilde{x} = \tilde{y}$. By Theorem 10.19(i), $h = 1_{\tilde{X}}$.

Definition. Let G be a group acting on a space X. An open set V in X is **proper** if $gV \cap V = \varnothing$ for every $g \in G - \{1\}$. One says that G acts **properly** on X if every point in X has a proper open neighborhood.

Our preliminary discussion shows that $\operatorname{Cov}(\tilde{X}/X)$ acts properly on $\tilde{X}$.

Theorem 10.54. *Let X be a connected locally path connected space, let G be a group acting properly on X, and let $p: X \to X/G$ be the natural map.*

(i) *(X, p) is a regular covering space of X/G.*
(ii) *If X is semilocally 1-connected, then* $\operatorname{Cov}(X/(X/G)) \cong G$.
(iii) *If X is simply connected, then* $\pi_1(X/G, *) \cong G$.

Proof. (i) The natural map p is an identification. If U is any open set in X, then

$$p^{-1}(p(U)) = \bigcup_{g \in G} gU$$

is open; it follows that $p(U)$ is open, hence p is an open map. Let $\xi \in X/G$, let $x \in X$ be such that $p(x) = \xi$, and let U be a proper open neighborhood of x.

We claim that $p(U)$ is evenly covered by p (we do know that $p(U)$ is an open neighborhood of ξ). If g, h are distinct elements of G, then $gU \cap hU = \varnothing$ (lest $g^{-1}hU \cap U \neq \varnothing$). It remains to prove that $p|gU: gU \to p(U)$ is a bijection (for we already know that $p|gU$ is an open continuous map). If $u \in U$, then u and gu lie in the same orbit, for every $g \in G$, and so $p(gu) = p(u)$; hence $p|gU$ is surjective. If $p(gu) = p(gv)$, where $u, v \in U$, then there exists $h \in G$ with $gu = hgv$; hence $gU \cap hgU \neq \varnothing$, a contradiction. Therefore $p|gU$ is an injection. We have proved that (X, p) is a covering space of X/G.

Now $G \subset \text{Cov}(X/(X/G))$ because each $g \in G$ may be regarded as a homeomorphism of X with $pg = p$. As the fiber over ξ is $\{gx: g \in G\}$ (where $p(x) = \xi$), it follows that G, hence $\text{Cov}(X/(X/G))$, acts transitively on the fiber. By Theorem 10.18, (X, p) is regular.

(ii) This follows at once from Theorem 10.52.

(iii) This follows at once from Corollary 10.29. □

EXERCISES

10.29. Let X be connected, locally path connected, and semilocally 1-connected, let $(\tilde{X}, p)$ be its universal covering space, and let G and H be subgroups of $\text{Cov}(\tilde{X}/X)$. Prove that $G \subset H$ if and only if $(\tilde{X}/G, \nu)$ is a covering space of $\tilde{X}/H$, where ν sends G-orbits of elements of $\tilde{X}$ into H-orbits.

10.30. (i) If a group G acts properly on a space X, then G acts **without fixed points**; that is, if $g \in G$ and $g \neq 1$, then g has no fixed points.
(ii) If X is Hausdorff, G is finite, and G acts on X without fixed points, then G acts properly on X.

10.31. If G is a topological group, then every subgroup H acts on G by left translation: if $h \in H$ and $x \in G$, then $h: x \mapsto hx$. Prove that if H is a discrete subgroup, then H acts properly on G. (*Hint*: See the proof of Theorem 10.2.)

*10.32. (i) For every $p \geq 2$, show that the action of $\mathbf{Z}/p\mathbf{Z}$ on S^3 giving lens spaces $L(p, q)$ (Example 8.22) is proper.
(ii) Show that S^3 is a universal covering space of $L(p, q)$ for all q and that $\pi_1(L(p, q)) \cong \mathbf{Z}/p\mathbf{Z}$.
(iii) Show that $L(p, q)$ is a compact connected 3-manifold.

10.33. Let G be a group. If there exists a tree T on which G acts properly, then G is free. (*Hint*: $T \to T/G$ is a universal covering space.)

Remark. The theory of groups acting on spaces in a rich one; we recommend [Bredon] to the interested reader.

CHAPTER 11

Homotopy Groups

Since a closed path $f: (\mathbf{I}, \dot{\mathbf{I}}) \to (X, x_0)$ can be viewed as a map $(S^1, 1) \to (X, x_0)$, one may view $\pi_1(X, x_0)$ as (pointed) homotopy classes of (pointed) maps from S^1 into X. It is thus quite natural to consider (pointed) maps of S^n into a space X; their homotopy classes will be elements of the *homotopy group* $\pi_n(X, x_0)$. This chapter gives the basic properties of the homotopy groups; in particular, it will be seen that they satisfy every Eilenberg–Steenrod axiom save excision.

Function Spaces

We shall soon be examining subspaces of the space of all paths in a space, so let us begin by looking at function spaces.

Definition. If X and Y are topological spaces, then X^Y is the set of all continuous functions from Y into X. The **compact-open topology** on X^Y is the topology having a sub-basis consisting of all subsets $(K; U)$, when K is a compact subset of Y, U is an open subset of X, and

$$(K; U) = \{f \in X^Y: f(K) \subset U\}.$$

A typical open set in X^Y is thus an arbitrary union of finite intersections of sets of the form $(K; U)$.

Although there are other topologies one can give X^Y, we shall always consider it topologized with the compact-open topology. We remark that the compact-open topology does arise naturally. For example, if X is a metric space, then the compact-open topology on X^Y, for any space Y, is precisely the topology given by uniform convergence on compact subsets (see [Munkres (1975), p. 286]).

Let X, Y, and Z be sets, and let $F\colon Z \times Y \to X$ be a function of two variables. If we fix the first variable, then $F(z, \ )\colon Y \to X$ is a function of the other variable; let us write F_z instead of $F(z, \ )$. Thus F determines a one-parameter family of functions $F_z\colon Y \to X$; better, F determines a function $F^{\#}\colon Z \to \operatorname{Hom}(Y, X)$ by $F^{\#}(z) = F_z$ (where $\operatorname{Hom}(Y, X)$ denotes the set of *all* functions from Y into X).

Definition. If $F\colon Z \times Y \to X$ is a function, then its **associate** is the function $F^{\#}\colon Z \to \operatorname{Hom}(Y, X)$ defined by $F^{\#}(z) = F_z$ (where $F_z\colon y \mapsto F(z, y)$).

Note that F can be recaptured from its associate $F^{\#}$: if $G\colon Z \to \operatorname{Hom}(Y, X)$, define $G^{\flat}\colon Z \times Y \to X$ by $G^{\flat}(z, y) = G(z)(y)$. Indeed $F \mapsto F^{\#}$ is a bijection $\operatorname{Hom}(Z \times Y, X) \to \operatorname{Hom}(Z, \operatorname{Hom}(Y, X))$ with inverse $G \mapsto G^{\flat}$ (this is called the *exponential law* for sets because it becomes $X^{Z \times Y} = (X^Y)^Z$ if one uses exponential notation). A decent topology on function spaces (the set of all *continuous* functions) should give analogous results.

There is another obvious function in this context.

Definition. If X and Y are sets, then the **evaluation map** $e\colon \operatorname{Hom}(Y, X) \times Y \to X$ is defined by

$$e(f, y) = f(y).$$

Theorem 11.1. *Let X and Z be topological spaces, let Y be a locally compact Hausdorff space, and let X^Y have the compact-open topology* (as usual).

(i) *The evaluation map $e\colon X^Y \times Y \to X$ is continuous.*
(ii) *A function $F\colon Z \times Y \to X$ is continuous if and only if its associate $F^{\#}\colon Z \to X^Y$ is continuous.*

Proof. (i) Let $(f, y) \in X^Y \times Y$, and let V be an open neighborhood of $f(y)$ in X. Since f is continuous, there is an open neighborhood W of y with $f(W) \subset V$; since Y is locally compact Hausdorff, there is an open set U with $\bar{U}$ compact such that $x \in U \subset \bar{U} \subset W$. Now $(\bar{U}; V) \times U$ is an open neighborhood of (f, y). If $(f', y') \in (\bar{U}; V) \times U$, then $e(f', y') = f'(y') \in f'(U) \subset f'(\bar{U}) \subset V$, as desired. Therefore e is continuous.

(ii) Assume that $F^{\#}\colon Z \to X^Y$ is continuous. It is easy to check that F is the composite

$$Z \times Y \xrightarrow{F^{\#} \times 1} X^Y \times Y \xrightarrow{\ e\ } X;$$

since e is continuous, it follows that F is continuous.

Conversely, assume that F is continuous. Observe first that if $z \in Z$, then $y \mapsto (z, y)$ is a continuous map $1_z\colon Y \to Z \times Y$ and $F_z = F \circ 1_z$; it follows that each F_z is continuous and that the target of $F^{\#}$ is indeed X^Y (not merely $\operatorname{Hom}(Y, X)$).

It suffices to prove that if $z \in Z$ and $(K; U)$ is any sub-basic open neighborhood of $F^{\#}(z) = F_z$, then there exists an open neighborhood V of z with $F^{\#}(V) \subset (K; U)$. Now $F_z \in (K; U)$ means that $F(z, y) \in U$ for every $y \in K$; equivalently, $F(\{z\} \times K) \subset U$; continuity of F says that $F^{-1}(U)$ is an open subset of $Z \times Y$. Hence $F^{-1}(U) \cap (Z \times K)$ is an open subset of $Z \times K$ containing $\{z\} \times K$, and the tube lemma (Lemma 8.9′) gives an open neighborhood V of z with $V \times K \subset F^{-1}(U)$. It follows that $F^{\#}(V) \subset (K; U)$, as desired. □

Corollary 11.2. *Let X and Z be spaces, and let Y be locally compact Hausdorff. A function $g: Z \to X^Y$ is continuous if and only if the composite $e \circ (g \times 1)$ is continuous.*

$$Z \times Y \xrightarrow{g \times 1} X^Y \times Y \xrightarrow{e} X.$$

PROOF. If this composite is denoted by F, then g is just its associate $F^{\#}$. □

Remember the following commutative diagram:

$$\begin{array}{ccc} Z \times Y & \xrightarrow{F^{\#} \times 1} & X^Y \times Y \\ & \searrow^{F} & \downarrow e \\ & & X. \end{array}$$

A thorough treatment of function spaces proves the **exponential law**: if X, Y, Z are spaces with Y locally compact Hausdorff, then $F \mapsto F^{\#}$ is a bijection $X^{Z \times Y} \to (X^Y)^Z$ with inverse $G \mapsto G^{\flat}$; indeed this bijection is a homeomorphism.

Homotopy fits nicely into this setting. Assume that $f, g: Y \to X$ are homotopic maps, where Y is locally compact Hausdorff. For this remark, let a homotopy be a continuous function $F: \mathbf{I} \times Y \to X$ with $F_0 = f$ and $F_1 = g$ (usually, the domain of F is $Y \times \mathbf{I}$). The associate $F^{\#}$ of F is a continuous map $F^{\#}: \mathbf{I} \to X^Y$; that is, $F^{\#}$ is a path in X^Y from f to g. Conversely, every path in X^Y determines a homotopy. It follows that the homotopy classes are the path components of X^Y: $[Y, X] = \pi_0(X^Y)$.

Group Objects and Cogroup Objects

From concrete point-set topology, we now pass to categories. As we are interested in the homotopy category (actually, $\mathbf{hTop}_*$), a category with complicated morphisms, this abstract approach is probably the simplest.

Definition. An object A in a category $\mathscr{C}$ is an **initial object** if, for each object X in $\mathscr{C}$, there exists a unique morphism $A \to X$. An object Z in $\mathscr{C}$ is a **terminal object** if, for each object X in $\mathscr{C}$, there exists a unique morphism $X \to Z$.

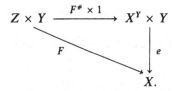

It is plain that any two initial objects in a category, if such exist, are equivalent; similarly, terminal objects are unique. One can thus speak of *the* initial object and *the* terminal object (if either exists).

EXAMPLE 11.1. In the category **Sets**, the empty set $\varnothing$ is the initial object and a singleton set is the terminal object.

EXAMPLE 11.2. The category of nonempty sets has no initial object.

EXAMPLE 11.3. Let $\mathbf{Sets}_*$ be the category of pointed sets. If $\{*\}$ is a singleton, then $A = (\{*\}, *)$ is both an initial object and a terminal object. (An object that is both an initial object and a terminal object is called a **zero object**.)

EXAMPLE 11.4. In **Groups**, the group of order 1 is a zero object.

One can give a formal definition of *duality* in a category (we shall not do so). Suffice it to say that the dual of a commutative diagram is the commutative diagram obtained by (formally) reversing each of its arrows; the dual of an object that is defined by diagrams is the object defined by the dual diagrams. Thus initial and terminal objects are dual; another pair of dual notions is product and coproduct.

Definition. If C_1 and C_2 are objects in a category, then their **product** is an object $C_1 \times C_2$ together with morphisms $p_i: C_1 \times C_2 \to C_i$, for $i = 1, 2$, called **projections**, such that, for every object X with morphisms $q_i: X \to C_i$, there exists a unique morphism $\theta: X \to C_1 \times C_2$ making the following diagram commute:

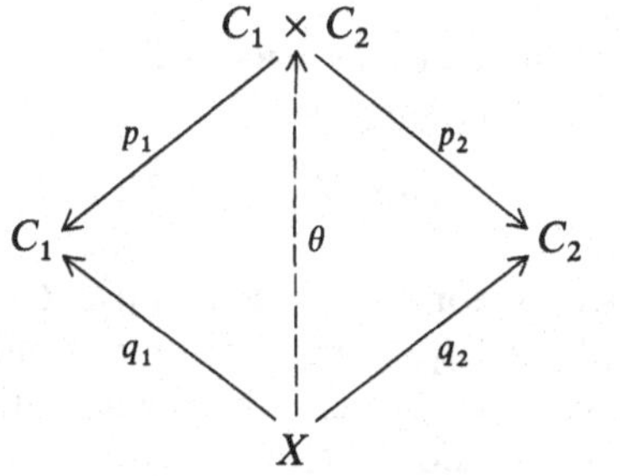

The map θ is denoted by (q_1, q_2).

In the case of **Sets**, products are the usual cartesian products equipped with the usual projections onto the factors, and $(q_1, q_2): x \mapsto (q_1(x), q_2(x))$. In particular, if $C_1 = C_2 = C$, say, then $(1_C, 1_C): C \to C \times C$ is the diagonal $x \mapsto (x, x)$. In general, define the **diagonal** $\Delta_C: C \to C \times C$ by $\Delta_C = (1_C, 1_C)$. Also, note that $(p_1, p_2) = 1_{C_1 \times C_2}$.

Definition. If C_1 and C_2 are objects in a category, then their **coproduct** is an object $C_1 \coprod C_2$ together with morphisms $j_i: C_i \to C_1 \coprod C_2$, for $i = 1, 2$, called

injections, such that, for every object X with morphisms $k_i: C_i \to X$, there exists a unique morphism $\theta: C_1 \amalg C_2 \to X$ making the following diagram commute:

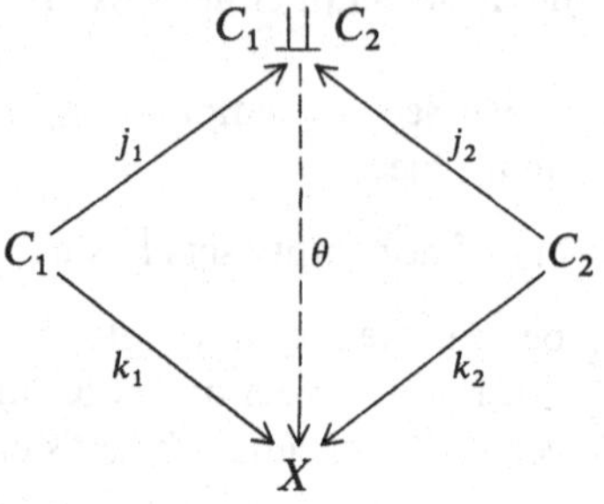

The map θ is denoted by (k_1, k_2).

In the case of **Sets**, coproducts are just disjoint unions equipped with the usual inclusions into the separate pieces, and (k_1, k_2) is the function whose restriction to C_i is k_i for $i = 1, 2$. In particular, if $C_1 = C_2 = C$, say, then $(1_C, 1_C): C \amalg C \to C$ maps each of the two copies of any $c \in C$ to itself; this map is often called the **folding map**. In general, define the **codiagonal** $\nabla_C: C \amalg C \to C$ by $\nabla_C = (1_C, 1_C)$. Also, note that $(j_1, j_2) = 1_{C_1 \amalg C_2}$.

Theorem 11.3.

(i) *Let C_1, C_2 be objects in a category in which $C_1 \times C_2$ exists. Then, for every object X, there is a natural bijection*

$$\text{Hom}(X, C_1) \times \text{Hom}(X, C_2) \xrightarrow{\sim} \text{Hom}(X, C_1 \times C_2).$$

(ii) *Let C_1, C_2 be objects in a category in which $C_1 \amalg C_2$ exists. Then, for every object X, there is a natural bijection*

$$\text{Hom}(C_1, X) \times \text{Hom}(C_2, X) \xrightarrow{\sim} \text{Hom}(C_1 \amalg C_2, X).$$

PROOF. (i) Define a function $\text{Hom}(X, C_1) \times \text{Hom}(X, C_2) \to \text{Hom}(X, C_1 \times C_2)$ by sending the ordered pair (f_1, f_2) into the unique morphism (also denoted by (f_1, f_2)!) which is guaranteed to exist by the definition of product. Define a function in the reverse direction as follows: to $g: X \to C_1 \times C_2$ associate the ordered pair $(p_1 g, p_2 g)$, where $p_i: C_1 \times C_2 \to C_i$ is the projection (for $i = 1, 2$). It is easy to check that these functions are inverse, hence both are bijections. The check of naturality is also left to the reader: if $h: X \to Y$ is any morphism, then the following diagram commutes:

$$\text{Hom}(X, C_1) \times \text{Hom}(X, C_2) \to \text{Hom}(X, C_1 \times C_2)$$

$$h^* \times h^* \Big\uparrow \qquad\qquad\qquad\qquad \Big\uparrow h^*$$

$$\text{Hom}(Y, C_1) \times \text{Hom}(Y, C_2) \to \text{Hom}(Y, C_1 \times C_2).$$

(ii) This proof is dual to that in the first part. $\square$

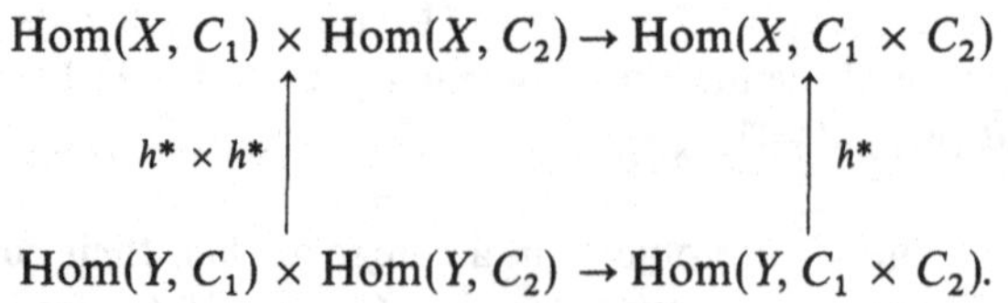

The notation (f_1, f_2) for the morphism $X \to C_1 \times C_2$ (or for $C_1 \amalg C_2 \to X$), though awkward in the proof of Theorem 11.3, is now seen to be convenient.

EXERCISES

11.1. (i) Let C_1, C_2 be sets. Define a new category $\mathscr{C}$ as follows: its objects are all ordered triples (X, q_1, q_2), where X is a set and $q_i: X \to C_i$ (for $i = 1, 2$) is a function; a morphism $\theta: (X, q_1, q_2) \to (Y, r_1, r_2)$ is a function $\theta: X \to Y$ making the following diagram commute:

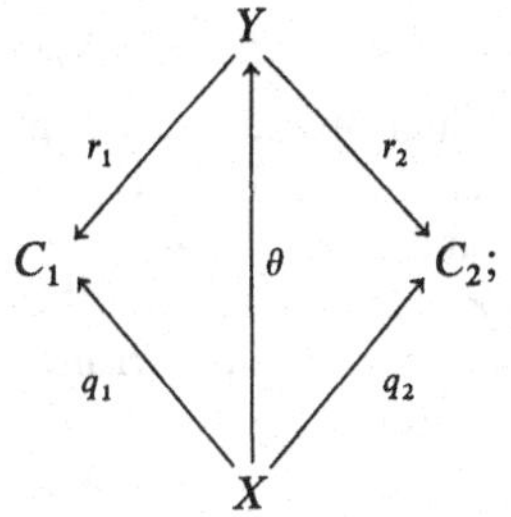

composition is ordinary composition of functions. Prove that $(C_1 \times C_2, p_1, p_2)$ is a terminal object in $\mathscr{C}$.
      (ii) Given sets C_1, C_2, construct a category in which their coproduct is an initial object.

11.2. In **Ab**, show that product and coproduct coincide ($C_1 \times C_2 = C_1 \oplus C_2 = C_1 \amalg C_2$). In **Groups**, show that product is direct product and that coproduct is free product (and so product and coproduct are distinct).

*11.3. (i) In **Sets**$_*$ and **Top**$_*$, consider the objects (A_i, a_i) for $i = 1, 2$. Their product is $(A_1 \times A_2, (a_1, a_2))$ and their coproduct is the wedge $(A_1 \vee A_2, *)$, where $*$ is the pair $\{a_1, a_2\}$ identified to a point.
      (ii) In **Top**$_*$, show that $A_1 \vee A_2$ is homeomorphic to the subset $(A_1 \times \{a_2\}) \cup (\{a_1\} \times A_2)$ of $A_1 \times A_2$. (In general, the coproduct cannot be imbedded in the product; for example, if A_1 and A_2 are finite groups with more than one element, then their free product (coproduct in **Groups**) is infinite while their direct product (product in **Groups**) is finite.)

11.4. If products exist, then the associative and commutative laws hold; similarly for coproducts. (*Warning*: One needs an extra diagrammatic axiom to deduce the generalized associative law from the associative law involving three terms.)

*11.5. (i) If $C_1 \times C_2$ and $D_1 \times D_2$ exist, and if $f_i: C_i \to D_i$ are morphisms for $i = 1, 2$, then there is a unique morphism $f_1 \times f_2: C_1 \times C_2 \to D_1 \times D_2$ making the diagrams (for $i = 1, 2$) commute (unlabeled arrows are projections):

$$
\begin{array}{ccc}
C_1 \times C_2 & \xrightarrow{\,f_1 \times f_2\,} & D_1 \times D_2 \\
\downarrow & & \downarrow \\
C_i & \xrightarrow[\;f_i\;]{} & D_i.
\end{array}
$$

      (ii) There is a dual construction $f_1 \amalg f_2: C_1 \amalg C_2 \to D_1 \amalg D_2$.

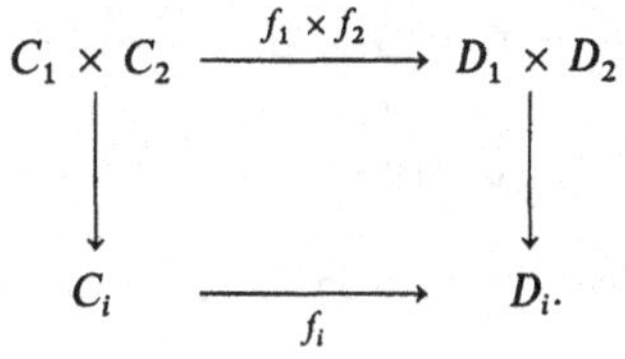

*11.6. (i) If $q_i\colon X \to C_i$ are morphisms for $i = 1, 2$, prove that

$$(q_1, q_2) = (q_1 \times q_2)\Delta_X.$$

(ii) If $k_i\colon C_i \to X$ are morphisms for $i = 1, 2$, prove that

$$(k_1, k_2) = \nabla_X(k_1 \coprod k_2).$$

(iii) If A and B are abelian groups and $f, g \in \mathrm{Hom}(A, B)$, then

$$A \times B = A \oplus B = A \coprod B.$$

and

$$f + g = \nabla_B(f \times g)\Delta_A.$$

*11.7. (i) If Z is the terminal object in a category $\mathscr{C}$ and if X is any object in $\mathscr{C}$, then $X \times Z$ is equivalent to X via the projection $X \times Z \to X$. (*Hint*: Let $\lambda\colon X \times Z \to X$ and $q\colon X \times Z \to Z$ be the projections, and let $\theta = (1_X, \omega)$, where $\omega\colon X \to Z$ is the unique morphism in $\mathrm{Hom}(X, Z)$. Then $\lambda\theta = 1_X$ and $\theta\lambda = 1_{X \times Z}$, the latter equality arising from the fact that both morphisms complete the diagram

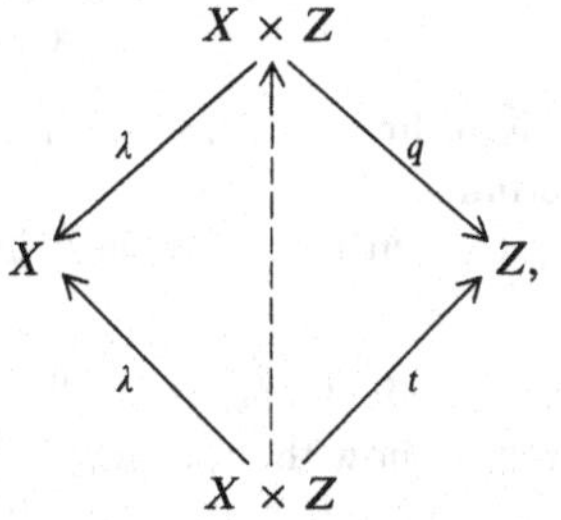

where t is the unique morphism $X \times Z \to Z$.

(ii) If A is an initial object, then $A \coprod X$ is equivalent to X via the injection $X \to A \coprod X$.

The axioms in the definition of a group can be rewritten so that they become assertions that certain diagrams commute! There are two reasons for doing this: one can consider "group-like" objects in a category; one can reverse arrows to obtain the dual notion of "cogroup".

Definition. Let $\mathscr{C}$ be a category having (finite) products and a terminal object Z. A **group object** in $\mathscr{C}$ is an object G and morphisms $\mu\colon G \times G \to G$ (called **multiplication**), $\eta\colon G \to G$, and $\varepsilon\colon Z \to G$ such that the following diagrams commute (the morphisms $f \times g$ and (f, g) are defined in Exercises 11.5 and 11.6).

(i) **Associativity**:

$$\begin{array}{ccc} G \times G \times G & \xrightarrow{1 \times \mu} & G \times G \\ {\scriptstyle \mu \times 1}\downarrow & & \downarrow{\scriptstyle \mu} \\ G \times G & \xrightarrow[\mu]{} & G. \end{array}$$

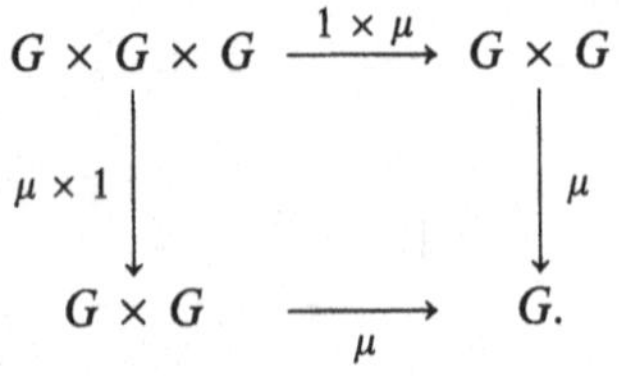

(ii) **Identity**:

$$\begin{array}{ccccc} G \times Z & \xrightarrow{1 \times \varepsilon} & G \times G & \xleftarrow{\varepsilon \times 1} & Z \times G \\ & {\scriptstyle\lambda} \searrow & \downarrow {\scriptstyle\mu} & \swarrow {\scriptstyle\rho} & \\ & & G & & \end{array}$$

(λ and ρ are the equivalences of Exercise 11.7(i) namely, the projections $\lambda: G \times Z \to G$ and $\rho: Z \times G \to G$).

(iii) **Inverse**:

$$\begin{array}{ccccc} G & \xrightarrow{(1, \eta)} & G \times G & \xleftarrow{(\eta, 1)} & G \\ {\scriptstyle\omega} \downarrow & & \downarrow {\scriptstyle\mu} & & \downarrow {\scriptstyle\omega} \\ Z & \xrightarrow[\varepsilon]{} & G & \xleftarrow[\varepsilon]{} & Z, \end{array}$$

where $\omega: G \to Z$ is the unique morphism to the terminal object.

It is easy to see that a group object in **Sets** is a group and that a group object in **Top** is a topological group. In **hTop**, the weaker notion of a space X equipped with μ and ε satisfying condition (ii) is an H-space.

Here is the dual of a group object.

Definition. Let $\mathscr{C}$ be a category having (finite) coproducts and an initial object A. A **cogroup object** in $\mathscr{C}$ is an object C and morphisms $m: C \to C \amalg C$ (called **comultiplication**), $h: C \to C$, and $e: C \to A$, such that the following diagrams commute (the morphism $f \amalg g$ is defined in Exercise 11.5).

(i) **Co-associativity**:

$$\begin{array}{ccc} C & \xrightarrow{m} & C \amalg C \\ {\scriptstyle m} \downarrow & & \downarrow {\scriptstyle 1 \amalg m} \\ C \amalg C & \xrightarrow[m \amalg 1]{} & C \amalg C \amalg C. \end{array}$$

(ii) **Co-identity**:

$$\begin{array}{ccccc} C \amalg A & \xleftarrow{1 \amalg e} & C \amalg C & \xrightarrow{e \amalg 1} & A \amalg C. \\ & \nwarrow & \uparrow {\scriptstyle m} & \nearrow & \\ & & C & & \end{array}$$

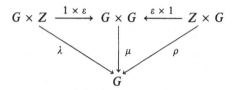

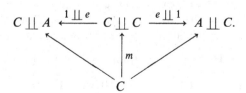

(iii) **Co-inverse:**

$$\begin{array}{ccccc}
C & \xleftarrow{(1,\,h)} & C \coprod C & \xrightarrow{(h,\,1)} & C \\
\uparrow{\scriptstyle \alpha} & & \uparrow{\scriptstyle m} & & \uparrow{\scriptstyle \alpha} \\
A & \xleftarrow[e]{} & C & \xrightarrow[e]{} & A,
\end{array}$$

where $\alpha: A \to C$ is the unique morphism from A to C.

We shall see that suspensions lead to cogroup objects in $\mathbf{hTop}_*$.

Recall that if G is an object in a category $\mathscr{C}$, then the (contravariant) functor $\mathrm{Hom}(\ \ , G): \mathscr{C} \to$ **Sets** is defined on morphisms $g: X \to Y$ as $g^*: \mathrm{Hom}(Y, G) \to \mathrm{Hom}(X, G)$, where $g^*: f \mapsto fg$. Similarly, (covariant) functor $\mathrm{Hom}(G, \ \ ): \mathscr{C} \to$ **Sets** is defined on g as $g_*: \mathrm{Hom}(G, X) \to \mathrm{Hom}(G, Y)$, where $g_*: f \mapsto gf$.

When we say that $\mathrm{Hom}(\ \ , G)$ takes values in **Groups**, then it follows, of course, that $\mathrm{Hom}(X, G)$ is a group for every object X and g^* is a homomorphism for every morphism g; a similar remark holds if $\mathrm{Hom}(G, \ \ )$ takes values in **Groups**.

Theorem 11.4. *Let $\mathscr{C}$ be a category with (finite) products and a terminal object. An object G in $\mathscr{C}$ is a group object in $\mathscr{C}$ if and only if* $\mathrm{Hom}(\ \ , G)$ *takes values in* **Groups**.

In this case, the multiplication

$$M_X: \mathrm{Hom}(X, G) \times \mathrm{Hom}(X, G) \to \mathrm{Hom}(X, G)$$

is given by

$$(f, g) \mapsto \mu(f, g),$$

where μ is the multiplication on G and $(f, g): X \to G \times G$ is the morphism of Theorem 11.3(i).

Proof. Assume that G is a group object in $\mathscr{C}$. By Theorem 11.3(i), we identify $\mathrm{Hom}(X, G) \times \mathrm{Hom}(X, G)$ with $\mathrm{Hom}(X, G \times G)$. Define M_X as in the statement. For every fixed object X, apply $\mathrm{Hom}(X, \ \ )$ to each of the three diagrams in the definition of group object. It follows that $\mathrm{Hom}(X, G)$ is a group object in **Sets**, hence is a group. It remains to show that if $h: X \to Y$, then $h^*: \mathrm{Hom}(Y, G) \to \mathrm{Hom}(X, G)$ is a homomorphism. If $f, g \in \mathrm{Hom}(Y, G)$, then $h^* M_Y(f, g) = h^*(\mu(f, g)) = \mu(f, g)h = \mu(fh, gh) = \mu(h^*f, h^*g) = M_X(h^*f, h^*g)$. Therefore, $\mathrm{Hom}(\ \ , G)$ takes values in **Groups**.

Conversely, for each object X, assume that there is some group operation

$$M_X: \mathrm{Hom}(X, G) \times \mathrm{Hom}(X, G) \to \mathrm{Hom}(X, G).$$

Again, identify $\mathrm{Hom}(X, G) \times \mathrm{Hom}(X, G)$ with $\mathrm{Hom}(X, G \times G)$, and now specialize X to $G \times G$. Thus

$$M_{G \times G}\colon \operatorname{Hom}(G \times G, G \times G) \to \operatorname{Hom}(G \times G, G).$$

Define $\mu \in \operatorname{Hom}(G \times G, G)$ as the image of the identity $1_{G \times G}$ under the function $M_{G \times G}$. If $\eta_X\colon \operatorname{Hom}(X, G) \to \operatorname{Hom}(X, G)$ is inversion, set $X = G$ and define $\eta \in \operatorname{Hom}(G, G)$ as $\eta_G(1_G)$. Define $\varepsilon \in \operatorname{Hom}(Z, G)$, where Z is the terminal object in $\mathscr{C}$, as the identity element of the group $\operatorname{Hom}(Z, G)$. One can also view ε as an image of an identity morphism. For each object X, $\operatorname{Hom}(X, Z)$ is a singleton and there is a function $\varepsilon_X\colon \operatorname{Hom}(X, Z) \to \operatorname{Hom}(X, G)$ whose (unique) value is the identity element of the group $\operatorname{Hom}(X, G)$. Set $X = Z$, so the unique element of $\operatorname{Hom}(Z, Z)$ is 1_Z; then $\varepsilon = \varepsilon_Z(1_Z)$.

That G so equipped is a group object in $\mathscr{C}$ can be seen using the Yoneda lemma, Exercise 9.7. We prove associativity, but the similar proofs of the commutativity of the identity and inverse diagrams are left to the reader.

By hypothesis, each $\operatorname{Hom}(X, G)$ is a group, and the associative law holds for its multiplication: there is a commutative diagram

$$\begin{array}{ccc}
\operatorname{Hom}(X, G) \times \operatorname{Hom}(X, G) \times \operatorname{Hom}(X, G) & \xrightarrow{M_X \times 1} & \operatorname{Hom}(X, G) \times \operatorname{Hom}(X, G) \\
{\scriptstyle 1 \times M_X}\downarrow & & \downarrow{\scriptstyle M_X} \\
\operatorname{Hom}(X, G) \times \operatorname{Hom}(X, G) & \xrightarrow[M_X]{} & \operatorname{Hom}(X, G).
\end{array}$$

By Theorem 11.3, we may rewrite this diagram as

$$\begin{array}{ccc}
\operatorname{Hom}(X, G \times G \times G) & \xrightarrow{M_X \times 1} & \operatorname{Hom}(X, G \times G) \\
{\scriptstyle 1 \times M_X}\downarrow & & \downarrow{\scriptstyle M_X} \\
\operatorname{Hom}(X, G \times G) & \xrightarrow[M_X]{} & \operatorname{Hom}(X, G).
\end{array}$$

One checks easily that there is a natural transformation $M\colon \operatorname{Hom}(\ \ , G \times G) \to \operatorname{Hom}(\ \ , G)$ with $M = (M_X)$, that is, the appropriate diagrams commute. Write $\mu = M_{G \times G}(1_{G \times G}) \in \operatorname{Hom}(G \times G, G)$. By the Yoneda lemma, Exercise 9.7(iv), for every object X and every morphism $f\colon X \to G \times G$, one has

$$M_X(f) = \mu \circ f. \tag{$*$}$$

The associativity diagram above can be used to show

$$M(M \times 1) = M(1 \times M)\colon \operatorname{Hom}(\ \ , G \times G \times G) \to \operatorname{Hom}(\ \ , G)$$

is a natural transformation. If $h\colon X \to G \times G \times G$, then the Yoneda lemma gives

$$M_X(M_X \times 1)(h) = u \circ h,$$

where $u = M_{G \times G}(M_{G \times G} \times 1)(1_{G \times G} \times 1_G) \in \operatorname{Hom}(G \times G \times G, G)$. Since

$1_{G\times G} \times 1_G = 1_{G\times G\times G}$, one has

$$u = M_{G\times G}(M_{G\times G}(1_{G\times G}) \times 1_G)$$
$$= M_{G\times G}(\mu \times 1_G) = \mu(\mu \times 1_G), \quad \text{by Eq. } (*).$$

Similarly, $[M(1 \times M)]_X(h) = v \circ h$, where $v = \mu(1_G \times \mu)$. Taking $X = G \times G \times G$ and $h = 1_{G\times G\times G}$, we see that $u = v$, i.e, $\mu(\mu \times 1) = \mu(1 \times \mu)$, as desired. □

Remark. Recall that if $p_1, p_2\colon G \times G \to G$ are the projections, then $1_{G\times G} = (p_1, p_2)$. Therefore, $\mu = M_{G\times G}(1_{G\times G}) = M_{G\times G}(p_1, p_2)$; that is, μ is the product (in the group $\mathrm{Hom}(G \times G, G)$) of the morphisms p_1 and p_2.

There is a dual result.

Theorem 11.4′. *Let $\mathscr{C}$ be a category with (finite) coproducts and an initial object. An object G in $\mathscr{C}$ is a cogroup object in $\mathscr{C}$ if and only if* $\mathrm{Hom}(G, \quad)$ *takes values in* **Groups**.

In this case, the multiplication

$$P_X\colon \mathrm{Hom}(G, X) \times \mathrm{Hom}(G, X) \to \mathrm{Hom}(G, X)$$

is given by

$$(f, g) \mapsto (f, g)m,$$

where m is the comultiplication of G and $(f, g)\colon G \coprod G \to X$ is the morphism of Theorem 11.3(ii).

Proof. The argument is similar (dual) to the one just given, but let us describe the comultiplication of G when $\mathrm{Hom}(G, \quad)$ takes values in **Groups**. For each object X, there is a multiplication

$$P_X\colon \mathrm{Hom}(G, X) \times \mathrm{Hom}(G, X) \to \mathrm{Hom}(G, X).$$

Identify $\mathrm{Hom}(G, X) \times \mathrm{Hom}(G, X)$ with $\mathrm{Hom}(G \coprod G, X)$ as in Theorem 11.3(ii). Now set $X = G \coprod G$, so that

$$P_{G\coprod G}\colon \mathrm{Hom}(G \coprod G, G \coprod G) \to \mathrm{Hom}(G, G \coprod G).$$

Then the comultiplication $m\colon G \to G \coprod G$ is the image of $1_{G\coprod G}$ under $P_{G\coprod G}$. But $(j_1, j_2) = 1_{G\coprod G}$, where j_1 and j_2 are the injections of the coproduct $G \coprod G$. Therefore, m is the product of j_1 and j_2 in the group $\mathrm{Hom}(G, G \coprod G)$. □

Exercises

11.8. Prove that a group object in **Groups** is an abelian group.

11.9. In **Sets** and in **Top**, the only cogroup object is $\varnothing$.

11.10. In **Sets**$_*$ and in **Top**$_*$, the only cogroup object is $*$.

*11.11. (i) Let $\mathscr{C}$ be a category with (finite) products and a terminal object. If G and H are group objects in $\mathscr{C}$, call a morphism $f: G \to H$ **special** if the following diagram commutes:

$$\begin{array}{ccc} G \times G & \xrightarrow{f \times f} & H \times H \\ \downarrow & & \downarrow \\ G & \xrightarrow[f]{} & H, \end{array}$$

where the vertical arrows are the multiplications in G and in H. Show that all group objects in $\mathscr{C}$ and all special morphisms form a category. State and prove the analogous result for cogroups.

(ii) If G and H are group objects in $\mathscr{C}$, then $\mathrm{Hom}(X, G)$ and $\mathrm{Hom}(X, H)$ are groups. Show that if $f: G \to H$ is special, then $f_*: \mathrm{Hom}(X, G) \to \mathrm{Hom}(X, H)$ is a homomorphism.

11.12. Prove that every abelian group is both a group object in **Ab** and a cogroup object in **Ab**.

11.13. Prove that every f.g. free group is a cogroup object in **Groups**. (One can dispense with the finiteness hypothesis.)

11.14. Every topological group (with its identity element as basepoint) is a group object in $\mathbf{hTop}_*$ and in $\mathbf{Top}_*$.

Loop Space and Suspension

The homotopy category **hTop** is the interesting category for us. It is easy to see that the empty set $\varnothing$ is an initial object; because there are no (continuous) functions $X \to \varnothing$ when X is nonempty, there are no cogroups in **hTop**. If we consider pointed spaces, however, then we shall see that there are interesting cogroups in $\mathbf{hTop}_*$.

Lemma 11.5. *The category* $\mathbf{hTop}_*$ *has a terminal object and an initial object (indeed it has a zero object), and it has (finite) products and (finite) coproducts.*

Proof. Let $*$ be a singleton, and let $A = (*, *)$. If X is a pointed space (we do not display the basepoint), then there is a morphism from X to A, namely, $[f]$, where $f: X \to *$ is the constant map. This morphism is unique, for any morphism $[g]: X \to A$ is a (pointed) homotopy class of pointed maps $g: X \to A$; but the only such pointed map is the constant map. A similar argument shows that A is also an initial object, for the only pointed map $A \to (X, x_0)$ is the function taking $*$ to x_0.

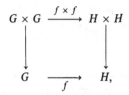

If C_1 and C_2 are pointed spaces, let $C_1 \times C_2$ (with projections p_i) be their product in $\mathbf{Top}_*$. We claim that $C_1 \times C_2$ with projections $[p_i]$ is their product in $\mathbf{hTop}_*$. Let $[f_i]: X \to C_i$ be morphisms. In $\mathbf{Top}_*$, one can complete the appropriate diagram with (f_1, f_2). Suppose that $F_i: f_i \simeq f_i'$ is a pointed homotopy $X \times \mathbf{I} \to C_i$, for $i = 1, 2$. Then $(F_1, F_2): X \times \mathbf{I} \to C_1 \times C_2$ is a pointed homotopy $(f_1, f_2) \simeq (f_1', f_2')$. It follows that $[(f_1, f_2)]: X \to C_1 \times C_2$ completes the appropriate diagram in $\mathbf{hTop}_*$. We let the reader prove uniqueness of this morphism. A similar argument shows that $C_1 \vee C_2$ is the coproduct in $\mathbf{hTop}_*$. □

Notation. A pointed space (X, x_0) may be denoted by X if there is no need to display the basepoint. In particular, if $*$ is a singleton, then the pointed space $(*, *)$ may be denoted by $*$.

Definition. A pointed space (X, x_0) is an ***H*-group** if there are continuous pointed maps $\mu: X \times X \to X$ and $\eta: X \to X$, and pointed homotopies:

$$\mu(1_X \times \mu) \simeq \mu(\mu \times 1_X) \quad \text{(associativity)};$$

$$\mu j_1 \simeq 1_X \simeq \mu j_2,$$

where $j_1, j_2: X \to X \times X$ are "injections" defined by $j_1(x) = (x, x_0)$ and $j_2(x) = (x_0, x)$;

$$\mu(1_X, \eta) \simeq c \simeq \mu(\eta, 1_X),$$

where $c: X \to X$ is the constant map at x_0.

Before giving the dual definition, let us agree on notation. As in Exercise 11.3(ii), the wedge $X \vee X$ is viewed as the subspace $X \times \{x_0\} \cup \{x_0\} \times X$ of the product $X \times X$. If $p_i: X \times X \to X$, for $i = 1, 2$, are the usual projections onto the first or second coordinates, respectively, then define "projections" $q_i: X \vee X \to X$, for $i = 1, 2$, by $q_i = p_i|X \vee X$; each q_i sends the appropriate copy of $x \in X$, namely, (x, x_0) or (x_0, x), into itself.

Definition. A pointed space (X, x_0) is an ***H'*-group** if there are continuous pointed maps $m: X \to X \vee X$ and $h: X \to X$, and there are pointed homotopies:

$$(1_X \vee m)m \simeq (m \vee 1_X)m \quad \text{(co-associativity)};$$

$$q_1 m \simeq 1_X \simeq q_2 m;$$

$$(1_X, h)m \simeq c \simeq (h, 1_X)m,$$

where $c: X \to X$ is the constant map at x_0.

Lemma 11.6.

(i) *For every pointed space (Z, z_0), the maps (j_1, j_2) and $(q_1, q_2): Z \vee Z \to Z \times Z$ are equal to k, the inclusion*

$$Z \times \{z_0\} \cup \{z_0\} \times Z \hookrightarrow Z \times Z.$$

(ii) *If* (X, x_0) *is an* H'*-group with comultiplication* m*, then the following diagram commutes to homotopy (i.e., commutes in* **hTop$_*$**):

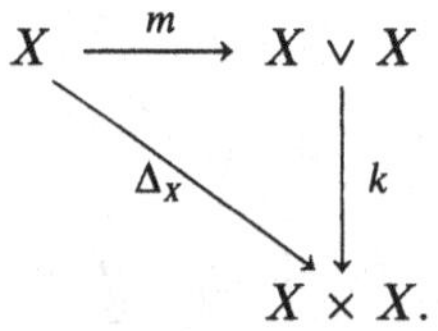

(iii) *If* (Y, y_0) *is an H-group, with multiplication* μ*, then the following diagram commutes to homotopy*:

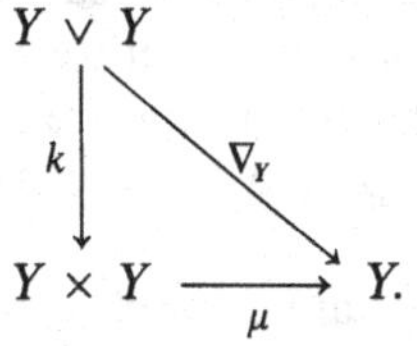

PROOF. (i) Both (j_1, j_2) and k make the following diagram commute

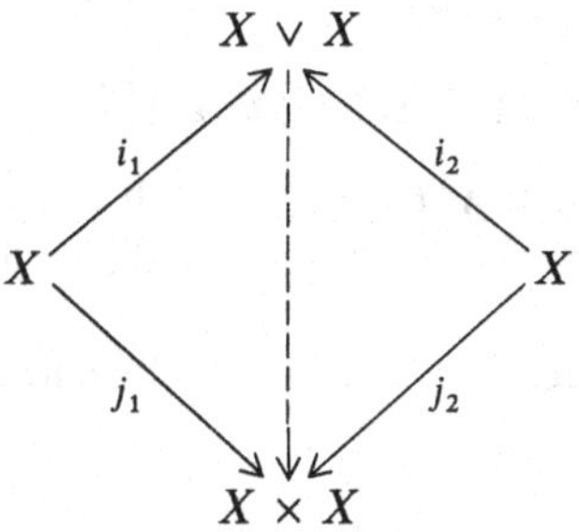

where i_1, i_2 are the injections; uniqueness gives $(j_1, j_2) = k$.

A similar argument gives $(q_1, q_2) = k$, using the diagram

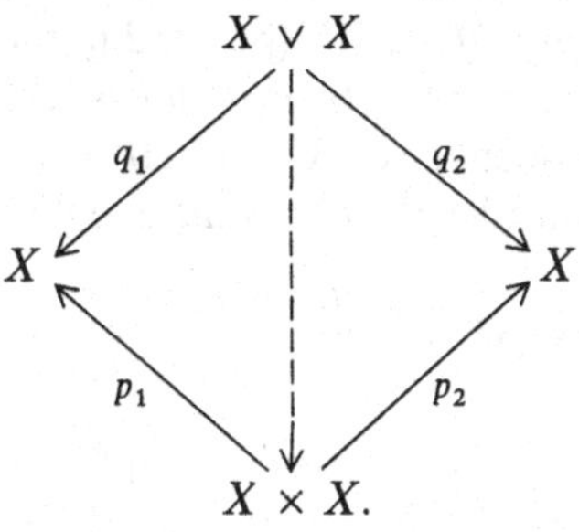

(ii) This follows from $q_1 m \simeq 1_X \simeq q_2 m$.

(iii) This follows from $\mu j_1 \simeq 1_Y \simeq \mu j_2$. □

Lemma 11.7. *The group objects in* **hTop$_*$** *are the H-groups, and the cogroup objects in* **hTop$_*$** *are the* H'*-groups.*

Proof. Suppose that (G, g_0) is a group object in **hTop**$_*$. There is a commutative diagram that surmounts the diagram of the identity axiom:

$$\begin{array}{ccc} & & G \\ & \swarrow^{[\lambda^{-1}]} & \big\downarrow{\scriptstyle [j_1]} \\ G \times \{*\} & \xrightarrow[{[1 \times \varepsilon]}]{} & G \times G, \end{array}$$

where $\lambda^{-1}: g \mapsto (g, *)$. Lifting to **Top**$_*$, $(1 \times \varepsilon)\lambda^{-1} \simeq j_1$, and so $\mu j_1 \simeq \mu(1 \times \varepsilon)\lambda^{-1} \simeq \lambda\lambda^{-1} \simeq 1_G$. The third homotopy in the definition exists because the composite $\varepsilon\omega$ in the diagram for inverse must be the constant map $G \to G$ at g_0. Since associativity holds, by hypothesis, G is an H-group.

The routine argument that H-groups are group objects in **hTop**$_*$ (as well as the dual result for cogroups and H'-groups) is left to the reader. □

At last, here are the important examples.

Definition. If (X, x_0) is a pointed space, then its **loop space**, denoted by $\Omega(X, x_0)$, is the function space

$$\Omega(X, x_0) = (X, x_0)^{(\mathbf{I}, \dot{\mathbf{I}})},$$

topologized as a subspace of $X^{\mathbf{I}}$ (equipped with the compact-open topology). One usually chooses ω_0, the constant path at x_0, as the basepoint of $\Omega(X, x_0)$.

Although the loop space does depend on the choice of basepoint, we often write ΩX instead of $\Omega(X, x_0)$.

Theorem 11.8. *Loop space defines a functor* Ω: **hTop**$_* \to$ **hTop**$_*$.

Proof. By Theorem 0.4, it suffices to prove that there is a functor Ω: **Top**$_* \to$ **Top**$_*$ with $f_0 \simeq f_1$ implying $\Omega f_0 \simeq \Omega f_1$ (pointed homotopies). If $f: X \to Y$ is a pointed map, define $\Omega f: \Omega X \to \Omega Y$ by $\omega \mapsto f\omega$, where ω is a loop in X (at the basepoint). As ΩX is a subspace of $X^{\mathbf{I}}$ and $\Omega f = f_* | \Omega X$, it suffices to show that f_* is continuous. Consider the commutative diagram

$$\begin{array}{ccc} X^{\mathbf{I}} \times \mathbf{I} & \xrightarrow{f_* \times 1} & Y^{\mathbf{I}} \times \mathbf{I} \\ \big\downarrow{\scriptstyle e} & & \big\downarrow{\scriptstyle e} \\ X & \xrightarrow[f]{} & Y, \end{array}$$

where the maps e are evaluations. Since $\mathbf{I}$ is compact Hausdorff, e and hence fe are continuous, by Theorem 11.1(i). Therefore $e(f_* \times 1)$ is continuous, and so f_* is continuous, by Corollary 11.2. That Ω so defined on objects and morphisms of **Top**$_*$ is a functor is left as a routine exercise.

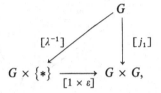

Suppose that $F: X \times \mathbf{I} \to Y$ is a pointed homotopy with $F_0 = f_0$ and $F_1 = f_1$. Define $\Phi: \Omega X \times \mathbf{I} \to \Omega Y$ by $(\omega, t) \mapsto F_t\omega$. It suffices to prove that Φ is continuous. Define $u: X^{\mathbf{I}} \times \mathbf{I} \times \mathbf{I} \to X^{\mathbf{I}} \times \mathbf{I} \times \mathbf{I}$ by $(\omega, t, s) \mapsto (\omega, s, t)$; of course, u is continuous. Consider the commutative diagram:

$$\begin{array}{ccc} X^{\mathbf{I}} \times \mathbf{I} \times \mathbf{I} & \xrightarrow{\Phi \times 1} & Y^{\mathbf{I}} \times \mathbf{I} \\ \Big\downarrow{\scriptstyle (e \times 1)u} & & \Big\downarrow{\scriptstyle e} \\ X \times \mathbf{I} & \xrightarrow[F]{} & Y. \end{array}$$

The counterclockwise composite is continuous (since the evaluation is), hence $e(\Phi \times 1)$ is continuous; Corollary 11.2 now gives the result. □

Theorem 11.9. *If (X, x_0) is a pointed space, then ΩX is an H-group.*

Proof. Define $\mu: \Omega X \times \Omega X \to \Omega X$ by

$$(\omega, \omega') \mapsto \omega * \omega',$$

where, as usual,

$$(\omega * \omega')(t) = \begin{cases} \omega(2t) & \text{if } 0 \le t \le \frac{1}{2} \\ \omega'(2t - 1) & \text{if } \frac{1}{2} \le t \le 1. \end{cases}$$

To see that μ is continuous, consider the composite

$$\Omega X \times \Omega X \times \mathbf{I} \xrightarrow{\mu \times 1} \Omega X \times \mathbf{I} \xrightarrow{e} X$$

(remember that ΩX is a subspace of $X^{\mathbf{I}}$). On $\Omega X \times \Omega X \times [0, \frac{1}{2}]$, this composite is equal to

$$\Omega X \times \Omega X \times [0, \tfrac{1}{2}] \xrightarrow{p_1 \times q} \Omega X \times \mathbf{I} \xrightarrow{e} X,$$

where p_1 is the first projection $\Omega X \times \Omega X \to \Omega X$ and $q: t \mapsto 2t$. Since this latter map is continuous, so is $e(\mu \times 1)$; by Corollary 11.2, $\mu | \Omega X \times \Omega X \times [0, \frac{1}{2}]$ is continuous. A similar argument shows that $\mu | \Omega X \times \Omega X \times [\frac{1}{2}, 1]$ is continuous, hence μ is continuous (because the two restrictions agree on the overlap).

Let us prove homotopy associativity. To define $G: \Omega X \times \Omega X \times \Omega X \times \mathbf{I} \to \Omega X$, it suffices to define $F: \Omega X \times \Omega X \times \Omega X \times \mathbf{I} \times \mathbf{I} \to X$, to set $G = F^\#$, and to check that the image of $F^\#$ in $X^{\mathbf{I}}$ actually lies in ΩX. Let

$$F(\omega, \omega', \omega'', t, s) = \begin{cases} \omega(4s/t + 1) & \text{if } 0 \le s \le (t + 1)/4 \\ \omega'(4s - t - 1) & \text{if } (t + 1)/4 \le s \le (t + 2)/4 \\ \omega''((4s - 2 - t)/(2 - t)) & \text{if } (t + 2)/4 \le s \le 1. \end{cases}$$

Again, Corollary 11.2 shows that G is continuous and hence is a (pointed) homotopy $\mu(\mu \times 1) \simeq \mu(1 \times \mu)$. (These formulas are, of course, similar to those that show homotopy associativity of paths.)

Let ω_0 be the constant map at x_0; we must show that the maps $\mu j_1: \omega \mapsto \omega * \omega_0$ and $\mu j_2: \omega \mapsto \omega_0 * \omega$ are each homotopic to the identity on ΩX. Define $F: \Omega X \times \mathbf{I} \to \Omega X$ by $(\omega, t) \mapsto \bar{\omega}_t$, where

$$\bar{\omega}_t(s) = \begin{cases} \omega(2s/(t+1)) & \text{if } 0 \le s \le (t+1)/2 \\ x_0 & \text{if } (t+1)/2 \le s \le 1. \end{cases}$$

Now F is continuous because $e(F \times 1)$ is, and so F is the desired homotopy. The argument for $\omega \mapsto \omega_0 * \omega$ is similar.

Finally, define $\eta: \Omega X \to \Omega X$ by $\omega(t) \mapsto \omega(1-t)$; again, Corollary 11.2 can be used to prove continuity. Define a homotopy $H: \Omega X \times \mathbf{I} \to \Omega X$ by defining $K: \Omega X \times \mathbf{I} \times \mathbf{I} \to X$; let

$$K(\omega, t, s) = \begin{cases} x_0 & \text{if } 0 \le s \le t/2 \\ \omega(2s - t) & \text{if } t/2 \le s \le \frac{1}{2} \\ \omega(2 - 2s - t) & \text{if } \frac{1}{2} \le s \le (2-t)/2 \\ x_0 & \text{if } (2-t)/2 \le s \le 1. \end{cases}$$

Again, continuity is proved by Corollary 11.2; details of the (now familiar) proofs are left to the reader. Hence $\omega \mapsto \mu(\omega, \eta(\omega))$ is nullhomotopic, and a similar argument shows that $\omega \mapsto \mu(\eta(\omega), \omega)$ is nullhomotopic. □

The homotopies are just those that arose in Theorem 3.2; the extra feature is their continuity as maps of function spaces. Note that we use both $*$ and **h**: $\mathbf{Top}_*$ (not merely **Top**) is needed so that μ is defined (the loops must be loops at the same point); $\mathbf{hTop}_*$ is needed so that the axioms for a group object are satisfied.

Recall that $\text{Hom}(X, Y)$ in the homotopy category is denoted by $[X, Y]$. We use the same notation for $\text{Hom}(X, Y)$ in $\mathbf{hTop}_*$ when X and Y are pointed spaces.

Corollary 11.10. *For any pointed space X, $[\ \ , \Omega X]$ is a (contravariant) functor from $\mathbf{hTop}_*$ into* **Groups**. *If Y is a pointed space, and if $[f], [g] \in [Y, \Omega X]$, then their product is $\mu([f], [g]) = [f * g]$.*

Proof. By Lemma 11.7, ΩX is a group object in $\mathbf{hTop}_*$. Theorem 11.4 shows that $[\ \ , \Omega X]$ is group valued; the proof of this last theorem also exhibits the multiplication in $[Y, \Omega X]$. □

Here is a related construction. Recall that if $G: Z \to X^Y$, then $G^\flat: Z \times Y \to X$ is defined by $G^\flat(z, y) = G(z)(y)$. In particular, $G \mapsto G^\flat$ is a function (even a bijection) $\text{Hom}(Z, X^{\mathbf{I}}) \to \text{Hom}(Z \times \mathbf{I}, X)$. Let X and Z be pointed spaces (with respective basepoints x_0, z_0), and replace $X^{\mathbf{I}}$ by its subspace $\Omega(X, x_0)$. This means that we restrict attention to those G such that $G(z_0)$ is the constant loop at x_0 and such that $G(z)(0) = x_0 = G(z)(1)$. Therefore, for all $z \in Z$ and all $t \in \mathbf{I}$,

$$G^\flat(z, 0) = G^\flat(z, 1) = x_0 = G^\flat(z_0, t);$$

that is, $G^\flat$ sends $(Z \times \dot{\mathbf{I}}) \cup (\{z_0\} \times \mathbf{I})$ into x_0. These remarks suggest the following definition.

Definition. If (Z, z_0) is a pointed space, then the **suspension** of Z, denoted by ΣZ, is the quotient space

$$\Sigma Z = (Z \times \mathbf{I})/((Z \times \dot{\mathbf{I}}) \cup (\{z_0\} \times \mathbf{I})),$$

where the identified subset is regarded as the basepoint of ΣZ.

There is another notion of suspension in topology, namely, the *double cone*: the quotient space of $Z \times \mathbf{I}$ in which $Z \times \{0\}$ is identified to a point and $Z \times \{1\}$ is identified to another point.

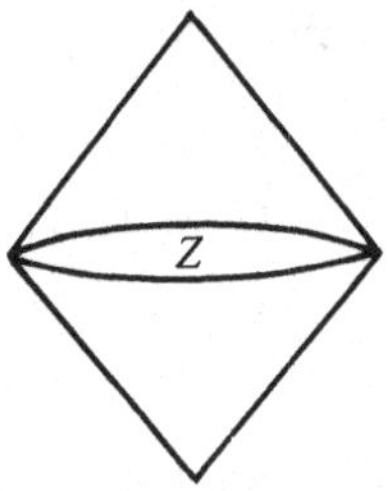

double cone on Z.

The suspension ΣZ just defined is often called the *reduced suspension* to distinguish it from the double cone. The picture of ΣZ is thus the following one with all points on the dashed line identified.

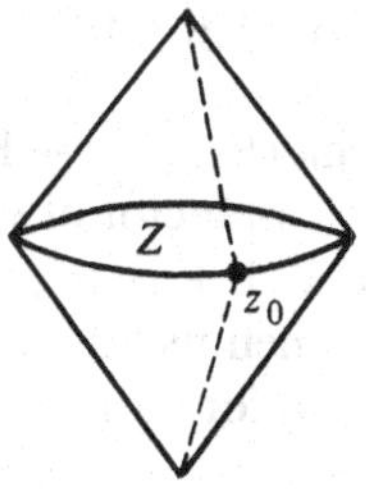

(reduced) suspension ΣZ

If $(z, t) \in Z \times \mathbf{I}$, denote the corresponding element of ΣZ by $[z, t]$. Abuse notation and write $z_0 = [z, 0] = [z, 1] = [z_0, t]$ for all $z \in Z$ and $t \in \mathbf{I}$.

Theorem 11.11. *Suspension defines a functor* Σ: $\mathbf{hTop_*} \to \mathbf{hTop_*}$.

Proof. By Theorem 0.4, it suffices to show that Σ is a functor on $\mathbf{Top_*}$ for which $f_0 \simeq f_1$ implies $\Sigma f_0 \simeq \Sigma f_1$ (pointed homotopies). It is routine to show that Σ is a functor on $\mathbf{Top}_*$ if, for $f: Z \to Y$, we define $\Sigma f: \Sigma Z \to \Sigma Y$ by $[z, t] \mapsto [f(z), t]$. That $\Sigma f_0 \simeq \Sigma f_1$ follows easily from Corollary 8.10. □

One can define a comultiplication, a co-identity, and a co-inverse on every $\Sigma(Z, z_0)$, and one can show that ΣZ is always an H'-group. However, there is an intimate relationship between Σ and Ω (they form an adjoint pair of functors) that will allow us to see this painlessly.

If $\mathscr{A}$ is a category, we sometimes write $\mathrm{Hom}_{\mathscr{A}}(A, A')$ to denote the Hom set in $\mathscr{A}$.

Definition. Let $F: \mathscr{A} \to \mathscr{C}$ and $G: \mathscr{C} \to \mathscr{A}$ be functors. The ordered pair (F, G) is an **adjoint pair** if, for each object A in $\mathscr{A}$ and each object C in $\mathscr{C}$, there is a bijection

$$\tau = \tau_{AC}: \mathrm{Hom}_{\mathscr{C}}(FA, C) \to \mathrm{Hom}_{\mathscr{A}}(A, GC),$$

which is natural in each variable; that is, the following diagrams commute for all $f: A' \to A$ in $\mathscr{A}$ and $g: C \to C'$ in $\mathscr{C}$:

$$\begin{array}{ccc} \mathrm{Hom}_{\mathscr{C}}(FA, C) & \xrightarrow{(Ff)^*} & \mathrm{Hom}_{\mathscr{C}}(FA', C) \\ \downarrow \tau & & \downarrow \tau \\ \mathrm{Hom}_{\mathscr{A}}(A, GC) & \xrightarrow[f^*]{} & \mathrm{Hom}_{\mathscr{A}}(A', GC); \end{array}$$

$$\begin{array}{ccc} \mathrm{Hom}_{\mathscr{C}}(FA, C) & \xrightarrow{g_*} & \mathrm{Hom}_{\mathscr{C}}(FA, C') \\ \downarrow \tau & & \downarrow \tau \\ \mathrm{Hom}_{\mathscr{A}}(A, GC) & \xrightarrow[(Gg)_*]{} & \mathrm{Hom}_{\mathscr{A}}(A, GC'). \end{array}$$

In short, τ is a natural equivalence $\mathrm{Hom}_{\mathscr{C}}(F\_\_, \_\_) \to \mathrm{Hom}_{\mathscr{A}}(\_\_, G\_\_)$ (if one makes the only reasonable definition of a functor of two variables). The reason for the name "adjoint" is quite formal. If V is an inner product space and if $f: V \to V$ is a linear transformation, then its adjoint is a linear transformation $g: V \to V$ such that $(fv, w) = (v, gw)$ for all vectors $v, w \in V$.

EXAMPLE 11.5. Let $\mathscr{A} = \mathscr{C} = \mathbf{Sets}$, and let Y be a fixed set. Define $F: \mathbf{Sets} \to \mathbf{Sets}$ by $F = \_\_ \times Y$, and define $G: \mathbf{Sets} \to \mathbf{Sets}$ by $G = \mathrm{Hom}(Y, \quad)$. For sets A, C, define $\tau_{AC}: \mathrm{Hom}(A \times Y, C) \to \mathrm{Hom}(A, \mathrm{Hom}(Y, C))$ by $G \mapsto G^\#$, the associate of G. It is routine to check that $(\_\_ \times Y, \mathrm{Hom}(Y, \quad))$ is an adjoint pair.

EXAMPLE 11.6. Let $\mathscr{A} = \mathscr{C} = \mathbf{Top}$. If Y and C are spaces, then $\mathrm{Hom}(Y, C) = C^Y$, and one can use the exponential law to show that $(\_\_ \times Y, (\quad)^Y)$ is an adjoint pair when Y is locally compact Hausdorff.

EXERCISES

*11.15. Let $\mathscr{A} = \mathscr{C} = \mathbf{Ab}$. For any abelian group Y, show that $(\_\_ \otimes Y, \mathrm{Hom}(Y, \quad))$ is an adjoint pair.

11.16. Let $F: \mathbf{Ab} \to \mathbf{Sets}$ be the forgetful functor (Example 0.8), and let $G: \mathbf{Sets} \to \mathbf{Ab}$ be the "free" functor: if X is a set, then GX is the free abelian group having basis X; if $f: X \to Y$ is a function, then $Gf: GX \to GY$ is the homomorphism obtained from f by extending by linearity. Prove that (G, F) is an adjoint pair.

*11.17. If (F, G) is an adjoint pair, then F preserves coproducts and G preserves products.

Exercise 11.17 is a special case of the main property of adjoint pairs: there is a notion of *limit* (inverse limit) and *colimit* (direct limit); if (F, G) is an adjoint pair, then F preserves colimits and G preserves limits (see [Rotman (1979), pp. 47, 55]). Examples of limits are products, pullbacks (defined below), kernels, nested intersections, and completions; examples of colimits are coproducts, pushouts, cokernels, and ascending unions.

Once we recall how suspension arose, the next result is almost obvious.

Theorem 11.12. *(Σ, Ω) is an adjoint pair of functors on* $\mathbf{hTop}_*$.

PROOF. If X and Y are pointed spaces, define

$$\tau_{XY}: [\Sigma X, Y] \to [X, \Omega Y]$$

by $[F] \mapsto [F^\#]$, where $F^\#$ is the associate of F. Now τ_{XY} is well defined, because if $H: \Sigma X \times \mathbf{I} \to Y$ is a (pointed) homotopy from F_0 to F_1, say, then $H^\#: X \times \mathbf{I} \to \Omega Y$, if continuous, is a (pointed) homotopy from $F_0^\#$ to $F_1^\#$. But Theorem 11.1(ii) shows that continuity of H implies that of $H^\#$. Each τ_{XY} is a bijection (its inverse is $[G] \mapsto [G^\flat]$); we leave the routine check that the required diagrams commute to the reader. □

As we remarked earlier, there are various consequences of adjointness; for example, Exercise 11.17 gives $\Sigma(X \vee Y) = \Sigma X \vee \Sigma Y$ and $\Omega(X \times Y) = \Omega X \times \Omega Y$.

Corollary 11.13. *If X is a pointed space, then ΣX is a cogroup object in* $\mathbf{hTop}_*$.

PROOF. For every pointed space Y, adjointness gives a bijection $\tau = \tau_{XY}$: $[\Sigma X, Y] \to [X, \Omega Y]$, namely, $[f] \mapsto [f^\#]$, where $f^\#$ is the associate of f. Since ΩY is a group object in $\mathbf{hTop}_*$, $[X, \Omega Y]$ is a group. We use τ to define a group structure on $[\Sigma X, Y]$: if $[f], [g] \in [\Sigma X, Y]$, then their product is

$$[\mu(f^\#, g^\#)] = [f^\# * g^\#].$$

Note that τ is now an isomorphism of groups.

We claim that the functor $[\Sigma X, \quad]$ is group valued; if so, then the result follows from Theorem 11.4. It remains to prove that if $\varphi: Y \to Y'$ is a pointed map, then $[\varphi]_*$ is a homomorphism. Consider the commutative diagram (in the definition of adjointness):

$$\begin{array}{ccc} [\Sigma X, Y] & \xrightarrow{[\varphi]_*} & [\Sigma X, Y'] \\ \Big\downarrow \tau & & \Big\downarrow \tau \\ [X, \Omega Y] & \xrightarrow[(\Omega[\varphi])_*]{} & [X, \Omega Y']. \end{array}$$

Now $\Omega[\varphi]\colon \Omega Y \to \Omega Y'$ is a special map (see Exercise 11.11) because $\varphi_*(\omega * \omega') = \varphi\omega * \varphi\omega'$ for all $\omega, \omega' \in \Omega X$. It follows from Exercise 11.11(ii) that $(\Omega[\varphi])_*$ is a homomorphism, and the diagram above now shows that $[\varphi]_*$ is a homomorphism (because the vertical maps are isomorphisms). □

Remark. Here is an explicit formula for the comultiplication $m\colon \Sigma X \to \Sigma X \vee \Sigma X$. In the proof of Theorem 11.4′, we saw that m is the product of j_1 and j_2 in the group $[\Sigma X, \Sigma X \vee \Sigma X]$, where j_1 and j_2 are the injections $\Sigma X \to \Sigma X \vee \Sigma X$. If we regard $\Sigma X \vee \Sigma X$ as a subspace of $\Sigma X \times \Sigma X$ (as in Exercise 11.3), then $j_1([x, t]) = ([x, t], *)$ and $j_2([x, t]) = (*, [x, t])$. But it was shown in Corollary 11.13 that the product of j_1 and j_2 is $[j_1^\# * j_2^\#]$. Recall that $j_1^\#\colon X \to \Omega(\Sigma X \vee \Sigma X)$ is given by $j_1^\#(x) = ([x, \quad], *)$; there is a similar formula for $j_2^\#$. Therefore,

$$m([x, t]) = \begin{cases} ([x, 2t], *) & \text{if } 0 \le t \le \frac{1}{2} \\ (*, [x, 2t - 1]) & \text{if } \frac{1}{2} \le t \le 1. \end{cases}$$

The comultiplication on the suspension ΣX is thus obtained by "pinching".

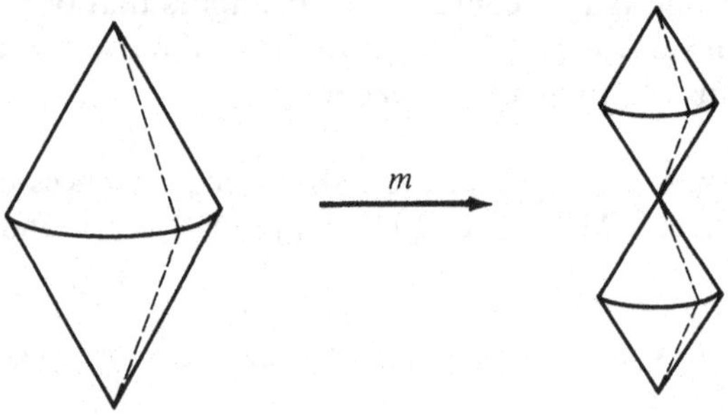

In particular, for $X = S^n$, the comultiplication $m\colon \Sigma S^n \to \Sigma S^n \vee \Sigma S^n$ may be viewed as the map $S^{n+1} \to S^{n+1} \vee S^{n+1}$ which identifies the equator to a point.

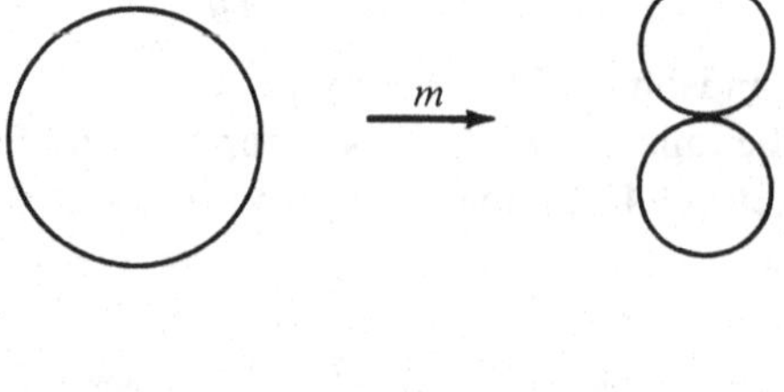

The result linking this discussion to homotopy groups is that $\Sigma S^n \approx S^{n+1}$ for all $n \geq 0$. Explicit homeomorphisms can be given: see [Spanier, p. 42] or [Whitehead, p. 107], but we prefer another proof that we learned from M. Ramachandran.

Definition. Let X be a locally compact Hausdorff space, and let ∞ denote a point outside X. Then the **one-point compactification** of X, denoted by X^∞, is the set $X^\infty = X \cup \{\infty\}$ equipped with the topology consisting of all open sets of X together with all sets of the form $(X - K) \cup \{\infty\}$, where K is a compact subset of X. We choose ∞ as the basepoint of X^∞.

Exercises

*11.18. The one-point compactification of $\mathbf{R}^n$ is S^n for all $n \geq 1$. (*Hint*: Stereographic projection.)

*11.19. If $J = [0, 1)$, then $(J^n)^\infty \approx \mathbf{I}^n$.

*11.20. If X is a compact Hausdorff space and A is a closed subset, then, in $\mathbf{Top}_*$,

$$X/A \approx (X - A)^\infty.$$

Definition. If X and Y are pointed spaces, then their **smash product**, denoted by $X \wedge Y$ (some authors write $X \# Y$), is the pointed space

$$X \wedge Y = (X \times Y)/(X \vee Y)$$

(where $X \vee Y$ is regarded as a subspace of $X \times Y$ as in Exercise 11.3).

The smash product does depend on the basepoint. For example, if 0 is chosen as basepoint of $\mathbf{I}$, then $\mathbf{I} \wedge \mathbf{I}$ is homeomorphic to $\mathbf{I} \times \mathbf{I}$ (one identifies two adjacent sides of $\mathbf{I} \times \mathbf{I}$). On the other hand, if $\frac{1}{2}$ is chosen as basepoint of $\mathbf{I}$, then $\mathbf{I} \wedge \mathbf{I}$ is homeomorphic to the wedge of four copies of $\mathbf{I} \times \mathbf{I}$.

Lemma 11.14. *If X is a locally compact Hausdorff pointed space, then*

$$\Sigma X \approx X \wedge S^1.$$

Proof. Since X is locally compact Hausdorff, the map $1 \times \exp\colon X \times \mathbf{I} \to X \times S^1$ is an identification, by Lemma 8.9. If $\nu\colon X \times S^1 \to X \wedge S^1$ is the natural map, then $h = \nu(1 \times \exp)$ is also an identification (Exercise 1.10). But it is easy to check that $(X \times \mathbf{I})/\ker h = \Sigma X$, and so the result follows from Corollary 1.10. □

Lemma 11.15. *If X and Y are locally compact Hausdorff spaces, then*

$$X^\infty \wedge Y^\infty \approx (X \times Y)^\infty,$$

where, in each case, ∞ is chosen as basepoint of the one-point compactification.

Proof. By definition, $X^\infty \wedge Y^\infty = (X^\infty \times Y^\infty)/(X^\infty \vee Y^\infty)$. Since the numerator is compact and the denominator is closed, Exercise 11.20 shows that

$X^\infty \wedge Y^\infty$ is the one-point compactification of $X^\infty \times Y^\infty - X^\infty \vee Y^\infty$. But

$$X^\infty \times Y^\infty = (X \times Y) \cup (\{\infty\} \times Y^\infty) \cup (X^\infty \times \{\infty\}),$$

while Exercise 11.3 shows that

$$X^\infty \vee Y^\infty = (\{\infty\} \times Y^\infty) \cup (X^\infty \times \{\infty\}).$$

Their difference is thus $X \times Y$, as desired. □

Theorem 11.16. $\Sigma S^n \approx S^{n+1}$ *for all* $n \geq 0$.

PROOF (M. Ramachandran). If $n = 0$, the result is easy and is left to the reader. If $n \geq 1$, then Exercise 11.18 and the above lemma give

$$\Sigma S^n = S^n \wedge S^1 = (\mathbf{R}^n)^\infty \wedge \mathbf{R}^\infty = (\mathbf{R}^n \times \mathbf{R})^\infty = (\mathbf{R}^{n+1})^\infty = S^{n+1}. \quad \square$$

Corollary 11.17. S^n *is a cogroup object in* **hTop**$_*$ *for all* $n \geq 1$.

PROOF. Each such sphere is a suspension. □

EXERCISES

11.21. Prove that $S^m \wedge S^n \approx S^{m+n}$ for all $m, n \geq 1$.

11.22. Prove that $\mathbf{I}^n \wedge \mathbf{I} \approx \mathbf{I}^{n+1}$, where the origin is taken as the basepoint of $\mathbf{I}^n$ and 0 is the basepoint of $\mathbf{I}$. (*Hint*: Use Exercise 11.19.)

Homotopy Groups

For each pointed space X, we know that

$$\pi_1(X) = [S^1, X],$$

where $(1, 0)$ is the basepoint of S^1.

Convention. For every $n \geq 0$, regard $s_n = (1, 0, \ldots, 0) \in \mathbf{R}^{n+1}$ as the basepoint of S^n.

Definition. For every pointed space (X, x_0) and every $n \geq 0$,

$$\pi_n(X, x_0) = [(S^n, s_n), (X, x_0)].$$

We shall usually abbreviate $\pi_n(X, x_0)$ to $\pi_n(X)$. When $n \geq 2$, $\pi_n(X)$ is called a (higher) **homotopy group**. Of course, π_n is a functor with domain **hTop**$_*$.

Theorem 11.18. *For every pointed space* X, $\pi_0(X)$ *is a pointed set, and* $\pi_n(X)$ *is a group for all* $n \geq 1$.

PROOF. That $[(S^0, 1), (X, x_0)]$ coincides with $\pi_0(X)$ as defined in Chapter 1 is left to the reader; the basepoint of $\pi_0(X)$ is the path component containing the basepoint of X. If $n \geq 1$, then the result is immediate from Corollary 11.17 and Theorem 11.4′. □

What is the product of $[f], [g] \in \pi_n(X)$? If $m: S^n \to S^n \vee S^n$ is the comultiplication (pinching), then

$$[f] * [g] = ([f], [g])[m] = [(f, g)m].$$

Suppose that X happens to be an H-group, with multiplication $\mu: X \times X \to X$. Then Theorem 11.4 also equips $\pi_n(X)$ with a group structure, namely,

$$[f] \circ [g] = [\mu]([f], [g]) = [\mu(f, g)].$$

Theorem 11.19. *If Q is an H'-group and P is an H-group, then the group operations on $[Q, P]$ determined by the comultiplication m of Q and by the multiplication μ of P coincide.*

PROOF. Let $f, g: Q \to P$. By Lemma 11.6, the following diagram commutes to homotopy:

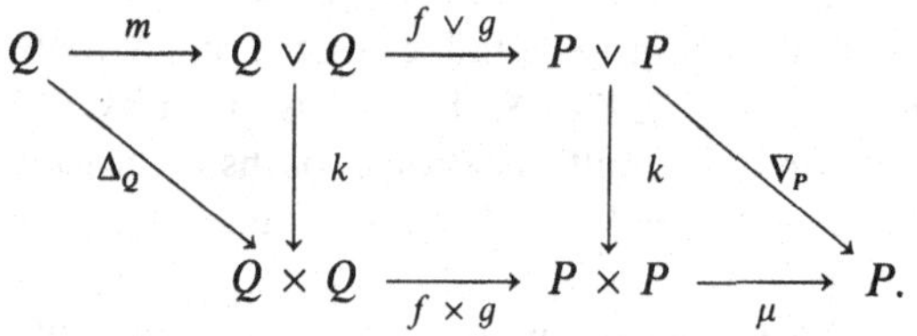

But the multiplication determined by m is $[f] * [g] = [(f, g)m] = [\nabla(f \vee g)m]$ (Exercise 11.6), and that determined by μ is $[f] \circ [g] = [\mu(f, g)] = [\mu(f \times g)\Delta]$. Hence $[f] * [g] = [f] \circ [g]$. □

Theorem 11.20. *If X is a pointed space, then*

$$\pi_n(X) \cong \pi_{n-k}(\Omega^k X)$$

for all $1 \leq k \leq n - 1$ (where Ω^k is the composite of Ω with itself k times). In particular, if $n \geq 2$,

$$\pi_n(X) \cong \pi_1(\Omega^{n-1} X).$$

PROOF. $\pi_n(X) = [S^n, X] = [\Sigma^n S^0, X]$

$$= [\Sigma^{n-k} S^0, \Omega^k X] = [S^{n-k}, \Omega^k X] = \pi_{n-k}(\Omega^k X).$$ □

Corollary 11.21. *If X is a pointed space, then $\pi_n(X)$ is abelian for all $n \geq 2$.*

PROOF. By the theorem, $\pi_n(X) \cong \pi_1(\Omega^{n-1} X)$ if $n \geq 2$. But $\Omega^{n-1} X$ is a loop space, hence is an H-group, and hence is an H-space. By Theorem 3.20, $\pi_1(\Omega^{n-1} X)$ is abelian. □

The functors π_n are defined on **hTop**$_*$ (with values in **Sets**$_*$ when $n = 0$, in **Groups** when $n = 1$, and in **Ab** when $n \geq 2$). Such functors can be viewed as functors on **Top**$_*$, which satisfy the (pointed) homotopy axiom: if there is a pointed homotopy $f \simeq g$ of pointed maps, then the induced maps f_* and g_* are equal.

If $n \geq 2$, it is plain that $\pi_n(X) = 0$ means that $\pi_n(X)$ is the trivial group; we extend this notation to the case $n = 1$. Also, we write $\pi_0(X) = 0$ to mean that $\pi_0(X)$ has only one element (i.e., X is path connected).

The following result is important even though it is easy.

Theorem 11.22 (Dimension Axiom). *If X is a singleton, then $\pi_n(X) = 0$ for all $n \geq 0$.*

PROOF. There is only one function from S^n into X, namely, the constant function, and so $[S^n, X]$ has only one element. □

There is a down-to-earth description of the multiplication in $\pi_n(X)$, eschewing functors and cogroup objects, which is a straightforward generalization of the multiplication in $\pi_1(X)$. We have already proved (in Corollary 11.13) that if $[f], [g] \in [\Sigma X, Y]$, then their product is $[f^\# * g^\#]$, where $f^\#: X \to \Omega Y$ is the associate of f. In more detail, elements of ΣX have the form $[x, t]$, where $x \in X$ and $t \in \mathbf{I}$; if $f: \Sigma X \to Y$, then $f^\#$ is given by $f^\#(x) = f([x, \quad])$; the star multiplication is the multiplication of paths in the loop space ΩY. This discussion applies to $\pi_n(X) = [S^n, X]$ upon recalling that $S^n = \Sigma S^{n-1}$.

Definition. Let $\mathbf{I}^n$ be the cartesian product of n copies of $\mathbf{I}$, and let $\dot{\mathbf{I}}^n = \{(t_1, \ldots, t_n) \in \mathbf{I}^n: \text{some } t_i \in \dot{\mathbf{I}}\}$.

If $n \geq 1$, then Exercise 11.20 shows that $\mathbf{I}^n/\dot{\mathbf{I}}^n \approx (\mathbf{I}^n - \dot{\mathbf{I}}^n)^\infty$ (one-point compactification). But $\mathbf{I}^n - \dot{\mathbf{I}}^n \approx \mathbf{R}^n$, and $(\mathbf{R}^n)^\infty \approx S^n$; therefore $\mathbf{I}^n/\dot{\mathbf{I}}^n \approx S^n$; choose homeomorphisms $\theta = \theta_n: \mathbf{I}^n/\dot{\mathbf{I}}^n \simeq S^n$. If $n \geq 2$, we prove that there is a homeomorphism $\varphi = \varphi_n: \mathbf{I}^n/\dot{\mathbf{I}}^n \simeq \Sigma S^{n-1}$ with

$$\varphi_n: [t_1, \ldots, t_n] \mapsto [\theta_{n-1}[t_1, \ldots, t_{n-1}], t_n],$$

where $[t_1, \ldots, t_n]$ is the image of $(t_1, \ldots, t_n)$ in $\mathbf{I}^n/\dot{\mathbf{I}}^n$. Recall the identities $\dot{\mathbf{I}}^n = (\dot{\mathbf{I}}^{n-1} \times \mathbf{I}) \cup (\mathbf{I}^{n-1} \times \dot{\mathbf{I}})$ and $(\mathbf{I}^{n-1} \times \mathbf{I})/(\dot{\mathbf{I}}^{n-1} \times \mathbf{I}) = (\mathbf{I}^{n-1}/\dot{\mathbf{I}}^{n-1}) \times \mathbf{I}$, and consider the diagram

$$\begin{array}{ccc}
(\mathbf{I}^{n-1}/\dot{\mathbf{I}}^{n-1}) \times \mathbf{I} = \dfrac{\mathbf{I}^{n-1} \times \mathbf{I}}{\dot{\mathbf{I}}^{n-1} \times \mathbf{I}} & \xrightarrow{\nu} & \dfrac{\mathbf{I}^{n-1} \times \mathbf{I}}{(\dot{\mathbf{I}}^{n-1} \times \mathbf{I}) \cup (\mathbf{I}^{n-1} \times \dot{\mathbf{I}})} = \mathbf{I}^n/\dot{\mathbf{I}}^n \\
\Big\downarrow{\scriptstyle \theta \times 1} & & \\
S^{n-1} \times \mathbf{I} = \dfrac{S^{n-1} \times \mathbf{I}}{\{s_{n-1}\} \times \mathbf{I}} & \xrightarrow[\xi]{} & \dfrac{S^{n-1} \times \mathbf{I}}{(\{s_{n-1}\} \times \mathbf{I}) \cup (S^{n-1} \times \dot{\mathbf{I}})} = \Sigma S^{n-1},
\end{array}$$

where ν and ξ are the natural maps. Define $h = \xi(\theta \times 1)$. Now $\theta \times 1$ is an identification, by Lemma 8.9, and so h is an identification, by Exercise 1.10. Corollary 1.10 thus gives a homeomorphism φ making the following diagram commute

$$\begin{array}{ccc} (\mathbf{I}^{n-1} \times \mathbf{I})/(\dot{\mathbf{I}}^{n-1} \times \mathbf{I}) & \xrightarrow{h} & \Sigma S^{n-1} \\ & \searrow^{\nu} & \nearrow_{\varphi} \\ & & \mathbf{I}^n/\dot{\mathbf{I}}^n \end{array}$$

(because $\mathbf{I}^n/\dot{\mathbf{I}}^n = (\mathbf{I}^{n-1} \times \mathbf{I}/\dot{\mathbf{I}}^{n-1} \times \mathbf{I})/\ker h$), and this says that $\varphi\colon [t_1, \ldots, t_n] \mapsto [\theta[t_1, \ldots, t_{n-1}], t_n]$, as desired.

Each function $f\colon (\mathbf{I}^n, \dot{\mathbf{I}}^n) \to (X, x_0)$ induces a pointed map $\bar{f}\colon \mathbf{I}^n/\dot{\mathbf{I}}^n \to X$; moreover, Corollary 8.10 shows that if f, g are two such maps which are homotopic rel $\dot{\mathbf{I}}^n$, then there is a pointed homotopy $\bar{f} \simeq \bar{g}$. Therefore, there is a bijection

$$b\colon [(\mathbf{I}^n, \dot{\mathbf{I}}^n), (X, x_0)] \to [(\Sigma S^{n-1}, *), (X, x_0)] = \pi_n(X, x_0)$$

given by $[f] \mapsto [\bar{f}\varphi^{-1}]$. This bijection equips $[(\mathbf{I}^n, \dot{\mathbf{I}}^n), (X, x_0)]$ with a group structure: if $f, g\colon (\mathbf{I}^n, \dot{\mathbf{I}}^n) \to (X, x_0)$, define $f + g\colon (\mathbf{I}^n, \dot{\mathbf{I}}^n) \to (X, x_0)$ by

$$(f + g)(t_1, \ldots, t_n) = \begin{cases} f(t_1, \ldots, t_{n-1}, 2t_n) & \text{if } 0 \le t_n \le \frac{1}{2} \\ g(t_1, \ldots, t_{n-1}, 2t_n - 1) & \text{if } \frac{1}{2} \le t_n \le 1. \end{cases}$$

To see that $[f] + [g]$ defined as $[f + g]$ actually gives a group isomorphic to $\pi_n(X, x_0)$, it suffices to show that $b([f + g]) = b([f]) * b([g])$. But $b([f]) = [\bar{f}\varphi^{-1}]$, where $\bar{f}\varphi^{-1}$ is defined on all $[\theta[t_1, \ldots, t_{n-1}], t_n] \in \Sigma S^{n-1}$. Our earlier discussion therefore shows that $[\bar{f}\varphi^{-1}] * [\bar{g}\varphi^{-1}]$ corresponds to $[(\bar{f}\varphi^{-1})^{\#} * (\bar{g}\varphi^{-1})^{\#}]$, and

$$\begin{aligned} &(\bar{f}\varphi^{-1})^{\#} * (\bar{g}\varphi^{-1})^{\#}(\theta[t_1, \ldots, t_{n-1}]) \\ &\quad = \bar{f}\varphi^{-1}([\theta[t_1, \ldots, t_{n-1}], \text{—}]) * \bar{g}\varphi^{-1}(\theta[t_1, \ldots, t_{n-1}], \text{—}]) \\ &\quad = \bar{f}([t_1, \ldots, t_{n-1}, \text{—}]) * \bar{g}([t_1, \ldots, t_{n-1}, \text{—}]) \\ &\quad = (f + g)(t_1, \ldots, t_{n-1}, \text{—}) \end{aligned}$$

as desired.

If $n = 2$, we may picture $f + g$ schematically:

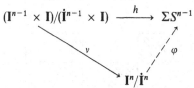

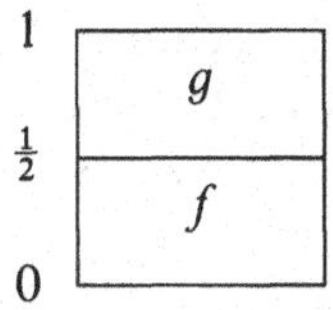

The following figure in which the shaded regions are constant (from [Whitehead, p. 125]) suggests a direct argument, using the above formula for $f + g$, that multiplication in $\pi_2(X)$ (and in all higher $\pi_n(X)$) is abelian.

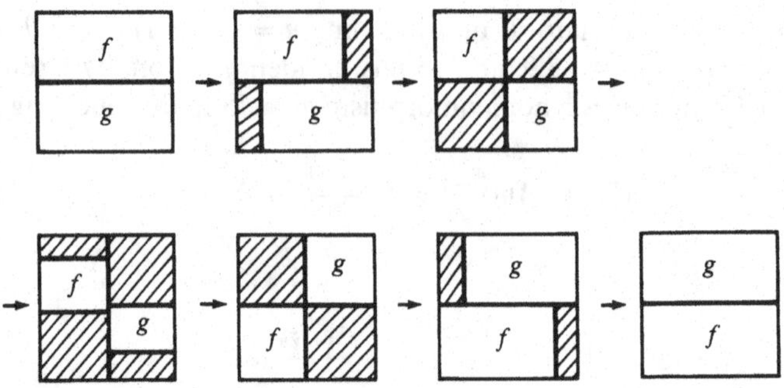

There are two obvious questions. If X is path connected, are its homotopy groups independent of the choice of basepoint? If X and Y are path connected spaces having the same homotopy type, do X and Y have the same homotopy groups? As is true for fundamental groups, the answers are positive, but the proofs are more involved here.

Definition. If X is a space, then a **local system** of groups is a family of groups $\{T(x): x \in X\}$ and a family of homomorphisms $\{T(\varphi): T(\varphi(0)) \to T(\varphi(1)) |$ every path $\varphi\}$ such that:

(i) if $\varphi \simeq \varphi'$ rel $\dot{\mathbf{I}}$, then $T(\varphi) = T(\varphi')$;
(ii) if i_x is the constant path at x, then $T(i_x)$ is the identity map on $T(x)$;
(iii) if φ, ψ are paths in X with $\varphi(1) = \psi(0)$, then

$$T(\varphi * \psi) = T(\psi)T(\varphi).$$

Denote the fundamental groupoid of X (see Theorem 3.2) by $\Pi(X)$. In Exercise 3.9, this groupoid was made into a category: define objects to be the points of X, and define morphisms to be the path classes. A local system is just a functor $T: \Pi(X) \to$ **Groups** (condition (i) guarantees that T is well defined on morphisms). Since every morphism in $\Pi(X)$ is an equivalence (the inverse of $[\varphi]$ is $[\varphi^{-1}]$), every $T(\varphi)$ is an isomorphism. We are going to see that if X is path connected, then there is a local system on X with $T(x) = \pi_n(X, x)$ for all $x \in X$.

Definition. Let $F: \mathbf{I}^n \times \mathbf{I} \to X$ be a free homotopy. If $0 = (0, \ldots, 0)$ denotes the origin in $\mathbf{I}^n$, then $\varphi = F(0, \quad)$ is a path in X; we say that F is a homotopy **along** φ. If $F(u, t) = \varphi(t)$ for all $u \in \dot{\mathbf{I}}^n$, then we say that F is a **level homotopy** along φ.

There is a retraction

$$r: \mathbf{I}^n \times \mathbf{I} \to (\mathbf{I}^n \times \{0\}) \cup (\dot{\mathbf{I}}^n \times \mathbf{I})$$

(when $n = 2$, the right side is a box without a top). Regard $\mathbf{I}^n \times \mathbf{I}$ as imbedded in $\mathbf{R}^{n+1}$, and let p be a point above an interior point of $\mathbf{I}^n \times \{1\}$; for example, let $p = (\frac{1}{2}, \ldots, \frac{1}{2}, 2)$. If $y \in \mathbf{I}^n \times \mathbf{I}$, define $r(y)$ to be the point where the line from p to y intersects $(\mathbf{I}^n \times \{0\}) \cup (\dot{\mathbf{I}}^n \times \mathbf{I})$. Call r a **stereographic retraction.**

In a similar manner, one sees that there is a retraction

$$R: \mathbf{I}^n \times \mathbf{I} \times \mathbf{I} \to (\mathbf{I}^n \times \mathbf{I} \times \{0\}) \cup (\dot{\mathbf{I}}^n \times \mathbf{I} \times \mathbf{I}).$$

Notation. Let $Q(X, x_0)$ denote the set of all maps $\alpha: (\mathbf{I}^n, \dot{\mathbf{I}}^n) \to (X, x_0)$.

If φ is a path in X from x_0 to x_1, and if $\alpha \in Q(X, x_0)$, define $L': (\mathbf{I}^n \times \{0\}) \cup (\dot{\mathbf{I}}^n \times \mathbf{I}) \to X$ by

$$L'(u, 0) = \alpha(u) \quad \text{if } u \in \mathbf{I}^n$$

and

$$L'(u, t) = \varphi(t) \quad \text{if } u \in \dot{\mathbf{I}}^n \text{ and } t \in \mathbf{I}$$

(the formulas agree on $\dot{\mathbf{I}}^n \times \{0\}$ because $\alpha \in Q(X, x_0)$). Then

$$L = L'r$$

is a level homotopy along φ with $L(\ \ , 0) = \alpha$ (where $r: \mathbf{I}^n \times \mathbf{I} \to (\mathbf{I}^n \times \{0\}) \cup (\dot{\mathbf{I}}^n \times \mathbf{I})$ is a stereographic retraction).

Definition. If φ is a path in X from x_0 to x_1, then

$$\varphi_{\#}: Q(X, x_0) \to Q(X, x_1)$$

is defined by

$$\varphi_{\#}: \alpha \mapsto L(\ \ , 1).$$

Lemma 11.23. *Let $\alpha, \beta \in Q(X, x_0)$.*

(i) *Let φ be a closed path in X at x_0, and let $F: \alpha \simeq \beta$ be a level homotopy along φ. If φ is nullhomotopic* rel $\dot{\mathbf{I}}$, *then*

$$\alpha \simeq \beta \text{ rel } \dot{\mathbf{I}}^n.$$

(ii) *Assume that φ, φ' are paths in X from x_0 to x_1 and that $F: \alpha \simeq \beta$ is a level homotopy along φ'. If $\varphi \simeq \varphi'$* rel $\dot{\mathbf{I}}$, *then*

$$\varphi_{\#}(\alpha) \simeq \beta \text{ rel } \dot{\mathbf{I}}^n.$$

Proof. (i) Let $\mu: \mathbf{I} \times \mathbf{I} \to X$ be a homotopy rel $\dot{\mathbf{I}}$ showing that φ is nullhomotopic; thus, for all $t, s \in \mathbf{I}$,

$$\mu(t, 0) = \varphi(t) \quad \text{and} \quad \mu(t, 1) = \mu(0, s) = \mu(1, s) = x_0.$$

Define $h: (\mathbf{I}^n \times \mathbf{I} \times \{0\}) \cup (\dot{\mathbf{I}}^n \times \mathbf{I} \times \mathbf{I}) \to X$ by

$$h(u, t, 0) = F(u, t) \quad \text{if } u \in \mathbf{I}^n \text{ and } t \in \mathbf{I};$$

$$h(u, t, s) = \mu(t, s) \quad \text{if } u \in \dot{\mathbf{I}}^n \text{ and } t, s \in \mathbf{I}$$

(note that these formulas agree on the overlap $\dot{\mathbf{I}}^n \times \mathbf{I} \times \{0\}$ because F is a level homotopy).

Define $H: \mathbf{I}^n \times \mathbf{I} \times \mathbf{I} \to X$ by $H = hR$ (where $R: \mathbf{I}^n \times \mathbf{I} \times \mathbf{I} \to (\mathbf{I}^n \times \mathbf{I} \times \{0\}) \cup (\dot{\mathbf{I}}^n \times \mathbf{I} \times \mathbf{I})$ is the retraction defined before this lemma), and define $K: \mathbf{I}^n \times \mathbf{I} \to X$ by

$$K(u, t) = \begin{cases} H(u, 0, 4t) & \text{if } 0 \le t \le \frac{1}{4} \\ H(u, 4t - 1, 1) & \text{if } \frac{1}{4} \le t \le \frac{1}{2} \\ H(u, 1, 2 - 2t) & \text{if } \frac{1}{2} \le t \le 1. \end{cases}$$

One checks easily that the formulas agree on the overlaps $\mathbf{I}^n \times \{\frac{1}{4}\}$ and $\mathbf{I}^n \times \{\frac{1}{2}\}$. Moreover, for $u \in \mathbf{I}^n$,

$$K(u, 0) = H(u, 0, 0) = F(u, 0) = \alpha(u)$$

and

$$K(u, 1) = H(u, 1, 0) = F(u, 1) = \beta(u),$$

and for $u \in \dot{\mathbf{I}}^n$,

$$K(u, t) = x_0.$$

(ii) Define $K: \mathbf{I}^n \times \mathbf{I} \to X$ by

$$K(u, t) = \begin{cases} L(u, 1 - 2t) & \text{if } 0 \le t \le \frac{1}{2} \\ F(u, 2t - 1) & \text{if } \frac{1}{2} \le t \le 1 \end{cases}$$

(both formulas agree on $(u, \frac{1}{2})$ with common value $\alpha(u)$; the map $L = L'r$ has been defined before this lemma). Now $K: \varphi_{\#}(\alpha) \simeq \beta$ is a level homotopy along $\varphi^{-1} * \varphi'$. Since $\varphi^{-1} * \varphi'$ is nullhomotopic rel $\dot{\mathbf{I}}$, the first part gives $\varphi_{\#}(\alpha) \simeq \beta$ rel $\dot{\mathbf{I}}^n$. □

Theorem 11.24. *If X is path connected, then there is a local system $T: \Pi(X) \to$* **Groups** *with $T(x) = \pi_n(X, x)$ for all $x \in X$, hence*

$$\pi_n(X, x_0) \cong \pi_n(X, x_1)$$

for all $x_0, x_1 \in X$.

Proof. We have already observed that every homomorphism $T(\varphi)$ in a local system must be an isomorphism; therefore, if φ is a path in X from x_0 to x_1, then $T(\varphi): \pi_n(X, x_0) \xrightarrow{\simeq} \pi_n(X, x_1)$.

Define $T(\varphi): \pi_n(X, x_0) \to \pi_n(X, x_1)$ by

$$[\alpha] \mapsto [\varphi_{\#}(\alpha)].$$

To see that $T(\varphi)$ is a well defined function, assume that $F: \alpha \simeq \beta$ is a homotopy rel $\dot{\mathbf{I}}^n$; that is, F is a level homotopy along the constant path e at x_0. Combining

F with the level homotopy $L: \beta \simeq \varphi_{\#}(\beta)$ along φ gives a level homotopy $\alpha \simeq \varphi_{\#}(\beta)$ along $e * \varphi$. Since $e * \varphi \simeq \varphi$ rel $\dot{\mathbf{I}}$, Lemma 11.23(ii) gives $\varphi_{\#}(\alpha) \simeq \varphi_{\#}(\beta)$ rel $\dot{\mathbf{I}}^n$.

To see that $T(\varphi)$ is a homomorphism, it suffices to prove that $\varphi_{\#}(\alpha * \beta) \simeq \varphi_{\#}(\alpha) * \varphi_{\#}(\beta)$ rel $\dot{\mathbf{I}}^n$, where $\alpha, \beta \in Q(X, x_0)$ and φ is a path with $\varphi(0) = x_0$; by Lemma 11.23(ii), it suffices to prove that there is a level homotopy $\alpha * \beta \simeq \varphi_{\#}(\alpha) * \varphi_{\#}(\beta)$ along φ. If $L_1: \alpha \simeq \varphi_{\#}(\alpha)$ and $L_2: \beta \simeq \varphi_{\#}(\beta)$ are the level homotopies along φ, define $G: \mathbf{I}^n \times \mathbf{I} \to X$ by

$$G(t_1, \ldots, t_n, t) = \begin{cases} L_1(t_1, \ldots, t_{n-1}, 2t_n, t) & \text{if } 0 \le t_n \le \frac{1}{2} \\ L_2(t_1, \ldots, t_{n-1}, 2t_n - 1, t) & \text{if } \frac{1}{2} \le t_n \le 1. \end{cases}$$

If $t_n = \frac{1}{2}$, then $(t_1, \ldots, t_{n-1}, 2t_n) = (t_1, \ldots, t_{n-1}, 1) \in \dot{\mathbf{I}}^n$ and $(t_1, \ldots, t_{n-1}, 2t_n - 1) = (t_1, \ldots, t_{n-1}, 0) \in \dot{\mathbf{I}}^n$; it follows that both formulas give the same value for each $t \in \mathbf{I}$, namely, $\varphi(t)$, and so G is well defined. But $G: \alpha * \beta \simeq \varphi_{\#}(\alpha) * \varphi_{\#}(\beta)$ is a level homotopy along φ, as desired.

We now verify the conditions in the definition of local system. If $\varphi \simeq \varphi'$ rel $\dot{\mathbf{I}}$, then, for all $\alpha \in Q(X, x_0)$, we have $\alpha \simeq \varphi'_{\#}(\alpha)$ along φ'; Lemma 11.23(ii) gives $\varphi_{\#}(\alpha) \simeq \varphi'_{\#}(\alpha)$ rel $\dot{\mathbf{I}}^n$, that is, $T(\varphi) = T(\varphi')$. If e is the constant path at x_0, then $e_{\#}(\alpha) \simeq \alpha$, hence $T(e)$ is the identity. Finally, if ψ is a path in X with $\psi(0) = \varphi(1)$, then there are level homotopies $\alpha \simeq \varphi_{\#}(\alpha)$ along φ and $\varphi_{\#}(\alpha) \simeq \psi_{\#}(\varphi_{\#}(\alpha))$ along ψ. Together, there is a level homotopy $\alpha \simeq \psi_{\#}(\varphi_{\#}(\alpha))$ along $\varphi * \psi$. Lemma 11.23(ii) gives $(\varphi * \psi)_{\#}(\alpha) \simeq \psi_{\#}(\varphi_{\#}(\alpha))$ rel $\dot{\mathbf{I}}^n$; that is, $T(\varphi * \psi) = T(\psi)T(\varphi)$. □

Lemma 11.25. *Let $f, g: X \to X$ be maps, and let $F: f \simeq g$ be a free homotopy; if $x_0 \in X$, denote the path $F(x_0, \ )$ from $f(x_0)$ to $g(x_0)$ by φ. Then there is a commutative diagram*

$$\begin{array}{ccc} \pi_n(X, x_0) & \xrightarrow{f_*} & \pi_n(X, f(x_0)) \\ & {}_{g_*}\searrow & \Big\downarrow T(\varphi) \\ & & \pi_n(X, g(x_0)). \end{array}$$

Proof. If $\alpha \in Q(X, x_0)$, then $G: \mathbf{I}^n \times \mathbf{I} \to X$ defined by

$$G(u, t) = F(\alpha(u), t)$$

is a level homotopy along φ with $G: f\alpha \simeq g\alpha$. By Lemma 11.23(ii), $\varphi_{\#}(f\alpha) \simeq g\alpha$ rel $\dot{\mathbf{I}}^n$, and this says that the diagram commutes. □

Corollary 11.26. *If $f: X \to Y$ is a homotopy equivalence, then $f_*: \pi_n(X, x_0) \to \pi_n(Y, f(x_0))$ is an isomorphism.*

Proof. Repeat the argument of Theorem 3.10, using Lemma 11.25. □

Corollary 11.27. *Let X and Y be path connected spaces having the same homotopy type. Then, for every $x_0 \in X$ and $y_0 \in Y$,*

$$\pi_n(X, x_0) \cong \pi_n(Y, y_0).$$

Proof. If $f: X \to Y$ is a homotopy equivalence, then $f_*: \pi_n(X, x_0) \to \pi_n(Y, f(x_0))$ is an isomorphism, by Corollary 11.26. But $\pi_n(Y, f(x_0)) \cong \pi_n(Y, y_0)$, by Theorem 11.24. □

Corollary 11.28. *If X is contractible, then $\pi_n(X, x_0) = 0$ for all $n \geq 0$.*

Proof. Immediate from Corollary 11.27 and the dimension axiom, Theorem 11.22. □

Here is a direct proof of this last result. If X is contractible, then Theorem 1.13 says that every map $f: S^n \to X$ is (freely) nullhomotopic; in particular, every pointed map $f: (S^n, s_n) \to (X, x_0)$ is nullhomotopic. By Theorem 1.6, there is a pointed homotopy from f to the constant map at x_0, hence $[f] = 0$ in $\pi_n(X, x_0)$.

Remark. The fundamental group acts on the homotopy groups. If (X, x_0) is a pointed space, if $[\varphi] \in \pi_1(X, x_0)$, and if $[\alpha] \in \pi_n(X, x_0)$, then define $[\varphi].[\alpha] = [\varphi_\#(\alpha)] \in \pi_n(X, x_0)$, where $\varphi_\#$ is the map occurring in the local system of Theorem 11.24. If $n \geq 2$ (so that $\pi_n(X, x_0)$ is an abelian group), then this action shows that $\pi_n(X, x_0)$ is a $\mathbf{Z}\pi_1(X, x_0)$-module, where $\mathbf{Z}\pi_1(X, x_0)$ denotes the integral group ring of the fundamental group. If $n = 1$, this action is conjugation by $[\varphi]$.

Call a space X ***n*-simple** if the action of $\pi_1(X, x_0)$ on $\pi_n(X, x_0)$ is trivial, i.e., each $[\varphi] \in \pi_1(X, x_0)$ acts as the identity; simply connected spaces and H-spaces are n-simple for every n. If X is n-simple, then $[(S^n, s_n), (X, x_0)] = [S^n, X]$, i.e., the pointed homotopy classes in $\mathbf{hTop}_*$ coincide with the (unpointed) homotopy classes in **hTop**. See [Maunder, p. 266].

Theorem 11.29. *If $(\tilde{X}, p)$ is a covering space of X, then*

$$p_*: \pi_n(\tilde{X}) \to \pi_n(X)$$

is an isomorphism for all $n \geq 2$.

Proof. Recall that if $[\tilde{f}] \in \pi_n(\tilde{X}) = [S^n, \tilde{X}]$, then $p_*[\tilde{f}] = [p\tilde{f}]$. To see that p_* is surjective, take $[f] \in \pi_n(X)$, and consider the diagram

$$\begin{array}{ccc} & & \tilde{X} \\ & & \downarrow p \\ S^n & \xrightarrow{f} & X. \end{array}$$

Since $n \geq 2$, S^n is simply connected (Corollary 7.6), and so the lifting criterion (more precisely, Corollary 10.14) provides a pointed map $\tilde{f}: S^n \to \tilde{X}$ with $p\tilde{f} = f$; therefore $p_*[\tilde{f}] = [f]$.

To see that p_* is injective, suppose that $[p\tilde{f}] = [p\tilde{f}_1]$, where $\tilde{f}, \tilde{f}_1: S^n \to \tilde{X}$ are pointed maps. Then $p\tilde{f} \simeq p\tilde{f}_1$, and the covering homotopy theorem (Theorem 10.5) says that their liftings are homotopic, that is, $[\tilde{f}] = [\tilde{f}_1]$. □

Theorem 11.30. $\pi_n(S^1) = 0$ *for all* $n \geq 2$.

PROOF. Since $(\mathbf{R}, \exp)$ is a covering space of S^1, Theorem 11.29 applies and gives $\pi_n(\mathbf{R}) \cong \pi_n(S^1)$ for all $n \geq 2$. But $\mathbf{R}$ is contractible, so that the result follows from Corollary 11.28. □

Since $\pi_1(S^1) = \mathbf{Z}$, all the homotopy groups of S^1 are known. (One also knows that $\pi_n(S^n) = \mathbf{Z}$ for every $n \geq 1$ (see [Maunder, p. 288]).) This is the exception; one does not even know all $\pi_p(S^n)$ for $n \geq 2$! (It is a theorem of Serre that when n is odd, $\pi_p(S^n)$ is finite for $p \neq n$, and when n is even, $\pi_p(S^n)$ is finite except for $p = n$ and $p = 2n - 1$; moreover, $\pi_{2n-1}(S^n)$ is a f.g. abelian group of rank 1.) The only finite simply connected CW complexes all of whose homotopy groups are known are contractible.

Theorem 11.31. *If* $0 < q < n$, *then* $\pi_q(S^n) = 0$, *and* $\pi_n(S^n) \neq 0$.

PROOF. By Theorem 7.5, every continuous map $f: S^q \to S^n$ is (freely) nullhomotopic; now apply Theorem 1.6.

If $f: X \to Y$ is (freely) homotopic to a constant, then $f_*: H_n(X) \to H_n(Y)$ (homology!) is the zero map for every $n \geq 0$. Now the identity map $1 = 1_{S^n}$ induces 1_*, the identity isomorphism on $H_n(S^n)$; since the latter group is nonzero, $1_* \neq 0$, and so 1 is not (freely) homotopic to a constant; *a fortiori*, there is no pointed homotopy from 1 to a constant. Therefore $[1] \in [S^n, S^n] = \pi_n(S^n)$ is nontrivial. □

EXAMPLE 11.7. There are path connected spaces X and Y having the same homotopy groups that are not of the same homotopy type; indeed X and Y can have different homology groups. Let $m > n > 1$, let $X = \mathbf{R}P^m \times S^n$, and let $Y = \mathbf{R}P^n \times S^m$. Now $\pi_1(X) \cong \mathbf{Z}/2\mathbf{Z} \cong \pi_1(Y)$ (since $\pi_1(\mathbf{R}P^m) \cong \mathbf{Z}/2\mathbf{Z}$, by Corollary 10.11), while $\pi_q(X) \cong \pi_q(Y)$ for all $q \geq 2$ because $S^m \times S^n$ is a universal covering space of each (so that Theorem 11.29 applies). On the other hand, if m is even and n is odd, then the Künneth formula (Theorem 9.37) shows that $H_{m+n}(X) = 0 \neq H_{m+n}(Y)$. (In Example 11.14, we shall exhibit two spaces with the same homology groups and with different homotopy groups.)

EXERCISES

11.23. If $\beta: (X, x_0) \to (Y, y_0)$ is freely nullhomotopic, then the induced homomorphism $\beta_*: \pi_n(X, x_0) \to \pi_n(Y, y_0)$ is trivial.

*11.24. Prove that if X and Y are pointed spaces, then, for all $n \geq 2$,

$$\pi_n(X \times Y) \cong \pi_n(X) \oplus \pi_n(Y)$$

(direct sum of abelian groups). Conclude that the higher homotopy groups of the torus are trivial. (*Hint*: Use Exercise 11.17: $\Omega(X \times Y) \approx \Omega X \times \Omega Y$.)

11.25. Prove that $\pi_q(S^n) \cong \pi_q(\mathbf{R}P^n)$ for all $q \geq 2$.

11.26. Let X be a contractible locally path connected space, and let G be a group acting properly on X. Prove that $\pi_n(X/G) = 0$ for all $n \geq 2$. (*Hint*: Use Theorem 10.54.)

*11.27. Let X and Y be objects in a category, and let $*$ and $\circ$ be binary operations on $\text{Hom}(X, Y)$ such that:

(i) there is a common two-sided identity $e \in \text{Hom}(X, Y)$, that is, for all $f \in \text{Hom}(X, Y)$,

$$e * f = f * e = f = e \circ f = f \circ e;$$

(ii) for all $a, b, c, d \in \text{Hom}(X, Y)$,

$$(a * b) \circ (c * d) = (a \circ c) * (b \circ d).$$

Prove that $*$ and $\circ$ coincide and that each is commutative. (*Hint*: To show that $f * g = f \circ g$, evaluate $(f \circ e) * (e \circ g)$ in two ways; to prove commutativity, evaluate $(e \circ g) * (f \circ e)$.) Recall that identity (ii) arose in the proof of Theorem 10.42.

11.28. (i) If Q is an H'-group and P is an H-space, then $[Q, P]$ is an abelian group. (*Hint*: If $\mu: P \times P \to P$ is the multiplication and $m: Q \to Q \vee Q$ is the comultiplication, then define $[f] * [g] = [\mu(f, g)]$ and $[f] \circ [g] = [(f, g)m]$. Show that $[e]$ is a common two-sided identity (where e is the constant map), and verify condition (ii) of Exercise 11.27 by evaluating both sides on elements.)

(ii) If X and Y are pointed spaces, prove that $[\Sigma^2 X, Y]$ is an abelian group. (*Remark*: This gives a second proof of Corollary 11.21. Groups of the form $[\Sigma X, Y]$ are called **track groups**.)

11.29. For every pointed space X, show that there is a homomorphism $\Sigma_n: \pi_n(X) \to \pi_{n+1}(X)$ given by $[f] \mapsto [\Sigma f]$. (Σ_n is called the **suspension homomorphism**.)

11.30. If X and Y are compact pointed polyhedra, then $[Y, X]$ is countable. (*Hint*: Use the simplicial approximation theorem.) Conclude that for every compact polyhedron X, $\pi_n(X)$ is countable for every $n \geq 0$.

Exact Sequences

Let (X, x_0) be a pointed space, and let A be a subspace of X containing x_0 (so that there is an inclusion $(A, x_0) \hookrightarrow (X, x_0)$, which is a pointed map). As in homology, there are relative homotopy groups $\pi_n(X, A)$, connecting homomorphisms $\pi_n(X, A) \to \pi_{n-1}(A)$, and an exact sequence

$$\cdots \to \pi_n(A) \to \pi_n(X) \to \pi_n(X, A) \to \pi_{n-1}(A) \to \cdots.$$

For small n, $\pi_n(X)$ and $\pi_n(X, A)$ are merely pointed sets (not groups), so that we must define exactness again.

Definition. A sequence of pointed sets and pointed functions

$$(X', x_0') \xrightarrow{f} (X, x_0) \xrightarrow{g} (X'', x_0'')$$

is **exact in Sets$_*$** if $\operatorname{im} f = \ker g$, where $\ker g = g^{-1}(x_0'')$.

If the pointed sets are groups (with identity elements as basepoints) and if the pointed maps are homomorphisms, then this definition is the usual definition of exactness. The basepoint, which is often an annoyance, is now essential.

There are computational proofs of the exact homotopy sequence using the description of π_n as homotopy classes of maps with domain $(\mathbf{I}^n, \dot{\mathbf{I}}^n)$ (see [Fuks and Rokhlin, Chap. 5], [Hilton (1953), Chap. IV], or [Whitehead, p. 162]). We present a proof in the categorical style (elaborating the proof in [Dold (1966)]), which simultaneously gives the exact sequence of a fibration.

The appropriate notion of exactness in **hTop$_*$** corresponds to a familiar result in algebra. A sequence

$$0 \to A' \to A \to A''$$

of abelian groups and homomorphisms is exact if and only if the sequence

$$0 \to \operatorname{Hom}(G, A') \to \operatorname{Hom}(G, A) \to \operatorname{Hom}(G, A'')$$

is exact for every abelian group G.

Definition. A sequence of pointed spaces and pointed maps

$$\cdots \to X_{n+1} \to X_n \to X_{n-1} \to \cdots$$

is **exact in hTop$_*$** if the induced sequence

$$\cdots \to [Z, X_{n+1}] \to [Z, X_n] \to [Z, X_{n-1}] \to \cdots$$

is an exact sequence in **Sets$_*$** for every pointed space Z.

Definition. If $f: (X, x_0) \to (Y, y_0)$ is a pointed map, then its **mapping fiber** is the pointed space

$$Mf = \{(x, \omega) \in X \times Y^{\mathbf{I}}: \omega(0) = y_0 \text{ and } \omega(1) = f(x)\}$$

(the basepoint is (x_0, ω_0), where ω_0 is the constant path at y_0).

The elements of Mf are ordered pairs (x, ω), where ω is a path in Y from y_0 to $f(x)$. The subspace of Mf consisting of all such ordered pairs of the form (x_0, ω) is just the loop space $\Omega(Y, y_0)$; more precisely, there is an injection $k: \Omega(Y, y_0) \to Mf$ given by $\omega \mapsto (x_0, \omega)$. There is also an obvious map

$f'\colon Mf \to X$, namely, the projection $(x, \omega) \mapsto x$. Both k and f' are pointed maps, and there is thus a sequence (which will be seen to be exact in **hTop**$_*$)

$$\Omega X \xrightarrow{\Omega f} \Omega Y \xrightarrow{k} Mf \xrightarrow{f'} X \xrightarrow{f} Y.$$

The construction of Mf can be repeated for f'; it is a certain subspace of $Mf \times X^{\mathbf{I}}$, namely,

$$Mf' = \{(x, \omega, \beta) \in X \times Y^{\mathbf{I}} \times X^{\mathbf{I}}\colon \omega(0) = y_0, \omega(1) = f(x),$$
$$\beta(0) = x_0, \beta(1) = f'(x, \omega) = x\};$$

there is an injection $j\colon \Omega(Y, y_0) \to Mf'$ given by

$$j\colon \omega \mapsto (x_0, \omega, \beta_0),$$

where β_0 is the constant path at x_0.

Notation. If $\beta\colon \mathbf{I} \to X$ is a path and if $s \in \mathbf{I}$, then β_s is the path defined by

$$\beta_s(t) = \beta(st).$$

In particular, if β is a path in X, then

$$(\beta^{-1})_s(t) = \beta^{-1}(st) = \beta(1 - st).$$

(Note that $(\beta_s)^{-1}(t) = \beta_s(1 - t) = \beta(s(1 - t))$, so that $(\beta^{-1})_s \neq (\beta_s)^{-1}$; however, we shall use only the former construction $(\beta^{-1})_s$.)

Lemma 11.32. *Let $f\colon (X, x_0) \to (Y, y_0)$ be a pointed map, let $f'\colon Mf \to X$ be the pointed map $(x, \omega) \mapsto x$, and let $j\colon \Omega(Y, y_0) \to Mf'$ be the pointed injection $\omega \mapsto (x_0, \omega, \beta_0)$. Then $\Omega(Y, y_0)$ is a pointed deformation retract of Mf', hence $[j]\colon \Omega(Y, y_0) \to Mf'$ is an equivalence in* **hTop**$_*$.

PROOF. We define a continuous map $F\colon Mf' \times \mathbf{I} \to Mf'$ such that

$$F(x, \omega, \beta, 0) = (x, \omega, \beta),$$
$$F(x, \omega, \beta, 1) \in j\Omega(Y, y_0),$$
$$F(x_0, \omega_0, \beta_0, s) = (x_0, \omega_0, \beta_0)$$

for all $(x, \omega, \beta) \in Mf'$ and all $s \in \mathbf{I}$. Build F in two stages. The first stage F_1 merely begins at (x, ω, β) and ends at $(x, \omega * \omega_1, \beta)$, where ω_1 is the constant path at $f(x)$. The second stage F_2 is given by the formula

$$F_2(x, \omega, \beta, s) = (\beta(1 - s), \omega * f(\beta^{-1})_s, \beta_{1-s}).$$

It is easy to see that im $F_2 \subset Mf'$, that $F_2(x, \omega, \beta, 0) = (x, \omega * \omega_1, \beta)$, and that $F_2(x, \omega, \beta, 1) = (x_0, \omega * f\beta^{-1}, \beta_0) \in j\Omega(Y, y_0)$. Hence F defined by $F = F_1 * F_2$ has all the desired properties; in particular, (x_0, ω_0, β_0) does stay fixed throughout the homotopy because $\omega_0 * \omega_0 = \omega_0$ (equal, not merely homotopic). □

Lemma 11.33. *If $f: X \to Y$ is a pointed map, then the following diagram is commutative in* **hTop$_*$**:

$$\begin{array}{ccccccccc}
\Omega X & \xrightarrow{\Omega f} & \Omega Y & \xrightarrow{k} & Mf & \xrightarrow{f'} & X & \xrightarrow{f} & Y \\
\downarrow{\scriptstyle j'i} & & \downarrow{\scriptstyle j} & & \downarrow{\scriptstyle 1} & & \downarrow{\scriptstyle 1} & & \downarrow{\scriptstyle 1} \\
Mf'' & \xrightarrow[f''']{} & Mf' & \xrightarrow[f'']{} & Mf & \xrightarrow[f']{} & X & \xrightarrow[f]{} & Y
\end{array}$$

where $i: \Omega X \to \Omega X$ is the homeomorphism $\beta \mapsto \beta^{-1}$, and j' is defined below.

PROOF. The second square commutes in **Top$_*$** (and hence commutes in **hTop$_*$**): recall that $k: \omega \mapsto (x_0, \omega)$, where ω is a loop in Y at y_0, that $j: \omega \mapsto (x_0, \omega, \beta_0)$ for any path ω in Y with $\omega(0) = y_0$, and that $f'': (x, \omega, \beta) \mapsto (x, \omega)$. A simple evaluation shows that $f''j = k$.

Here are the definitions of the maps in the first square. If β is a loop in X at x_0, then $\Omega f: \beta \mapsto f\beta$, $j': \beta \mapsto (x_0, \omega_0, \beta, \gamma_0)$ (where γ_0 is the constant loop at the basepoint of Mf), and $f''': (x, \omega, \beta, \gamma) \mapsto (x, \omega, \beta)$ (where γ is a suitable path in Mf). Hence $j(\Omega f): \beta \mapsto (x_0, f\beta, \beta_0)$ and $f'''j'i: \beta \mapsto (x_0, \omega_0, \beta^{-1})$.

Define $F: \Omega(X, x_0) \times \mathbf{I} \to Mf'$ by

$$F(\beta, s) = (\beta(1 - s), f\beta_{1-s}, (\beta^{-1})_s),$$

where $\beta_{1-s}(t) = \beta((1 - s)t)$ and $(\beta^{-1})_s(t) = \beta(1 - st)$. Note that F is a continuous map taking values in Mf', that $F(\beta, 0) = (x_0, f\beta, \beta_0)$, and that $F(\beta, 1) = (x_0, \omega_0, \beta)$, as desired. □

Remark. Note that $[j]$ and $[j']$ are equivalences in **hTop$_*$**, by Lemma 11.32; since i is a homeomorphism, $[j'i] = [j'][i]$ is also an equivalence in **hTop$_*$**.

The next result will be used in proving that the rows in the diagram of Lemma 11.33 are exact in **hTop$_*$**.

Lemma 11.34. *Let $f(X, x_0) \to (Y, y_0)$ be a pointed map, and let $q: Mf \to Y$ be defined by $q: (x, \omega) \mapsto \omega(1)$. Then f is nullhomotopic* rel x_0 *if and only if there exists a pointed map φ making the following diagram commute*:

$$\begin{array}{ccc}
 & Mf & \\
{\scriptstyle \varphi}\nearrow & & \searrow{\scriptstyle q} \\
X & \xrightarrow[f]{} & Y.
\end{array}$$

PROOF. If f is nullhomotopic rel x_0, then there is a continuous map $F: X \times \mathbf{I} \to Y$ with $F(x, 0) = y_0$ for all $x \in X$, $F(x, 1) = f(x)$ for all $x \in X$, and $F(x_0, t) = y_0$ for all $t \in \mathbf{I}$. Define $\varphi: X \to Mf$ by

$$\varphi(x) = (x, F_x),$$

where $F_x\colon \mathbf{I} \to Y$ is given by $F_x(t) = F(x, t)$. It is a simple matter to see that φ is a pointed map with $q\varphi = f$.

Conversely, assume that such a map φ exists; thus $\varphi(x) = (\lambda(x), \omega_x) \in Mf \subset X \times Y^{\mathbf{I}}$; that φ is a pointed map gives $\varphi(x_0) = (x_0, \omega_0)$, so that $\omega_{x_0} = \omega_0$, the constant path at y_0; commutativity of the diagram gives $\omega_x(1) = f(x)$. Define $F\colon X \times \mathbf{I} \to Y$ by $F(x, t) = \omega_x(t)$. Another simple check shows that F is a pointed homotopy $\omega_0 \simeq f$. □

Lemma 11.35. *If $f\colon X \to Y$ is a pointed map, then the sequence*

$$Mf \xrightarrow{f'} X \xrightarrow{f} Y$$

is exact in $\mathbf{hTop}_*$.

Proof. Consider the sequence in $\mathbf{Sets}_*$ (where Z is any pointed space):

$$[Z, Mf] \xrightarrow{f'_*} [Z, X] \xrightarrow{f_*} [Z, Y].$$

The basepoint in $[Z, Y]$ is the class of the constant map, so that the "kernel" of f_* consists of all maps $h\colon Z \to X$ with fh nullhomotopic.

$\operatorname{im} f'_* \subset \ker f_*$: Define $\varphi\colon Mf \to M(ff')$ by $\varphi\colon (x, \omega) \mapsto (x, \omega, \omega)$ ($M(ff') \subset Mf \times Y^{\mathbf{I}} \subset X \times Y^{\mathbf{I}} \times Y^{\mathbf{I}}$ because $ff'\colon Mf \to Y$). It is easy to see that the diagram

$$\begin{array}{ccc} & M(ff') & \\ {}^{\varphi}\nearrow & & \searrow^{q} \\ Mf & \xrightarrow[ff']{} & Y \end{array}$$

commutes, hence ff' is nullhomotopic, by Lemma 11.34. It follows that $ff'g$ is nullhomotopic for every $[g] \in [Z, Mf]$, as desired.

$\ker f_* \subset \operatorname{im} f'_*$: Assume that $[g] \in [Z, X]$ and that fg is nullhomotopic, say, $F\colon fg \simeq c \text{ rel } x_0$, where c is the constant map at x_0. The map $\varphi\colon Z \to M(fg)$ in the proof of Lemma 11.34, namely, $\varphi(z) = (z, F_z)$, makes the following diagram commute:

$$\begin{array}{ccc} & M(fg) & \\ {}^{\varphi}\nearrow & & \searrow^{q} \\ Z & \xrightarrow[fg]{} & Y. \end{array}$$

Now $M(fg) \subset Z \times Y^{\mathbf{I}}$, and it is easy to see that the restriction, call it r, of $g \times 1\colon Z \times Y^{\mathbf{I}} \to X \times Y^{\mathbf{I}}$ is a map $M(fg) \to M(f)$. Thus $r\varphi\colon Z \to Mf$, and one sees at once that $f'r\varphi = g$. Hence $[g] \in \operatorname{im} f'_*$, as desired. □

Corollary 11.36. *If $f\colon X \to Y$ is a pointed map, then the sequence*

$$\cdots \longrightarrow Mf'' \xrightarrow{f'''} Mf' \xrightarrow{f''} Mf \xrightarrow{f'} X \xrightarrow{f} Y$$

is exact in $\mathbf{hTop}_*$.

PROOF. Iterate Lemma 11.35. □

Corollary 11.37. *If $f\colon X \to Y$ is a pointed map, then the sequence*

$$\Omega X \xrightarrow{\Omega f} \Omega Y \xrightarrow{k} Mf \xrightarrow{f'} X \xrightarrow{f} Y$$

is exact in **hTop**$_*$.

PROOF. Consider the diagram in **hTop**$_*$ (of Lemma 11.33):

$$\begin{array}{ccccccccc}
\Omega X & \longrightarrow & \Omega Y & \longrightarrow & Mf & \longrightarrow & X & \longrightarrow & Y \\
\downarrow & & \downarrow & & \downarrow & & \downarrow & & \downarrow \\
Mf'' & \longrightarrow & Mf' & \longrightarrow & Mf & \longrightarrow & X & \longrightarrow & Y.
\end{array}$$

This diagram commutes (Lemma 11.33), the vertical maps are equivalences (Lemma 11.32), and the bottom row is exact (Corollary 11.36). Apply the functor $[Z, \quad]$ to this diagram (for any pointed space Z) to obtain a similar diagram in **Sets**$_*$. A diagram chase shows that the top row is exact in **Sets**$_*$, hence the top row of the original diagram is exact in **hTop**$_*$. □

The next lemma will allow us to extend the sequence of Corollary 11.37 to the left.

Lemma 11.38. *If $X' \to X \to X''$ is an exact sequence in* **hTop**$_*$, *then so is the "looped" sequence*

$$\Omega X' \to \Omega X \to \Omega X''.$$

PROOF. Use the adjointness of (Σ, Ω): for every pointed space Z, there is a commutative diagram in which the vertical functions are pointed bijections:

$$\begin{array}{ccccc}
[\Sigma Z, X'] & \longrightarrow & [\Sigma Z, X] & \longrightarrow & [\Sigma Z, X''] \\
\downarrow & & \downarrow & & \downarrow \\
[Z, \Omega X'] & \longrightarrow & [Z, \Omega X] & \longrightarrow & [Z, \Omega X''].
\end{array}$$

The top row is exact, by hypothesis, and so it follows that the bottom row is exact as well. □

Theorem 11.39 (Puppe Sequence). *If $f\colon X \to Y$ is a pointed map, then the following sequence is exact in* **hTop**$_*$:

$$\cdots \xrightarrow{\Omega^2 k} \Omega^2(Mf) \xrightarrow{\Omega^2 f'} \Omega^2 X \xrightarrow{\Omega^2 f} \Omega^2 Y \xrightarrow{\Omega k} \Omega(Mf) \xrightarrow{\Omega f'}$$
$$\Omega X \xrightarrow{\Omega f} \Omega Y \xrightarrow{k} Mf \xrightarrow{f'} X \xrightarrow{f} Y$$

(*of course*, $\Omega^0 X = X$ *and* $\Omega^{n+1} X = \Omega(\Omega^n X)$).

PROOF. By Corollary 11.37, the sequence

$$\Omega X \to \Omega Y \to Mf \to X \to Y$$

is exact in **hTop**$_*$, and by Lemma 11.38, the looped sequence

$$\Omega^2 X \to \Omega^2 Y \to \Omega(Mf) \to \Omega X \to \Omega Y$$

is exact in **hTop**$_*$. Since these sequences overlap, they may be spliced together to form a longer exact sequence. The result now follows by induction. □

Remark. There is another Puppe sequence, dual to this one. A sequence of pointed spaces and pointed maps

$$\cdots \to X_{n+1} \to X_n \to X_{n-1} \to \cdots$$

is called **coexact** in **hTop**$_*$ if the induced (reversed) sequence

$$\cdots \to [X_{n-1}, Z] \to [X_n, Z] \to [X_{n+1}, Z] \to \cdots$$

is exact in **Sets**$_*$ for every pointed space Z. In place of the mapping fiber Mf of a pointed map $f: X \to Y$, one works with the **mapping cone** Cf defined as follows. First define the **(reduced) cone** cX as the smash product $X \wedge \mathbf{I}$, and note that X can be identified with the closed subspace $\{[x, 1]: x \in X\}$; then Cf is defined as the space obtained from Y by attaching cX via f: $Cf = cX \coprod_f Y$. One pictures Cf as a (creased) witch's hat:

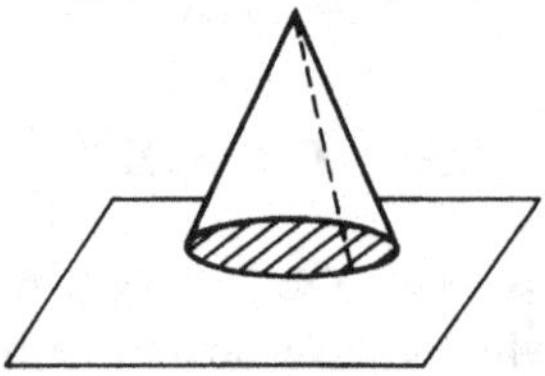

the cone cX surmounts the "brim" Y, and points in the shaded area are identified by $[x, 1] = f(x)$ for all $x \in X$. (One can show that this geometric construction corresponds to the algebraic mapping cone given in Chapter 9.) Using suspension in place of loop space, one obtains the **coexact Puppe sequence** (see [Atiyah], [Dyer], or [Spanier, p. 369]):

$$X \xrightarrow{f} Y \to Cf \to \Sigma X \to \Sigma Y \to \Sigma(Cf) \to \Sigma^2 X \to \Sigma^2 Y \to \Sigma^2(Cf) \to \cdots.$$

This sequence is important, but it is less convenient for us than the sequence we have presented: its various constructions involve quotient spaces instead of subspaces, and so all maps and homotopies require more scrutiny to ensure that they are well defined and continuous.

Corollary 11.40. *Let (X, x_0) be a pointed space, let A be a subspace of X containing x_0, and let $i: A \hookrightarrow X$ be the inclusion. Then there is an exact sequence* *in* **Sets$_*$**:

$$\cdots \to \pi_{n+1}(A) \xrightarrow{(\Omega^n i)_*} \pi_{n+1}(X) \xrightarrow{(\Omega^n k)_*} [S^0, \Omega^n(Mi)] \xrightarrow{(\Omega^n i')_*} \pi_n(A) \to \pi_n(X) \to \cdots$$
$$\cdots \to \pi_1(A) \to \pi_1(X) \to [S^0, Mi] \to \pi_0(A) \to \pi_0(X).$$

PROOF. Apply the functor $[S^0, \ ]$ to the Puppe sequence of the inclusion $i: A \hookrightarrow X$, and recall that $\pi_{n+1}(A) = \pi_1(\Omega^n A)$. □

This corollary is actually the long homotopy sequence once we replace the terms $[S^0, \Omega^n Mi] = [S^n, Mi] = \pi_n(Mi)$, for $n \geq 0$, by something more manageable. We also want a good formula for the "connecting homomorphism" $[S^0, \Omega^n Mi] \to \pi_n(A)$.

Definition. Let (X, x_0) be a pointed space. A **pointed pair** is an ordered pair (X, A) (often written (X, A, x_0)) in which A is a subspace of X that contains x_0.

Of course, the inclusion $A \hookrightarrow X$ is a pointed map when (X, A) is a pointed pair.

Definition. Let (X, A, x_0) and (Y, B, y_0) be pointed pairs. A **pointed pair map** $f: (X, A) \to (Y, B)$ is a pointed map $f: X \to Y$ with $f(A) \subset B$. If $f, g: (X, A) \to (Y, B)$, then a **pointed pair homotopy** $F: f \simeq g$ is a continuous map $F: X \times \mathbf{I} \to Y$ with

$$F(x, 0) = f(x) \quad \text{and} \quad F(x, 1) = g(x) \quad \text{for all } x \in X,$$
$$F(x_0, t) = y_0 \quad \text{for all } t \in \mathbf{I},$$
$$F(A \times \mathbf{I}) \subset B.$$

Definition. If (Y, B) and (X, A) are pointed pairs, then

$$[(Y, B, y_0), (X, A, x_0)]$$

is the set of all (pointed pair) homotopy classes of pointed pair maps β: $(Y, B, y_0) \to (X, A, x_0)$. We often suppress basepoints and write $[(Y, B), (X, A)]$.

There is an obvious basepoint in $[(Y, B), (X, A)]$, namely, the class of the constant map at x_0; thus $[(Y, B), (X, A)]$ may be regarded as a pointed set.

Definition. Let $s_n = (1, \ldots, 0, 0) \in S^n$ be the common basepoint of S^n and of D^{n+1}. For $n \geq 1$, the **relative homotopy group** of the pointed pair (X, A) is

$$\pi_n(X, A, x_0) = [(D^n, S^{n-1}, s_{n-1}), (X, A, x_0)]$$

(we usually abbreviate $\pi_n(X, A, x_0)$ to $\pi_n(X, A)$).

This definition reminds us of characteristic maps and suggests that CW complexes are convenient for homotopy theory. Note that $\pi_1(X, A, x_0)$ does have genuine interest; for example, $\pi_1(X, A, x_0) = 0$ means that every path ω in X with $\omega(1) = x_0$ and $\omega(0) \in A$ is nullhomotopic in X (by a pointed pair homotopy).

Since there is a homeomorphism $(D^n, S^{n-1}) \to (\mathbf{I}^n, \dot{\mathbf{I}}^n)$, one can also describe $\pi_n(X, A)$ as $[(\mathbf{I}^n, \dot{\mathbf{I}}^n), (X, A)]$. Moreover, using Corollary 8.10, one sees at once that "absolute" homotopy groups are special cases of relative ones:

$$[(D^n, S^{n-1}), (X, x_0)] = [(D^n/S^{n-1}, *), (X, x_0)] = [(S^n, *), (X, x_0)] = \pi_n(X, x_0).$$

Therefore one can identify the absolute group $\pi_n(X, x_0)$ with the relative group $\pi_n(X, x_0, x_0)$.

Calling $\pi_n(X, A)$ a group does not make it one; indeed $\pi_1(X, A)$ has no obvious group structure and it is merely a pointed set (with basepoint the class of the constant function). The next lemma will be used to identify $[S^0, \Omega^n Mi] = [S^n, Mi]$ with $\pi_{n+1}(X, A)$, where $i: A \hookrightarrow X$ is the inclusion.

Lemma 11.41.[2] *Let* $s_n = (1, \ldots, 0, 0)$ *and* $0 = (0, \ldots, 0)$ *be points of* D^{n+1}. *There is a continuous map* $F: D^{n+1} \times \mathbf{I} \to D^{n+1}$ *such that*

$$F(z, 0) = z \quad \text{for all } z \in D^{n+1},$$

$$F(u, t) = u \quad \text{for all } u \in S^n \text{ and all } t \in \mathbf{I},$$

$$F(0, 1) = s_n.$$

Remark. Thus F is a pointed pair homotopy $1_{D^{n+1}} \simeq \xi$, where $\xi(z) = F(z, 1)$.

PROOF. Regard each point in S^n as being connected to 0 by an elastic radius. The homotopy consists of pulling 0 toward s_n (along the radius). The picture at time t is thus

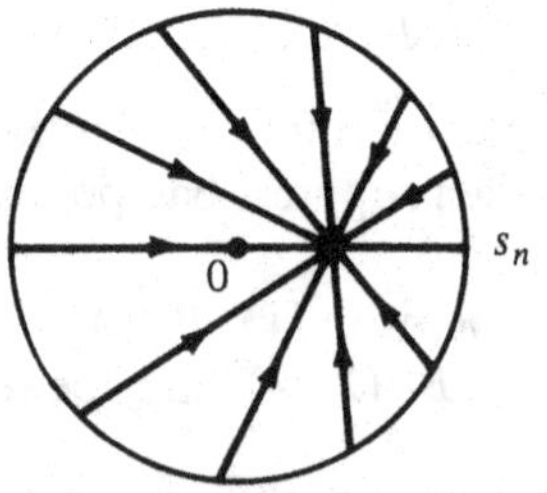

□

Lemma 11.42. *Let* (X, A) *be a pointed pair, and let* $i: A \hookrightarrow X$ *be the inclusion. Then there is a bijection* θ *and a commutative diagram*

[2] Cogniscenti will note that this lemma allows us to avoid reduced cones cS^n.

$$
\begin{array}{ccccc}
 & & [S^0, \Omega^n(Mi)] & & \\
 & \overset{(\Omega^n k)_*}{\nearrow} & & \overset{(\Omega^n i')_*}{\searrow} & \\
\pi_{n+1}(X) & & \downarrow \theta & & \pi_n(A), \\
 & \underset{j_*}{\searrow} & & \underset{d}{\nearrow} & \\
 & & \pi_{n+1}(X, A) & &
\end{array}
$$

where j_* is induced by the inclusion $j: (X, x_0, x_0) \to (X, A, x_0)$ (after identifying the absolute group $\pi_{n+1}(X, x_0)$ with the relative group $\pi_{n+1}(X, x_0, x_0)$) and $d: [g] \mapsto [g|S^n]$.

PROOF. First, adjointness of (Σ, Ω) allows us to replace $[S^0, \Omega^n Mi]$ with $[S^n, Mi]$. Next, if $h: S^n \to Mi$ we must define a map $\bar{h}: (D^{n+1}, S^n) \to (X, A)$. Now $Mi = \{(a, \omega) \in A \times X^{\mathbf{I}}: \omega(0) = x_0 \text{ and } \omega(1) = a\}$. Hence for each $u \in S^n$, $h(u) = (a_u, \omega_u)$, where $\omega_u \in X^{\mathbf{I}}$ is such that $\omega_u(0) = x_0$ and $\omega_u(1) = a_u \in A$; also, if $*$ $(= s_n)$ is the basepoint of S^n, then $h(*) = (x_0, \omega_0)$, where ω_0 is the constant path at x_0. If $p: Mi \to X^{\mathbf{I}}$ is the projection $(a, \omega) \mapsto \omega$, then $ph: S^n \to X^{\mathbf{I}}$ is a continuous map; by Theorem 11.1(ii), the map $S^n \times \mathbf{I} \to X$ defined by $(u, t) \mapsto \omega_u(t)$ is continuous. But each $z \in D^{n+1}$ can be written $z = tu$, where $t \in \mathbf{I}$ and $u \in S^n$, and this factorization is unique for $t \neq 0$. It follows that there is a continuous map $\bar{h}: D^{n+1} \to X$ defined by $\bar{h}(tu) = \omega_u(t)$ (note that $\bar{h}(0)$ is defined and is x_0, because $\bar{h}(0) = \omega_u(0) = x_0$ for all $u \in S^n$). Now $\bar{h}(u) = \omega_u(1) = a_u \in A$, and $\bar{h}(*) = \omega_0(1) = x_0$, so that $\bar{h}: (D^{n+1}, S^n) \to (X, A)$ is a map of pointed pairs. Finally, define $\theta: [S^n, Mi] \to \pi_{n+1}(X, A)$ by $\theta([h]) = [\bar{h}]$.

We claim that θ does not depend on the choice of $h \in [h]$. Suppose that $h' \in [h]$ and that $F: S^n \times \mathbf{I} \to Mi$ is a pointed homotopy displaying $h \simeq h'$. Thus

$$F(u, 0) = h(u) \quad \text{and} \quad F(u, 1) = h'(u) \quad \text{for all } u \in S^n;$$

$$F(*, s) = (x_0, \omega_0) \quad \text{for all } s \in \mathbf{I}.$$

For each $u \in S^n$ and $s \in \mathbf{I}$, let the second coordinate of $F(u, s)$ be denoted by $\omega_{u,s}$. As above, each $F_s: S^n \to Mi$ defines a continuous map $\bar{F}_s: (D^{n+1}, S^n) \to (X, A)$, and hence a continuous map $G: D^{n+1} \times \mathbf{I} \to X$, namely, $G(z, s) = \bar{F}_s(z)$; hence $G(z, s) = G(tu, s) = \omega_{u,s}(t)$. It is routine to check that G is a pointed pair homotopy $\bar{h} \simeq \bar{h}'$. Therefore θ is a well defined function.

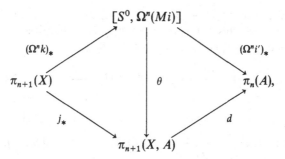

To show that θ is a bijection, we construct its inverse. Let $\beta: (D^{n+1}, S^n) \to (X, A)$ be a pointed map, and assume further that $\beta(0) = x_0$ (the basepoint of D^{n+1} is not 0 but $s_n \in S^n$). If $u \in S^n$, define $\omega_u \in X^{\mathbf{I}}$ by $\omega_u(t) = \beta(tu)$. Now $\omega_u(0) = \beta(0) = x_0$, by our assumption, while $\omega_u(1) = \beta(u) \in A$; thus $(\beta(u), \omega_u) \in Mi$. It is routine to check that $\tilde{\beta}: S^n \to Mi$, defined by $u \mapsto (\beta(u), \omega_u)$, is a continuous pointed map. Next, if $\gamma: (D^{n+1}, S^n) \to (X, A)$ is any pointed map, then Lemma 11.41 shows that there is a pointed pair homotopy $\gamma \simeq \gamma\xi$, and $\gamma\xi(0) = x_0$. We leave as an exercise that $[\beta] \mapsto [\tilde{\beta}]$ does not depend on the choice of β in $[\beta]$, and that both composites of this function with θ are identities.

Adjointness of (Σ, Ω) gives a commutative diagram:

$$\begin{array}{ccccc} [S^0, \Omega^n(\Omega X)] & \xrightarrow{(\Omega^n k)_*} & [S^0, \Omega^n(Mi)] & \xrightarrow{(\Omega^n i')_*} & [S^0, \Omega^n A] \\ \downarrow & & \downarrow & & \downarrow \\ [S^n, \Omega X] & \xrightarrow[k_*]{} & [S^n, Mi] & \xrightarrow[i'_*]{} & [S^n, A]. \end{array}$$

As there are now explicit (and simple) formulas for each function, it is straightforward to see that the diagram in the statement commutes. □

Theorem 11.43 (Homotopy Sequence of a Pair). *If (X, A) is a pointed pair, then there is an exact sequence*

$$\cdots \to \pi_{n+1}(A) \to \pi_{n+1}(X) \to \pi_{n+1}(X, A) \xrightarrow{d} \pi_n(A) \to \pi_n(X) \to \cdots$$
$$\cdots \to \pi_1(A) \to \pi_1(X) \to \pi_1(X, A) \xrightarrow{d} \pi_0(A) \to \pi_0(X).$$

Moreover, $d\colon \pi_{n+1}(X, A) \to \pi_n(A)$ is the map $[\beta] \mapsto [\beta | S^n]$, while the other maps are induced by inclusions.

PROOF. Immediate from Corollary 11.40 and Lemma 11.42. □

Corollary 11.44. *$\pi_n(X, A)$ is a group for all $n \geq 2$, and it is an abelian group for all $n \geq 3$.*

PROOF. The bijection $\theta\colon [S^n, Mi] \to \pi_{n+1}(X, A)$ is used to equip $\pi_{n+1}(X, A)$ with a group structure when $[S^n, Mi]$ is a group. But $[S^n, Mi] = \pi_n(Mi)$ is a group for $n \geq 1$, and it is an abelian group for $n \geq 2$. □

What is the group multiplication in the relative homotopy group $\pi_n(X, A)$? Recall that $S^n \approx \mathbf{I}^n/\dot{\mathbf{I}}^n$, and we saw (just after Theorem 11.22) that one can view the elements of the "absolute" homotopy group $\pi_n(X) = [S^n, X]$ as being represented by continuous maps $f\colon (\mathbf{I}^n, \dot{\mathbf{I}}^n) \to (X, x_0)$. Now $D^{n+1} \approx \mathbf{I}^{n+1}$, and one can show that elements of $\pi_{n+1}(X, A)$ can be represented by continuous maps

$$f\colon (\mathbf{I}^{n+1}, \dot{\mathbf{I}}^{n+1}, (\dot{\mathbf{I}}^n \times \mathbf{I}) \cup (\mathbf{I}^n \times \{1\}) \to (X, A, x_0);$$

moreover, the multiplication (really, addition, since most homotopy groups are abelian) is the same as in the absolute case:

$$(f + g)(t_1, \ldots, t_{n+1}) = \begin{cases} f(t_1, \ldots, t_n, 2t_{n+1}) & \text{if } 0 \leq t_{n+1} \leq \frac{1}{2} \\ g(t_1, \ldots, t_n, 2t_{n+1} - 1) & \text{if } \frac{1}{2} \leq t_{n+1} \leq 1. \end{cases}$$

Theorem 11.45. *Let $f\colon (X, A) \to (Y, B)$ be a map of pointed pairs. Then there is a commutative diagram with exact rows:*

$$\begin{array}{ccccccccccc}
\cdots \to \pi_2(X, A) & \to & \pi_1(A) & \to & \pi_1(X) & \to & \pi_1(X, A) & \to & \pi_0(A) & \to & \pi_0(X) \\
\downarrow & & \downarrow & & \downarrow & & \downarrow & & \downarrow & & \downarrow \\
\cdots \to \pi_2(Y, B) & \to & \pi_1(B) & \to & \pi_1(Y) & \to & \pi_1(Y, B) & \to & \pi_0(B) & \to & \pi_0(Y).
\end{array}$$

PROOF. The easy verification is left to the reader. □

EXERCISES

11.31. If $r: X \to A$ is a retraction, then there are isomorphisms, for all $n \geq 2$,

$$\pi_n(X) \cong \pi_n(A) \oplus \pi_n(X, A).$$

(*Hint*: See Exercise 5.14(ii).)

11.32. Let $B \subset A \subset X$ be pointed spaces. Then there is an **exact sequence of the triple** (X, A, B):

$$\cdots \to \pi_{n+1}(X, A) \to \pi_n(A, B) \to \pi_n(X, B) \to \pi_n(X, A) \to \pi_{n-1}(A, B) \to \cdots.$$

(*Hint*: Use remark (3) after Theorem 5.9.)

*11.33. For every pointed space X, $\pi_n(X, X) = 0$ for all $n \geq 1$.

Fibrations

Covering spaces arose from examining the proof that $\pi_1(S^1) = \mathbf{Z}$; fibrations arise from examining a key property of covering spaces (which occurs in other interesting contexts). It will be seen that fibrations determine exact homotopy sequences (the proof of exactness is an application of the Puppe sequence). A theorem of Milnor states that there is an analogue of the Eilenberg–Steenrod axioms for homology that characterizes the homotopy groups.

Definition. Let E and B be topological spaces (without chosen basepoints). A map $p: E \to B$ has the **homotopy lifting property** with respect to a space X if, for every two maps $\tilde{f}: X \to E$ and $G: X \times \mathbf{I} \to B$ for which $p\tilde{f} = Gi$ (where $i: X \to X \times \mathbf{I}$ is the map $x \mapsto (x, 0)$), there exists a continuous map $\tilde{G}: X \times \mathbf{I} \to E$ making both triangles below commute.

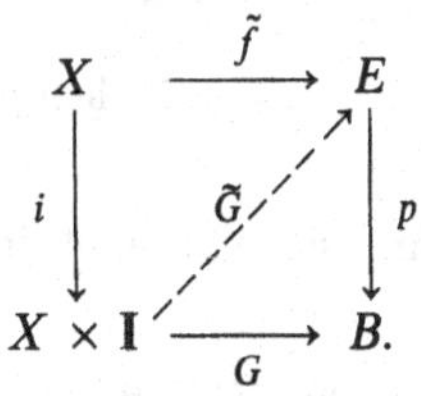

If one defines $f: X \to B$ by $f(x) = G(x, 0)$, then $\tilde{f}$ is a lifting of f; if one defines $g: X \to B$ by $g(x) = G(x, 1)$, then G is a homotopy $f \simeq g$. The map $\tilde{G}$ is a homotopy $\tilde{f} \simeq \tilde{g}$, where $\tilde{g} = \tilde{G}(x, 1)$ is a lifting of g. Thus, if $f \simeq g$ and if

f has a lifting $\tilde{f}$, then the homotopy can be lifted, hence g has a lifting $\tilde{g}$ with $\tilde{f} \simeq \tilde{g}$.

Definition. A map $p: E \to B$ is called a **fibration** (or **Hurewicz fiber space**) if it has the homotopy lifting property with respect to every space X. If $b_0 \in B$, then $p^{-1}(b_0) = F$ is called the **fiber**.

We do not assert that different fibers of a fibration are homeomorphic, because this is not true (Exercise 11.38); however, Theorem 11.47 shows that all fibers do have the same homotopy type.

EXAMPLE 11.8. Every covering projection $p: \tilde{X} \to X$ is a fibration (Theorem 10.5) having a discrete fiber.

EXAMPLE 11.9. If $E = B \times F$, then the projection $p: E \to B$ defined by $(b, x) \mapsto b$ (where $b \in B$ and $x \in F$) is a fibration with fiber F. To see this, consider the commutative diagram

$$\begin{array}{ccc} X & \xrightarrow{\tilde{f}} & B \times F \\ {\scriptstyle i}\downarrow & & \downarrow{\scriptstyle p} \\ X \times \mathbf{I} & \xrightarrow[G]{} & B, \end{array}$$

and define $\tilde{G}: X \times \mathbf{I} \to B \times F$ by $\tilde{G}(x, t) = (G(x, t), q\tilde{f}(x))$, where $q: B \times F \to F$ is the projection $(b, x) \mapsto x$.

EXAMPLE 11.10. A fiber bundle $p: E \to B$ with B paracompact is a fibration (see [Spanier, p. 96] for definitions and proof).

EXERCISES

11.34. If B is a singleton, then every map $p: E \to B$ is a fibration.

11.35. If $p: E \to B$ has the homotopy lifting property with respect to a singleton, then every path ω in B with $\omega(0) \in \operatorname{im} p$ can be lifted to E.

11.36. If $p: E \to B$ and $q: B \to B'$ are fibrations, then $qp: E \to B'$ is a fibration.

11.37. If $p_i: E_i \to B_i$ is a fibration for $i = 1, 2$, then $p_1 \times p_2: E_1 \times E_2 \to B_1 \times B_2$ is a fibration.

*11.38. (i) Let E be the (two-dimensional) triangle in $\mathbf{R}^2$ having vertices (0, 0), (0, 1), and (1, 0):

$$E = \{(x, y) \in \mathbf{R}^2: x \in \mathbf{I} \text{ and } 0 \le y \le 1 - x\}.$$

Show that $p: E \to \mathbf{I}$, defined by $(x, y) \mapsto x$, is a fibration. (*Hint*: If $\tilde{f}: X \to E$ and $G: X \times \mathbf{I} \to \mathbf{I}$ satisfy $p\tilde{f} = Gi$, where $i: x \mapsto (x, 0)$, define $\tilde{G}: X \times \mathbf{I} \to E$ by

$$\tilde{G}(x, t) = (G(x, t), \min\{1 - G(x, t), q\tilde{f}(x)\}),$$

where $q: E \to \mathbf{I}$ is the map $(x, y) \mapsto y$.)

(ii) Show that the fibers in this case are not homeomorphic.

We are going to use the Puppe sequence to show that every fibration gives rise to an exact sequence of homotopy groups, for virtually all the work has already been done. Afterward, however, we shall weaken the notion of fibration, and we shall give a functor-free proof (independent of the next proof) that there is also an exact sequence in this more general case.

If B is a pointed space with basepoint b_0, then every map $p: E \to B$ can be viewed as a pointed map if the basepoint of E is any point in the fiber over b_0.

Lemma 11.46. *Let $p: (E, x_0) \to (B, b_0)$ be a fibration with fiber $F = p^{-1}(b_0)$. Then F and the mapping fiber Mp have the same homotopy type.*

PROOF.[3] Recall that $Mp = \{(x, \omega) \in E \times B^{\mathbf{I}}: \omega(0) = b_0 \text{ and } \omega(1) = p(x)\}$, and there is a commutative diagram

$$\begin{array}{ccc} Mp & \xrightarrow{q} & B^{\mathbf{I}} \\ {\scriptstyle p'}\downarrow & & \downarrow{\scriptstyle d} \\ E & \xrightarrow[p]{} & B, \end{array}$$

where $p': (x, \omega) \mapsto x$, $q: (x, \omega) \mapsto \omega$, and $d: \omega \mapsto \omega(1)$.

If $x \in F$ and ω_0 is the constant path at b_0, then $(x, \omega_0) \in Mp$; define $\lambda: F \to Mp$ by $x \mapsto (x, \omega_0)$. We now construct a homotopy inverse of λ. Consider the map $G: Mp \times \mathbf{I} \to B$ defined by

$$G(x, \omega, t) = \omega(1 - t)$$

(G is continuous, being the composite of the continuous maps $(x, \omega, t) \mapsto (x, \omega, 1 - t) \mapsto (\omega, 1 - t) \mapsto \omega(1 - t)$; indeed G shows that pp' is nullhomotopic.) Since $p: E \to B$ is a fibration, there is a map $\tilde{G}: Mp \times \mathbf{I} \to E$ making the following diagram commute:

$$\begin{array}{ccc} Mp & \xrightarrow{p'} & E \\ {\scriptstyle i}\downarrow & \nearrow{\scriptstyle \tilde{G}} & \downarrow{\scriptstyle p} \\ Mp \times \mathbf{I} & \xrightarrow[G]{} & B. \end{array}$$

Hence $\tilde{G}(x, \omega, 0) = p'(x, \omega) = x$ and $p\tilde{G}(x, \omega, s) = G(x, \omega, s) = \omega(1 - s)$ for

---

[3] The proof shows that the conclusion holds if $p: E \to B$ has the homotopy lifting property with respect to the mapping fiber Mp.

all $(x, \omega) \in Mp$ and all $s \in \mathbf{I}$. In particular, $p\tilde{G}(x, \omega, 1) = \omega(0) = b_0$, so that $(x, \omega) \mapsto \tilde{G}(x, \omega, 1)$ defines a continuous map $\gamma: Mp \to F$.

It is easy to see that $\tilde{G}(\lambda \times 1)$ is a map $F \times \mathbf{I} \to F$ (because $p\tilde{G}(\lambda \times 1)$: $(x, s) \mapsto (x, \omega_0, s) \mapsto \omega_0(1 - s) = b_0$) that is a homotopy $1_F \simeq \gamma\lambda$. For the other composite $\lambda\gamma$, suppose that there were a map $J: Mp \times \mathbf{I} \to B^{\mathbf{I}}$ such that, for all $(x, \omega) \in Mp$ and $s \in \mathbf{I}$, one has $J(x, \omega, s): 1 \mapsto \omega(1 - s)$, $J(x, \omega, 0) = \omega$, and $J(x, \omega, 1) = \omega_0$. Then $(x, \omega, s) \mapsto (\tilde{G}(x, \omega, s), J(x, \omega, s))$ is a map $Mp \times \mathbf{I} \to Mp$ that is a homotopy $1_{Mp} \simeq \lambda\gamma$. Finally, one such J is given by $J(x, \omega, s): t \mapsto \omega(t(1 - s))$. □

Theorem 11.47. *Let $p: E \to B$ be a fibration and let $b_0, b_1 \in B$. If B is path connected, then the fibers $p^{-1}(b_0)$ and $p^{-1}(b_1)$ have the same homotopy type.*

PROOF. For $i = 0, 1$, let $M_i p$ denote the mapping fiber of p for the basepoint $b_i \in B$ (our previous notation does not display the dependence on the basepoint). Since B is path connected, there is a path λ in B from b_1 to b_0, and it is easy to see that $(e, \omega) \mapsto (e, \lambda * \omega)$ is a homotopy equivalence $M_0 p \to M_1 p$. The result now follows from the lemma. □

Theorem 11.48 (Homotopy Sequence of a Fibration). *If $p: E \to B$ is a fibration with fiber F, then there is an exact sequence*

$$\cdots \to \pi_2(E) \xrightarrow{p_*} \pi_2(B) \to \pi_1(F) \to \pi_1(E) \xrightarrow{p_*} \pi_1(B) \to \pi_0(F) \to \pi_0(E) \xrightarrow{p_*} \pi_0(B).$$

PROOF. By Lemma 11.46, Mp and F have the same homotopy type; by Corollary 11.26, $[S^n, Mp] \cong [S^n, F]$ for all $n \geq 0$. The result now follows by applying $[S^0, \ ]$ to the Puppe sequence of p (and using adjointness of (Σ, Ω)). □

Remarks. (1) Theorem 11.48 implies Theorem 11.29, for a covering projection is a fibration having a discrete fiber F, hence $\pi_n(F) = 0$ for all $n \geq 1$.

(2) In view of Exercise 11.24, the exact sequence arising from the projection of a product onto a factor is not interesting.

There is an unpointed version of the mapping fiber which is useful.

Definition. Let $p: E \to B$ be a map. Then the **fiber product** is the space

$$Fp = \{(x, \omega) \in E \times B^{\mathbf{I}}: \omega(1) = p(x)\}.$$

Of course, the mapping fiber Mp is a subspace of Fp.

The fiber product and the mapping fiber are special cases of a general (categorical) construction.

Definition. Let $p: E \to B$ and $q: D \to B$ be morphisms in a category. A **solution** is an ordered triple (X, f, g), where $f: X \to E$ and $g: X \to D$ are morphisms such that $qg = pf$; that is, the following diagram commutes:

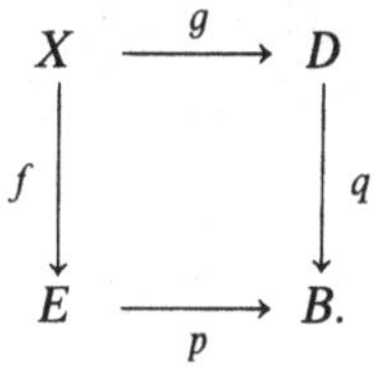

A **pullback** is a solution (Z, r, s) that is "best" in the following sense: for any solution (X, f, g), there exists a unique morphism $\theta: X \to Z$ giving commutativity of the diagram

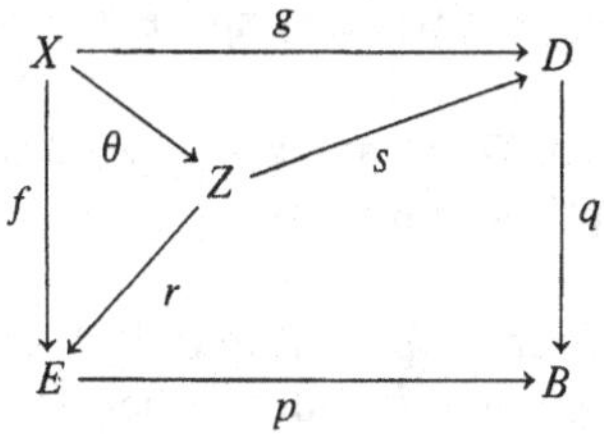

Pullback is the dual of pushout.

EXERCISES

11.39. If the pullback of two morphisms $p: E \to B$ and $q: D \to B$ exists, then it is unique to equivalence.

11.40. In **Top**, the pullback of $p: E \to B$ and $q: B^{\mathbf{I}} \to B$ (where $q: \omega \mapsto \omega(1)$) is Fp. (*Hint*: Define $r: Fp \to E$ by $(x, \omega) \mapsto x$ and $s: Fp \to B^{\mathbf{I}}$ by $(x, \omega) \mapsto \omega$.)

11.41. For every pointed map $p: E \to B$, show that Mp is a pullback in $\mathbf{Top}_*$.

11.42. Define $\lambda: E \to Fp$ by $\lambda(x) = (x, \omega_x)$, where ω_x is the constant path at $p(x)$; define $\mu: Fp \to E$ by $(x, \omega) \mapsto x$. Show that $\mu\lambda = 1$ and that $\lambda\mu \simeq 1$, hence λ is a homotopy equivalence.

11.43. (**Hurewicz**). Let $p: (E, e_0) \to (B, b_0)$ be continuous, and define $\pi: Fp \to B$ by $(x, \omega) \mapsto \omega(0)$. Show that π is a fibration with fiber Mp. (*Hint*: To construct a map $\tilde{G}: X \times \mathbf{I} \to Fp$, it suffices to find a commutative diagram

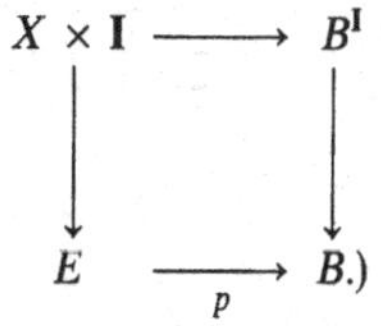

11.44. Every map $h: X \to Y$ is the composite $h = \pi\lambda$, where λ is an injection that is a homotopy equivalence and π is a fibration. (*Hint*: Consider $X \to Fh \to Y$.)

We merely mention a dual notion (see Dold (1966) for a discussion of the duality).

Definition. A pair (X, A) has the **homotopy extension property** with respect to a space Y if, for every map $f: X \times \{0\} \to Y$ and every map $G: A \times \mathbf{I} \to Y$ with

$G(a, 0) = f(a, 0)$ for every $a \in A$, there exists a map $F: X \times \mathbf{I} \to Y$ making the following diagram commute:

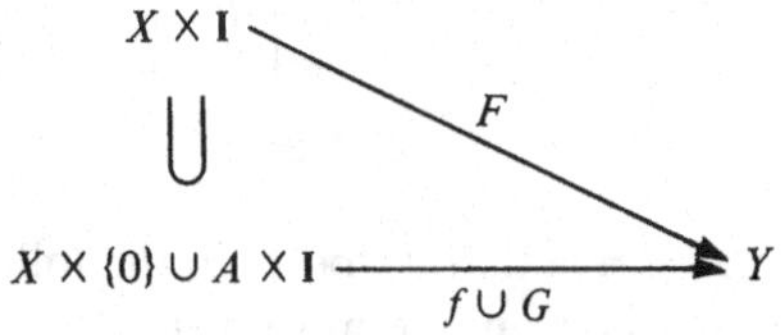

The inclusion $i: A \hookrightarrow X$ is called a **cofibration**[4] if (X, A) has the homotopy extension property with respect to every space Y.

EXAMPLE 11.11. If X is a CW complex and A is a CW subcomplex, then $i: A \hookrightarrow X$ is a cofibration (Theorem 8.33).

It can be shown (see [Spanier, p. 97]) that if $g: A \to X$ is a cofibration and if A and X are locally compact Hausdorff, then for every space Y, the map $g^*: Y^X \to Y^A$ is a fibration. In particular, if X is a locally compact CW complex and A is a CW subcomplex, then the restriction map $i^*: Y^X \to Y^A$ is a fibration.

We now proceed to the generalized notion of fibration mentioned earlier.

Definition. A map $p: E \to B$ is a **weak fibration** (or **Serre fiber space**) if it has the homotopy lifting property with respect to every cube $\mathbf{I}^n$, $n \geq 0$ (by definition, $\mathbf{I}^0$ is a singleton).

EXERCISE

11.45. Let L be the portion of the graph $y = x - 1$ for $x \in \mathbf{I}$. If $\mathbf{Z}^+$ is the set of positive integers, define

$$E = L \cup \bigcup_{n \in \mathbf{Z}^+} (\mathbf{I} \times \{1/n\})$$

$E =$

(i) Show that $p: E \to \mathbf{I}$, defined by $(x, y) \mapsto x$, is a weak fibration. (*Hint*: One can cover homotopies $G: X \times \mathbf{I} \to \mathbf{I}$ for every path connected space X.)

(ii) Show that $p^{-1}(1)$ and $p^{-1}(0)$ do not have the same homotopy type (Exercise 1.5). Use Theorem 11.47 to conclude that $p: E \to \mathbf{I}$ is not a fibration. (Here

[4] More generally, one says that any map $g: A \to X$ (where A is not necessarily a subspace of X) is a *cofibration* if the definition above is modified to read "$G(a, 0) = f(g(a), 0)$ for every $a \in A$".

is an explicit homotopy that cannot be covered. Let $X = p^{-1}(1)$, let $\tilde{f}: X \to E$ be the inclusion, and let $G: X \times \mathbf{I} \to \mathbf{I}$ be a homotopy from the constant function at 1 to the constant function at 0.)

Theorem 11.49. *A weak fibration $p: E \to B$ has the homotopy lifting property with respect to every CW complex X.*

PROOF. Since a CW complex has the weak topology determined by its skeletons, it suffices to prove that there exists a map $\tilde{G}_n$, for every $n \geq 0$, making the following diagram commute:

$$\begin{array}{ccc} X^{(n)} & \xrightarrow{\tilde{f}_n} & E \\ \downarrow & \overset{\tilde{G}_n}{\nearrow} & \downarrow p \\ X^{(n)} \times \mathbf{I} & \xrightarrow[G_n]{} & B, \end{array}$$

where $\tilde{f}: X \to E$ and $G: X \times \mathbf{I} \to B$ are given, and $\tilde{f}_n$ and G_n are appropriate restrictions. We prove this by induction on n. Let $n = 0$. For each $x \in X^{(0)}$, there exists a continuous map $h_x: \{x\} \times \mathbf{I} \to E$ with $h_x(x, 0) = \tilde{f}_0(x)$ and $ph_x = G_0$, because $p: E \to B$ is a weak fibration. Because $X^{(0)}$ is discrete, the function $\tilde{G}_0: X^{(0)} \times \mathbf{I} \to E$ given by $\tilde{G}_0(x, t) = h_x(x, t)$ is continuous. Assume now that $n > 0$ and that $\tilde{G}_{n-1}: X^{(n-1)} \times \mathbf{I} \to E$ exists; let e be an n-cell in X, and let $\Phi_e: (D^n, S^{n-1}) \to (e \cup X^{(n-1)}, X^{(n-1)})$ be the characteristic map of e. Consider the diagram

$$\begin{array}{ccc} (D^n \times \{0\}) \cup (S^{n-1} \times \mathbf{I}) & \xrightarrow{h} & E \\ \downarrow & \nearrow & \downarrow p \\ D^n \times \mathbf{I} & \xrightarrow[G(\Phi_e \times 1)]{} & B, \end{array}$$

where $h|D^n \times \{0\} = \tilde{f}\Phi_e$ and $h|S^{n-1} \times \mathbf{I} = \tilde{G}_{n-1}(\Phi_e \times 1)$ (note that h is well defined because the two functions agree on the overlap $S^{n-1} \times \{0\}$). There is a homeomorphism of the pairs $(\mathbf{I}^{n+1}, \mathbf{I}^n)$ and $(D^n \times \mathbf{I}, D^n \times \{0\} \cup S^{n-1} \times \mathbf{I})$; therefore the given homotopy lifting property provides a continuous map $\gamma_e: D^n \times \mathbf{I} \to E$ making the above diagram commute. It is now routine to check that $g_e: \bar{e} \times \mathbf{I} \to E$ defined by $g_e(x, t) = \gamma_e(u, t)$, where $x \in \bar{e}$ and $u \in D^n$ satisfies $\Phi_e(u) = x$, is a well defined continuous function giving commutativity of the diagram

$$\begin{array}{ccc} \bar{e} & \xrightarrow{\tilde{f}} & E \\ \downarrow & \overset{g_e}{\nearrow} & \downarrow p \\ \bar{e} \times \mathbf{I} & \xrightarrow[G]{} & B. \end{array}$$

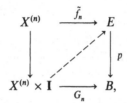

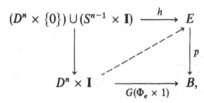

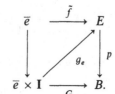

All the maps g_e, as e varies over all n-cells in X, may be assembled[5] to form a function $\tilde{G}_n: X^{(n)} \times \mathbf{I} \to E$ with $\tilde{G}_n|\bar{e} = g_e$. It is easy to see that $\tilde{G}_n$ extends $\tilde{G}_{n-1}$ and that $\tilde{G}_n$ makes the appropriate diagram commute; finally, $\tilde{G}_n$ is continuous because its restriction to every closed cell in $X^{(n)} \times \mathbf{I}$ (namely, $\bar{e} \times \{0\}$, $\bar{e} \times \{1\}$, and $\bar{e} \times \mathbf{I}$) is continuous. □

The proof of the exactness of the homotopy sequence of a fibration $p: E \to B$ was based on the Puppe sequence of p. The coming proof of the exactness of the homotopy sequence of a weak fibration $p: E \to B$ with fiber F is based on the sequence of the pair (E, F), that is, on the Puppe sequence of the inclusion $F \hookrightarrow E$.

Theorem 11.50 (Serre). *Let $p: E \to B$ be a weak fibration with fiber $F = p^{-1}(b_0)$ for some $b_0 \in B$. Then $p'_*: \pi_n(E, F) \to \pi_n(B, b_0)$ is a bijection for all $n \geq 1$, where $p'j = p$ and $j: (E, x_0) \hookrightarrow (E, F)$ is the inclusion.*

Remark. If $n \geq 2$, p_* is an isomorphism because p_* is a homomorphism; if $n = 1$, however, $\pi_1(E, F)$ has no obvious group structure.

Proof. An easy induction on n shows that the dashed arrow exists making both triangles commute (because $p: E \to B$ is a weak fibration)

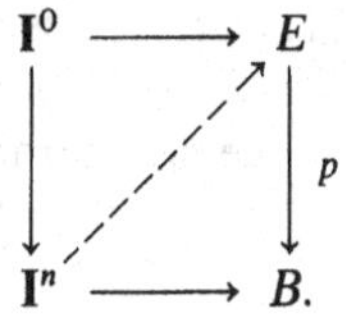

Suppose that $[g] \in \pi_n(B, b_0)$; we may regard g as a map of pairs $g: (\mathbf{I}^n, \dot{\mathbf{I}}^n) \to (B, b_0)$. Choose $e_0 \in F$ and define $\tilde{f}: \mathbf{I}^0 \to E$ by $\tilde{f}(*) = e_0$. The first paragraph shows that there exists a map $G: \mathbf{I}^n \to E$ with $pG = g$. Since $g(\dot{\mathbf{I}}^n) = \{b_0\}$, it follows that $G(\dot{\mathbf{I}}^n) \subset F$, hence $G: (\mathbf{I}^n, \dot{\mathbf{I}}^n) \to (E, F)$. Therefore $[G] \in \pi_n(E, F)$, $p_*([G]) = [g]$, and p_* is surjective.

Assume that $f: (D^n, S^{n-1}) \to (E, F)$ is such that pf (more precisely, the map $D^n/S^{n-1} \to B$ induced by pf) is nullhomotopic; we claim that f is null-homotopic. There is a homotopy of pointed pairs $G: (D^n \times \mathbf{I}, S^{n-1} \times \mathbf{I}) \to (B, b_0)$ with $G(z, 0) = pf(z)$ and $G(z, 1) = b_0$ for all $z \in D^n$. Consider the diagram

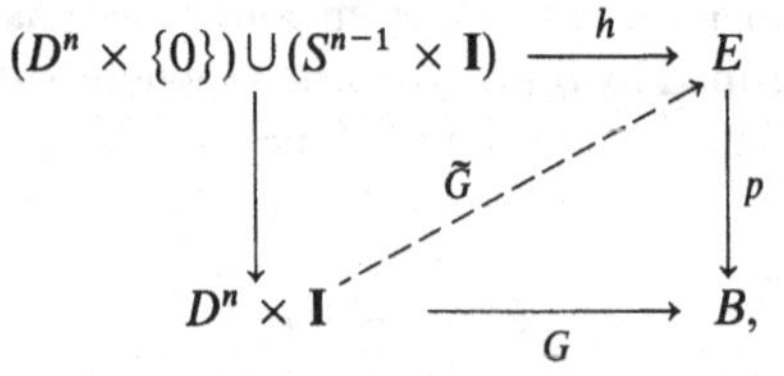

[5] A family $\{B_i: i \in I\}$ of subsets of a topological space X is **locally finite** if each $x \in X$ has a neighborhood meeting only finitely many B_i; if each B_i is closed, then it is easy to see that $\bigcup_{i \in I} B_i$ is also closed; moreover there is a gluing lemma for a locally finite closed cover of a space.

where $h(z, 0) = f(z)$ for all $z \in D^n$ and $h(u, t) = x_0$, where $x_0 \in F$ is the basepoint of E, $u \in S^{n-1}$, and $t \in \mathbf{I}$. Note that h is well defined (i.e., $h(u, 0) = x_0$ for all $u \in S^{n-1}$) and that the diagram commutes. Since $(D^n \times \mathbf{I}, (D^n \times \{0\}) \cup (S^{n-1} \times \mathbf{I}))$ is homeomorphic to $(\mathbf{I}^{n+1}, \mathbf{I}^n)$, there is a map $\tilde{G}: D^n \times \mathbf{I} \to E$ making both triangles commute. The map $\tilde{G}$ is easily seen to be a homotopy of pointed pairs $\tilde{G}: f \simeq g$, where $g: D^n \to F$. But this says that $[f] = [g]$ is in the image $\pi_n(F, F) \to \pi_n(E, F)$ under the map induced by inclusion. Since $\pi_n(F, F) = 0$, by Exercise 11.33, it follows that $[f] = 0$, as desired.

If $n \geq 2$, then $\pi_n(E, F)$ is a group and p_* is a homomorphism. The argument above shows that $\ker p_*$ is trivial and so p_* is injective. Finally, let $n = 1$, and assume that $[f_1], [f_2] \in \pi_1(E, F)$ and that $[pf_1] = [pf_2]$ in $\pi_1(B)$. For $i = 1, 2$, $f_i: (\mathbf{I}, \dot{\mathbf{I}}) \to (E, F)$ is a path with $f_i(0) = x_0$, the basepoint in E; let $h = f_1^{-1} * f_2$. Since ph is nullhomotopic, it follows that h is nullhomotopic. But $f_2 \simeq f_1 * h$, hence $f_2 \simeq f_1$, as desired. □

Theorem 11.51. (Homotopy Sequence of a Weak Fibration). *Let $p: E \to B$ be a weak fibration. Choose basepoints $x_0 \in E$ and $b_0 = p(x_0) \in B$, so that $F = p^{-1}(b_0)$ is the fiber. Then there is an exact sequence*

$$\cdots \to \pi_2(E) \xrightarrow{p_*} \pi_2(B) \to \pi_1(F) \to \pi_1(E) \xrightarrow{p_*} \pi_1(B) \to \pi_0(F) \to \pi_0(E).$$

PROOF. In the exact sequence of the pair (E, F), replace the relative homotopy group $\pi_n(E, F)$ with $\pi_n(B)$ for all $n \geq 1$; the map $\pi_n(E) \to \pi_n(B)$ is the composite $p'_* j_* = p_*$, where $j: (E, x_0) \to (E, F)$ is the inclusion and $p': (E, F) \to (B, b_0)$ is the map p regarded as a map of pairs. □

One needs examples of weak fibrations in order to use this last result. Here is the most useful generalization of covering spaces.

Definition. A **locally trivial bundle** with **fiber** F is a map $p: E \to B$ for which there is an open cover $\mathscr{V}$ of B and homeomorphisms

$$\varphi_V: V \times F \to p^{-1}(V)$$

for all $V \in \mathscr{V}$ such that

$$p\varphi_V(v, x) = v \quad \text{for all } (v, x) \in V \times F.$$

The open sets $V \in \mathscr{V}$ are called **coordinate neighborhoods**.

In a locally trivial bundle, all fibers (i.e., all subspaces of E of the form $p^{-1}(b)$) are homeomorphic to F.

EXAMPLE 11.12. Every covering space $p: \tilde{X} \to X$ is a locally trivial bundle. Note first that the fibers $p^{-1}(x_0)$, as x_0 varies over X, are homeomorphic discrete spaces, any one of which we may denote by F. Define the coordinate neighborhoods to be the admissible open sets. Thus, if V is admissible and $x_0 \in V$, then

$$p^{-1}(V) = \bigcup_{y \in F} S_y,$$

where $F = p^{-1}(x_0)$ and S_y is the sheet over V containing y. Finally, define $\varphi_V\colon V \times F \to p^{-1}(V)$ by $(v, y) \mapsto (p|S_y)^{-1}(v)$; it is easy to see that φ_V is a homeomorphism and that $p\varphi_V(v, y) = v$.

Example 11.13. Let B and F be topological spaces, let $E = B \times F$, and let $p\colon E \to B$ be the projection $(b, x) \mapsto b$, where $b \in B$ and $x \in F$. Then $p\colon E \to B$ is a locally trivial bundle. Indeed, $p\colon E \to B$ is called a **trivial bundle**.

Theorem 11.52. *A locally trivial bundle $p\colon E \to B$ with fiber F is a weak fibration.*[6]

Proof. Consider the commutative diagram

$$\begin{array}{ccc} \mathbf{I}^n & \xrightarrow{\tilde{f}} & E \\ \downarrow & & \downarrow p \\ \mathbf{I}^{n+1} & \xrightarrow[G]{} & B. \end{array}$$

The family of open sets of the form $G^{-1}(V)$, where V ranges over the coordinate neighborhoods, is an open cover of the compact metric space $\mathbf{I}^{n+1}$. If λ is the Lebesgue number of this cover, then any subset A of $\mathbf{I}^{n+1}$ of diameter $< \lambda$ lies in some $G^{-1}(V)$, that is, $G(A) \subset V$. Triangulate $\mathbf{I}^n$ (say, by iterated barycentric subdivision) so that every simplex σ has diameter $< \lambda/2$; choose points $0 = t_0 < t_1 < \cdots < t_m = 1$ so that $t_{j+1} - t_j < \lambda/2$ for all $0 \le j < m$; it is easy to see that $\operatorname{diam}(\sigma \times [t_j, t_{j+1}]) < \lambda$ for every σ and every j. It follows that, for each σ and j, there exists a coordinate neighborhood $V = V_{\sigma,j}$ with $G(\sigma \times [t_j, t_{j+1}]) \subset V$.

Let L denote the simplicial complex of the triangulation of $\mathbf{I}^n$, and let $L^{(k)}$ denote its k-skeleton. We prove by induction on $k \ge 0$ that there exist continuous maps $\tilde{h}_k$ giving commutativity of

$$\begin{array}{ccc} L^{(k)} & \xrightarrow{\tilde{f}} & E \\ {\scriptstyle i}\downarrow & \overset{\tilde{h}_k}{\nearrow} & \downarrow {\scriptstyle p} \\ L^{(k)} \times [0, t_1] & \xrightarrow[G]{} & B \end{array}$$

and with $\tilde{h}_{k+1}$ extending $\tilde{h}_k$ (we abuse notation and denote restrictions of i, $\tilde{f}$, and G by the same letters).

Let the projections $V \times F \to V$ and $V \times F \to F$ be denoted by α_V and β_V, respectively. If $e \in p^{-1}V$ (for a coordinate neighborhood V). then $\varphi_V^{-1}e = (\alpha_V\varphi_V^{-1}e, \beta_V\varphi_V^{-1}e)$. Since $p\varphi_V(v, x) = v$, however, if follows that $\alpha_V\varphi_V^{-1}e = pe$;

[6] We remind the reader that a fiber bundle $p\colon E \to B$ with B paracompact is a fibration (Example 11.10).

hence, for all $e \in p^{-1}V$,

$$\varphi_V(pe, \beta_V \varphi_V^{-1} e) = e.$$

If $u \in \mathrm{Vert}(L) = L^{(0)}$, then there is a coordinate neighborhood V containing $G(\{u\} \times [0, b_1])$. Define $h^u: \{u\} \times [0, t_1] \to p^{-1}V \subset E$ by $h^u(u, t) = \varphi_V(G(u, t), \beta_V \varphi_V^{-1} \tilde{f}(u))$. Now $ph^u(u, t) = G(u, t)$, while $h^u(u, 0) = \varphi_V(G(u, 0), \beta_V \varphi_V^{-1} \tilde{f}(u)) = \varphi_V(p\tilde{f}(u), \beta_V \varphi_V^{-1} \tilde{f}(u)) = \tilde{f}(u)$. Since $L^{(0)}$ is discrete, one may glue these maps h^u together to obtain a continuous map $\tilde{h}_0: L^{(0)} \times [0, t] \to E$; the induction begins.

For the inductive step, let σ be a $(k + 1)$-simplex in L, and let V be a coordinate neighborhood containing $G(\sigma \times [0, t_1])$. Since $\sigma \approx \mathbf{I}^{k+1}$, an obvious modification of a stereographic retraction $\mathbf{I}^{k+1} \times \mathbf{I} \to (\mathbf{I}^{k+1} \times \{0\}) \cup (\dot{\mathbf{I}}^{k+1} \times \mathbf{I})$ gives a retraction $r_\sigma: \sigma \times [0, t_1] \to (\sigma \times \{0\}) \cup (\dot{\sigma} \times [0, t_1])$. Define

$$\tilde{v}_\sigma: (\sigma \times \{0\}) \cup (\dot{\sigma} \times [0, t_1]) \to p^{-1}V$$

by $\tilde{v}_\sigma | \sigma \times \{0\} = \tilde{f}i | \sigma \times \{0\}$ and $\tilde{v}_\sigma | \dot{\sigma} \times [0, t_1] = \tilde{h}_k | \dot{\sigma} \times [0, t_1]$, the latter map existing by induction. Finally, define

$$h^\sigma: \sigma \times [0, t_1] \to p^{-1}V \subset E$$

by

$$h^\sigma(u, t) = \varphi_V(G(u, t), \beta_V \varphi_V^{-1} \tilde{v}_\sigma r_\sigma(u, t)),$$

for $u \in \sigma$ and $t \in [0, t_1]$. We claim that $h^\sigma | \dot{\sigma} \times [0, t_1] = \tilde{h}_k | \dot{\sigma} \times [0, t_1]$ and that the following diagram commutes

$$\begin{array}{ccc} \sigma & \xrightarrow{\tilde{f}} & p^{-1}V \\ {\scriptstyle i}\downarrow & \overset{h^\sigma}{\nearrow} & \downarrow{\scriptstyle p} \\ \sigma \times [0, t_1] & \xrightarrow[G]{} & V. \end{array}$$

Clearly, $ph^\sigma(u, t) = p\varphi_V(G(u, t), \text{stuff}) = G(u, t)$, so that the lower triangle commutes. If $(u, t) \in (\sigma \times \{0\}) \cup (\dot{\sigma} \times [0, t_1])$, then $r_\sigma(u, t) = (u, t)$. If $(u, t) \in \sigma \times \{0\}$, then $G(u, t) = G(u, 0) = p\tilde{f}(u)$ and $\tilde{v}_\sigma(u, t) = \tilde{f}(u)$; hence $h\sigma(u, t) = \varphi_V(p\tilde{f}(u), \beta_V \varphi_V^{-1} \tilde{f}(u)) = \tilde{f}(u)$. If $(u, t) \in \dot{\sigma} \times [0, t_1]$, then $G(u, t) = p\tilde{h}_k(u, t)$ and $\tilde{v}_\sigma(u, t) = \tilde{h}_k(u, t)$; hence, $h^\sigma(u, t) = \varphi_V(p\tilde{h}_k(u, t), \beta_V \varphi_V^{-1} \tilde{h}_k(u, t)) = \tilde{h}_k(u, t)$, as desired. Since simplexes in L intersect in lower dimensional faces, the gluing lemma allows us to assemble all the maps h^σ to obtain a continuous map $\tilde{h}_{k+1}: L^{(k+1)} \times [0, t_1] \to E$, as desired. In particular, for $k = n$, there is a continuous map $\tilde{G}_1 = \tilde{h}_n$ making the following diagram commute.

$$\begin{array}{ccc} \mathbf{I}^n & \xrightarrow{\tilde{f}} & E \\ \downarrow & \overset{\tilde{G}_1}{\nearrow} & \downarrow{\scriptstyle p} \\ \mathbf{I}^n \times [0, t_1] & \xrightarrow[G]{} & B. \end{array}$$

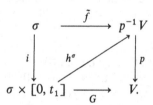

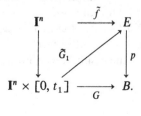

Now repeat this construction with $[t_1, t_2]$ playing the role of $[t_0, t_1]$ (for $t_0 = 0$) to obtain a map $\mathbf{I}^n \times [t_1, t_2] \to E$ agreeing with $\tilde{G}_1$ on $\mathbf{I}^n \times \{t_1\}$. These maps can be glued together to obtain a map $\tilde{G}_2: \mathbf{I}^n \times [0, t_2] \to E$ making the appropriate diagram commute. Iterate to obtain $\tilde{G} = \tilde{G}_m$ defined on $\mathbf{I}^n \times [0, t_m] = \mathbf{I}^n \times \mathbf{I}$, as desired. □

Recall the constructions of complex and quaternionic projective spaces (Exercises 8.6 and 8.7). There is a map $p: S^{2n+1} \to \mathbf{C}P^n$ given by

$$(x_1, \ldots, x_{2n+2}) = (z_0, \ldots, z_n) \mapsto [z_0, \ldots, z_n],$$

where $z_j = x_{2j+1} + ix_{2j+2}$. Also, there is a map $q: S^{4n+3} \to \mathbf{H}P^n$ given by

$$(x_1, \ldots, x_{4n+4}) = (h_0, \ldots, h_n) \mapsto [h_0, \ldots, h_n],$$

where h_ν is the quaternion $x_{4\nu+1} + x_{4\nu+2}i + x_{4\nu+3}j + x_{4\nu+4}k$.

Remark. The maps p and q are called **Hopf fibrations**.

Theorem 11.53. *The Hopf fibrations $p: S^{2n+1} \to \mathbf{C}P^n$ and $q: S^{4n+3} \to \mathbf{H}P^n$ are locally trivial bundles with fiber S^1 and S^3, respectively.*

Proof. We show that p is a locally trivial bundle; the proof for q is similar. For each j with $0 \leq j \leq n$, define

$$V_j = \{[z_0, \ldots, z_n] \in \mathbf{C}P^n: z_j \neq 0\};$$

each V_j is open because its complement is the image of a (closed, hence) compact subset of S^{2n+1}. Define $\varphi_j: V_j \times S^1 \to p^{-1}V_j \subset S^{2n+1}$ by

$$\varphi_j([z_0, \ldots, z_n], u) = (\|z_j\|u/\sqrt{\Sigma\|z_k\|^2})(z_j^{-1}z_0, \ldots, z_j^{-1}z_n).$$

A short computation shows that if $\lambda \in \mathbf{C}$ and $\lambda \neq 0$, then $\varphi_j([\lambda z_0, \ldots, \lambda z_n], u) = \varphi_j([z_0, \ldots, z_n], u)$, hence φ_j is well defined. Since the inverse of φ_j is (easily seen to be) the map $p^{-1}V_j \to V_j \times S^1$ given by $(z_0, \ldots, z_n) \mapsto ([z_0, \ldots, z_n], z_j/\|z_j\|)$, it follows that φ_j is a homeomorphism. Since $p\varphi_j([z_0, \ldots, z_n], u) = [z_0, \ldots, z_n]$, it follows that p is a locally trivial bundle. □

Corollary 11.54.

(i) $\pi_n(S^3) \cong \pi_n(S^2)$ *for all* $n \geq 3$.
(ii) $\pi_n(S^4) \cong \pi_{n-1}(S^3)$ *for* $1 \leq n \leq 6$.

Proof. (i) Recall Exercise 8.2: $\mathbf{C}P^1 \approx S^2$. Thus $p: S^3 \to S^2$ is a locally trivial bundle with fiber S^1 and hence is a weak fibration. Therefore, by Theorem 11.30, $\pi_n(S^1) = 0$ for all $n \geq 2$, so that the exact homotopy sequence of a weak fibration gives $\pi_n(S^3) \cong \pi_n(S^2)$ for all $n \geq 3$.

(ii) Recall Exercise 8.2: $\mathbf{H}P^1 \approx S^4$. Thus $q: S^7 \to S^4$ is a locally trivial bundle with fiber S^3 and hence is a weak fibration. Therefore, by Theorem 11.31, $\pi_n(S^7) = 0$ for all $n \leq 6$, and the result now follows from the exact homotopy sequence. □

Corollary 11.55. $\pi_3(S^2) \neq 0$.

PROOF. $\pi_3(S^3) \cong \pi_3(S^2)$, by Corollary 11.54, and $\pi_3(S^3) \neq 0$, by Theorem 11.31. □

One knows that $\pi_n(S^n) \cong \mathbf{Z}$, so that $\pi_3(S^2) \cong \mathbf{Z}$.

Remark. There is a nonassociative real division algebra called the **Cayley numbers** with additive group $\mathbf{R}^8$; it can be used to construct a locally trivial bundle $h\colon S^{15} \to S^8$ with fiber S^7 (see [Hilton (1953), p. 54] or [Steenrod, p. 108]) (this map is also called a **Hopf fibration**).

EXAMPLE 11.14. There are spaces X and Y having isomorphic homology groups but different homotopy groups (see also Example 11.7).

Let $X = S^2 \vee S^4$ and let $Y = \mathbf{C}P^2$. It is easy to see that $H_n(X) \cong H_n(Y)$ for all $n \geq 0$ ($H_n = \mathbf{Z}$ for $n = 0, 2, 4$ and $H_n = 0$ otherwise). Since S^4 is a retract of X, $\pi_4(S^4)$ is a retract (direct summand) of $\pi_4(X)$, and so $\pi_4(X) \neq 0$. On the other hand, since there is a locally trivial bundle $S^{2n+1} \to \mathbf{C}P^n$ with fiber S^1, Theorem 11.51 yields $\pi_q(S^{2n+1}) \cong \pi_q(\mathbf{C}P^n)$ for all $q \geq 3$ (for $\pi_q(S^1) = 0$ for all $q \geq 2$). In particular, $\pi_4(\mathbf{C}P^2) \cong \pi_4(S^5) = 0$ (Theorem 11.31). Therefore $\pi_4(X) \ncong \pi_4(Y)$.

Homotopy groups do not behave like homology groups: they need not vanish in degrees above the dimension of the space. Indeed there is no exact homology sequence of a fibration, and there is no excision for homotopy groups. For example, $\mathbf{R}^2$ is a covering space of $S^1 \times S^1$ with discrete fiber $\mathbf{Z} \times \mathbf{Z}$, and exactness of

$$\cdots \to H_2(\mathbf{Z} \times \mathbf{Z}) \to H_2(\mathbf{R}^2) \to H_2(S^1 \times S^1) \to H_1(\mathbf{Z} \times \mathbf{Z}) \to \cdots$$

would give $0 = H_2(\mathbf{R}^2) \cong H_2(S^1 \times S^1)$, which contradicts the fact that $H_2(S^1 \times S^1) \cong \mathbf{Z}$. To see that excision fails for homotopy, it suffices to show that the Mayer–Vietoris sequence is not exact for homotopy groups. Write $S^2 = A^\circ \cup B^\circ$, where A is the complement of the north pole and B is the complement of the south pole. Note that the equator S^1 is a pointed deformation retract of $A \cap B$ (deform along longitudes; as usual, the basepoint is $(1, 0, 0)$); it follows that $\pi_n(A \cap B) \cong \pi_n(S^1)$ for all n. The Mayer–Vietoris sequence in homotopy would say that there is an exact sequence

$$\cdots \to \pi_3(A \cap B) \to \pi_3(A) \oplus \pi_3(B) \to \pi_3(A \cup B) \to \pi_2(A \cap B) \to \cdots,$$

that is, an exact sequence

$$\cdots \to \pi_3(S^1) \to \pi_3(A) \oplus \pi_3(B) \to \pi_3(S^2) \to \pi_2(S^1) \to \cdots.$$

Since $\pi_3(S^1) = 0 = \pi_2(S^1)$, there is an isomorphism $\pi_3(A) \oplus \pi_3(B) \cong \pi_3(S^2)$. But both A and B are contractible, hence $\pi_3(A) \oplus \pi_3(B) = 0$ (Corollary 11.28), and this contradicts $\pi_3(S^2) \neq 0$.

[Milnor (1956), p. 279] shows that there is an analogue for homotopy groups of the Eilenberg–Steenrod theorem: there is a unique sequence of functors $\pi_n\colon \mathscr{C} \to \textbf{Sets}$, where $\mathscr{C}$ is the category of all pointed pairs of topological spaces, that is, all (X, A, x_0), where $x_0 \in A \subset X$, which satisfies the Eilenberg–Steenrod axioms with excision replaced by the exact sequence of a weak fibration, and with $\pi_0(X, x_0)$ the set of path components of X. Two comments are needed. First, Milnor calls $A_0 \to A_1 \xrightarrow{f} A_2 \to A_3$ exact in **Sets** if A_0 and A_3 singletons implies that f is a bijection (this definition is much weaker than our definition "exact in $\textbf{Sets}_*$"). Second, the precise statement of the fibration axiom is: if B_0 is a path component of B, if $p\colon E \to B_0$ is a weak fibration, and if (B, A) is a pointed pair, then $p_*\colon \pi_n(E, p^{-1}(A)) \to \pi_n(B, A)$ is a bijection for all $n \geq 1$ (Theorem 11.51 is the special case $B = B_0$ and $A = \{x_0\}$).

Exercises

11.46. If $p\colon E \to B$ is a weak fibration with fiber F, then $\pi_2(E, F)$ is an abelian group.

11.47. (i) If $q \geq 3$, then $\pi_q(\mathbf{C}P^n) \cong \pi_q(S^{2n+1})$.
(ii) If $q \leq 4n + 2$, then $\pi_q(\mathbf{H}P^n) \cong \pi_{q-1}(S^3)$.

11.48. If $p\colon E \to B$ is a weak fibration with simply connected fibers, then $p_*\colon \pi_1(E) \to \pi_1(B)$ is an isomorphism.

11.49. Prove that $\pi_2(S^2) \cong \mathbf{Z}$. (*Hint*: Use the Hopf fibration $p\colon S^3 \to S^2$.)

11.50. Let $O(n)$ denote the **orthogonal group** consisting of all $n \times n$ real matrices A with $A^tA = E$ (A^t is the transpose of A and E is the identity matrix). If e_n is the (column) vector $(0, \ldots, 0, 1)$, then Ae_n is a unit vector in $\mathbf{R}^n$ and hence lies in S^{n-1}. Using the fact (see [Gray, p. 89]) that $p\colon O(n) \to S^{n-1}$ (defined by $A \mapsto Ae_n$) is a locally trivial bundle with fiber $O(n-1)$, show that $\pi_q(O(n-1)) \cong \pi_q(O(n))$ for $q < n - 2$. Conclude, for fixed q and $m, n \geq q + 2$, that $\pi_q(O(n)) \cong \pi_q(O(m))$.

*11.51. If X is a convergent sequence with its limit and Y is a countable discrete space, then X and Y do not have the same homotopy type (Exercise 1.5), $H_n(X) \cong H_n(Y)$ for all $n \geq 0$, and $\pi_n(X, x_0) \cong \pi_n(Y, y_0)$ for all $n \geq 0$ and all basepoints $x_0 \in X$ and $y_0 \in Y$.

A Glimpse Ahead

In Chapter 4, we constructed the Hurewicz map

$$\varphi\colon \pi_1(X, x_0) \to H_1(X)$$

(singular homology), and we proved in Theorem 4.29 that φ induces an isomorphism $\pi_1(X, x_0)_{\text{ab}} \xrightarrow{\sim} H_1(X)$, where $\pi_1(X, x_0)_{\text{ab}}$ denotes the quotient group of $\pi_1(X, x_0)$ modulo its commutator subgroup.

Let $h_n\colon \Delta^n \to S^n$ be the natural map, where we identify S^n with the quotient space $\Delta^n/\dot{\Delta}^n$. Then h_n is an n-cycle and cls h_n is a generator of $H_n(S^n) = \mathbf{Z}$. If we regard the elements of $\pi_n(X, x_0)$ as pointed homotopy classes $[S^n, X]$, then the **Hurewicz map**

$$\varphi_n\colon \pi_n(X, x_0) \to H_n(X)$$

is defined by $[\alpha] \mapsto \text{cls}\, \alpha h_n$. Replacing the homology group $H_n(X)$ by the reduced homology group $\tilde{H}_n(X) = H_n(X, x_0)$, one sees that φ_n is a natural transformation $\pi_n \to \tilde{H}_n$ (both are functors $\mathbf{Top}_* \to \mathbf{Ab}$). Call a space X ***n*-connected** if $\pi_q(X, x_0) = 0$ for all $q \leq n$.

Hurewicz Theorem. *If X is an n-connected space with $n \geq 2$, then $\tilde{H}_q(X) = 0$ for all $q \leq n$ and the Hurewicz map is an isomorphism*

$$\pi_{n+1}(X, x_0) \xrightarrow{\simeq} \tilde{H}_{n+1}(X).$$

A complete proof can be found in [Spanier, pp. 387–400]; indeed a more general version of the Hurewicz theorem for relative homotopy groups and relative homology groups is proved there. A shorter proof, but only for CW complexes, can be found in [Bott and Tu, p. 225], [Maunder, pp. 322–328], or [Whitehead, pp. 174–180]. Since S^n is $(n-1)$-connected, it follows at once that $\pi_n(S^n) \cong \mathbf{Z}$ (this last isomorphism can be established without the Hurewicz theorem: e.g., see [Maunder, p. 288]).

Suppose that X is a CW complex, A is a CW subcomplex, and $f\colon A \to Y$ is a continuous map. If e is an n-cell in X with $\dot{e} \subset A$, then $f|\dot{e}$ defines a certain element $c(f, e)$ of $\pi_{n-1}(Y)$, called its **obstruction**, and f can be extended to $A \cup e$ if and only if $c(f, e) = 0$. If one knew everything about homotopy groups, then one could see whether f extends to X by checking one cell at a time. This same problem leads to cohomology groups $H^n(X, A; \pi_{n-1}(Y))$ whose coefficient groups are homotopy groups! See [Hu (1959)] or [Spanier, Chap. 8].

A point $x_0 \in X$ is called **nondegenerate** if the inclusion $\{x_0\} \hookrightarrow X$ is a cofibration. By Theorem 8.33, every 0-cell of a CW complex is nondegenerate; indeed Lemma 8.30 shows that every point in a CW complex is nondegenerate.

Freudenthal Suspension Theorem. *Let X be an $(n-1)$-connected space having a nondegenerate basepoint. Then the suspension homomorphism $\pi_q(X) \to \pi_{q+1}(\Sigma X)$ is an isomorphism for all $q \leq 2n-2$ and is a surjection for $q = 2n-1$.*

For a proof, see [Gray, p. 145] or [Whitehead, p. 369]. Specializing to $X = S^n$ thus relates low-dimensional homotopy groups of S^n and S^{n+1}.

Here is a theorem with a similar conclusion; it also gives a condition for excision to hold for homotopy groups.

Blakers–Massey Theorem. *Let $X = X_1^\circ \cup X_2^\circ$, and let $i\colon (X_1, X_1 \cap X_2) \hookrightarrow (X_1, X_2)$ be the (excision) inclusion. If $(X_1, X_1 \cap X_2)$ is $(n-1)$-connected and $(X_2, X_1 \cap X_2)$ is $(m-1)$-connected, then $i_*\colon \pi_q(X_1, X_1 \cap X_2) \to \pi_q(X, X_2)$ is an isomorphism for $q < m + n - 2$ and is a surjection for $q = m + n - 2$.*

A proof can be found in [Gray, p. 143].

There is a homotopy analogue of Theorem 8.41. Let X be a path connected CW complex with CW subcomplex A. If A is k-connected and $\pi_1(X, A) = 0$, then $\pi_i(X, A) \cong \pi_i(X/A)$ for all i with $1 \leq i \leq k - 1$. For a proof of a more general result, see [Gray, p. 144].

The following theorem indicates the convenience of CW complexes (one should expect good results because homotopy groups are concerned with maps $S^n \to X$, relative homotopy groups are concerned with maps $(D^n, S^{n-1}) \to (X, A)$, and CW complexes are constructed from relative homeomorphisms $(D^n, S^{n-1}) \to (X, A)$).

Whitehead Theorem.[7] *If X and Y are connected CW complexes, and if $f: X \to Y$ is a continuous map such that $f_*: \pi_n(X, x_0) \to \pi_n(Y, f(x_0))$ is an isomorphism for all n, then f is a homotopy equivalence (so that X and Y have the same homotopy type).*

Corollary. *A connected CW complex X is contractible if and only if $\pi_n(X) = 0$ for all n.*

If Y is a one-point space, then the (constant) map $f: X \to Y$ induces isomorphisms between the trivial groups $\pi_n(X)$ and $\pi_n(Y)$.

A proof of Whitehead's theorem is in [Maunder, p. 300]. Note that one must assume that the isomorphisms are all induced by one continuous map lest Example 11.7 give a counterexample. One must also assume that both spaces X and Y are CW complexes: there is an example ([Maunder, p. 301]) of a path connected space X (a certain subspace of $\mathbf{R}^2$) that is not contractible and for which $\pi_n(X) = 0$ for all n; also see Exercise 11.51.

There is an inductive way of "killing" homotopy groups that is an iterative version of the construction of a universal covering space.

Theorem. *Given a CW complex X and an integer n, there exists a CW complex Y containing X as a CW subcomplex such that $\pi_q(X) \cong \pi_q(Y)$ for all $q < n$ and $\pi_q(Y) = 0$ for all $q \geq n$.*

A proof can be found in [Maunder, p. 303]. Using this theorem, one can prove that if $\pi_1, \pi_2, \ldots$ is a sequence of groups with π_n abelian for all $n \geq 2$ (actually, π_n must be a $\mathbf{Z}\pi_1$-module for all $n \geq 2$), then there exists a connected CW complex X with $\pi_n(X) \cong \pi_n$ for all $n \geq 1$ (see [Whitehead, p. 216]). Thus there are simply connected spaces whose higher homotopy groups can be any preassigned abelian groups. Also, given any integer $n \geq 1$ and any group π (abelian if $n \geq 2$), there exists a connected CW complex K with $\pi_n(K) = \pi$ and

---

[7] After J. H. C. Whitehead, who invented CW complexes and proved many of the fundamental theorems about them.

$\pi_q(K) = 0$ for all $q \neq n$. Such a space K is called an **Eilenberg–Mac Lane space**, and it is denoted by $K(\pi, n)$. If X is a CW complex having a contractible universal covering space, then Theorem 11.29 show that X is a $K(\pi, 1)$, where π is the fundamental group of X; Riemann surfaces are examples of such spaces X. For a fixed π and n, the Whitehead theorem can be used to prove that any two $K(\pi, n)$'s have the same homotopy type. These spaces arise in studying (contravariant) cohomology theory, for if X is a CW complex and G is an abelian group, then $\tilde{H}^n(X; G) \cong [X, K(G, n)]$. Now it follows from Theorem 11.20 that $\Omega K(\pi, n)$ "is" $K(\pi, n + 1)$. One defines an **Ω-spectrum** as a sequence of pointed spaces E_n, $n \geq 0$, and pointed maps $\varepsilon_n \colon \Omega E_n \to E_{n+1}$ such that $\varepsilon_{n*} \colon [X, E_{n+1}] \to [X, \Omega E_n]$ is a bijection for all CW complexes X. Since ΩE_n is an H-group, $[X, \Omega E_n]$ and hence $[X, E_{n+1}]$ can be regarded as a group (indeed as an abelian group). The sequence of contravariant functors $[\quad, E_n]$ is called a **generalized cohomology theory**; it satisfies all the Eilenberg–Steenrod axioms save the dimension axiom, so that it is an extraordinary cohomology theory. A theorem of E. H. Brown states that almost any extraordinary cohomology theory arises from some Ω-spectrum. For a discussion of these ideas, see [Atiyah], [Dyer], and [Maunder, §8.4].

The homology groups of the Eilenberg–Mac Lane spaces $K(\pi, 1)$ are the object of study of **cohomology of groups** (see [Brown]). The universal covering space $\tilde{X}$ of $K(\pi, 1)$ exists (Theorem 10.38) and is a CW complex (Theorem 10.43). By definition, $\pi_1(\tilde{X}) = 0$, while $\pi_n(\tilde{X}) \cong \pi_n(K(\pi, 1)) = 0$ for all $n \geq 2$ (Theorem 11.29). It follows from the corollary to Whitehead's theorem that $\tilde{X}$ is contractible. Now $\pi \cong \pi_1(K(\pi, 1))$ acts properly on $\tilde{X}$, and the orbit space $\tilde{X}/\pi$ is homeomorphic to $K(\pi, 1)$ (Lemma 10.49). But if G is an abelian group, then there are isomorphisms

$$H^n(\tilde{X}/\pi; G) \cong H^n(\pi; G)$$

(see [Mac Lane, p. 136]; the groups on the right-hand side are the cohomology groups of the group π, and these are defined purely algebraically; the groups on the left-hand side are cohomology groups of the space $\tilde{X}/\pi$ (with coefficients in G) and are discussed in Chapter 12).

If X and Y are CW complexes with the same homology groups ($H_n(X) \cong H_n(Y)$ for all n) and the same homotopy groups ($\pi_n(X) \cong \pi_n(Y)$ for all n), then do X and Y have the same homotopy type? The answer is "no". However, two CW complexes have the same homotopy type if and only if they have the same homology groups, the same homotopy groups, and the same **Postnikov invariants**.

After studying the fundamental group and computing $\pi_1(S^1)$, one can prove the fundamental theorem of algebra; after studying homology, one can, after computing $H_*(S^n)$, prove interesting results about euclidean space. In Chapter 12, we shall study cohomology; after computing the cohomology of $\mathbf{R}P^n$ mod 2, we shall prove more nice theorems about euclidean space. What are applications of homotopy groups?

Homotopy theory enters into solutions of problems, but usually not as the only ingredient of a proof. A **real division algebra** is a finite-dimensional real vector space equipped with a bilinear multiplication having a two-sided identity element and such that each nonzero element has a two-sided multiplicative inverse. Examples of real division algebras are the real numbers, the complex numbers, the real quaternions, and an eight-dimensional non-associative algebra called the Cayley numbers. It is a theorem of J. F. Adams that there are no other examples. To each element of $\pi_{2n-1}(S^n)$, one can associate an integer called its **Hopf invariant**. It was known that if $\pi_{2n-1}(S^n)$ contains no elements of Hopf invariant one, then there is no real division algebra of dimension n; Adams proved that such elements exist only if $n = 1$, 2, 4, 8; see [Atiyah, p. 137] for a much simpler proof using K-theory. (Unfortunately for our point, there is an equivalent statement involving "cohomology operations", and Adams' proof contains only a bit of homotopy theory.)

Algebraic K-theory is an area using homotopy theory in an essential way; here is one version of it. Let $\mathscr{C}$ be an exact category; that is, $\mathscr{C}$ is a full subcategory of **Ab** that is closed under extensions: if

$$0 \to A \to B \to C \to 0 \tag{1}$$

is an exact sequence with A, $C \in \operatorname{Obj}(\mathscr{C})$, then $B \in \operatorname{Obj}(\mathscr{C})$. Grothendieck defined an abelian group $K_0(\mathscr{C})$ as the abelian group having generators $\operatorname{Obj}(\mathscr{C})$ and relations $A + C = B$ if there is an exact sequence (1). Later, in analogy with topological K-theory, Bass invented a group $K_1(\mathscr{C})$, the **Whitehead group**, and he constructed a 5-term exact sequence involving $K_0(\mathscr{C})$ and $K_1(\mathscr{C})$. Quillen then constructed groups $K_i(\mathscr{C})$ for all $i \geq 0$ agreeing with the earlier groups when $i = 0$ and $i = 1$. To $\mathscr{C}$ he first associated a new category $\mathbf{Q}\mathscr{C}$, the **Q-construction**, he then took its classifying space $\mathbf{BQ}\mathscr{C}$, which is a functorial CW-complex, and then defined

$$K_i(\mathscr{C}) = \pi_{i+1}(\mathbf{BQ}\mathscr{C}).$$

CHAPTER 12

Cohomology

Cohomology is a contravariant version of homology. Although it is not difficult to define, let us first give some background for it.

Differential Forms

Throughout this section, X shall denote an open connected subset of $\mathbf{R}^n$. Recall that a function $\alpha: X \to \mathbf{R}$ is a **C^∞-function** if its partial derivatives $\partial^k\alpha/\partial x_1^k, \ldots, \partial^k\alpha/\partial x_n^k$ exist for all $k \geq 1$. The family $A(X)$ of all C^∞-functions on X is a commutative ring under pointwise operations: if $\alpha, \beta \in A(X)$, then $\alpha + \beta: x \mapsto \alpha(x) + \beta(x)$ and $\alpha\beta: x \mapsto \alpha(x)\beta(x)$; the unit is the constant function $\alpha(x) \equiv 1$.

Definition. If A is a commutative ring with 1, an **A-module** is an abelian group M equipped with a **scalar multiplication** $A \times M \to M$, denoted by $(a, m) \mapsto am$, such that the following identities hold for all $m, m' \in M$ and $a, a', 1 \in A$:

(i) $a(m + m') = am + am'$;
(ii) $(a + a')m = am + a'm$;
(iii) $(aa')m = a(a'm)$;
(iv) $1m = m$.

If $A = \mathbf{Z}$, then an A-module is merely an abelian group; if A is a field, then an A-module is a vector space over A. The ring A itself can be regarded as an A-module by taking scalar multiplication to be the given multiplication of A.

Given A-modules $M_1, \ldots, M_n$, forget the scalar multiplication for a moment, and form the direct sum of the abelian groups $M_1, \ldots, M_n$. Then $M_1 \oplus \cdots \oplus M_n$ is an A-module, called the **direct sum**, if one defines $a(m_1, \ldots, m_n) = (am_1, \ldots, am_n)$. In particular, the direct sum of n copies of A, denoted by $A^{(n)}$, is called a **free A-module**. If $e_i \in A^{(n)}$ is the n-tuple having 1 in the ith position and zeros elsewhere, then it is easy to see that every $m \in A^{(n)}$ has a unique expression of the form

$$m = \sum a_i e_i, \qquad a_i \in A.$$

Definition. A subset $\{b_1, \ldots, b_n\}$ of A^n is an **A-basis** if each $m \in A^{(n)}$ has a unique expression of the form $m = \sum a_i b_i$ with $a_i \in A$.

Thus $\{e_1, \ldots, e_n\}$ is an A-basis of $A^{(n)}$. Let $A = A(X)$, the ring of C^∞-functions on $X \subset \mathbf{R}^n$. In $A(X)^{(n)}$, rename e_i as dx_i, so that each $m \in A(X)^{(n)}$ has a unique expression of the form

$$m = \sum \alpha_i \, dx_i, \qquad \alpha_i \in A(X).$$

For integration, one needs expressions of the form $dx_1 \, dx_2 \cdots dx_p$; moreover, an expression of the form $dx \, dx$ should be zero.

Definition. If M is an A-module and $p \geq 0$, then the pth **exterior power** of M, denoted by $\bigwedge^p M$, is the abelian group with the following presentation:

Generators: $A \times M \times \cdots \times M$ (p factors M).
Relations: For all $a, a' \in A$ and $m_i, m_i' \in M$,

$$(a, m_1, \ldots, m_i + m_i', \ldots, m_p) = (a, m_1, \ldots, m_i, \ldots, m_p) + (a, m_1, \ldots, m_i', \ldots, m_p) \quad \text{for all } i;$$

$$(a + a', m_1, \ldots, m_p) = (a, m_1, \ldots, m_p) + (a', m_1, \ldots, m_p);$$

$$(aa', m_1, \ldots, m_i, \ldots, m_p) = (a, m_1, \ldots, a'm_i, \ldots, m_p) \quad \text{for all } i;$$

$$(a, m_1, \ldots, m_p) = 0 \quad \text{if } m_i = m_j \text{ for some } i \neq j.$$

If $p = 0$, then $\bigwedge^0 M = A$, and if $p = 1$, then $\bigwedge^1 M \cong M$. If F is the free abelian group with basis $A \times M \times \cdots \times M$ and if S is the subgroup of F generated by the relations, then the coset $(a, m_1, \ldots, m_p) + S$ is denoted by $am_1 \wedge \cdots \wedge m_p$. Thus every element of $\bigwedge^p M$ has an expression (not necessarily unique) of the form $\sum_j a_j m_1^j \wedge \cdots \wedge m_p^j$, where $a_j \in A$ and $m_i^j \in M$. It is now plain that $\bigwedge^p M$ is an A-module, because we can multiply any element by $a \in A$ (and the axioms will be satisfied).

Observe that $m \wedge m = 0$ for all $m \in M$. Hence, for $m, m' \in M$,

$$\begin{aligned} 0 &= (m + m') \wedge (m + m') \\ &= m \wedge m + m \wedge m' + m' \wedge m + m' \wedge m' \\ &= m \wedge m' + m' \wedge m. \end{aligned}$$

Therefore $m' \wedge m = -m \wedge m'$ for all $m, m' \in M$. A similar argument when $p \geq 2$ shows that interchanging two factors of $m_1 \wedge \cdots \wedge m_p$ reverses the sign.

If $M = A^{(n)}$ is a free A-module, then $\bigwedge^p M$ is also a free A-module; indeed, if $\{e_1, \ldots, e_n\}$ is an A-basis of M, then $\{e_{i_1} \wedge \cdots \wedge e_{i_p} : 1 \leq i_1 < i_2 < \cdots < i_p \leq n\}$ is an A-basis of $\bigwedge^p M$ (see [Greub, p. 105]). Each element of $\bigwedge^p(A^{(n)})$ thus has a unique expression

$$\sum a_{i_1 \ldots i_p} e_{i_1} \wedge \cdots \wedge e_{i_p},$$

where $a_{i_1 \ldots i_p} \in A$ and $1 \leq i_1 < i_2 < \cdots < i_p \leq n$. Furthermore, if $p > n$, then $\bigwedge^p M = 0$. It follows that a basis of $\bigwedge^p(A^{(n)})$ has binomial coefficient $\binom{n}{p}$ elements.

Definition. If X is a connected open subset of $\mathbf{R}^n$ and if $p \geq 0$, write $\Omega^p(X) = \bigwedge^p(A(X)^{(n)})$; an element $\omega \in \Omega^p(X)$ is called a **differential p-form** on X.

Definition. The **exterior derivative** $d^p \colon \Omega^p(X) \to \Omega^{p+1}(X)$ is defined inductively. If $\alpha \in \Omega^0(X) = A(X)$, then α is a C^∞-function, and

$$d^0(\alpha) = \sum_{j=1}^{n} \left(\frac{\partial \alpha}{\partial x_j}\right) dx_j;$$

if $p \geq 1$, then $\omega \in \Omega^p(X)$ has the form $\omega = \sum \alpha_{i_1 \ldots i_p} dx_{i_1} \wedge \cdots \wedge dx_{i_p}$, and

$$d^p(\omega) = \sum d^0(\alpha_{i_1 \ldots i_p}) \wedge dx_{i_1} \wedge \cdots \wedge dx_{i_p}.$$

Note that one can rewrite $d^p(\omega)$ with subscripts in ascending order by repeated use of the identities $dx_i \wedge dx_i = 0$ and $dx_j \wedge dx_i = -dx_i \wedge dx_j$.

A connected open set X in $\mathbf{R}^n$ thus determines a sequence of homomorphisms

$$0 \longrightarrow \Omega^0(X) \xrightarrow{d^0} \Omega^1(X) \xrightarrow{d^1} \cdots \xrightarrow{d^{n-1}} \Omega^n(X) \longrightarrow 0;$$

moreover, there is a straightforward computation showing that $dd = 0$. In other words, this sequence is a complex; its homology groups are called the **de Rham cohomology** of X (this discussion can be extended to differentiable n-manifolds X; see [Bott and Tu] or [Warner]).

Consider the special case $n = 3$, so that the complex is

$$0 \to \Omega^0(X) \to \Omega^1(X) \to \Omega^2(X) \to \Omega^3(X) \to 0.$$

If $\omega \in \Omega^0(X)$, then $\omega = \alpha(x, y, z)$, a C^∞-function on X, and

$$d^0\omega = \frac{\partial \alpha}{\partial x} dx + \frac{\partial \alpha}{\partial y} dy + \frac{\partial \alpha}{\partial z} dz,$$

a 1-form resembling the gradient, grad α. If $\omega \in \Omega^1(X)$, then $\omega = \alpha\, dx + \beta\, dy + \gamma\, dz$, and a simple calculation (using $dx_i \wedge dx_i = 0$ and $dx_j \wedge dx_i =$

$-dx_i \wedge dx_j$) gives

$$d^1\omega = \left(\frac{\partial\beta}{\partial x} - \frac{\partial\alpha}{\partial y}\right) dx \wedge dy + \left(\frac{\partial\gamma}{\partial y} - \frac{\partial\beta}{\partial z}\right) dy \wedge dz + \left(\frac{\partial\alpha}{\partial z} - \frac{\partial\gamma}{\partial x}\right) dz \wedge dx,$$

a 2-form resembling curl ω. If $\omega \in \Omega^2(X)$, then $\omega = \lambda\, dy \wedge dz + \mu\, dz \wedge dx + \nu\, dx \wedge dy$, and

$$d^2\omega = \left(\frac{\partial\lambda}{\partial x} + \frac{\partial\mu}{\partial y} + \frac{\partial\nu}{\partial z}\right) dx \wedge dy \wedge dz,$$

a 3-form resembling the divergence, div ω.

These are not mere resemblances. Since $\Omega^1(X)$ has an $A(X)$-basis $\{dx, dy, dz\}$, $d^0\omega$ is grad ω when ω is a 0-form; since $\Omega^2(X)$ has an $A(X)$-basis $\{dx \wedge dy, dy \wedge dz, dz \wedge dx\}$, $d^1\omega$ is curl ω when ω is a 1-form; since $\Omega^3(X)$ has an $A(X)$-basis $\{dx \wedge dy \wedge dz\}$, $d^2\omega$ is div ω when ω is a 2-form. That $d^1d^0 = 0$ and $d^2d^1 = 0$ are therefore the familiar identities curl grad $= 0$ and div curl $= 0$.

In advanced calculus, a 1-form ω is called **closed** if $d\omega = 0$, and it is called **exact** if $\omega =$ grad α for some C^∞-function α. In the language of cohomology, closed 1-forms are 1-cocycles and exact 1-forms are 1-coboundaries. The name "exact sequence" was suggested by this context, because every closed form is exact if and only if the corresponding cohomology group is zero. Similar remarks hold for 2-forms and 3-forms.

Consider the special case $n = 2$; the complex of differentials is

$$0 \to \Omega^0(X) \to \Omega^1(X) \to \Omega^2(X) \to 0.$$

If ω is a 0-form, then ω is a C^∞-function $\alpha(x, y)$, and

$$d^0\omega = \frac{\partial\alpha}{\partial x} dx + \frac{\partial\alpha}{\partial y} dy.$$

If $\omega = P\, dx + Q\, dy$ is a 1-form, then

$$d^1\omega = \left(\frac{\partial Q}{\partial x} - \frac{\partial P}{\partial y}\right) dx \wedge dy.$$

The special case $n = 1$ is also of interest. If $\omega = \alpha(x)$ is a 0-form, then $d^0\omega = (\partial\alpha/\partial x)\, dx = (d\alpha/dx)\, dx$.

Each differential p-form ω on X has a unique expression

$$\omega = \sum \alpha_{i_1 \ldots i_p} dx_{i_1} \wedge \cdots \wedge dx_{i_p},$$

where $1 \le i_1 < i_2 < \cdots < i_p \le n$ and $\alpha_{i_1 \ldots i_p}$ is a C^∞-function on X. Differential forms are required for integration. A singular p-simplex $\sigma: \Delta^p \to X$ determines n coordinate functions σ_i (if $y \in \Delta^p$, then $\sigma(y) = (\sigma_1(y), \ldots, \sigma_n(y)) \in X \subset \mathbf{R}^n$). Given a p-form ω and a singular p-simplex σ, define

$$\int_\sigma \omega = \int_{\sigma(\Delta^p)} \sigma_\# \omega,$$

where $\sigma_{\#}\omega = \sum \alpha_{i_1 \ldots i_p} \sigma J\, dx_{i_1} \ldots dx_{i_p}$ if σ is differentiable and J is the Jacobian $\det(\partial\sigma_{i_j}/\partial x_{i_k})$, and $\sigma_{\#}\omega = 0$ otherwise (if $p \geq 2$, the right side is a multiple integral). More generally, if $c = \sum k_i \sigma_i$ is a p-chain ($k_i \in \mathbf{Z}$), that is, $c \in S_p(X)$, then

$$\int_c \omega = \sum k_i \int_{\sigma_i} \omega.$$

Thus every differential p-form ω on X defines, via integration, a real-valued function on $S_p(X)$. In fact, integration gives a homomorphism $\Omega^p(X) \to \operatorname{Hom}(S_p(X), \mathbf{R})$, namely, a p-form ω defines $c \mapsto \int_c \omega$ in $\operatorname{Hom}(S_p(X), \mathbf{R})$.

There is a generalized Stokes theorem (see [Bott and Tu, p. 31]).

Theorem. *If c is a $(p + 1)$-chain and ω is a differential p-form, then*

$$\int_c d\omega = \int_{\partial c} \omega.$$

The classical Stokes theorem is the special case $n = 3$ and $p = 2$; Green's theorem is the case $n = 2$ and $p = 1$; the fundamental theorem of calculus is the case $n = 1$ and $p = 0$.

Cohomology Groups

For a fixed abelian group G, recall that $\operatorname{Hom}(\ \ , G)$: $\mathbf{Ab} \to \mathbf{Ab}$ is a contravariant functor: if $\varphi: A \to B$ is a homomorphism, then $\varphi^{\#}: \operatorname{Hom}(B, G) \to \operatorname{Hom}(A, G)$ is defined by $\varphi^{\#}: f \mapsto f\varphi$ (we have modified our usual notation by using superscript $\#$ instead of $*$). Also, $\operatorname{Hom}(\ \ , G)$ is an additive functor, hence $\varphi^{\#}$ is the zero map whenever φ is (Exercise 9.12). Recall that differential forms suggest the functor $\operatorname{Hom}(\ \ , \mathbf{R})$, because, as we observed earlier, integration defines a homomorphism $\Omega^p(X) \to \operatorname{Hom}(S_p(X), \mathbf{R})$.

Lemma 12.1. *If $(S_*(X), \partial)$ is the singular complex of a space X, then, for every abelian group G,*

$$0 \longrightarrow \operatorname{Hom}(S_0(X), G) \xrightarrow{\partial_1^{\#}} \operatorname{Hom}(S_1(X), G) \xrightarrow{\partial_2^{\#}} \operatorname{Hom}(S_2(X), G) \longrightarrow \cdots$$

is a complex (denoted by $\operatorname{Hom}(S_*(X), G)$*).*

Proof. For every $n \geq 1$,

$$\partial_{n+1}^{\#} \partial_n^{\#} = (\partial_n \partial_{n+1})^{\#} = 0^{\#} = 0. \qquad \square$$

Of course, the lemma holds if one begins with simplicial chains or with cellular chains.

When F is a free abelian group with basis B, then the elements $\varphi \in \operatorname{Hom}(F, G)$ are easy to describe; they correspond to functions $B \to G$. Plainly, every homomorphism $\varphi: F \to G$ determines the function $\varphi|B$; conversely,

Theorem 4.1 shows that every such function determines a unique homomorphism. In particular, $\text{Hom}(S_p(X), G)$ corresponds to G-valued functions on the p-simplexes in X.

Some notational changes are needed because of the contravariance of $\text{Hom}(\ , G)$, for the differentiations in a complex (A_*, d) must lower degrees by 1; for all n, $d_n: A_n \to A_{n-1}$. Applying $\text{Hom}(\ , G)$ to $(S_*(X), \partial)$ gives

$$\cdots \leftarrow \text{Hom}(S_{n+1}(X), G) \xleftarrow{\partial^\#_{n+1}} \text{Hom}(S_n(X), G) \xleftarrow{\partial^\#_n} \text{Hom}(S_{n-1}(X), G) \leftarrow \cdots.$$

If we write $\text{Hom}(S_n(X), G) = A_{-n}$ and $\partial^\#_{n+1} = d_{-n}$, then our notation is consistent with the definition of complex:

$$\cdots \longrightarrow A_{-n+1} \xrightarrow{d_{-n+1}} A_{-n} \xrightarrow{d_{-n}} A_{-n-1} \longrightarrow \cdots,$$

and (A_*, d) is a complex all of whose nonzero terms have nonpositive degree. It is now clear how to define cycles, boundaries, and homology:

$$H_{-n}(\text{Hom}(S_*(X), G)) = H_{-n}(A_*) = \ker d_{-n}/\text{im}\, d_{-n+1} = \ker \partial^\#_{n+1}/\text{im}\, \partial^\#_n.$$

However, negative indices are inconvenient, and one eliminates signs by raising indices. Thus we set

$$A^n = A_{-n} \quad \text{and} \quad \delta^n = d_{-n};$$

that is,

$$A^n = \text{Hom}(S_n(X), G), \quad H^n(S_*(X), G) = \ker \partial^\#_{n+1}/\text{im}\, \partial^\#_n, \quad \text{and} \quad \delta^n = \partial^\#_{n+1}.$$

We repeat the definition of δ: if $f: S_n(X) \to G$ is a homomorphism, then

$$\delta^n(f) = f\partial_{n+1}.$$

Because the complex $(\text{Hom}(S_*(X), G), \delta)$ involves contravariant functors, all the usual terms acquire the prefix "co".

Definition. Let G be an abelian group and let X be a space. If $n \geq 0$, then the group of (singular) ***n*-cochains** in X with **coefficients** G is $\text{Hom}(S_n(X), G)$. The group of ***n*-cocycles** is $\ker \delta^n$ and is denoted by $Z^n(X; G)$; the group of ***n*-coboundaries** is $\text{im}\, \delta^{n-1}$ and is denoted by $B^n(X; G)$. The ***n*th cohomology group** of X with **coefficients** G is

$$H^n(X; G) = Z^n(X; G)/B^n(X; G) = \ker \delta^n/\text{im}\, \delta^{n-1} = \ker \partial^\#_{n+1}/\text{im}\, \partial_n{}^\#.$$

An element of $H^n(X; G)$ is a coset $\zeta + B^n(X; G)$, where ζ is an n-cocycle; it is called a **cohomology class** and it is denoted by $\text{cls}\, \zeta$.

Theorem 12.2. *For each fixed $n \geq 0$ and each abelian group G, cohomology is a contravariant functor*

$$H^n(\ ; G): \textbf{Top} \to \textbf{Ab}.$$

PROOF. We have already defined H^n on objects. If $f: X \to Y$ is continuous, then $f_\#: S_*(X) \to S_*(Y)$ is a chain map; that is, the following diagram commutes:

$$\begin{array}{ccccccc}
\cdots & \longrightarrow & S_n(X) & \xrightarrow{\partial} & S_{n-1}(X) & \longrightarrow & \cdots \\
 & & \Big\downarrow f_\# & & \Big\downarrow f_\# & & \\
\cdots & \longrightarrow & S_n(Y) & \xrightarrow{\partial'} & S_{n-1}(Y) & \longrightarrow & \cdots .
\end{array}$$

Applying the contravariant functor Hom(, G) gives the commutative diagram

$$\begin{array}{ccccccc}
\cdots & \longleftarrow & \mathrm{Hom}(S_n(X), G) & \xleftarrow{\delta} & \mathrm{Hom}(S_{n-1}(X), G) & \longleftarrow & \cdots \\
 & & \Big\uparrow f^\# & & \Big\uparrow f^\# & & \\
\cdots & \longleftarrow & \mathrm{Hom}(S_n(Y), G) & \xleftarrow{\delta'} & \mathrm{Hom}(S_{n-1}(Y), G) & \longleftarrow & \cdots,
\end{array}$$

where $f^\#: h \mapsto hf_\#$ for every $h: S_n(Y) \to G$. It is easy to see, as in Lemma 4.9, that $f^\#(Z^n(Y; G)) \subset Z^n(X; G)$ and $f^\#(B^n(Y; G)) \subset B^n(X; G)$. Hence $f^\#$ induces a homomorphism

$$f^*: H^n(Y; G) \to H^n(X; G)$$

by

$$\zeta + B^n(Y; G) \mapsto f^\#(\zeta) + B^n(X; G) = \zeta f_\# + B^n(X; G)$$

(where $\zeta: S_n(Y) \to G$ is a cocycle); that is, $\operatorname{cls} \zeta \mapsto \operatorname{cls}(\zeta f_\#)$.

It is routine to see that $(fg)^* = g^* f^*$ and $1^* = 1$. □

Theorem 12.3 (Dimension Axiom). *If X is a one-point space, then*

$$H^p(X; G) = \begin{cases} G & \text{if } p = 0 \\ 0 & \text{if } p > 0. \end{cases}$$

PROOF. We saw in Theorem 4.12 that every $S_n(X) \cong \mathbf{Z}$, that $\partial_n = 0$ when n is odd, and that ∂_n is an isomorphism when n is even and positive. It is now an easy exercise to prove that $H^p(X; G) = 0$ for all $p \geq 1$.

Let us compute $H^0(X; G)$. The end of the singular complex is

$$S_1(X) \xrightarrow{\partial_1} S_0(X) \xrightarrow{\partial_0} 0,$$

where $S_1(X) \cong \mathbf{Z} \cong S_0(X)$ and $\partial_1 = 0$. Applying Hom(, G) gives

$$0 \xrightarrow{\partial_0^\#} \mathrm{Hom}(S_0(X), G) \xrightarrow{\partial_1^\#} \mathrm{Hom}(S_1(X), G).$$

Therefore, since $\partial_0 = 0$,

$$H^0(X; G) = \ker \partial_1^\# / \mathrm{im}\, \partial_0^\# = \ker \partial_1^\# = \mathrm{Hom}(S_0(X), G).$$

But $\mathrm{Hom}(\mathbf{Z}, G) \cong G$ (Example 9.2), so that $H^0(X; G) \cong G$. □

Theorem 12.4 (Homotopy Axiom). *If $f, g: X \to Y$ are homotopic, then they induce the same homomorphisms $H^n(Y; G) \to H^n(X; G)$ for all $n \geq 0$.*

PROOF. In the proof of Theorem 4.23, the problem was normalized to show that $\lambda_0, \lambda_1: X \to X \times \mathbf{I}$ induce the same homomorphisms on homology groups (where $\lambda_0: x \mapsto (x, 0)$ and $\lambda_1: x \mapsto (x, 1)$). This last fact was established by constructing a chain homotopy $P = \{P_n: S_n(X) \to S_{n+1}(Y)\}$, that is,

$$\lambda_{1\#} - \lambda_{0\#} = \partial_{n+1} P_n + P_{n-1} \partial_n.$$

Applying the functor Hom(, G), however, shows that $P^\#$ is a chain homotopy for $\lambda_1^\#$ and $\lambda_0^\#$; the easy details are left to the reader. □

We must do a bit of algebra before we can define relative cohomology groups. The next property of the functor Hom(, G) is called **left exactness**.

Lemma 12.5. *Let G be an abelian group. If $A' \xrightarrow{i} A \xrightarrow{p} A'' \to 0$ is an exact sequence of abelian groups, then there is an exact sequence*

$$0 \longrightarrow \operatorname{Hom}(A'', G) \xrightarrow{p^\#} \operatorname{Hom}(A, G) \xrightarrow{i^\#} \operatorname{Hom}(A', G).$$

PROOF. $p^\#$ is injective. Assume that $f: A'' \to G$ satisfies $0 = p^\#(f) = fp$; thus f annihilates im p. Since p is surjective, $f = 0$.

im $p^\# \subset$ ker $i^\#$. If $f: A'' \to G$, then $i^\# p^\#(f) = fpi = 0$ because $pi = 0$.

ker $i^\# \subset$ im $p^\#$. Assume that $g: A \to G$ satisfies $0 = i^\#(g) = gi$. Define $\hat{g}: A'' \to G$ by $\hat{g}(a'') = g(a)$ if $p(a) = a''$. Now $\hat{g}$ is well defined, because if $p(a_1) = a''$, then $a - a_1 \in \ker p = \operatorname{im} i$, and so $a - a_1 = i(a')$ for some $a' \in A'$. Therefore $g(a - a_1) = gi(a') = 0$, and so $g(a) = g(a_1)$. But $p^\#(\hat{g}) = \hat{g}p = g$, since $\hat{g}p(a) = g(a)$ for all $a \in A$. □

Even if we assume that i is injective, it does not follow that $i^\#$ is surjective, that is, applying Hom to a short exact sequence yields another short exact sequence. If the short exact sequence is split, however, then Exercise 9.13 shows that it does remain (split) exact after applying Hom(, G).

EXAMPLE 12.1. Let $G = \mathbf{Z}$, and consider the exact sequence

$$0 \to \mathbf{Z} \overset{i}{\hookrightarrow} \mathbf{Q} \to \mathbf{Q}/\mathbf{Z} \to 0.$$

Now $i^\#: \operatorname{Hom}(\mathbf{Q}, \mathbf{Z}) \to \operatorname{Hom}(\mathbf{Z}, \mathbf{Z})$ cannot be surjective, because $\operatorname{Hom}(\mathbf{Q}, \mathbf{Z}) = 0$ and $\operatorname{Hom}(\mathbf{Z}, \mathbf{Z}) = \mathbf{Z} \neq 0$.

Corollary 12.6. *Let G be an abelian group.*

(i) $\operatorname{Hom}(\mathbf{Z}, G) \cong G$.
(ii) $\operatorname{Hom}(\mathbf{Z}/m\mathbf{Z}, G) \cong G[m] = \{x \in G: mx = 0\}$.
(iii) $\operatorname{Hom}(\mathbf{Z}/m\mathbf{Z}, \mathbf{Z}/n\mathbf{Z}) \cong \mathbf{Z}/d\mathbf{Z}$, *where $d = \gcd\{m, n\}$.*

PROOF. (i) In Example 9.2, we saw that $\varepsilon: f \mapsto f(1)$ is an isomorphism $\operatorname{Hom}(\mathbf{Z}, G) \xrightarrow{\sim} G$ (which is a constituent of a natural equivalence).

(ii) Apply Hom(, G) to the exact sequence

$$0 \to \mathbf{Z} \xrightarrow{m} \mathbf{Z} \xrightarrow{p} \mathbf{Z}/m\mathbf{Z} \to 0,$$

where the first map is multiplication by m, and the second map p is the natural map. There is an exact sequence

$$0 \longrightarrow \operatorname{Hom}(\mathbf{Z}/m\mathbf{Z}, G) \xrightarrow{p^\#} \operatorname{Hom}(\mathbf{Z}, G) \xrightarrow{m^\#} \operatorname{Hom}(\mathbf{Z}, G).$$

Hence $\operatorname{Hom}(\mathbf{Z}/m\mathbf{Z}, G) \cong \operatorname{im} p^\# = \ker m^\#$. But $m^\#$ is also multiplication by m, and so the result follows from the commutative diagram

$$\begin{array}{ccccc} \operatorname{Hom}(\mathbf{Z}/m\mathbf{Z}, G) & \xrightarrow{p^\#} & \operatorname{Hom}(\mathbf{Z}, G) & \xrightarrow{m^\#} & \operatorname{Hom}(\mathbf{Z}, G) \\ & \searrow^{\varepsilon p^\#} & \downarrow \varepsilon & & \downarrow \varepsilon \\ & & G & \xrightarrow{m} & G. \end{array}$$

(iii) This is a consequence of part (ii), because $(\mathbf{Z}/n\mathbf{Z})[m] \cong \mathbf{Z}/d\mathbf{Z}$. □

Corollary 12.7. *If A and B are known f.g. abelian groups, then* $\operatorname{Hom}(A, B)$ *is also known.*

Proof. The fundamental theorem of f.g. abelian groups says that such a group G is a direct sum

$$G = F \oplus C_1 \oplus \cdots \oplus C_k,$$

where F is free abelian of finite rank, C_i is cyclic of order m_i, and $m_1 | m_2 | \cdots | m_k$; moreover, these summands are uniquely determined to isomorphism. By Exercise 9.13, both functors Hom(, B) and Hom(A,) preserve finite direct sums, hence the determination of $\operatorname{Hom}(A, B)$ is reduced to the special case when both A and B are cyclic, namely, Corollary 12.6. □

Lemma 12.8. *Let G be an abelian group and let A be a subspace of a space X. For every $n \geq 0$, there is an exact sequence of abelian groups*

$$0 \to \operatorname{Hom}(S_n(X)/S_n(A), G) \to \operatorname{Hom}(S_n(X), G) \to \operatorname{Hom}(S_n(A), G) \to 0.$$

Hence there is a short exact sequence of complexes

$$0 \to \operatorname{Hom}(S_*(X)/S_*(A), G) \to \operatorname{Hom}(S_*(X), G) \to \operatorname{Hom}(S_*(A), G) \to 0.$$

Proof. By Exercise 5.13, $S_n(X)/S_n(A)$ is a free abelian group. Hence $0 \to S_n(A) \to S_n(X) \to S_n(X)/S_n(A) \to 0$ is a split short exact sequence (Corollary 9.2 and Exercise 9.10), and so Exercise 9.13(i) shows that the sequence remains exact after applying Hom(, G). Finally, Exercise 5.8 shows that the sequence of complexes is exact. □

Definition. If A is a subspace of X and if G is an abelian group, then the nth **relative cohomology group** with **coefficients** G is

$$H^n(X, A; G) = H_{-n}\,\mathrm{Hom}(S_*(X)/S_*(A), G)).$$

Recall that $\bar{\partial}_{n+1}: S_{n+1}(X)/S_{n+1}(A) \to S_n(X)/S_n(A)$ is defined by $c + S_{n+1}(A) \mapsto \partial_{n+1}c + S_n(A)$. Therefore

$$H^n(X, A; G) = \ker(\bar{\partial}_{n+1})^{\#}/\mathrm{im}(\bar{\partial}_n)^{\#}.$$

Since there is a short exact sequence of complexes

$$0 \to \mathrm{Hom}(S_*(X)/S_*(A), G) \to \mathrm{Hom}(S_*(X), G) \to \mathrm{Hom}(S_*(A), G) \to 0,$$

Lemma 5.5 applies at once to give a **connecting homomorphism**

$$d: H^n(A; G) \to H^{n+1}(X, A; G).$$

Theorem 12.9 (Long Exact Sequence). *If A is a subspace of X and if G is an abelian group, there is an exact sequence*

$$0 \to H^0(X, A; G) \to H^0(X; G) \to H^0(A; G) \xrightarrow{d} H^1(X, A; G) \to H^1(X; G) \to \cdots.$$

Moreover, the connecting homomorphisms are natural.

Proof. Theorems 5.6 and 5.7. □

Theorem 12.10 (Excision). *Let X_1 and X_2 be subspaces of X with $X = X_1^\circ \cup X_2^\circ$. Then the inclusion $j: (X_1, X_1 \cap X_2) \hookrightarrow (X, X_2)$ induces isomorphisms for all $n \geq 0$,*

$$j^*: H^n(X, X_2; G) \xrightarrow{\cong} H^n(X_1, X_1 \cap X_2; G).$$

Proof. The straightforward adaptation of the proof of Theorem 6.17 is left to the reader. □

It has now been shown that all the obvious analogues of the Eilenberg–Steenrod axioms hold for cohomology.

Exercises

*12.1. If G is an abelian group, then $\mathrm{Hom}(\sum A_\lambda, G) \cong \prod \mathrm{Hom}(A_\lambda, G)$, where the group on the right consists of all elements in the cartesian product under coordinatewise addition. (Hint: If the projection $\sum A_\lambda \to A_\lambda$ is denoted by p_λ and if $f: \sum A_\lambda \to G$, then $f \mapsto (p_\lambda f)$ is an isomorphism.)

12.2. (i) If $\{X_\lambda : \lambda \in \Lambda\}$ is the set of path components of X, prove that, for every $n \geq 0$,

$$H^n(X; G) \cong \prod_\lambda H^n(X_\lambda; G).$$

(*Hint*: Use Exercise 12.1.)

(ii) If X is a nonempty path connected space, then $H^0(X; G) \cong G$.

12.3. If X and Y have the same homotopy type, then

$$H^n(X; G) \cong H^n(Y; G) \text{ for all } n \geq 0.$$

12.4. State and prove the Mayer–Vietoris theorem for cohomology.

12.5. Compute $H^p(S^n; G)$.

If X is a simplicial complex, then the two algebraic modifications (cohomology and homology with coefficients) are defined as for the singular theory; merely replace the complex of singular chains $S_*(X)$ by the complex of simplicial chains. Similarly, if X is a CW complex, replace $S_*(X)$ by the complex of cellular chains.

Universal Coefficients Theorems for Cohomology

The homology groups of a space X are defined in two steps: a topological step that involves setting up a chain complex ($S_*(X)$ in the singular case) and an algebraic step that associates homology groups to a complex. Since cohomology has been defined by modifying the algebraic half of the construction, it should not be surprising that there is an algebraic way of relating homology groups and cohomology groups.

There are two universal coefficients theorems for cohomology: the first (Theorem 12.11) shows how $H^n(X; G)$ is determined by $H_*(X)$; the second (Theorem 12.15) shows how $H^n(X; G)$ is determined by $H^*(X) = H^*(X; \mathbf{Z})$.

Just as an investigation of $\ker(A' \otimes G \to A \otimes G)$ yields Tor, so does investigation of $\operatorname{coker}(\operatorname{Hom}(A, G) \to \operatorname{Hom}(A', G))$ yield Ext.

Definition. For each abelian group A, choose an exact sequence $0 \to R \xrightarrow{i} F \to A \to 0$ with F (and hence R) free abelian. For any abelian group G, define

$$\operatorname{Ext}(A, G) = \operatorname{coker} i^\# = \operatorname{Hom}(R, G)/i^\# \operatorname{Hom}(F, G).$$

Now Ext is actually a functor of two variables (having the same variances as Hom), and it is independent of the choice of presentation $0 \to R \to F \to A \to 0$. The "sophisticated" way we viewed Tor (in Chapter 9) can be adapted to give the definition of Ext on morphisms; in the discussion there, apply the functor $\operatorname{Hom}(\ \ , G)$ instead of the tensor product functor.

(The reader may want a less sophisticated description of $\operatorname{Ext}(A, G)$. Let $Z(A, G)$ be the abelian group of all functions $f: G \times G \to A$ (under pointwise addition) satisfying the following identities for all $x, y, z \in G$:

$$f(x, 0) = 0 = f(0, x);$$

$$f(y, z) - f(x + y, z) + f(x, y + z) - f(x, y) = 0;$$

$$f(x, y) = f(y, x).$$

Let $B(A, G)$ be the set of all functions $g\colon G \times G \to A$ of the form $g(x, y) = \alpha(y) - \alpha(x + y) + \alpha(x)$, where $\alpha\colon G \to A$ is a function with $\alpha(0) = 0$. Then $B(A, G)$ is a subgroup of $Z(A, G)$, and it can be shown that $\text{Ext}(A, G) = Z(A, G)/B(A, G)$.)

Before we give the basic properties of Ext, we introduce a class of groups.

Definition. An abelian group G is **divisible** if, for every $x \in G$ and every integer $n > 0$, there exists $y \in G$ with $ny = x$.

EXAMPLE 12.2. The following groups are divisible:

$$\mathbf{Q};\quad \mathbf{R};\quad \mathbf{C};\quad S^1;\quad \mathbf{Q}/\mathbf{Z};\quad \mathbf{R}/\mathbf{Z}.$$

Proofs of the following facts can be found in any book on homological algebra.

[Ext 1]. If $0 \to A' \to A \to A'' \to 0$ is a short exact sequence, then there is an exact sequence

$$0 \to \text{Hom}(A'', G) \to \text{Hom}(A, G) \to \text{Hom}(A', G) \to \text{Ext}(A'', G) \to \text{Ext}(A, G) \to \text{Ext}(A', G) \to 0.$$

[Ext 1′]. If $0 \to G' \to G \to G'' \to 0$ is a short exact sequence, then there is an exact sequence

$$0 \to \text{Hom}(A, G') \to \text{Hom}(A, G) \to \text{Hom}(A, G'') \to \text{Ext}(A, G') \to \text{Ext}(A, G) \to \text{Ext}(A, G'') \to 0.$$

[Ext 2]. If F is free abelian, then

$$\text{Ext}(F, G) = 0.$$

[Ext 2′]. If D is divisible, then

$$\text{Ext}(A, D) = 0.$$

If $\{A_j\colon j \in J\}$ is a family of abelian groups, then $\prod A_j$ is the abelian group whose elements are all J-tuples (a_j) under coordinatewise addition (thus, $\sum A_j$ is the subgroup of $\prod A_j$ consisting of all J-tuples with only finitely many nonzero coordinates). When the index set J is finite, $\sum A_j = \prod A_j$.

[Ext 3]. $\text{Ext}(\sum A_j, G) \cong \prod \text{Ext}(A_j, G)$.
[Ext 3′]. $\text{Ext}(A, \prod G_j) \cong \prod \text{Ext}(A, G_j)$.
[Ext 4]. $\text{Ext}(\mathbf{Z}/m\mathbf{Z}, G) \cong G/mG$.

Using these properties, one can compute $\text{Ext}(A, G)$ whenever A and G are f.g. abelian groups.

Remark. The analogue of [Tor 5] ($\text{Tor}(A, B) \cong \text{Tor}(B, A)$) is false for Ext; it is easy to see that $\text{Ext}(\mathbf{Z}/m\mathbf{Z}, \mathbf{Z}) \ncong \text{Ext}(\mathbf{Z}, \mathbf{Z}/m\mathbf{Z})$, for example.

Let A be an abelian group, and let $0 \to R \to F \to A \to 0$ be an exact sequence with F free abelian. Since $\text{Ext}(F, G) = 0$, by [Ext 2], the exact sequence given in [Ext 1] shows that $\text{Ext}(A, G) \cong \text{coker}(\text{Hom}(F, G) \to \text{Hom}(R, G))$; we have recaptured the definition of Ext.

There is a cohomology version of the universal coefficients theorem (Theorem 9.32) that shows that the homology groups of a space determine its cohomology groups; there is also a cohomology version of the Künneth formula (Theorem 9.37).

We begin by constructing the "obvious" homomorphism $\beta: H^n(\text{Hom}(S_*, G)) \to \text{Hom}(H_n(S_*), G)$, where (S_*, ∂) is a chain complex and G is an abelian group. Let φ be an n-cocycle, that is, $\varphi \in \text{Hom}(S_n, G)$ and $0 = \delta(\varphi) = \varphi\partial_{n+1}$. Now $\varphi: S_n \to G$ and $0 = \varphi(\text{im } \partial_{n+1}) = \varphi(B_n)$. Thus φ induces a homomorphism $S_n/B_n \to G$, and hence a homomorphism $\varphi': H_n(S_*) = Z_n/B_n \to G$, namely, $z_n + B_n \mapsto \varphi(z_n)$. Moreover, if φ is an n-coboundary, that is, $\varphi = \delta(\psi) = \psi\partial$, then φ induces the map $z_n + B_n \mapsto \varphi(z_n) = \psi\partial(z_n) = 0$ (because z_n is a cycle). Thus there is a natural map

$$\beta: H^n(\text{Hom}(S_*, G)) \to \text{Hom}(H_n(S_*), G))$$

defined by

$$\beta: \text{cls } \varphi \mapsto \varphi',$$

where $\varphi'(z_n + B_n) = \varphi(z_n)$.

Theorem 12.11 (Dual Universal Coefficients).

(i) *For every space X and every abelian group G, there are exact sequences for all $n \geq 0$:*

$$0 \to \text{Ext}(H_{n-1}(X), G) \to H^n(X; G) \xrightarrow{\beta} \text{Hom}(H_n(X), G) \to 0,$$

where β is the map defined above.

(ii) *This sequence splits; that is, there are isomorphisms for all $n \geq 0$,*

$$H^n(X; G) \cong \text{Hom}(H_n(X), G) \oplus \text{Ext}(H_{n-1}(X), G).$$

Proof. One proves a more general result: if (C_*, ∂) is a free chain complex, then

$$H^n(\text{Hom}(C_*, G)) \cong \text{Hom}(H_n(C_*), G) \oplus \text{Ext}(H_{n-1}(C_*), G);$$

the theorem follows by specializing C_* to $S_*(X)$.

The proof of Theorem 9.32 can be adapted here: every occurrence there of the (covariant) functor $\_\_ \otimes G$ should be replaced by the contravariant functor $\text{Hom}(\quad, G)$. The appearances of [Tor 1] and [Tor 2] are replaced by [Ext 1] and [Ext 2], respectively. □

Corollary 12.12. *If F is a field of characteristic zero (e.g., **Q**, **R**, or **C**), then, for all $n \geq 0$,*

$$H^n(X; F) \cong \text{Hom}(H_n(X), F).$$

PROOF. The additive group of a field F of characteristic zero is divisible. By [Ext 2′], $\operatorname{Ext}(H_{n-1}(X), F) = 0$. □

Remark. One often abbreviates the notation for **integral cohomology** $H^*(X; \mathbf{Z})$ to $H^*(X)$.

EXERCISES

*12.6. Let K be a finite simplicial complex, and let $C_*(K)$ be its simplicial chain complex.
(i) Show that $C_n(K)/B_n(K) \cong H_n(K) \oplus$ (free abelian group).
(*Hint*: The exact sequence $0 \to Z_n/B_n \to C_n/B_n \to C_n/Z_n \to 0$ splits because C_n/Z_n is isomorphic to the free abelian group B_{n-1}.)
(ii) Show that $\operatorname{Ext}(C_n(K)/B_n(K), \mathbf{Z}) \cong \operatorname{Ext}(H_n(K), \mathbf{Z})$.
(iii) Consider the diagram

$$\begin{array}{ccc} C_n(K) & \xrightarrow{\partial_n} & C_{n-1}(K) \\ & {\scriptstyle p_n} \searrow \quad \nearrow {\scriptstyle i_{n-1}} & \\ & B_{n-1}(K), & \end{array}$$

where p_n differs from ∂_n only in its target, and where i_{n-1} is inclusion. If $H_{n-1}(K)$ is free abelian, prove that

$$B^n(K, \mathbf{Z}) = \operatorname{im} p_n^\#.$$

(*Hint*: $B^n = \operatorname{im} \partial_n^\# = \operatorname{im} p_n^\# i_{n-1}^\#$; but $i_{n-1}^\#$ is surjective because its cokernel is isomorphic to $\operatorname{Ext}(H_{n-1}(K), \mathbf{Z})$, which is zero here.)

12.7. If A is an abelian group, then $\operatorname{Hom}(A, \mathbf{Z}) \neq 0$ if and only if A has an infinite cyclic direct summand. (*Hint*: Corollary 9.2.)

12.8. Show that $H^p(S^n; G) \cong H_p(S^n; G)$ for all $p \geq 0$ and all $n \geq 0$.

12.9. (i) Prove that the direct sum and the direct product of (possibly infinitely many) divisible groups is divisible.
(ii) Prove that a quotient group of a divisible group is divisible.

12.10. Define the **character group** of an abelian group G, denoted by $G^\perp$, by

$$G^\perp = \operatorname{Hom}(G, \mathbf{R}/\mathbf{Z}).$$

Prove that $H^n(X; \mathbf{R}/\mathbf{Z}) \cong (H_n(X))^\perp$.

12.11. If X and Y are finite CW complexes, find $H^n(X \times Y)$. (*Hint*: Use the Künneth formula (Theorem 9.37) and the adjoint isomorphism (Exercise 11.15).)

*12.12. (i) Prove that, when n is even,

$$H^p(\mathbf{R}P^n) = \begin{cases} \mathbf{Z} & \text{if } p = 0 \\ \mathbf{Z}/2\mathbf{Z} & \text{if } p \text{ is even and } 2 \leq p \leq n \\ 0 & \text{otherwise.} \end{cases}$$

If n is odd, show that $H^p(\mathbf{R}P^n)$ is as above except for $p = n$, when $H^n(\mathbf{R}P^n) = \mathbf{Z}$. (*Hint*: Use Theorem 8.47.)

(ii) Prove that, for all $n \geq 1$,

$$H^p(\mathbf{R}P^n; \mathbf{Z}/2\mathbf{Z}) = \begin{cases} \mathbf{Z}/2\mathbf{Z} & \text{if } 0 \leq p \leq n \\ 0 & \text{otherwise.} \end{cases}$$

(This result does not depend on the parity of n.)

Theorem 12.11, the dual universal coefficients theorem, can be extended to an algebraic Künneth theorem (if S_* and T_* are chain complexes, there is a standard way of constructing a chain complex $\mathrm{Hom}(S_*, T_*)$), but this is not so interesting for us because the Eilenberg–Zilber theorem has no analogue. Instead we present a purely cohomological universal coefficients theorem (i.e., there is no mixture of homology and cohomology, as in Theorem 12.11) and a Künneth formula based on it.

Definition. A chain complex C_* is of **finite type** if each of its terms C_n is f.g. A space X is of **finite type** if each of its homology groups $H_n(X)$ is f.g.

Every compact polyhedron, more generally, every compact CW complex, is a space of finite type; $\mathbf{R}P^\infty$ is a space of finite type that is not compact.

Lemma 12.13. *If X is a space of finite type, then there exists a free chain complex C_* of finite type such that C_* is chain equivalent to $S_*(X)$.*

Proof. Let $\nu_n: Z_n(X) \to H_n(X)$ be the natural map. Since $H_n(X)$ is f.g., there is a f.g. subgroup of $Z_n(X)$, say F_n, necessarily free abelian, with $\nu_n | F_n: F_n \to H_n(X)$ surjective; let F_n' denote $\ker(\nu_n | F_n)$. Define

$$C_n = F_n \oplus F_{n-1}',$$

and define $d_n: C_n \to C_{n-1}$ by

$$d_n(\alpha, \alpha') = (\alpha', 0)$$

for $\alpha \in F_n$ and $\alpha' \in F_{n-1}'$. For each n, C_n is a free abelian group of finite rank; moreover,

$$H_n(C_*) = \ker d_n / \mathrm{im}\, d_{n+1} = F_n / F_{n-1}' = H_n(X).$$

Let us construct a chain map $f: C_* \to S_*(X)$. Since F_n' is free abelian, Theorem 9.1 provides a homomorphism $h_n: F_n' \to S_{n+1}(X)$ with $\partial_{n+1} h_n(\alpha') = \alpha'$ for all $\alpha' \in F_n'$. Define $f_n: C_n \to S_n(X)$ by

$$f_n(\alpha, \alpha') = \alpha + h_{n-1}(\alpha'),$$

where $\alpha \in F_n$ and $\alpha' \in F_n'$. Now f_n is a chain map:

$$\partial f(\alpha, \alpha') = \partial(\alpha + h_{n-1}(\alpha')) = \partial\alpha + \partial h_{n-1}(\alpha') = \alpha',$$

because $\alpha \in F_n \subset Z_n(X)$ and the definition of h_{n-1}. On the other hand,

$$fd(\alpha, \alpha') = f(\alpha', 0) = \alpha'.$$

It follows from Theorem 9.8 that f is a chain equivalence. □

Lemma 12.14. *If C_* is a free chain complex of finite type, then there is an isomorphism of chain complexes*

$$\text{Hom}(C_*, \mathbf{Z}) \otimes G \xrightarrow{\simeq} \text{Hom}(C_*, G)$$

for every abelian group G.

PROOF. For each n, define

$$\mu_n: \text{Hom}(C_n, \mathbf{Z}) \otimes G \to \text{Hom}(C_n, G)$$

by

$$\mu_n(f \otimes g): c \mapsto f(c)g$$

(note that $f(c) \in \mathbf{Z}$, so that the right side makes sense). It is clear that μ is a chain map. One proves that μ_n is an isomorphism by induction on rank C_n. If this rank is 1, then $C_n \cong \mathbf{Z}$, and the result follows from the identities $\mathbf{Z} \otimes G = G$ and $\text{Hom}(\mathbf{Z}, G) = G$. The inductive step follows from the identities $\text{Hom}(A \oplus B, G) = \text{Hom}(A, G) \oplus \text{Hom}(B, G)$ and $(A \oplus B) \otimes G = (A \otimes G) \oplus (B \otimes G)$. □

Before we proceed, note that if S_* and C_* are chain equivalent chain complexes, then for every abelian group G, the chain complexes $S_* \otimes G$ and $C_* \otimes G$ are chain equivalent, as are the chain complexes $\text{Hom}(S_*, G)$ and $\text{Hom}(C_*, G)$.

Theorem 12.15 (Universal Coefficients Theorems for Cohomology).

(i) *If X is a space of finite type and if G is an abelian group, then there is an exact sequence for every $n \geq 0$:*

$$0 \to H^n(X) \otimes G \xrightarrow{\alpha} H^n(X; G) \to \text{Tor}(H^{n+1}(X), G) \to 0,$$

where

$$\alpha: (\text{cls } z) \otimes g \mapsto \text{cls } zg,$$

where $zg: \sigma \mapsto z(\sigma)g$ for an n-simplex σ in X (recall that $z(\sigma) \in \mathbf{Z}$).

(ii) *This sequence splits; that is,*

$$H^n(X; G) \cong H^n(X) \otimes G \oplus \text{Tor}(H^{n+1}(X), G).$$

PROOF. Since X has finite type, Lemma 12.13 provides a free chain complex C_* of finite type with $H_*(C_*) = H_*(X)$. If $A^* = \text{Hom}(C_*, \mathbf{Z})$, then Theorem 9.32, the universal coefficients theorem for homology, applies because A^* is a free chain complex. (The device of raising indices and changing their sign converts the nonpositive chain complex $\text{Hom}(C_*, \mathbf{Z})$ into a nonnegative one, A^*; indices on homology groups are similarly changed, giving cohomology groups.) There is thus a split short exact sequence

$$0 \to H^n(A^*) \otimes G \xrightarrow{\alpha} H^n(A^* \otimes G) \to \operatorname{Tor}(H^{n+1}(A^*), G) \to 0.$$

Now $H^n(A^*) = H^n(\operatorname{Hom}(C_*, \mathbf{Z})) = H^n(\operatorname{Hom}(S_*(X), \mathbf{Z})) = H^n(X)$. Moreover, by Lemma 12.14,

$$A^* \otimes G = \operatorname{Hom}(C_*, \mathbf{Z}) \otimes G \cong \operatorname{Hom}(C_*, G) \cong \operatorname{Hom}(S_*(X), G),$$

so that $H^n(A^* \otimes G) \cong H^n(X; G)$, by Theorem 5.3. □

Recall from Exercise 9.28 that $S_i(X) \otimes S_j(Y)$ is free abelian with basis all symbols $\sigma_i \otimes \tau_j$, where σ_i is an i-simplex in X and τ_j is a j-simplex in Y.

Definition. Let R be a commutative ring. If $\varphi \in \operatorname{Hom}(S_m(X), R)$ and $\theta \in \operatorname{Hom}(S_n(Y), R)$, then define $\varphi \otimes \theta \in \operatorname{Hom}(S_*(X) \otimes S_*(Y), R)$ by

$$(\varphi \otimes \theta)(\sigma_i \otimes \tau_j) = \begin{cases} \varphi(\sigma_m)\theta(\tau_n) & \text{if } i = m \text{ and } j = n \\ 0 & \text{otherwise,} \end{cases}$$

where the right side is the product of two elements in the ring R.

Theorem 12.16 (Künneth Formula for Cohomology). *If X and Y are spaces of finite type, then there is a split short exact sequence*

$$0 \to \sum_{i+j=n} H^i(X) \otimes H^j(Y) \xrightarrow{\alpha'} H^n(X \times Y) \to \sum_{p+q=n+1} \operatorname{Tor}(H^p(X), H^q(Y)) \to 0,$$

where α': $\operatorname{cls} \varphi_i \otimes \operatorname{cls} \theta_j \mapsto \operatorname{cls} \zeta^\#(\varphi_i \otimes \theta_j)$ (ζ is an Eilenberg–Zilber chain equivalence $S_*(X \times Y) \to S_*(X) \otimes S_*(Y)$).

Proof. Since X and Y have finite type, Lemma 12.13 gives chain complexes C_* and E_* of finite type chain equivalent to $S_*(X)$ and $S_*(Y)$, respectively. We let the reader prove that there is a commutative diagram

$$\begin{array}{ccc} H^i(X) \otimes H^j(Y) & \xrightarrow{\alpha'} & H^{i+j}(X \times Y) \\ \downarrow & & \downarrow \\ H^i(C_*) \otimes H^j(E_*) & \xrightarrow{\alpha'} & H^{i+j}(\operatorname{Hom}(C_* \otimes E_*, \mathbf{Z})) \end{array}$$

with vertical map isomorphisms (note that $C_* \otimes E_*$ is chain equivalent to $S_*(X) \otimes S_*(Y)$); it follows that we may work with the bottom row. However, Theorem 9.36 applies at once, because both $\operatorname{Hom}(C_*, \mathbf{Z})$ and $\operatorname{Hom}(E_*, \mathbf{Z})$ are free chain complexes (because C_* and E_* are of finite type). □

Remark. If R is a commutative ring and A and B are R-modules, then there is a **tensor product over R**, denoted by $A \otimes_R B$; it is defined as the quotient of $A \otimes B$ by all relations of the form

$$(ra, b) = (a, rb) \quad \text{for all } r \in R, a \in A, b \in B.$$

The abelian group $A \otimes_R B$ is an R-module; in particular, $A \otimes_R B$ is a vector space over R when R is a field. There is also a version of $\text{Tor}(A, B)$, now denoted by $\text{Tor}_1^R(A, B)$, defined as $\ker i \otimes 1_B$, where $0 \to C \xrightarrow{i} F \to A \to 0$ is an exact sequence of R-modules and F is a free R-module.

These constructions arise in cohomology as follows. If R is a field, then each $\text{Hom}(S_p(X), R)$ is a vector space over R and the differentiations are R-linear transformations. It follows that cocycles and coboundaries are vector spaces and hence that each $H^p(X; R)$ is an R-vector space. It is natural to take this fact into account. For example, if R is any field and V and W are finite-dimensional R-vector spaces of dimensions m and n, respectively, then one can prove that $\dim(V \otimes_R W)$ is mn; on the other hand, $V \otimes W$ (no subscript R) is an infinite-dimensional R-vector space when R is the field of real numbers.

There is an analogue of Theorem 12.16 for any principal ideal domain R: if X and Y are spaces of finite type, then there is a split short exact sequence

$$0 \to \sum_{i+j=n} H^i(X; R) \otimes_R H^j(Y; R) \xrightarrow{\alpha'} H^n(X \times Y; R) \to$$

$$\to \sum_{p+q=n+1} \text{Tor}_1^R(H^p(X; R), H^q(Y; R)) \to 0.$$

If R is a field (of any characteristic), it is known that $\text{Tor}_1^R(V, W) = 0$ for any pair of R-vector spaces V and W; in this case, therefore, the homomorphism α' is an isomorphism.

Cohomology Rings

The direct sum of all the cohomology groups of a space X with coefficients in a commutative ring can be equipped with a functorial ring structure (this is not so for homology groups). Here are some algebraic preliminaries.

Definition. A ring R is a **graded ring** if there are additive subgroups R^n, $n \geq 0$, such that:

(i) $R = \sum_{n \geq 0} R^n$ (direct sum of additive groups);
(ii) $R^n R^m \subset R^{n+m}$ for all $n, m \geq 0$, that is, if $x \in R^n$ and $y \in R^m$, then $xy \in R^{n+m}$.

EXAMPLE 12.3. If A is a commutative ring, then the polynomial ring $R = A[x]$ is a graded ring if one sets $R^n = \{ax^n : a \in A\}$.

EXAMPLE 12.4. If $R = A[x_1, \ldots, x_p]$ is the polynomial ring in several variables, then R is a graded ring if one sets

$$R^n = \{\textstyle\sum a x_1^{e_1} \cdots x_p^{e_p} : a \in A \text{ and } \sum e_i = n\}.$$

Thus R^n is generated by all monomials of total degree n.

EXAMPLE 12.5. If M is an A-module, then $\sum_{p \geq 0} \bigwedge^p M$ is a graded ring if one defines

$$(m_1 \wedge \cdots \wedge m_p) \cdot (m_1' \wedge \cdots \wedge m_q') = m_1 \wedge \cdots \wedge m_p \wedge m_1' \wedge \cdots \wedge m_q'.$$

$\sum \bigwedge^p M$ is denoted by $\bigwedge M$ and is called the **exterior algebra** on M.

Definition. An element x in a graded ring $R = \sum R^n$ has **degree** n if $x \in R^n$; such elements are called **homogeneous**. A (two-sided) ideal I or a subring S is called **homogeneous** if it is generated by homogeneous elements.

It is easy to see that an ideal I (or a subring S) of a graded ring $R = \sum R^n$ is homogeneous if and only if $I = \sum (I \cap R^n)$ $(S = \sum (S \cap R^n))$.

Warning! The definition of degree in a graded ring differs from the usual definition in a polynomial ring. Both versions agree for monomials, but no other polynomial is provided with a degree in the sense of graded rings. Note also that the zero element has degree n for every $n \geq 0$.

The element 1 in $R = \sum R^n$ must be homogeneous of degree 0. If we assume $1 = e_0 + \cdots + e_k$, where $e_i \in R^i$, and if $a_n \in R^n$, then

$$a_n = e_0 a_n + \cdots + e_k a_n \in R^n \cap (R^n \oplus \cdots \oplus R^{n+k}) = R^n;$$

hence $e_i a_n = 0$ for all $i \geq 1$, and $a_n = e_0 a_n$; it follows that $a = e_0 a$ for all $a \in R$. A similar argument shows that $a = a e_0$ for every $a \in R$, so that e_0 is a two-sided identity in R. But two-sided identities in a ring are unique, hence $1 = e_0 \in R^0$.

Lemma 12.17. *If I is a homogeneous ideal in a graded ring $R = \sum R^n$, then R/I is a graded ring; indeed*

$$R/I = \sum (R^n + I)/I.$$

PROOF. Since I is homogeneous, $I = \sum (I \cap R^n)$. As abelian groups, $R/I = \sum R^n / \sum (I \cap R^n) \cong \sum (R^n / I \cap R^n) \cong \sum (R^n + I)/I$. Also, $(R^n + I)/I \cdot (R^m + I)/I \subset (R^n R^m + I)/I$ (because I is an ideal), and $(R^n R^m + I)/I \subset (R^{n+m} + I)/I$. □

Every (commutative) ring R is an abelian group under its addition, so that $H^n(X; R)$ makes sense. We are going to make $H^*(X; R) = \sum H^n(X; R)$ into a graded ring by equipping it with a multiplication, called *cup product*. The following technical lemma will be used in verifying elementary properties of this multiplication.

Definition. If $0 \leq i \leq d$, define (affine) maps $\lambda_i, \mu_i \colon \Delta^i \to \Delta^d$ by

$$\lambda_i \colon (t_0, \ldots, t_i) \mapsto (t_0, \ldots, t_i, 0, \ldots, 0)$$

and

$$\mu_i \colon (t_0, \ldots, t_i) \mapsto (0, \ldots, 0, t_0, \ldots, t_i).$$

One calls λ_i a **front face** and μ_i a **back face**.

A more complete notation for these maps, indicating their target, is λ_i^d and μ_i^d. Note that λ_d^d and μ_d^d are both identities, while λ_0^d has image $(1, 0, \ldots, 0) = e_0$ and μ_0^d has image $(0, \ldots, 0, 1) = e_d$; remember that $\Delta^d = [e_0, \ldots, e_d]$.

Recall the face maps $\varepsilon_i^{d+1}\colon \Delta^d \to \Delta^{d+1}$, where $0 \le i \le d+1$:

$$\varepsilon_0(t_0, \ldots, t_d) = (0, t_0, \ldots, t_d) = \mu_d^{d+1}(t_0, \ldots, t_d);$$

if $1 \le i \le d+1$, then

$$\varepsilon_i(t_0, \ldots, t_d) = (t_0, \ldots, t_{i-1}, 0, t_i, \ldots, t_d).$$

Lemma 12.18.

(i) *If* $\varepsilon_i^{d+1}\colon \Delta^d \to \Delta^{d+1}$ *is the* ith *face map, then*

$$\mu_d^{d+1} = \varepsilon_0^{d+1} \quad \textit{and} \quad \lambda_d^{d+1} = \varepsilon_{d+1}^{d+1}.$$

(ii) $\mu_{m+k}^d \mu_k^{m+k} = \mu_k^d$; $\lambda_{n+m}^d \lambda_n^{n+m} = \lambda_n^d$; $\mu_{m+k}^{n+m+k} \lambda_m^{m+k} = \lambda_{n+m}^{n+m+k} \mu_m^{n+m}$.

(iii)
$$\varepsilon_i^{d+1} \lambda_p^d = \begin{cases} \lambda_{p+1}^{d+1} \varepsilon_i^{p+1} & \textit{if } i \le p \\ \lambda_p^{d+1} & \textit{if } i \ge p+1; \end{cases}$$

$$\varepsilon_i^{d+1} \mu_q^d = \begin{cases} \mu_q^{d+1} & \textit{if } i \le d-q \\ \mu_{q+1}^{d+1} \varepsilon_{i+q-d}^{q+1} & \textit{if } i \ge d-q+1. \end{cases}$$

PROOF. Routine. Note, in the last identity in (iii), that the case $i = d - q + 1$ gives $\varepsilon_{d-q+1}\mu_q = \mu_q = \mu_{q+1}\varepsilon_0$. □

Notation. Given a space X and an abelian group G, write

$$S^n(X, G) = \operatorname{Hom}(S_n(X), G)$$

and

$$S^*(X, G) = \sum_{n \ge 0} S^n(X, G).$$

Notation. If $\varphi \in S^n(X, G)$ and $c \in S_n(X)$, write

$$(c, \varphi) = \varphi(c) \in G.$$

There are two important special cases. If $c' \in S_{n+1}(X)$, then

$$(c', \delta(\varphi)) = (\partial c', \varphi);$$

if $f\colon X \to Y$ is continuous and $\varphi \in S^n(Y, R)$, then

$$(c, f^{\#}(\varphi)) = (f_{\#}c, \varphi).$$

In particular, if c is an n-simplex σ, then

$$(\sigma, f^{\#}(\varphi)) = (f\sigma, \varphi).$$

Since $S_n(X)$ has a basis comprised of n-simplexes, $\varphi \in S^n(X, G)$ is determined by all (σ, φ) as σ ranges over the continuous maps $\Delta^n \to X$.

Definition. Let X be a space, and let R be a commutative ring. If $\varphi \in S^n(X, R)$ and $\theta \in S^m(X, R)$, define their **cup product** $\varphi \cup \theta \in S^{n+m}(X, R)$ by

$$(\sigma, \varphi \cup \theta) = (\sigma\lambda_n, \varphi)(\sigma\mu_m, \theta)$$

for every $(n + m)$-simplex σ in X, where the right side is the product of two elements in the ring R.

Of course, cup product defines a function

$$S^*(X, R) \times S^*(X, R) \to S^*(X, R)$$

by defining

$$(\sum \varphi_i) \cup (\sum \theta_j) = \sum_{i,j} \varphi_i \cup \theta_j,$$

where $\varphi_i \in S^i(X, R)$ and $\theta_j \in S^j(X, R)$.

Lemma 12.19. *If X is a space and R is a commutative ring, then $S^*(X, R) = \sum S^n(X, R)$ is a graded ring under cup product.*

PROOF. To prove left distributivity, it suffices to show that $\varphi \cup (\theta + \psi) = (\varphi \cup \theta) + (\varphi \cup \psi)$ when $\varphi \in S^n(X, R)$ and $\theta, \psi \in S^m(X, R)$. But if σ is an $(n + m)$-simplex,

$$\begin{aligned}(\sigma, \varphi \cup (\theta + \psi)) &= (\sigma\lambda_n, \varphi)(\sigma\mu_m, \theta + \psi)\\ &= (\sigma\lambda_n, \varphi)[(\sigma\mu_m, \theta) + (\sigma\mu_m, \psi)]\\ &= (\sigma, \varphi \cup \theta) + (\sigma, \varphi \cup \psi).\end{aligned}$$

A similar calculation proves right distributivity.

To prove associativity, let $\varphi \in S^n(X, R)$, $\theta \in S^m(X, R)$, and $\psi \in S^k(X, R)$. If σ is an $(n + m + k)$-simplex, then

$$(\sigma, \varphi \cup (\theta \cup \psi)) = (\sigma\lambda_n, \varphi)(\sigma\mu_{m+k}\lambda_m, \theta)(\sigma\mu_{m+k}\mu_k, \psi)$$

and

$$(\sigma, (\varphi \cup \theta) \cup \psi) = (\sigma\lambda_{n+m}\lambda_n, \varphi)(\sigma\lambda_{n+m}\mu_m, \theta)(\sigma\mu_k, \psi).$$

These two products are equal, by Lemma 12.18(ii).

Define $e \in S^0(X, R)$ by

$$(x, e) = 1$$

for all $x \in X$ (recall that 0-simplexes in X are identified with the points of X). It is easy to see that e is a (two-sided) identity in $S^*(X, R)$, hence $S^*(X, R)$ is a ring. It follows at once from the definition of cup product that $S^*(X, R)$ is a graded ring. □

The distributive laws give bilinearity of cup product $S^*(X, R) \times S^*(X, R) \to S^*(X, R)$; one may, therefore, regard cup product as a map

$$\cup: S^*(X, R) \otimes S^*(X, R) \to S^*(X, R).$$

Lemma 12.20. *If $f\colon X \to X'$ is a continuous map, then*

$$f^{\#}(\varphi \cup \theta) = f^{\#}(\varphi) \cup f^{\#}(\theta).$$

Moreover, if $e \in S^0(X, R)$ is the unit $((x, e) = 1$ for all $x \in X)$, and if $e' \in S^0(X', R)$ is defined by $(x', e') = 1$ for all $x' \in X'$, then

$$f^{\#}(e') = e.$$

Proof. It suffices to assume that $\varphi \in S^p(X', R)$ and $\theta \in S^q(X', R)$. Now if σ is a $(p+q)$-simplex in X,

$$\begin{aligned}(\sigma, f^{\#}(\varphi \cup \theta)) &= (f\sigma, \varphi \cup \theta)\\ &= (f\sigma\lambda_p, \varphi)(f\sigma\mu_q, \theta)\\ &= (\sigma\lambda_p, f^{\#}\varphi)(\sigma\mu_q, f^{\#}\theta) = (\sigma, f^{\#}\varphi \cup f^{\#}\theta).\end{aligned}$$

If $x \in X$, then $(x, f^{\#}(e')) = (f(x), e') = 1$. □

Corollary 12.21. *For a given commutative ring R, $S^*(\ \ , R)$ is a contravariant functor from* **Top** *to* **Graded Rings**.

Proof. Immediate from Lemmas 12.19 and 12.20. □

The ring $S^*(X, R)$ has several disadvantages: its enormous size makes it almost impossible to compute; it does not satisfy the homotopy axiom; and it is "very" noncommutative. We shall now see that the ring structure on $S^*(X, R)$ is inherited by $\sum_{n \geq 0} H^n(X; R)$ and that these defects of $S^*(X, R)$ disappear in passing to cohomology.

Lemma 12.22. *If $\varphi \in S^p(X, R)$ and $\theta \in S^q(X, R)$, then*

$$\delta(\varphi \cup \theta) = \delta\varphi \cup \theta + (-1)^p \varphi \cup \delta\theta.$$

Proof. Note that both sides have degree $d = p + q + 1$. If σ is a d-simplex, then

$$\begin{aligned}&(\sigma, \delta\varphi \cup \theta + (-1)^p \varphi \cup \delta\theta)\\ &= (\sigma\lambda_{p+1}, \delta\varphi)(\sigma\mu_q, \theta) + (-1)^p(\sigma\lambda_p, \varphi)(\sigma\mu_{q+1}, \delta\theta)\\ &= (\partial(\sigma\lambda_{p+1}), \varphi)(\sigma\mu_q, \theta) + (-1)^p(\sigma\lambda_p, \varphi)(\partial(\sigma\mu_{q+1}), \theta)\\ &= \sum_{i=0}^{p+1} (-1)^i(\sigma\lambda_{p+1}\varepsilon_i, \varphi)(\sigma\mu_q, \theta) + \sum_{j=0}^{q+1} (-1)^{j+p}(\sigma\lambda_p, \varphi)(\sigma\mu_{q+1}\varepsilon_j, \theta).\end{aligned}$$

By Lemma 12.18(i), $\sigma\lambda_{p+1}\varepsilon_{p+1} = \sigma\lambda_{p+1}\lambda_p = \sigma\lambda_p$ and $\sigma\mu_{q+1}\varepsilon_0 = \sigma\mu_{q+1}\mu_q = \sigma\mu_q$. It follows that term $p+1$ of the first sum cancels term 0 of the second sum, and so the two sums equal

$$\sum_{i=0}^{p}(-1)^i(\sigma\lambda_{p+1}\varepsilon_i,\varphi)(\sigma\mu_q,\theta)+\sum_{j=1}^{q+1}(-1)^{j+p}(\sigma\lambda_p,\varphi)(\sigma\mu_{q+1}\varepsilon_j,\theta).$$

On the other hand,

$$\begin{aligned}(\sigma,\delta(\varphi\cup\theta))&=(\partial\sigma,\varphi\cup\theta)\\&=\sum_{i=0}^{d}(-1)^i(\sigma\varepsilon_i,\varphi\cup\theta)\\&=\sum_{i=0}^{d}(-1)^i(\sigma\varepsilon_i\lambda_p,\varphi)(\sigma\varepsilon_i\mu_q,\theta)\\&=\sum_{i=0}^{p}(-1)^i(\sigma\varepsilon_i\lambda_p,\varphi)(\sigma\varepsilon_i\mu_q,\theta)+\sum_{i=p+1}^{d}(-1)^i(\sigma\varepsilon_i\lambda_p,\varphi)(\sigma\varepsilon_i\mu_q,\theta).\end{aligned}$$

Since $d-q=p+1$, Lemma 12.18(iii) shows that this equals

$$\sum_{i=0}^{p}(-1)^i(\sigma\lambda_{p+1}\varepsilon_i,\varphi)(\sigma\mu_q,\theta)+\sum_{i=p+1}^{d}(-1)^i(\sigma\lambda_p,\varphi)(\sigma\mu_{q+1}\varepsilon_{i-p},\theta).$$

But the index of summation in the second sum can be changed to $j=i-p$, giving $\sum_{j=1}^{q+1}(-1)^{j+p}(\sigma\lambda_p,\varphi)(\sigma\mu_{q+1}\varepsilon_j,\theta)$, as desired. □

Theorem 12.23. *For any commutative ring R, $H^*(\ ;R)=\sum_{p\geq 0}H^p(\ ;R)$ is a contravariant functor* **hTop → Graded Rings**.

PROOF. Let $Z^*(X,R)=\sum Z^p(X,R)$ and $B^*(X,R)=\sum B^p(X,R)$. If $\varphi\in Z^p$ and $\theta\in Z^q$, then $\delta\varphi=0=\delta\theta$, and

$$\delta(\varphi\cup\theta)=\delta\varphi\cup\theta+(-1)^p\varphi\cup\delta\theta=0;$$

hence $\varphi\cup\theta$ is a cocycle. It follows that Z^* is a (homogeneous) subring of $S^*(X,R)$.

If $\varphi\in Z^p$ and $\theta\in B^q$, then $\delta\varphi=0$ and $\theta=\delta\psi$ for some $\psi\in S^{q-1}(X,R)$. Hence

$$\begin{aligned}\varphi\cup\theta=\varphi\cup\delta\psi&=\pm(\delta(\varphi\cup\psi)-\delta\varphi\cup\psi)\\&=\pm\delta(\varphi\cup\psi),\end{aligned}$$

so that $\varphi\cup\theta$ is a coboundary; similarly, $\theta\cup\varphi$ is a coboundary. It follows that B^* is a two-sided homogeneous ideal in Z^*. By Lemma 12.17, $H^*(X;R)=Z^*/B^*$ is a graded ring. (Of course, multiplication in H^* is given by

$$\text{cls}\,\varphi\cup\text{cls}\,\theta=\text{cls}(\varphi\cup\theta).)$$

That a continuous map $f\colon X\to Y$ yields a ring homomorphism $f^*\colon H^*(Y;R)\to H^*(X;R)$, namely, $f^*\,\text{cls}\,\varphi=\text{cls}\,f^{\#}\varphi$, follows easily from Lemma 12.20. Indeed the homotopy axiom for cohomology, Theorem 12.4, shows that this ring map is independent of the choice of continuous map homotopic to f.

The reader may now easily show that H^* is a (contravariant) functor. □

Definition. The multiplication $H^*(X; R) \otimes H^*(X; R) \to H^*(X; R)$ is also called **cup product**,[1] and one defines

$$\text{cls } \varphi \cup \text{cls } \theta = \text{cls}(\varphi \cup \theta).$$

Definition. If X is a space and R is a commutative ring, then the **cohomology ring** with **coefficients** R is

$$H^*(X; R) = \sum_{p \geq 0} H^p(X; R).$$

The following discussion will show that cup product is essentially the only multiplication on $H^*(X; R)$ which extends the "obvious" ring structure on $H^0(X; R)$.

Definition. A **diagonal approximation** is an augmentation preserving natural chain map $\varkappa: S_*(X) \to S_*(X) \otimes S_*(X)$.

In more detail, the augmentation of $S_*(X)$ is the homomorphism $\varepsilon: S_0(X) \to \mathbf{Z}$ with $\varepsilon(x) = 1$ for all $x \in X$, and the augmentation of $S_*(X) \otimes S_*(X)$ is the homomorphism $\varepsilon': S_0(X) \otimes S_0(X) \to \mathbf{Z}$ with $\varepsilon'(x \otimes y) = 1$ for all $x, y \in X$; the condition is that $\varepsilon'\varkappa_0 = \varepsilon$.

Recall that an Eilenberg–Zilber natural chain map $\zeta: S_*(X \times X) \to S_*(X) \otimes S_*(X)$ satisfies $\zeta_0: (x, y) \mapsto x \otimes y$ for all $x, y \in X$. It follows easily that $\zeta d_\#$ is a diagonal approximation for $d: X \to X \times X$ the diagonal (augmentation preserving is thus a substitute for specifying ζ_0). The next result is that this example is essentially the only diagonal approximation.

Theorem 12.24. *Every two diagonal approximations are naturally chain homotopic, hence they induce the same homomorphisms in cohomology.*

PROOF. Let $\mathscr{M} = \{\Delta^p: p \geq 0\}$ be a family of models in **Top**. That the functor E: **Top** $\to$ **Comp** with $E(X) = S_*(X)$ is free with base in $\mathscr{M}$ is contained in Example 9.4; moreover, each Δ^p is totally E-acyclic because it is contractible and so all its reduced homology vanishes. In the proof of the Eilenberg–Zilber theorem, it was shown that the functor F: **Top** $\times$ **Top** $\to$ **Comp** with $F(X, Y) = S_*(X) \otimes S_*(Y)$ is free with base in the family of all models $\mathscr{M}' = \{(\Delta^p, \Delta^q): p \geq 0,$

[1] A geometric interpretation of cup product on manifolds as "intersection numbers" can be found in [Dold (1972), VII §4], [Greenberg and Harper, §31], [Munkres (1984), Chap. 8], or [Seifert and Threlfall, Chap. X]. There is an interpretation in terms of differential forms which is part of **de Rham's theorem**: if $\Omega^*(X)$ denotes the de Rham complex of a differentiable manifold X (see the first section of this chapter), then $H^p(\Omega^*(X)) \cong H^p(X; \mathbf{R})$ for all $p \geq 0$; moreover, if $\omega \in \Omega^p(X)$ and $\omega' \in \Omega^q(X)$ are closed differential forms, then

$$\text{cls } \omega \cup \text{cls } \omega' = \text{cls}(\omega \wedge \omega'),$$

where $\omega \wedge \omega'$ is the wedge product in the exterior algebra $\sum_{i \geq 0} \Omega^i(X)$ (see [Warner, pp. 211–214]).

$q \geq 0\}$; moreover, all such models are totally F-acyclic. It follows that the functor G: **Top** $\to$ **Comp** defined by $G(X) = F(X, X)$ is free with base in $\mathscr{M}$ and that all models in $\mathscr{M}$ are totally G-acyclic. Corollary 9.13(i) applies at once to show that every two diagonal approximations are naturally chain homotopic. □

Diagonal approximations $S_*(X) \to S_*(X) \otimes S_*(X)$ suggest multiplications on $S_*(X)$, better, on the homology groups of these chain complexes, but the arrow points in the wrong direction. Applying a contravariant functor corrects this, and we shall soon see cup product emerge. This discussion will then show that cup product is the unique multiplication on $H^*(X; R)$ arising from a diagonal approximation. (Incidentally, if a space X possesses a "nice" map $\mu: X \times X \to X$ (in place of the diagonal $d: X \to X \times X$), then $\mu_\# \psi: S_*(X) \otimes S_*(X) \to S_*(X)$ does lead to a product in homology, where ψ is a homotopy inverse of an Eilenberg–Zilber natural chain equivalence. In particular, for every H-space X, there is a graded ring structure on $\sum_{p \geq 0} H_p(X)$, called the **Pontrjagin product**.)

We seek a formula for an Eilenberg–Zilber map $\zeta: S_*(X \times Y) \to S_*(X) \otimes S_*(Y)$ (which will be specialized to the case $Y = X$). Now $\zeta_0: S_0(X \times Y) \to S_0(X) \otimes S_0(Y)$ is given by $(x, y) \mapsto x \otimes y$. To find ζ_1, let $\sigma: \Delta^1 \to X \times Y$ be continuous with $\sigma(e_0) = (x_0, y_0)$ and $\sigma(e_1) = (x_1, y_1)$. The map ζ_1 must make the following square commute:

$$\begin{array}{ccc} S_1(X \times Y) & \xrightarrow{\partial_1} & S_0(X \times Y) \\ \downarrow \zeta_1 & & \downarrow \zeta_0 \\ (S_1(X) \otimes S_0(Y)) \oplus (S_0(X) \otimes S_1(Y)) & \xrightarrow[D_1]{} & S_0(X) \otimes S_0(Y), \end{array}$$

where D_1 is the usual differentiation on the tensor product (if $\alpha \in S_i(X)$ and $\beta \in S_j(Y)$, then $D_{i+j}(\alpha \otimes \beta) = \partial\alpha \otimes \beta + (-1)^i \alpha \otimes \partial\beta$). Now we see that $\zeta_0 \partial_1 \sigma = \zeta_0(\sigma(e_1) - \sigma(e_0)) = \zeta_0((x_1, y_1) - (x_0, y_0)) = x_1 \otimes y_1 - x_0 \otimes y_0$. If π' and π'' are the projections of $X \times Y$ onto X and Y, respectively, let us write $\sigma' = \pi'\sigma$ and $\sigma'' = \pi''\sigma$ for these 1-simplexes in X and Y, respectively. A reasonable guess is to set

$$\zeta_1(\sigma) = \sigma' \otimes y + x \otimes \sigma''$$

for some $x \in X$ and $y \in Y$. Since $\partial y = 0 = \partial x$ (because all 0-chains are cycles),

$$\begin{aligned} D_1 \zeta_1(\sigma) &= (\sigma'(e_1) - \sigma'(e_0)) \otimes y + x \otimes (\sigma''(e_1) - \sigma''(e_0)) \\ &= (x_1 - x_0) \otimes y + x \otimes (y_1 - y_0). \end{aligned}$$

Hence, if we define

$$\zeta_1(\sigma) = \sigma' \otimes y_1 + x_0 \otimes \sigma'',$$

then $D_1 \zeta_1 = \zeta_0 \partial$. The computation of $\zeta_2(\sigma)$, for $\sigma: \Delta^2 \to X \times Y$, is more complicated. A reasonable guess is

$$\zeta_2(\sigma) = \sigma' \otimes y + \alpha \otimes \beta + x \otimes \sigma'',$$

where $x \in X$, $y \in Y$, and α, β are 1-simplexes, and one can choose these so that $D_2\zeta_2 = \zeta_1\partial_2$ (the reader is invited to do the calculation).

Theorem 12.25 (Alexander–Whitney). *The map* $\zeta\colon S_*(X \times Y) \to S_*(X) \otimes S_*(Y)$ *defined by*

$$\zeta_n(\sigma) = \sum_{i+j=n} \sigma'\lambda_i \otimes \sigma''\mu_j,$$

where $\sigma\colon \Delta^n \to X \times Y$ *and* $\sigma' = \pi'\sigma$, $\sigma'' = \pi''\sigma$ *(and where* π', π'' *are the projections of* $X \times Y$ *onto* X, Y, *respectively), is a natural chain equivalence over* $\zeta_0\colon (x, y) \mapsto x \otimes y$.

Proof. If ζ is, in fact, a natural chain map, then Corollary 9.13(ii) gives the result, for the hypotheses of that corollary were verified in the proof of the Eilenberg–Zilber theorem.

Recall that both chain complexes may be regarded as functors **Top** × **Top** → **Comp**, and morphisms in **Top** × **Top** are ordered pairs of continuous maps. It is routine to check naturality of ζ, that is, if $f\colon X \to X'$ and $g\colon Y \to Y'$, then the following diagram commutes:

$$\begin{array}{ccc} S_*(X \times Y) & \xrightarrow{\zeta} & S_*(X) \otimes S_*(Y) \\ {\scriptstyle (f \times g)_\#}\downarrow & & \downarrow{\scriptstyle f_\# \otimes g_\#} \\ S_*(X' \times Y') & \xrightarrow[\zeta]{} & S_*(X') \otimes S_*(Y'). \end{array}$$

It remains to show that ζ is a chain map. We normalize the problem. If $d\colon \Delta^n \to \Delta^n \times \Delta^n$ is the diagonal, and if $\sigma\colon \Delta^n \to X \times Y$ is an n-simplex, then

$$\sigma = (\sigma' \times \sigma'')d.$$

Suppose we prove that

$$D_n\zeta_n(d) = \zeta_{n-1}\partial(d). \qquad (*)$$

Then

$$\begin{aligned} D\zeta(\sigma) &= D\zeta((\sigma' \times \sigma'')d) \\ &= D\zeta(\sigma' \times \sigma'')_\#(d) \\ &= D(\sigma'_\# \otimes \sigma''_\#)\zeta(d) && \text{(by naturality)} \\ &= (\sigma'_\# \otimes \sigma''_\#)D\zeta(d) && (\sigma'_\# \otimes \sigma''_\# \text{ in a chain map}) \\ &= (\sigma'_\# \otimes \sigma''_\#)\zeta\partial(d) && \text{(by } (*)) \end{aligned}$$

$$= \zeta(\sigma' \times \sigma'')_\# \partial(d) \qquad \text{(by naturality)}$$
$$= \zeta\partial(\sigma' \times \sigma'')_\#(d) \qquad ((\sigma' \times \sigma'')_\# \text{ is a chain map})$$
$$= \zeta\partial((\sigma' \times \sigma'')d) = \zeta\partial(\sigma), \quad \text{as desired.}$$

Let us now verify (∗).

If $\alpha: \Delta^i \to \Delta^n$ is an affine map with $\alpha(e_k) \in \{e_0, \ldots, e_n\}$ for all k, we shall denote α by $(\alpha(e_0), \alpha(e_1), \ldots, \alpha(e_i))$. In particular, $\lambda_i = (e_0, \ldots, e_i)$ and $\mu_{n-i} = (e_i, e_{i+1}, \ldots, e_n)$.

Now

$$D\zeta(d) = D\sum_i (e_0, \ldots, e_i) \otimes (e_i, \ldots, e_n)$$
$$= \sum_i [\partial(e_0, \ldots, e_i) \otimes (e_i, \ldots, e_n) + (-1)^i(e_0, \ldots, e_i) \otimes \partial(e_i, \ldots, e_n)]$$
$$= \sum_{j \le i}\sum (-1)^j(e_0, \ldots, \hat{e}_j, \ldots, e_i) \otimes (e_i, \ldots, e_n)$$
$$+ \sum_{j \ge i}\sum (-1)^j(e_0, \ldots, e_i) \otimes (e_i, \ldots, \hat{e}_j, \ldots, e_n)$$

(note that the sign in the second sum is correct, because

$$\partial(e_i, \ldots, e_n) = \sum_{k=0}^{n-i} (-1)^k(e_i, \ldots, \hat{e}_{i+k}, \ldots, e_n) = \sum_{j=i}^{n} (-1)^{j-i}(e_i, \ldots, \hat{e}_j, \ldots, e_n)).$$

The portion of the first sum with $j = i$, namely,

$$\sum_{i=1}^{n} (-1)^i(e_0, \ldots, e_{i-1}) \otimes (e_i, \ldots, e_n),$$

cancels the portion of the second sum with $j = i$, namely,

$$\sum_{i=0}^{n-1} (-1)^i(e_0, \ldots, e_i) \otimes (e_{i+1}, \ldots, e_n).$$

Therefore

$$D\zeta(d) = \sum_{j<i} (-1)^j(e_0, \ldots, \hat{e}_j, \ldots, e_i) \otimes (e_i, \ldots, e_n)$$
$$+ \sum_{j>i} (-1)^j(e_0, \ldots, e_i) \otimes (e_i, \ldots, \hat{e}_j, \ldots, e_n).$$

On the other hand, the definition of $\zeta(\partial d)$ is

$$\zeta_{n-1}(\partial d) = \sum_{i=0}^{n-1} (\partial d)'\lambda_i \otimes (\partial d)''\mu_{n-1-i}.$$

Recall that $(\partial d)' = \pi'_\#(\partial d)$, where $\pi': \Delta^n \times \Delta^n \to \Delta^n$ is the projection on the first factor. But $\pi'_\#$ is a chain map, so that $\pi'_\#(\partial d) = \partial\pi'_\#(d) = \partial(\pi' d) = \partial(\delta^n)$, where δ^n is the identity map on Δ^n (for d is the diagonal map). Similar arguments show that $(\partial d)'' = \partial(\delta^n)$ and that $\varepsilon'_j = \varepsilon_j = \varepsilon''_j$ (where ε_j is the jth face map $\Delta^{n-1} \to \Delta^n$).

For each j, Lemma 12.18(iii) gives

$$\begin{aligned}\zeta(\varepsilon_j) &= \sum_i \varepsilon_j\lambda_i \otimes \varepsilon_j\mu_{n-1-i} \\ &= \sum_{i<j} \lambda_i \otimes \mu_{n-i}\varepsilon_{j-i-1} + \sum_{i>j} \lambda_{i+1}\varepsilon_j \otimes \mu_{n-i}.\end{aligned}$$

In our earlier notation,

$$\begin{aligned}\zeta(\varepsilon_j) = \sum_{i<j} &(e_0, \ldots, e_i) \otimes (e_i, \ldots, \hat{e}_j, \ldots, e_n) \\ &+ \sum_{i>j} (e_0, \ldots, \hat{e}_j, \ldots, e_i) \otimes (e_i, \ldots, e_n).\end{aligned}$$

Hence $\zeta(\partial d) = \sum(-1)^j\zeta(\varepsilon_j)$ equals $D\zeta(d)$ computed earlier. □

Definition. Define a function

$$\pi: S^*(X, R) \otimes S^*(Y, R) \to \operatorname{Hom}(S_*(X) \otimes S_*(Y), R)$$

as follows. If $\varphi \in S^n(X, R)$ and $\theta \in S^m(Y, R)$, then there is a function defined on $S^*(X, R) \times S^*(Y, R)$, namely, $(\varphi, \theta) \mapsto \varphi \otimes \theta$, where

$$(\sigma_i \otimes \tau_j, \varphi \otimes \theta) = \begin{cases}(\sigma_i, \varphi)(\tau_j, \theta) & \text{if } i = n \text{ and } j = m \\ 0 & \text{otherwise.}\end{cases}$$

Since this function is bilinear (the proof of Lemma 12.19), it defines a homomorphism π on the tensor product.

Definition. The (external) **cross product** is the map

$$\zeta^{\#}\pi: S^*(X, R) \otimes S^*(Y, R) \to S^*(X \times Y, R).$$

If $\varphi \in S^n(X, R)$ and $\theta \in S^m(Y, R)$, then their cross product is denoted by

$$\varphi \times \theta \in S^{n+m}(X \times Y, R).$$

Of course, the cross product may be regarded as a map in cohomology

$$H^*(X; R) \otimes H^*(Y; R) \to H^*(X \times Y; R).$$

It is the map α' of the Künneth formula, Theorem 12.16.

EXERCISES

*12.13. Let $f: X \to X'$ and $g: Y \to Y'$ be continuous. If $\operatorname{cls} \varphi \in H^p(X'; R)$ and $\operatorname{cls} \theta \in H^q(Y'; R)$, then $(f \times g)^*(\operatorname{cls} \varphi \times \operatorname{cls} \theta) = f^* \operatorname{cls} \varphi \times g^* \operatorname{cls} \theta$.

*12.14. Show that the cross product is associative.

Theorem 12.26. *Cup product is the composite* $d^{\#}\zeta^{\#}\pi$:

$$S^*(X, R) \otimes S^*(X, R) \to \operatorname{Hom}(S_*(X) \otimes S_*(X), R) \to S^*(X \times X, R) \to S^*(X, R).$$

PROOF. Let $\varphi \in S^n(X, R)$, let $\theta \in S^m(X, R)$, and let σ be an $(n + m)$-simplex in X. Then

$$(\sigma, d^{\#}\zeta^{\#}\varphi \otimes \theta) = (\zeta d_{\#}\sigma, \varphi \otimes \theta) = (\zeta(d\sigma), \varphi \otimes \theta).$$

Now the Alexander–Whitney formula gives

$$\zeta(d\sigma) = \sum_{i=0}^{n+m} (d\sigma)'\lambda_i \otimes (d\sigma)''\mu_{n+m-i}.$$

Recall that $(d\sigma)' = \pi' d\sigma$ and $(d\sigma)'' = \pi'' d\sigma$, where π' and π'' are projections of $X \times X$ onto the first and second factors, respectively. Since $d: X \to X \times X$ is the diagonal, however, both $\pi' d$ and $\pi'' d$ equal the identity on X. Hence

$$\zeta(d\sigma) = \sum_{i=0}^{n+m} \sigma\lambda_i \otimes \sigma\mu_{n+m-i}.$$

But $\varphi \otimes \theta$ vanishes off $S_n(X) \otimes S_m(X)$, so that

$$\begin{aligned}(\zeta(d\sigma), \varphi \otimes \theta) &= (\sigma\lambda_n \otimes \sigma\mu_m, \varphi \otimes \theta) \\ &= (\sigma\lambda_n, \varphi)(\sigma\mu_m, \theta) = (\sigma, \varphi \cup \theta).\end{aligned}$$ □

Corollary 12.27. *If $\varphi \in S^n(X, R)$ and $\theta \in S^m(X, R)$, then*

$$\varphi \cup \theta = d^{\#}(\varphi \times \theta).$$

PROOF. Immediate from the theorem and the definition of cross product. □

Lemma 12.28. *If (S_*, ∂) is a nonnegative chain complex, then the function $t: S_* \otimes S_* \to S_* \otimes S_*$ defined by*

$$t(\alpha \otimes \beta) = (-1)^{pq}\beta \otimes \alpha,$$

where $\alpha \in S_p$ and $\beta \in S_q$, is a natural chain equivalence.

PROOF. If D is the usual differentiation on the tensor product, then

$$Dt(\alpha \otimes \beta) = (-1)^{pq}D(\beta \otimes \alpha) = (-1)^{pq}\partial\beta \otimes \alpha + (-1)^{pq+q}\beta \otimes \partial\alpha.$$

On the other hand,

$$\begin{aligned}tD(\alpha \otimes \beta) &= t(\partial\alpha \otimes \beta + (-1)^p\alpha \otimes \partial\beta) \\ &= (-1)^{(p-1)q}\beta \otimes \partial\alpha + (-1)^{p+p(q-1)}\partial\beta \otimes \alpha \\ &= (-1)^{pq-q}\beta \otimes \partial\alpha + (-1)^{pq}\partial\beta \otimes \alpha.\end{aligned}$$

Since $(-1)^q = (-1)^{-q}$, it follows that $Dt = tD$, that is, t is a chain map. It is easy to see that t is a natural isomorphism. □

Theorem 12.29 (Anticommutativity)[2]. *If* cls $\varphi \in H^p(X; R)$ *and* cls $\theta \in H^q(X; R)$,

[2] This result is more natural in light of the de Rham theorem (see the previous footnote) which shows that, for differentiable manifolds, cup product and wedge product (in the exterior algebra) coincide.

then

$$\text{cls } \varphi \cup \text{cls } \theta = (-1)^{pq} \text{ cls } \theta \cup \text{cls } \varphi.$$

PROOF. Both $\zeta d_{\#}$ and $t\zeta d_{\#}$ are natural chain maps $S_*(X) \to S_*(X) \otimes S_*(X)$ over $\zeta_0 d_{\#0}$ (recall that $\zeta_0 d_{\#0}: x \mapsto x \otimes x$). Since $\zeta_0 d_{\#0}$ is a natural equivalence, acyclic models (Theorem 9.12(iii)) implies that $\zeta d_{\#}$ and $t\zeta d_{\#}$ are (naturally) chain equivalent. We conclude, after applying Hom(, R), that both chain maps induce the same map in cohomology: cls $d^{\#}\zeta^{\#}$ = cls $d^{\#}\zeta^{\#}t^{\#}$. The result now follows from Theorem 12.26, for cup product is just cls $d^{\#}\zeta^{\#}\pi$. □

EXERCISES

12.15. Show that every left or right homogeneous ideal in $H^*(X; R)$ is a two-sided ideal.

12.16. Show that the graded ring $S^*(X, R)$ is not anticommutative in the sense of Theorem 12.29.

*12.17. If the additive group of $H^*(X; R)$ has no elements of order 2, prove that if $\beta \in H^*(X; R)$ has odd degree, then $\beta \cup \beta = 0$.

12.18. Compute the ring $H^*(\mathbf{R}P^2) = H^*(\mathbf{R}P^2; \mathbf{Z})$.

12.19. Compute the ring $H^*(S^n)$.

Computations and Applications

There are not many general results helping one to compute cohomology rings; one such is Theorem 12.31 below.

Lemma 12.30.

(i) *If R and S are rings, then there is a ring structure on $R \otimes S$ with multiplication*

$$(r \otimes s)(r' \otimes s') = rr' \otimes ss',$$

where $r, r' \in R$ and $s, s' \in S$.

(ii) *If R and S are graded rings, then $R \otimes S$ is a graded ring with multiplication*

$$(r_i \otimes s_j)(r'_p \otimes s'_q) = (-1)^{jp} r_i r'_p \otimes s_j s'_q,$$

where $r_i \in R_i$, $r'_p \in R_p$, $s_j \in S_j$, and $s'_q \in S_q$ (of course,

$$(R \otimes S)_n = \sum_{i+j=n} R_i \otimes S_j).$$

PROOF. The formula for multiplication is well defined, since it is the composite

$$R \otimes S \otimes R \otimes S \xrightarrow{1 \otimes t \otimes 1} R \otimes R \otimes S \otimes S \xrightarrow{\mu \otimes \nu} R \otimes S,$$

where $t: S \otimes R \to R \otimes S$ is the map $s \otimes r \mapsto r \otimes s$, and μ and ν are the given multiplications in R and S, respectively. Verification of the ring axioms is left as a routine exercise. The graded case is also left to the reader; the sign is present because of Lemma 12.28. □

It can be shown that $R \otimes S$ is the coproduct of R and S in the category of rings (or the category of graded rings).

The sign in the definition of multiplication in the graded ring case forces anticommutativity; without the sign, the elements $r \otimes 1$ and $1 \otimes s$ commute.

EXAMPLE 12.6. If R and S are graded rings, then the two tensor products can differ.

Let $R = \mathbf{Z}[x]$, polynomials over $\mathbf{Z}$ in one variable x, and let $S = \mathbf{Z}[y]$. As (ungraded) rings, $R \otimes S \cong \mathbf{Z}[x, y]$, polynomials over $\mathbf{Z}$ in two (commuting) variables x and y (one identifies x with $x \otimes 1$ and y with $1 \otimes y$). As graded rings, however, $R \otimes S$ consists of all polynomials over $\mathbf{Z}$ in two variables x and y in which $xy = -yx$.

EXAMPLE 12.7. If M and N are abelian groups, then there is a graded ring isomorphism of exterior algebras:

$$\bigwedge(M \oplus N) \cong \bigwedge M \otimes \bigwedge N$$

(see [Greub, p. 121]).

Theorem 12.31.

(i) *If X and Y are spaces, then cross product $H^*(X) \otimes H^*(Y) \to H^*(X \times Y)$ is a homomorphism of graded rings.*
(ii) *If X and Y are spaces of finite type (for example, compact* CW *complexes) with $H_n(X)$ free abelian for all $n \geq 0$, then cross product is an isomorphism.*

Remark. Recall that $H^*(X) = H^*(X; \mathbf{Z})$.

PROOF. The Künneth formula (Theorem 12.16) gives an exact sequence

$$0 \to \sum_{i+j=n} H^i(X) \otimes H^j(Y) \xrightarrow{\alpha'} H^n(X \times Y) \to \sum_{p+q=n+1} \operatorname{Tor}(H^p(X), H^q(Y)) \to 0$$

in which the map α' is the cross product. Now (ii) follows from (i) as follows. If $H_n(X)$ is free abelian for all $n \geq 0$, then Theorem 12.11 shows that $H^p(X)$ is free abelian for all $p \geq 0$, and so the Tor term is zero.

To prove (i), let $\varphi \in Z^n(X, R)$, $\varphi' \in Z^p(X, R)$, $\theta \in Z^m(Y, R)$, and $\theta' \in Z^q(Y, R)$; let $u = \operatorname{cls} \varphi$, $u' = \operatorname{cls} \varphi'$, $v = \operatorname{cls} \theta$, and $v' = \operatorname{cls} \theta'$. It must be shown that

$$\alpha'((u \otimes v)(u' \otimes v')) = \alpha'(u \otimes v) \cup \alpha'(u' \otimes v').$$

To evaluate the left side, the definition of multiplication in tensor products of graded rings gives

$$(u \otimes v)(u' \otimes v') = (-1)^{mp}(u \cup u') \otimes (v \cup v').$$

Since $\alpha'(u \otimes v)$ is the cross product $u \times v$, it thus remains to prove that

$$(-1)^{mp}(u \cup u') \times (v \cup v') = (u \times v) \cup (u' \times v').$$

Now there is a commutative diagram

$$\begin{array}{ccc} X \times Y & \xrightarrow{d_X \times d_Y} & X \times (X \times Y) \times Y \\ & \searrow^{d_{X \times Y}} & \downarrow{\scriptstyle 1 \times t \times 1} \\ & & X \times (Y \times X) \times Y, \end{array}$$

where $d_X \colon X \to X \times X$ is the diagonal $x \mapsto (x, x)$, and $t \colon X \times Y \to Y \times X$ is the map given by $(x, y) \mapsto (y, x)$.

Using Corollary 12.27 and Exercise 12.14 (associativity of cross product), we have

$$\begin{aligned} (u \times v) \cup (u' \times v') &= d^*_{X \times Y}(u \times v \times u' \times v') \\ &= (d_X \times d_Y)^*(1 \times t \times 1)^*(u \times v \times u' \times v') \\ &= (d_X \times d_Y)^*(u \times t^*(v \times u') \times v') && \text{(Exercise 12.13)} \\ &= (-1)^{mp}(d_X \times d_Y)^*(u \times u' \times v \times v') && \text{(Lemma 12.28)} \\ &= (-1)^{mp} d^*_X(u \times u') \times d^*_Y(v \times v') \\ &= (-1)^{mp}(u \cup u') \times (v \cup v'). \end{aligned}$$

□

Remarks. (1) $H^*(X) \otimes H^*(Y)$ is always a subring of $H^*(X \times Y)$; if it is a proper subring, then it cannot be an ideal because it contains the unit.

(2) Let R be a field. The version of the Künneth formula described after the proof of Theorem 12.16 shows that if X and Y are of finite type, if cohomology groups over $\mathbf{Z}$ are replaced by coefficients R, and if $\otimes$ is replaced by $\otimes_R$, then cross product is necessarily an isomorphism.

Corollary 12.32. *If T^r is an r-torus, that is, the cartesian product of r copies of S^1, then the cohomology ring $H^*(T^r)$ is isomorphic to the exterior algebra $\bigwedge(\mathbf{Z}^{(r)})$, where $\mathbf{Z}^{(r)}$ denotes a free abelian group of rank r.*

PROOF. We do an induction on $r \geq 1$. When $r = 1$, then the additive structure of $H^*(S^1) = H^0(S^1) \oplus H^1(S^1) = \mathbf{Z} \oplus \mathbf{Z}$. Choose generators $1 \in H^0(S^1)$ and $a \in H^1(S^1)$; the multiplication is determined by 1 being the unit element and $a^2 = 0$. If $r > 1$, then $T^r = S^1 \times T^{r-1}$; Theorem 12.31 applies to give $H^*(T^r) \cong H^*(S^1) \otimes H^*(T^{r-1})$. By induction, $H^*(T^{r-1}) \cong \bigwedge(\mathbf{Z}^{(r-1)})$, and so the result follows from Example 12.7. □

It follows that $H^i(T^r)$ is free abelian of rank the binomial coefficient $\binom{r}{i}$.

Computation of cup products is difficult; let us therefore retreat from general spaces to polyhedra. Recall the construction of simplicial homology. If K is an *oriented* simplicial complex (there is a partial order on Vert(K) whose restriction to the vertex set of any simplex is a linear order), then $H_*(K) = H_*(C_*(K))$, where $C_*(K)$ is defined as follows. The qth term $C_q(K)$ is the free abelian group with basis all symbols $\langle p_0, \ldots, p_q \rangle$, where $\{p_0, \ldots, p_q\}$ spans a q-simplex of K and $p_0 < p_1 < \cdots < p_q$. The differentiations $\partial_q \colon C_q(K) \to C_{q-1}(K)$ are the usual alternating sums:

$$\partial\langle p_0, \ldots, p_q \rangle = \sum_{i=0}^{q} (-1)^i \langle p_0, \ldots, \hat{p}_i, \ldots, p_q \rangle.$$

In Theorem 7.22, we proved that $H_*(C_*(K)) \cong H_*(|K|)$. More precisely, if $j_q \colon C_q(K) \to S_q(|K|)$ is defined by

$$j_q(\langle p_0, \ldots, p_q \rangle) = \sigma,$$

where $\sigma \colon \Delta^q \to |K|$ is the affine map $\sum t_i e_i \mapsto \sum t_i p_i$, then j is a chain map and $j_* \colon H_*(C_*(K)) \to H_*(|K|)$ is an isomorphism. By Theorem 9.8, it follows that j is a chain equivalence.

Definition. If K is an oriented simplicial complex and G is an abelian group, then the **simplicial cohomology groups** of K with **coefficients** G are defined by

$$H^n(K; G) = H^n(\mathrm{Hom}(C_*(K), G)).$$

Since $C_*(K)$ and $S_*(|K|)$ are chain equivalent, it follows that $\mathrm{Hom}(C_*(K), G)$ and $\mathrm{Hom}(S_*(|K|), G)$ are chain equivalent and hence have the same cohomology groups (Exercise 9.14). Therefore simplicial cohomology groups are independent of orientation.

Notation. If K is an oriented simplicial complex and R is a commutative ring, define

$$C^n(K, R) = \mathrm{Hom}(C_n(K), R)$$

and

$$C^*(K, R) = \sum_{n \geq 0} C^n(K, R).$$

Definition. If K is an oriented simplicial complex and R is a commutative ring, define **cup product** as follows. If $\varphi \in C^n(K, R)$ and $\theta \in C^m(K, R)$, then

$$(\langle p_0, \ldots, p_{n+m} \rangle, \varphi \cup \theta) = (\langle p_0, \ldots, p_n \rangle, \varphi)(\langle p_n, \ldots, p_{n+m} \rangle, \theta).$$

Theorem 12.33. *Let K be an oriented simplicial complex and let R be a commutative ring.*

(i) *$H^*(K; R) = \sum_{n \geq 0} H^n(K; R)$ inherits a ring structure from the cup product on simplicial cochains.*
(ii) *The rings $H^*(K; R)$ and $H^*(|K|; R)$ are isomorphic (via the chain equivalence of Theorem 7.22).*

PROOF. (i) The argument of Lemma 12.19 shows that $C^*(K, R)$ is a graded ring; moreover, the argument of Lemma 12.22 carries over to $C^*(K, R)$. It follows that the cocycles form a graded subring, the coboundaries form a homogeneous two-sided ideal in the cocycles, and hence $H^*(K; R)$ is a graded ring.

(ii) The explicit formula for the chain equivalence $j: C_*(K) \to S_*(|K|)$ shows that the isomorphism $j^*: H^*(|K|; R) \to H^*(K; R)$ preserves cup products. □

Corollary 12.34. *The cohomology ring of an oriented simplicial complex K does not depend on the orientation.*

PROOF. The ring $H^*(|K|; R)$ does not depend on the orientation. □

EXERCISE

*12.20. (i) If $X \coprod Y$ is a disjoint union, then there is a ring isomorphism $H^*(X \coprod Y; R) \cong H^*(X; R) \times H^*(Y; R)$.

(ii) If X and Y are polyhedra with basepoints and $X \vee Y$ is their wedge, then there is a ring isomorphism

$$\tilde{H}^*(X \vee Y; R) \cong \tilde{H}^*(X; R) \times \tilde{H}^*(Y; R),$$

where $\tilde{H}^*$ is the ideal in H^* generated by all terms of degree >0 ($\tilde{H}^*$ is a ring without unit!).

EXAMPLE 12.8. Let $X = S^2 \vee S^1 \vee S^1$ and let $Y = S^1 \times S^1$. In Exercise 9.47, we saw that X and Y have the same homology groups (hence the same cohomology groups) and the same fundamental group. We prove that X and Y do not have the same homotopy type by showing that their integral cohomology rings are not isomorphic (one can also show that $\pi_2(X) \neq 0$ and $\pi_2(Y) = 0$).

By Exercise 12.20(ii), $\tilde{H}^*(X) \cong \tilde{H}^*(S^2) \times \tilde{H}^*(S^1) \times \tilde{H}^*(S^1)$ as graded rings. The elements of degree 1 lie in the direct product of the subrings $\tilde{H}^*(S^1) \times \tilde{H}^*(S^1)$ (because $H^1(S^2) = 0$); it follows that the cup product of any two elements of degree 1 is zero. On the other hand, $H^*(Y)$ is the exterior algebra $\bigwedge(\mathbf{Z}^{(2)})$; if $\{a, b\}$ is a basis of $\bigwedge^1(\mathbf{Z}^{(2)})$, then $a \wedge b \neq 0$. The graded rings $H^*(X)$ and $H^*(Y)$ are not isomorphic, and so X and Y do not have the same homotopy type.

A serious consequence of this example is that if X and Y have singular chain complexes which are chain equivalent, then they do have isomorphic homology and cohomology groups, but they may have nonisomorphic cohomology rings.

Remark. By Theorem 12.15, the (integral) cohomology *groups* $H^*(X)$ of a space X of finite type determine the cohomology groups $H^*(X; G)$ for every abelian group G. In contrast, the (integral) cohomology *ring* $H^*(X)$ does not determine the cohomology ring $H^*(X; R)$. If $X = \mathbf{R}P^3$ and $Y = \mathbf{R}P^2 \vee S^3$, then it is shown in [Hilton and Wylie, p. 151] that the cohomology rings $H^*(X)$ and $H^*(Y)$ are isomorphic, but the rings $H^*(X; \mathbf{Z}/2\mathbf{Z})$ and $H^*(Y; \mathbf{Z}/2\mathbf{Z})$ are not isomorphic (Exercise 12.17 suggests a reason for this).

EXERCISES

12.21. Show that $S^1 \vee S^2 \vee S^3$ and $S^1 \times S^2$ do not have the same homotopy type.
12.22. Show that $S^n \vee S^m$ is not a retract of $S^n \times S^m$, where $m, n \geq 1$.

There are many difficulties in computing cohomology and cup product. At the most basic level, it is not obvious how to construct cocycles (other than coboundaries). Let us give a negative example in this regard. If F is a free abelian group with a finite basis B, then there is a **dual basis** of $\mathrm{Hom}(F, \mathbf{Z})$ consisting of all b^* for $b \in B$, where $b^*: F \to \mathbf{Z}$ is defined by

$$b^*(b) = 1 \quad \text{and} \quad b^*(c) = 0 \quad \text{for all } c \in B - \{b\}.$$

If $c = \sum m_i b_i$ is a chain, then its dual is defined to be $c^* = \sum m_i b_i^*$. It is easy to see that the dual basis is a basis of the free abelian group $\mathrm{Hom}(F, \mathbf{Z})$; hence, every cochain has a unique expression of the form c^*. It is not true that the dual of a cycle is a cocycle. Consider the following simple example:

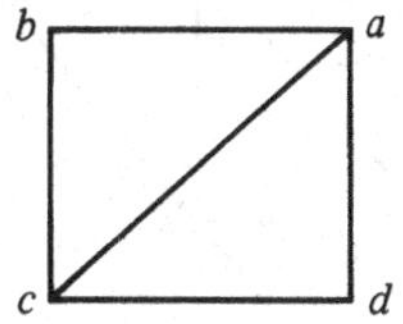

Clearly, $z = \langle a, b\rangle + \langle b, c\rangle - \langle a, c\rangle$ is a 1-cycle. On the other hand, if $\sigma = \langle a, c, d\rangle$, then

$$\begin{aligned}(\sigma, \delta z^*) &= (\partial\sigma, z^*)\\ &= (\langle c, d\rangle - \langle a, d\rangle + \langle a, c\rangle, z^*) = -1;\end{aligned}$$

therefore $(\sigma, \delta z^*) \neq 0$, so that $\delta z^* \neq 0$ and z^* is not a cocycle.

Let us illustrate how one can compute with simplicial cohomology.

EXAMPLE 12.9. As usual, triangulate the torus $T = S^1 \times S^1$ by first triangulating the square:

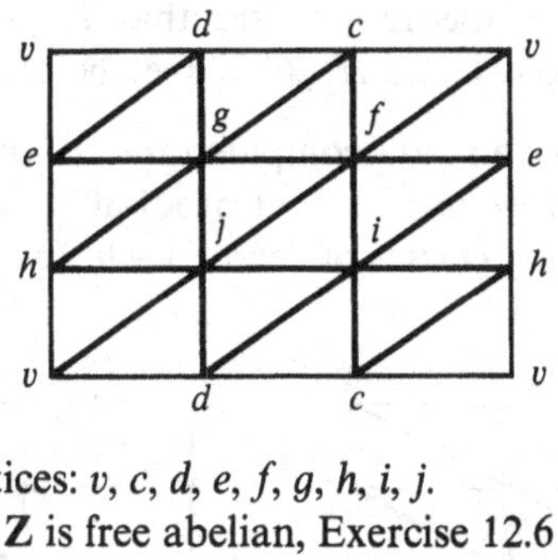

Linearly order the vertices: $v, c, d, e, f, g, h, i, j$.

Since $H_1(T) = \mathbf{Z} \oplus \mathbf{Z}$ is free abelian, Exercise 12.6 applies to give

$$\begin{aligned}B^2(T, \mathbf{Z}) &= \{\varphi\partial_2 : \varphi \in \mathrm{Hom}(B_1(T), \mathbf{Z})\}\\ &= \{c^*\partial_2 : c \in B_1(T)\}.\end{aligned}$$

Every 2-cochain γ in $C^2(T, \mathbf{Z})$ has the form $\gamma = \sum m_i \sigma_i^*$, where σ_i is a 2-simplex. We claim that $\gamma = \sum m_i \sigma_i^*$ is a 2-coboundary if its coefficient sum

$\sum m_i$ is zero. Let $\tau = \langle x, y, z\rangle$ be a 2-simplex (with vertices ordered $x < y < z$). By our observation above, a typical generator of $B^2(T, \mathbf{Z})$ is $\delta(\partial\tau)^* = (\langle y, z\rangle^* - \langle x, z\rangle^* + \langle x, y\rangle^*)\partial$. If $\sigma = \langle u, v, w\rangle$ (with $u < v < w$), then

$$\begin{aligned}(\sigma, \delta(\partial\tau)^*) &= (\partial\sigma, (\partial\tau)^*)\\ &= (\langle v, w\rangle - \langle u, w\rangle + \langle u, v\rangle, \langle y, z\rangle^* - \langle x, z\rangle^* + \langle x, y\rangle^*).\end{aligned}$$

It follows easily that $(\sigma, \delta(\partial\tau)^*) = 0$ unless σ and τ have a common edge. Write $\delta(\partial\tau)^* = \sum m_i\sigma_i^*$, where the σ_i are 2-simplexes. Since the dual basis behaves as an orthonormal basis, $m_i = (\sigma_i, \delta(\partial\tau)^*)$. But each edge of τ is an edge of exactly one other 2-simplex, and it occurs there with opposite orientation. Hence, there are only four (oriented) 2-simplexes involved in the expression for $(\partial\tau)^*$, namely, $\sigma_i = \langle x, y, z\rangle, \langle y, x, u\rangle, \langle z, y, w\rangle$, and $\langle x, z, v\rangle$.

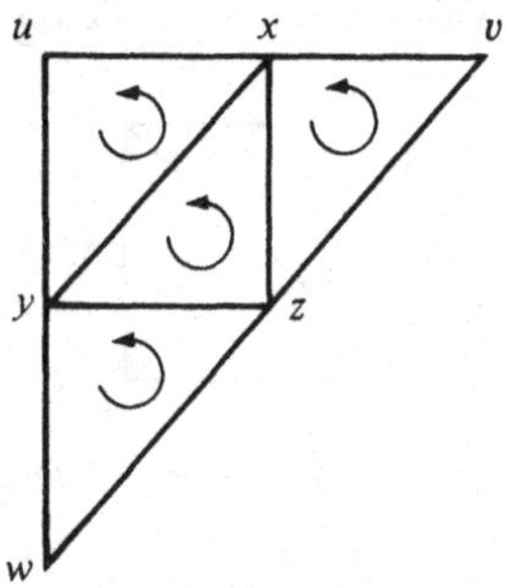

Evaluating gives values $m_i = 3, -1, -1$, and -1; the coefficient sum is thus 0. Therefore, the coefficient sum of every 2-coboundary is zero.

We conclude that, for every 2-simplex σ in T, the 2-cocycle σ^* is not a coboundary (all 2-cochains here are 2-cocycles); that is, cls $\sigma^* \neq 0$ in $H^2(T; \mathbf{Z})$. Indeed, one can show that cls σ^* is a generator of $H^2(T; \mathbf{Z})$. Note that if we are interested only in finding some generator of $H^2(T; \mathbf{Z})$, then we can invoke the universal coefficients theorem to see that $H^2(T; \mathbf{Z}) \cong \mathrm{Hom}(H_2(T), \mathbf{Z})$ (since $\mathrm{Ext}(H_1(T), \mathbf{Z}) = 0$ because $H_1(T)$ is free abelian).

EXAMPLE 12.10. We have already computed the cohomology ring of T (with much algebra). Let us now give another proof of its most important feature: there are two cohomology classes of degree 1 whose cup product is nonzero.

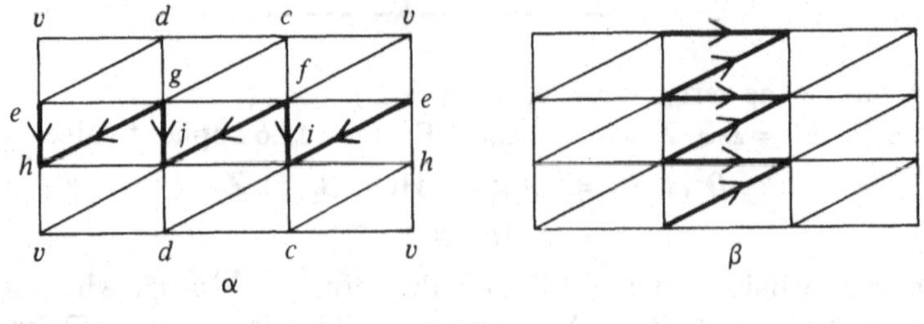

Define 1-chains

$$\alpha = \langle e, h\rangle + \langle g, h\rangle + \langle g, j\rangle + \langle f, j\rangle + \langle f, i\rangle + \langle e, i\rangle$$

and

$$\beta = -\langle c, d\rangle - \langle c, g\rangle - \langle f, g\rangle - \langle f, j\rangle - \langle i, j\rangle + \langle d, i\rangle.$$

Note that $(\partial\sigma, \alpha^*) = 0$ for all 2-simplexes σ having no edges in common with α; in fact, a simple calculation shows that $(\partial\sigma, \alpha^*) = 0$ for every 2-simplex σ, hence α^* is a cocycle. Similarly, one can show that β^* is a cocycle.

Another easy calculation shows that

$$\alpha^* \cup \beta^* = -\langle f, i, j\rangle^*:$$

for example,

$$(\langle f, g, j\rangle, \alpha^* \cup \beta^*) = (\langle f, g\rangle, \alpha^*)(\langle g, j\rangle, \beta^*) = 1;$$

that $(\sigma, \alpha^* \cup \beta^*) = 0$ for every other 2-simplex σ is left to the reader. By the previous example,

$$\text{cls}\,\alpha^* \cup \text{cls}\,\beta^* = \text{cls}\,\alpha^* \cup \beta^* = \text{cls}\langle f, g, j\rangle^* \neq 0 \text{ in } H^2(T; \mathbf{Z}).$$

as desired.

We saw in Chapter 11 that if $v: E \to B$ is a fibration with fiber F, then there is a relation between the homotopy groups of E, B, and F given by the exact sequence of a fibration. The cohomology rings of E, B, and F are also related, by the *Leray–Serre spectral sequence* (there is also a spectral sequence relating the homology groups). Specializing to fibrations with fiber S^q and coefficient ring $\mathbf{Z}/2\mathbf{Z}$, one obtains the following result.

Theorem (Gysin Sequence). *Let $v: E \to B$ be a fibration with fiber S^q, where $q \geq 0$. Denote $\mathbf{Z}/2\mathbf{Z}$ by* $\mathbf{2}$.

(i) *There is an exact sequence*

$$\cdots \longrightarrow H^k(B; \mathbf{2}) \xrightarrow{v^*} H^k(E; \mathbf{2}) \longrightarrow H^{k-q}(B; \mathbf{2}) \xrightarrow{\psi_k} H^{k+1}(B; \mathbf{2}) \xrightarrow{v^*}$$
$$\longrightarrow H^{k+1}(E; \mathbf{2}) \longrightarrow H^{k+1-q}(B; \mathbf{2}) \longrightarrow \cdots$$

which begins

$$0 \longrightarrow H^0(B; \mathbf{2}) \xrightarrow{\psi_0} H^{-q}(B; \mathbf{2}) \xrightarrow{v^*} H^1(E; \mathbf{2}) \longrightarrow H^1(B; \mathbf{2}) \longrightarrow \cdots.$$

(ii) *If* cls $e \in H^0(B; \mathbf{2})$ *is the unit of the cohomology ring* $H^*(B; \mathbf{2})$, *write*

$$\Omega_B = \psi_0(\text{cls } e) \in H^{q+1}(B; \mathbf{2})$$

(Ω_B is called the **characteristic class** *of the fibration).*

Then the map $\psi_k: H^{k-q}(B; \mathbf{2}) \to H^{k+1}(B; \mathbf{2})$, for all $k \geq 1$, is given by

$$\beta \mapsto \Omega_B \cup \beta.$$

(iii) *If $v'\colon E' \to B'$ is another fibration with fiber S^q and if the following diagram commutes*

$$\begin{array}{ccc} E' & \xrightarrow{\tilde{f}} & E \\ {\scriptstyle v'}\downarrow & & \downarrow{\scriptstyle v} \\ B' & \xrightarrow[f]{} & B, \end{array}$$

then $f^*(\Omega_B) = \Omega_{B'}$.

For a proof without spectral sequences, see [Spanier, p. 260]; for a proof with spectral sequences, see [Spanier, p. 499] or [McCleary, p. 134].

Recall from Exercise 12.12(ii) that $H^k(\mathbf{R}P^n; \mathbf{Z}/2\mathbf{Z}) \cong \mathbf{Z}/2\mathbf{Z}$ for all k with $0 \le k \le n$. Furthermore, the usual covering projection $v\colon S^n \to \mathbf{R}P^n$, which identifies antipodal points, is a fibration (Theorem 10.5) with fiber S^0.

Theorem 12.35. *The cohomology ring $H^*(\mathbf{R}P^n; \mathbf{Z}/2\mathbf{Z})$ is isomorphic to the polynomial ring $(\mathbf{Z}/2\mathbf{Z})[x]$ modulo the ideal (x^{n+1}). In particular, if Ω_n is the nonzero element of $H^1(\mathbf{R}P^n; \mathbf{Z}/2\mathbf{Z})$, then the nonzero element of $H^k(\mathbf{R}P^n; \mathbf{Z}/2\mathbf{Z})$ (where $1 \le k \le m$) is Ω_m^k, the cup product of Ω_n with itself k times.*

Proof[3]. The Gysin sequence of the fibration $v\colon S^n \to \mathbf{R}P^n$ with fiber S^0 is

$$\cdots \longrightarrow H^k(S^n; \mathbf{2}) \longrightarrow H^k(\mathbf{R}P^n; \mathbf{2}) \xrightarrow{\psi_k} H^{k+1}(\mathbf{R}P^n; \mathbf{2}) \longrightarrow H^{k+1}(S^n; \mathbf{2}) \longrightarrow \cdots$$

(again, we have denoted $\mathbf{Z}/2\mathbf{Z}$ by $\mathbf{2}$). Since $H^k(S^n; \mathbf{2}) = 0$ unless $k = 0$ or $k = n$, it follows that ψ_k is an isomorphism if $0 < k < n - 1$, ψ_0 is a surjection, and ψ_{n-1} is an injection. Now ψ_0 is always injective, so that ψ_0 is an isomorphism. To see that ψ_{n-1} is surjective (hence is an isomorphism), consider the "end" of the Gysin sequence:

$$H^{n-1}(\mathbf{R}P^n; \mathbf{2}) \xrightarrow{\psi_{n-1}} H^n(\mathbf{R}P^n; \mathbf{2}) \to H^n(S^n; \mathbf{2}) \xrightarrow{v^*} H^n(\mathbf{R}P^n; \mathbf{2}) \to H^{n+1}(\mathbf{R}P^n; \mathbf{2}).$$

Now $H^{n+1}(\mathbf{R}P^n; \mathbf{2}) = 0$ because $\mathbf{R}P^n$ is a polyhedron of dimension n; therefore $v^*\colon H^n(S^n; \mathbf{2}) \to H^n(\mathbf{R}P^n; \mathbf{2})$ is surjective. As both groups have order 2, the map v^* must be an isomorphism, hence v^* is an injection. Exercise 5.2 now implies that ψ_{n-1} is a surjection.

If Ω_n is the nonzero element of $H^1(\mathbf{R}P^n; \mathbf{2})$, then Ω_n is the characteristic class because ψ_0 is an injection. Now statement (ii) of the Gysin sequence is that $\psi_k(\beta) = \Omega_n \cup \beta$. It follows easily by induction that Ω_n^k is the nonzero element of H^k if $k \le n$. Since $\Omega_n^{n+1} = 0$, the structure of $H^*(\mathbf{R}P^n; \mathbf{2})$ is as stated. □

[3] There are proofs of this theorem avoiding the Gysin sequence (all are long): see [Wallace, p. 127], [Dold (1972), p. 223], [Maunder, p. 348], or [Munkres (1984), p. 403].

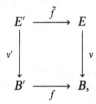

Corollary 12.36. *Let $n > m \geq 1$, and let $f\colon \mathbf{R}P^n \to \mathbf{R}P^m$ be continuous. If Ω_m is the nonzero element of $H^1(\mathbf{R}P^m; \mathbf{Z}/2\mathbf{Z})$, then $f^*(\Omega_m) = 0$.*

PROOF. Since f^* is a map of graded rings, it follows that $f^*(\Omega_m) = 0$ or $f^*(\Omega_m) = \Omega_n$ (for $H^1(\mathbf{R}P^n; \mathbf{Z}/2\mathbf{Z})$ has only two elements). In the latter case, $f^*(\Omega_m^{m+1}) = \Omega_n^{m+1}$, and this gives a contradiction because $f^*(\Omega_m^{m+1}) = 0$ and $\Omega_n^{m+1} \neq 0$. □

Theorem 12.37. *If $n > m \geq 1$, then $\mathbf{R}P^m$ is not a retract of $\mathbf{R}P^n$.*

PROOF. Let $i\colon \mathbf{R}P^m \hookrightarrow \mathbf{R}P^n$ be (any) injection, and assume that there is some retraction $r\colon \mathbf{R}P^n \to \mathbf{R}P^m$, that is, $ri = 1$, the identity on $\mathbf{R}P^m$. It follows that $i^*r^* = 1$, so that $r^*\colon H^*(\mathbf{R}P^m; 2) \to H^*(\mathbf{R}P^n; 2)$ is an injection, and this contradicts the corollary. □

If $n \geq m \geq 1$, the **usual imbedding** $i\colon \mathbf{R}P^m \hookrightarrow \mathbf{R}P^n$ is given by $[x_0, \ldots, x_m] \mapsto [x_0, \ldots, x_m, 0, \ldots, 0]$ (where $(x_0, \ldots, x_m) \in S^m$ and $[x_0, \ldots, x_m]$ is the equivalence class of $(x_0, \ldots, x_m)$ obtained by identifying antipodal points). The following diagram commutes when i is the usual imbedding:

$$\begin{array}{ccc} S^m & \xrightarrow{\ \tilde{i}\ } & S^n \\ {\scriptstyle v'}\downarrow & & \downarrow{\scriptstyle v} \\ \mathbf{R}P^m & \xrightarrow[\ i\ ]{} & \mathbf{R}P^n, \end{array}$$

where v', v are fibrations with fiber S^0, namely, $(x_0, \ldots, x_m) \mapsto [x_0, \ldots, x_m]$, and $\tilde{i}\colon S^m \hookrightarrow S^n$ is the imbedding $(x_0, \ldots, x_m) \mapsto (x_0, \ldots, x_m, 0, \ldots, 0)$. Part (iii) of the Gysin sequence thus says that

$$i^*(\Omega_n) = \Omega_m$$

whenever $i\colon \mathbf{R}P^m \hookrightarrow \mathbf{R}P^n$ is the usual imbedding.

EXERCISES

*12.23. If $n > m \geq 1$ and $i\colon \mathbf{R}P^m \hookrightarrow \mathbf{R}P^n$ is the usual imbedding, then it is true that $i^*\colon H^q(\mathbf{R}P^n; \mathbf{Z}/2\mathbf{Z}) \to H^q(\mathbf{R}P^m; \mathbf{Z}/2\mathbf{Z})$ is an isomorphism for all $q \leq m$.

*12.24. Let $i\colon \mathbf{R}P^1 \to \mathbf{R}P^n$ be the usual imbedding, where $n \geq 2$, and let $v\colon S^1 \to \mathbf{R}P^1$ be the fibration $(x_0, x_1) \mapsto [x_0, x_1]$. Show that $[i \circ v]$ is a nontrivial element of $\pi_1(\mathbf{R}P^n, *)$. (Recall Corollary 10.11 that $\pi_1(\mathbf{R}P^n, *) \cong \mathbf{Z}/2\mathbf{Z}$ and that $\mathbf{R}P^1 \approx S^1$.)

Lemma 12.38. *Let $n > m \geq 1$, and let $f\colon \mathbf{R}P^n \to \mathbf{R}P^m$ be a continuous map. If $v\colon S^m \to \mathbf{R}P^m$ is the covering projection identifying antipodal points, then there exists a lifting $\tilde{f}\colon \mathbf{R}P^n \to S^m$ (i.e., $v\tilde{f} = f$).*

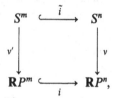

$$\begin{array}{ccc} & & S^m \\ & \overset{\tilde{f}}{\nearrow} & \downarrow{\scriptstyle \nu} \\ \mathbf{R}P^n & \xrightarrow[f]{} & \mathbf{R}P^m. \end{array}$$

PROOF. By the lifting criterion, Theorem 10.13, it is enough to show that the induced map $f_*: \pi_1(\mathbf{R}P^n, *) \to \pi_1(\mathbf{R}P^m, *)$ is zero.

If $m = 1$, then $\mathbf{R}P^m = \mathbf{R}P^1 = S^1$ and so $\pi_1(\mathbf{R}P^1, *) \cong \mathbf{Z}$; since $\pi_1(\mathbf{R}P^n, *) \cong \mathbf{Z}/2\mathbf{Z}$, it follows that $f_* = 0$ in this case. We may therefore assume that $m \geq 2$. Now f^*, the induced map in cohomology, satisfies $f^*(\Omega_m) = 0$, by Corollary 12.36. Hence, if $i: \mathbf{R}P^1 \hookrightarrow \mathbf{R}P^n$ is the usual imbedding,

$$0 = i^*f^*(\Omega_m) = (f \circ i)^*(\Omega_m).$$

On the other hand, if $j: \mathbf{R}P^1 \hookrightarrow \mathbf{R}P^m$ is the usual imbedding, then Exercise 12.23 shows that $j^*(\Omega_m) \neq 0$ (because $m \geq 2$). It follows that the maps j and $f \circ i$ from $\mathbf{R}P^1$ to $\mathbf{R}P^m$ are not homotopic. As $\mathbf{R}P^1 = S^1$, however, these maps represent elements of $\pi_1(\mathbf{R}P^m, *) \cong \mathbf{Z}/2\mathbf{Z}$ (again, we use $m \geq 2$). As j is not nullhomotopic ($j^*(\Omega_m) \neq 0$), it follows that $f \circ i$ is nullhomotopic. By Exercise 12.24, the nontrivial element of $\pi_1(\mathbf{R}P^m, *)$ is $[i \circ \nu']$, and so $f_*[i \circ \nu'] = [f \circ i \circ \nu'] = 0$, as desired (where $\nu': S^1 \to \mathbf{R}P^1$ is as in Exercise 12.24). □

Theorem 12.39. *If $n > m \geq 1$, then there is no continuous map $g: S^n \to S^m$ with $g(-x) = -g(x)$ for all $x \in S^n$.*

Remark. The special case $m = 1$ has been proved in Theorem 6.28.

PROOF. If such a map g exists, then there exists a continuous map f making the following diagram commute:

$$\begin{array}{ccc} S^n & \xrightarrow{g} & S^m \\ {\scriptstyle \nu'}\downarrow & & \downarrow{\scriptstyle \nu} \\ \mathbf{R}P^n & \xrightarrow[f]{} & \mathbf{R}P^m, \end{array}$$

where ν', ν are the covering projections. By the lemma, there is a lifting $\tilde{f}: \mathbf{R}P^n \to S^m$ with $\nu\tilde{f} = f$. Now consider the diagram

$$\begin{array}{ccc} & & S^m \\ & \nearrow & \downarrow{\scriptstyle \nu} \\ S^n & \xrightarrow[f\nu']{} & \mathbf{R}P^m. \end{array}$$

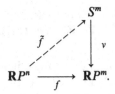

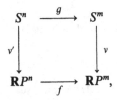

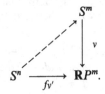

Commutativity of the original diagram shows that both g and $\widetilde{fv'}$ are liftings of fv':

$$vg = fv' = v\widetilde{fv'}.$$

Choose $x_0 \in S^n$; by definition, $v(g(x_0)) = v(-g(x_0))$, and so the (two-point) fiber over $v(g(x_0))$ consists of $\{\pm g(x_0)\}$. Now the point $\widetilde{fv'}(x_0)$ lies in this fiber (because $v\widetilde{fv'}(x_0) = fv'(x_0) = vg(x_0)$), so that either $\widetilde{fv'}(x_0) = g(x_0)$ or $\widetilde{fv'}(x_0) = -g(x_0)$. In the second case,

$$\widetilde{fv'}(-x_0) = \widetilde{fv'}(x_0) = -g(x_0) = g(-x_0).$$

Hence, in either case, the liftings g and $\widetilde{fv'}$ agree at a point. The uniqueness theorem, Lemma 10.3, gives $g = \widetilde{fv'}$. But this is a contradiction: for every $x \in S^n$, we have $v'(-x) = v'(x)$, and so $\widetilde{fv'}(-x) = \widetilde{fv'}(x)$; on the other hand, $g(-x) = -g(x)$. □

Corollary 12.40 (Borsuk–Ulam). *If $f: S^n \to \mathbf{R}^n$ is continuous and $n \geq 1$, then there exists $x \in S^n$ with $f(x) = f(-x)$.*

PROOF. The case $n = 1$ was proved in Exercise 6.15, and so we may assume that $n \geq 2$. If no such x exists, then the map $g: S^n \to S^{n-1}$ given by

$$g(x) = \frac{f(x) - f(-x)}{\|f(x) - f(-x)\|}$$

is a well defined continuous map, and $g(-x) = -g(x)$ for every $x \in S^n$, contradicting the theorem. □

EXERCISES

12.25. If $f: S^n \to \mathbf{R}^n$ satisfies $f(-x) = -f(x)$ for every $x \in S^n$, then there exists $x_0 \in S^n$ with $f(x_0) = 0$.

12.26. Prove that $\mathbf{R}^n$ contains no subspace homeomorphic to S^n.

12.27. Prove the Lusternik–Schnirelmann theorem: if S^n is the union of $n + 1$ closed subsets $F_1, \ldots, F_{n+1}$, then at least one F_i contains a pair of antipodal points. (See Corollary 6.30.)

Theorem 12.41 (Ham Sandwich Theorem). *Let $\mathbf{R}^n$ contain n bounded Lebesgue measurable subsets $A_1, \ldots, A_n$. Then there exists a hyperplane that bisects every A_j, $j = 1, \ldots, n$ (i.e., half the measure of each A_j lies on each side of the hyperplane).*

Remarks. (1) The name of the theorem comes from the case $n = 3$: given a piece of white bread, a piece of rye bread, and a piece of ham, one can slice the sandwich with one cut into two sandwiches, each having the same amounts of white bread, rye bread, and ham.

(2) The alimentary example above is misleading, because the subsets A_j may intersect.

Proof. We begin with some elementary geometry. As usual, regard $\mathbf{R}^n$ as imbedded in $\mathbf{R}^{n+1}$ as all $(n+1)$-tuples with last coordinate zero. Choose, once and for all, a point $y \in \mathbf{R}^{n+1} - \mathbf{R}^n$. If 0 is the origin and $x \in S^n$, let $x0$ denote the line determined by x and 0. There is a one-parameter family of hyperplanes perpendicular to $x0$; let $\pi(x)$ denote that hyperplane in the family containing y. Note that

$$\pi(x) = x^{\perp} + y,$$

where $x^{\perp}$ is the n-dimensional sub-vector-space of $\mathbf{R}^{n+1}$:

$$x^{\perp} = \{z \in \mathbf{R}^{n+1}: (z, x) = 0\}.$$

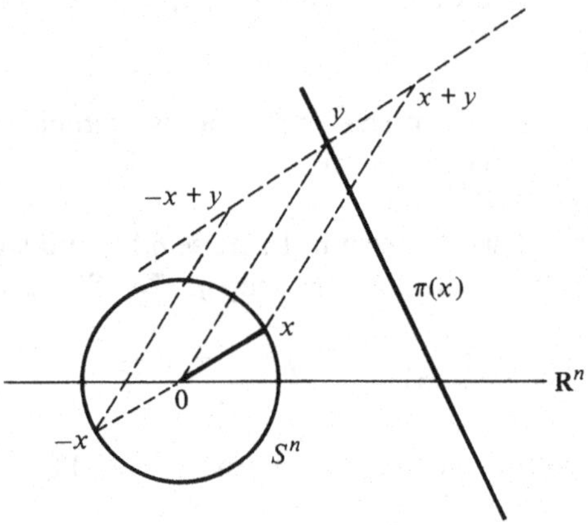

Here are three observations.

(1) $\pi(x) \neq \mathbf{R}^n$ (for $y \in \pi(x)$ and $y \notin \mathbf{R}^n$).
(2) $x + y$ and $-x + y$ lie on different sides of $\pi(x)$. First, $\pm x + y \notin \pi(x)$; otherwise $\pm x + y = z + y$ for some $z \in x^{\perp}$; but then $\pm x = z$ and $(x, x) = 0$, contradicting $(x, x) = 1$. But $y \in \pi(x)$ is the midpoint of the line segment joining $x + y$ and $-x + y$).
(3) $\pi(x) = \pi(-x)$ (the lines $x0$ and $-x0$ coincide).

We now begin the proof proper. For each j with $1 \leq j \leq n$, define $u_j: S^n \to \mathbf{R}$ by setting $u_j(x)$ to be the measure of that portion of A_j lying on the same side of $\pi(x)$ as $x + y$ (should all of A_j lie on the other side, then $u_j(x) = 0 = \mu(\varnothing)$, where μ is Lebesgue measure). That u_j is continuous follows from countable additivity of Lebesgue measure: if $x_m \to x$ and if s_m is the "slab" between $\pi(x_m)$ and $\pi(x_{m+1})$, then $\lim_{m\to\infty} \mu(A_j \cap s_m) = 0$.

Define $f: S^n \to \mathbf{R}^n$ by $f(x) = (u_1(x), \ldots, u_n(x))$; f is continuous because each u_j is continuous. By the Borsuk–Ulam theorem, there is $x_0 \in S^n$ with $f(x_0) = f(-x_0)$; that is, $u_j(x_0) = u_j(-x_0)$ for $j = 1, \ldots, n$. For every $x \in S^n$, we have

$$u_j(x) + u_j(-x) = \mu(A_j);$$

this follows from (2), (3), and the additivity of μ. Therefore $u_j(x_0) = \frac{1}{2}\mu(A_j)$ for $j = 1, \ldots, n$. Finally, the required hyperplane in $\mathbf{R}^n$ is just $\pi(x_0) \cap \mathbf{R}^n$ (which is a hyperplane in $\mathbf{R}^n$ by (1)). □

For an elementary proof of the Borsuk–Ulam theorem when $n = 2$, see [Kosniowski, p. 157]; for elementary proofs of the ham sandwich theorem for $n = 2$ and $n = 3$, see [Kosniowski, pp. 64, 159].

The last problem we consider is whether a sphere S^n can be an H-space.

Definition. A **graded co-ring** is a graded abelian group $B = \sum_{p \geq 0} B^p$ with a homomorphism $c: B \to B \otimes B$, called **comultiplication**, for which

$$c(B^p) \subset \sum_{i+j=p} B^i \otimes B^j.$$

Definition. A graded abelian group $B = \sum B^p$ is a **Hopf algebra** over $\mathbf{Z}$ if

(i) B is a graded ring;
(ii) B is a graded co-ring;
(iii) the comultiplication $c: B \to B \otimes B$ is a homomorphism of graded rings.

Thus a Hopf algebra combines the two "dual" notions of graded ring and graded co-ring, with axiom (iii) as a compatibility condition.

Definition. A **co-unit** of a graded co-ring B is a homomorphism $\varepsilon: B \to \mathbf{Z}$ making the following diagram commute:

$$\begin{array}{ccccc} B \otimes \mathbf{Z} & \xleftarrow{1 \otimes \varepsilon} & B \otimes B & \xrightarrow{\varepsilon \otimes 1} & \mathbf{Z} \otimes B. \\ & {\scriptstyle l} \nwarrow & \uparrow {\scriptstyle c} & \nearrow {\scriptstyle r} & \\ & & B & & \end{array}$$

where l and r are the isomorphisms defined by $l: b \mapsto b \otimes 1$ and $r: b \mapsto 1 \otimes b$.

In the language of Chapter 11, a co-unit is a co-identity in the category of graded rings (for it is easy to see that $\mathbf{Z}$, the graded ring having $\mathbf{Z}$ in degree 0 and zero elsewhere, is an initial object).

Definition. A Hopf algebra $B = \sum B^p$ is **connected** if

(iv) B^0 is infinite cyclic with generator the unit e;
(v) the map $\varepsilon: B \to \mathbf{Z}$, defined by

$$\varepsilon(e) = 1 \quad \text{and} \quad \varepsilon(b^p) = 0 \quad \text{for all } b^p \in B^p, \quad p \geq 1,$$

is a co-unit.

Connected Hopf algebras arise naturally.

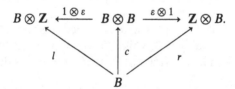

Theorem 12.42. *If X is a path connected H-space whose homology groups are f.g. free abelian groups, then $H^*(X)$ is a connected Hopf algebra over* $\mathbf{Z}$.

PROOF. As always, $H^*(X)$ is a graded ring under cup product. Since X is an H-space, there is a given continuous map $\mu: X \times X \to X$. By Theorem 12.31, the hypotheses on X imply that cross product $\alpha': H^*(X) \otimes H^*(X) \to H^*(X \times X)$ is an isomorphism of graded rings. If β is its inverse, define $c: H^*(X) \to H^*(X) \otimes H^*(X)$ as the composite

$$H^*(X) \xrightarrow{\mu^*} H^*(X \times X) \xrightarrow{\beta} H^*(X) \otimes H^*(X).$$

As μ^* is a map of graded rings (Lemma 12.20), it follows that c is a homomorphism of graded rings, and so $H^*(X)$ is a Hopf algebra over $\mathbf{Z}$. (Since every space X has continuous maps $X \times X \to X$, the connectedness of the Hopf algebra must be the crucial point where the hypothesis that X is an H-space is used.) Now X path connected implies that $H^0(X) \cong \mathbf{Z}$; it is easy to see that the unit e is a generator (e is defined by $(x, e) = 1$ for all $x \in X$).

Recall the definition of H-space. There is $x_0 \in X$ so that the following diagram commutes to homotopy:

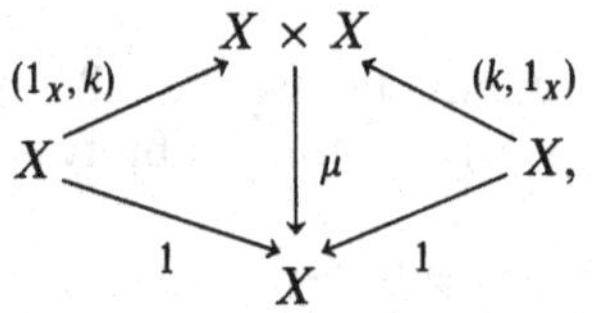

where $k: X \to X$ is the constant map at x_0, $(k, 1_X): x \mapsto (x_0, x)$, and $(1_X, k): x \mapsto (x, x_0)$. Let $i: \{x_0\} \hookrightarrow X$ be the inclusion.

Since $H^*(\{x_0\}) = H^0(\{x_0\})$, the map of graded rings $i^*: H^*(X) \to H^*(\{x_0\})$ carries $H^p(X)$ into 0 for all $p > 0$, while $i^*(\text{cls } e) = \text{cls } e_0$, the unit of $H^*(\{x_0\})$, by Lemma 12.20. We identify $H^*(\{x_0\})$ with $\mathbf{Z}$, and so i^* is the map that must be shown to be a co-unit.

Consider the following subdivision of the defining co-unit diagram:

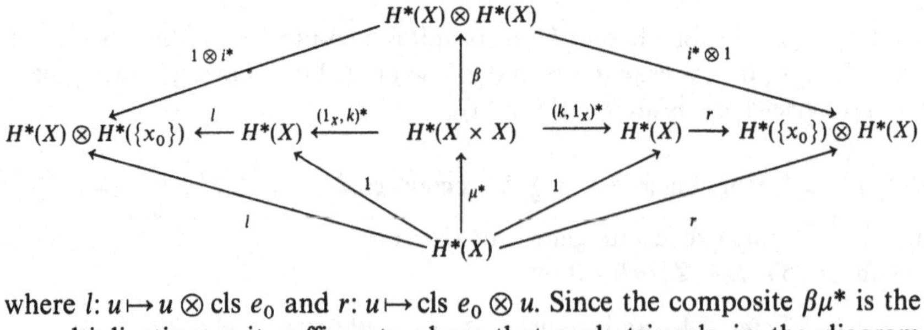

where $l: u \mapsto u \otimes \text{cls } e_0$ and $r: u \mapsto \text{cls } e_0 \otimes u$. Since the composite $\beta\mu^*$ is the comultiplication c, it suffices to show that each triangle in the diagram commutes (for then i^* is a co-unit). Commutativity of the lower outside triangles is plain, while commutativity of the other two lower triangles results from applying the functor H^* to the homotopy commutative diagram above.

It remains to prove that $l(1_X, k)^* = (1 \otimes i^*)\beta$ and $r(k, 1_X)^* = (i^* \otimes 1)\beta$; since β is the inverse of α', it suffices to prove that

$$l(1_X, k)^*\alpha' = 1 \otimes i^* \quad \text{and} \quad r(k, 1_X)^*\alpha' = i^* \otimes 1.$$

Now $i^* \otimes 1$ maps $H^p(X) \otimes H^q(X)$ into $H^p(\{x_0\}) \otimes H^q(X)$, and hence it is zero when $p > 0$. If $p = 0$, it suffices to look at $(i^* \otimes 1)(\text{cls } e \otimes \text{cls } \theta)$ (for X path connected implies that cls e generates $H^0(X)$), and

$$\begin{aligned}(i^* \otimes 1)(\text{cls } e \otimes \text{cls } \theta) &= \text{cls } i^{\#}e \otimes \text{cls } \theta \\ &= \text{cls } e_0 \otimes \text{cls } \theta.\end{aligned}$$

On the other hand, if φ is a p-cocycle and θ is a q-cocycle, then $r(k, 1_X)^*\alpha'(\text{cls } \varphi \otimes \text{cls } \theta) = r(k, 1_X)^*(\text{cls } \varphi \times \text{cls } \theta)$, by definition of cross product. But $(k, 1_X)^* = d^*(k \times 1_X)^*$, where $d: X \to X \times X$ is the diagonal, so that Exercise 12.13 gives

$$r(k, 1_X)^*(\text{cls } \varphi \times \text{cls } \theta) = rd^*(k^* \text{ cls } \varphi \times \text{cls } \theta).$$

It follows that this is zero for $p > 0$ (because k is a constant map). If $p = 0$, we may again assume that $\varphi = e$, and now

$$rd^*(k^* \text{ cls } e \times \text{cls } \theta) = rd^*(\text{cls } e \times \text{cls } \theta)$$

(for $k: X \to X$ implies that k^* cls e = cls e). By Corollary 12.27,

$$\begin{aligned}rd^*(\text{cls } e \times \text{cls } \theta) &= r(\text{cls } e \cup \text{cls } \theta) \\ &= r \text{ cls } \theta = \text{cls } e_0 \otimes \text{cls } \theta,\end{aligned}$$

as desired. A similar argument handles $1 \otimes i^*$. □

Remark. If one replaces **Z** by a field R throughout, then one obtains graded R-algebras instead of graded rings and connected Hopf algebras over R instead of over **Z**; the analogous result holds for any space X of finite type (whose homology groups need not be free abelian).

Theorem 12.43 (Hopf). *If $n > 0$ is even,*[4] *then S^n is not an H-space.*

PROOF. We know that $H^*(S^n) = H^0 \oplus H^n$, say, with generators e and x. It suffices to show that $H^*(S^n)$ cannot be a connected Hopf algebra; let us assume otherwise.

Now the comultiplication $c: H^* \to H^* \otimes H^*$ is a map of graded rings. Since x has degree n,

$$c(x) = re \otimes sx + ux \otimes ve, \qquad r, s, u, v \in \mathbf{Z}.$$

If $\varepsilon: H^* \to \mathbf{Z}$ is the co-unit, then $(\varepsilon \otimes 1)c(x) = 1 \otimes x$ and $(1 \otimes \varepsilon)c(x) = x \otimes 1$.

[4] It is known that S^n is an H-space only for $n = 0, 1, 3$, and 7.

Since $H^0 \otimes H^n \to \mathbf{Z}$ (given by $ae \otimes bx \mapsto ab$) and $H^n \otimes H^0 \to \mathbf{Z}$ (given by $bx \otimes ae \mapsto ab$) are isomorphisms of abelian groups, $1 = rs = uv$. Since $re \otimes sx = e \otimes rsx = rse \otimes x$, it follows that

$$c(x) = e \otimes x + x \otimes e.$$

Now $x \cup x = 0$ because its degree is $2n > n$, and so $c(x \cup x) = 0$. But c is multiplicative, so that

$$\begin{aligned} 0 = c(x \cup x) &= c(x)c(x) \\ &= (e \otimes x + x \otimes e)(e \otimes x + x \otimes e) \\ &= (e \otimes x)^2 + (e \otimes x)(x \otimes e) + (x \otimes e)(e \otimes x) + (x \otimes e)^2. \end{aligned}$$

Recall that multiplication in $H^* \otimes H^*$ satisfies

$$(a \otimes b)(y \otimes z) = (-1)^{\deg b \deg y}(a \cup y) \otimes (b \cup z).$$

It follows that

$$(e \otimes x)^2 = e \otimes (x \cup x) \quad \text{and} \quad (x \otimes e)^2 = (x \cup x) \otimes e,$$

and each of these is zero because $x \cup x = 0$. Also

$$(x \otimes e)(e \otimes x) = x \otimes x$$

because e has degree 0, while

$$(e \otimes x)(x \otimes e) = (-1)^{n^2} x \otimes x.$$

Since n is even, it follows from the above expansion of $0 = c(x \cup x)$ that $2x \otimes x = 0$ in $H^n \otimes H^n$, hence $x \otimes x = 0$. But $x \otimes x \neq 0$, by Exercise 9.34. This contradiction completes the proof. □

It follows, of course, that S^{2n} is never the underlying space of a topological group (when $n > 0$).

You are now in the hands of [J. F. Adams].

Bibliography

J. F. Adams, *Algebraic Topology: A Student's Guide*, Cambridge University Press, Cambridge, 1972.

P. S. Alexandroff and H. Hopf, *Topologie*, Springer-Verlag, New York, 1935.

E. Artin and H. Braun, *Introduction to Algebraic Topology*, Merrill, Westerville, OH, 1969.

M. F. Atiyah, *K-Theory*, Benjamin, Elmsford, NY, 1967.

R. H. Bing, The elusive fixed point property, *Am. Math. Monthly* **76**, 119–132 (1969).

R. Bott and L. W. Tu, *Differentiable Forms in Algebraic Topology*, Springer-Verlag, New York, 1982.

G. Bredon, *Introduction to Compact Transformation Groups*, Academic, Orlando, FL, 1972.

E. J. Brody, The topological classification of the lens spaces, *Ann. Math.* **71**, 163–184 (1963).

K. S. Brown, *Cohomology of Groups*. Springer-Verlag, New York, 1982.

R. Brown, *Topology: A Geometric Account of General Topology, Homotopy Types, and the Fundamental Groupoid*, Wiley, New York, 1988.

H. Cartan and S. Eilenberg, *Homological Algebra*, Princeton University Press, Princeton, 1956.

G. E. Cooke and R. L. Finney, *Homology of Cell Complexes*, Princeton University Press, Princeton, 1967.

A. Dold, Halbexakte Homotopiefunktoren, *Springer Lecture Notes* 12, 1966.

A. Dold, *Lectures on Algebraic Topology*, Springer-Verlag, New York, 1972.

C.W. Dowker, Topology of metric complexes, *Am. J. Math.* **74**, 555–577 (1952).

J. Dugundji, *Topology*, Allyn & Bacon, Newton, MA, 1966.

N. Dunford and J. Schwartz, *Linear Operators I*, Interscience, New York, 1958.

E. Dyer, *Cohomology Theories*, Benjamin, New York, 1969.

S. Eilenberg and N. Steenrod, *Foundations of Algebraic Topology*, Princeton University Press, Princeton, 1952.

G. Flores, Über n-dimensionale Komplexe die im R_{2n+1} absolut selbstverschlungen sind, *Ergeb. eines math. Kolloq.* **6**, 4–7 (1933–34).

D. B. Fuks and V. A. Rokhlin, *Beginner's Course in Topology*, Springer-Verlag, New York, 1984.

B. Gray, *Homotopy Theory, An Introduction to Algebraic Topology*, Academic, Orlando, FL, 1975.

M. Greenberg and J. R. Harper, *Algebraic Topology: A First Course*, Benjamin/Cummings, Menlo Park, CA, 1981.

W. H. Greub, *Multilinear Algebra*, Springer-Verlag, New York, 1967.

E. Hewitt and K. A. Ross, *Abstract Harmonic Analysis*, Springer-Verlag, New York, 1963.

P. J. Hilton, *An Introduction to Homotopy Theory*, Cambridge University Press, Cambridge, 1953.

P. J. Hilton and S. Wylie, *Homology Theory*, Cambridge University Press, Cambridge, 1960.

M. W. Hirsch, A proof of the nonretractibility of a cell onto its boundary, *Proc. Am. Math. Soc.* **14**, 364–365 (1963).

J. G. Hocking and G. S. Young, *Topology*, Addison-Wesley, Reading, MA, 1961.

S.-T. Hu, *Homotopy Theory*, Academic, Orlando, FL, 1959.

S.-T. Hu, *Homology Theory*, Holden Day, Oakland, CA, 1966.

W. Hurewicz and H. Wallman, *Dimension Theory*, Princeton University Press, Princeton, 1948.

D. Husemoller, *Fibre Bundles*, McGraw-Hill, New York, 1966.

N. Jacobson, *Basic Algebra I*, Freeman, New York, 1974.

J. L. Kelley, *General Topology*, Van Nostrand, New York, 1955.

C. Kosniowski, *A First Course in Algebraic Topology*, Cambridge University Press, Cambridge, 1980.

S. Lang, *Algebra*, Addison-Wesley, Reading, MA, 1965.

S. Lefschetz, *Introduction to Topology*, Princeton University Press, Princeton, 1949.

A. T. Lundell and S. Weingram, *Topology of CW Complexes*, van Nostrand-Reinhold, New York, 1968.

S. Mac Lane, *Homology*, Springer-Verlag, New York, 1963.

R. Maehara, The Jordan curve theorem via the Brouwer fixed point theorem, *Am. Math. Monthly* **91**, 641–644 (1984).

W. S. Massey, *Algebraic Topology, An Introduction*, Harcourt-Brace, Orlando, FL, 1967.

W. S. Massey, *Homology and Cohomology Theory*, Dekker, New York, 1978.

W. S. Massey, *Singular Homology Theory*, Springer-Verlag, New York, 1980.

C. R. F. Maunder, *Algebraic Topology*, Cambridge University Press, Cambridge, 1980.

J. McCleary, *User's Guide to Spectral Sequences*, Publish or Perish, Wilmington, DE, 1985.

J. Milnor, Construction of universal bundles I, *Ann. Math.* **63**, 272–284 (1956).

J. Milnor, *Morse Theory*, Princeton University Press, Princeton, 1963.

J. Milnor, Analytic proofs of the "hairy ball theorem" and the Brouwer fixed point theorem, *Am. Math. Monthly* **85**, 521–524 (1978).

J. R. Munkres, *Topology, A First Course*, Prentice-Hall, Englewood Cliffs, NJ, 1975.

J. R. Munkres, *Elements of Algebraic Topology*, Addison-Wesley, Reading, MA, 1984.

P. Olum, Non-abelian cohomology and van Kampen's theorem, *Ann. Math.* **68**, 658–668, (1958).

J. J. Rotman, *An Introduction to the Theory of Groups*, 3rd ed., Allyn & Bacon, Newton, MA, 1984.

J. J. Rotman, *An Introduction to Homological Algebra*, Academic, Orlando, FL, 1979.

H. Schubert, *Topology*, Allyn & Bacon, Newton, MA, 1968.

H. Seifert and W. Threlfall, *A Textbook of Topology*, Academic, Orlando, FL, 1980. (English translation of *Lehrbuch der Topology*, Teubner, Stuttgart, 1934.)

E. Spanier, *Algebraic Topology*, Springer-Verlag, New York, 1982 (McGraw-Hill, New York, 1966).

N. Steenrod, *The Topology of Fibre Bundles*, Princeton University Press, Princeton, 1951.

R. M. Switzer, *Algebraic Topology—Homotopy and Homology*, Springer-Verlag, New York, 1975.

J. W. Vick, *Homology Theory, An Introduction to Algebraic Topology*, Academic, Orlando, FL, 1973.

J. W. Walker, A homology version of the Borsuk–Ulam theorem, *Am. Math. Monthly* **90**, 466–468 (1983).

A. H. Wallace, *Algebraic Topology*, Benjamin, New York, 1970.

F. W. Warner, *Foundations of Differentiable Manifolds and Lie Groups*, Scott, Foresman, Glenview, IL, 1971.

G. W. Whitehead, *Elements of Homotopy Theory*, Springer-Verlag, New York, 1978.

Notation

$\mathbf{C}$	complex numbers
$\mathbf{I}$	closed unit interval $[0, 1]$
$\mathbf{Q}$	rational numbers
$\mathbf{R}$	real numbers
$\mathbf{Z}$	integers
$\bar{X}$	closure of X (as subspace of a larger space)
X°	interior of X (as subspace of a larger space)
$\dot{U}$	boundary of U (as subspace of a larger space)
$\|X\|$	cardinal of a set X
$\mathbf{R}^n$	euclidean n-space 1
S^n	n-sphere 1
D^n	n-disk 2
Δ^n	standard n-simplex 2
1_A	identity function or identity morphism on A 6
obj $\mathscr{C}$	objects in a category $\mathscr{C}$ 6
Groups	category of groups 7
Sets	category of sets 7
Top	category of topological spaces 7
Ab	category of abelian groups 8
$\mathbf{Top}^2$	category of pairs (X, A), where X is a topological space and A is a subspace of X 8
$\mathbf{Sets}_*$	category of pointed sets 8
$\mathbf{Top}_*$	category of pointed topological spaces 8
f_*	map $\mathrm{Hom}(M, A) \to \mathrm{Hom}(M, B)$ induced by $f\colon A \to B$ 11
f^* or $f^\#$	map $\mathrm{Hom}(B, M) \to \mathrm{Hom}(A, M)$ induced by $f\colon A \to B$ 12
hTop	homotopy category 16
X/A	quotient space 20

Index

Graduate Texts in Mathematics

continued from page ii

118 PEDERSEN. Analysis Now.
119 ROTMAN. An Introduction to Algebraic Topology.
120 ZIEMER. Weakly Differentiable Functions: Sobolev Spaces and Functions of Bounded Variation.
121 LANG. Cyclotomic Fields I and II. Combined 2nd ed.
122 REMMERT. Theory of Complex Functions. *Readings in Mathematics*
123 EBBINGHAUS/HERMES et al. Numbers. *Readings in Mathematics*
124 DUBROVIN/FOMENKO/NOVIKOV. Modern Geometry—Methods and Applications. Part III.
125 BERENSTEIN/GAY. Complex Variables: An Introduction.
126 BOREL. Linear Algebraic Groups. 2nd ed.
127 MASSEY. A Basic Course in Algebraic Topology.
128 RAUCH. Partial Differential Equations.
129 FULTON/HARRIS. Representation Theory: A First Course. *Readings in Mathematics*
130 DODSON/POSTON. Tensor Geometry.
131 LAM. A First Course in Noncommutative Rings.
132 BEARDON. Iteration of Rational Functions.
133 HARRIS. Algebraic Geometry: A First Course.
134 ROMAN. Coding and Information Theory.
135 ROMAN. Advanced Linear Algebra.
136 ADKINS/WEINTRAUB. Algebra: An Approach via Module Theory.
137 AXLER/BOURDON/RAMEY. Harmonic Function Theory.
138 COHEN. A Course in Computational Algebraic Number Theory.
139 BREDON. Topology and Geometry.
140 AUBIN. Optima and Equilibria. An Introduction to Nonlinear Analysis.
141 BECKER/WEISPFENNING/KREDEL. Gröbner Bases. A Computational Approach to Commutative Algebra.
142 LANG. Real and Functional Analysis. 3rd ed.
143 DOOB. Measure Theory.
144 DENNIS/FARB. Noncommutative Algebra.
145 VICK. Homology Theory. An Introduction to Algebraic Topology. 2nd ed.
146 BRIDGES. Computability: A Mathematical Sketchbook.
147 ROSENBERG. Algebraic *K*-Theory and Its Applications.
148 ROTMAN. An Introduction to the Theory of Groups. 4th ed.
149 RATCLIFFE. Foundations of Hyperbolic Manifolds.
150 EISENBUD. Commutative Algebra with a View Toward Algebraic Geometry.
151 SILVERMAN. Advanced Topics in the Arithmetic of Elliptic Curves.
152 ZIEGLER. Lectures on Polytopes.
153 FULTON. Algebraic Topology: A First Course.
154 BROWN/PEARCY. An Introduction to Analysis.
155 KASSEL. Quantum Groups.
156 KECHRIS. Classical Descriptive Set Theory.
157 MALLIAVIN. Integration and Probability.
158 ROMAN. Field Theory.
159 CONWAY. Functions of One Complex Variable II.
160 LANG. Differential and Riemannian Manifolds.
161 BORWEIN/ERDÉLYI. Polynomials and Polynomial Inequalities.
162 ALPERIN/BELL. Groups and Representations.
163 DIXON/MORTIMER. Permutation Groups.
164 NATHANSON. Additive Number Theory: The Classical Bases.
165 NATHANSON. Additive Number Theory: Inverse Problems and the Geometry of Sumsets.
166 SHARPE. Differential Geometry: Cartan's Generalization of Klein's Erlangen Program.
167 MORANDI. Field and Galois Theory.
168 EWALD. Combinatorial Convexity and Algebraic Geometry.
169 BHATIA. Matrix Analysis.
170 BREDON. Sheaf Theory. 2nd ed.
171 PETERSEN. Riemannian Geometry.
172 REMMERT. Classical Topics in Complex Function Theory.
173 DIESTEL. Graph Theory.
174 BRIDGES. Foundations of Real and Abstract Analysis.
175 LICKORISH. An Introduction to Knot Theory.
176 LEE. Riemannian Manifolds.
177 NEWMAN. Analytic Number Theory.
178 CLARKE/LEDYAEV/STERN/WOLENSKI. Nonsmooth Analysis and Control Theory.
179 DOUGLAS. Banach Algebra Techniques in Operator Theory. 2nd ed.
180 SRIVASTAVA. A Course on Borel Sets.
181 KRESS. Numerical Analysis.

182 WALTER. Ordinary Differential Equations.
183 MEGGINSON. An Introduction to Banach Space Theory.
184 BOLLOBÁS. Modern Graph Theory.
185 COX/LITTLE/O'SHEA. Using Algebraic Geometry.
186 RAMAKRISHNAN/VALENZA. Fourier Analysis on Number Fields.
187 HARRIS/MORRISON. Moduli of Curves.
188 GOLDBLATT. Lectures on Hyperreals.
189 LAM. Lectures on Rings and Modules.

Printed in the United States
by Baker & Taylor Publisher Services